Online Weather Studies

2nd Edition

Education Program

American Meteorological Society

The American Meteorological Society Education Program

The American Meteorological Society (AMS), founded in 1919, is a **scientific and professional** society. Interdisciplinary in its scope, the Society actively promotes the development and dissemination of information on the atmospheric and related oceanic and hydrologic sciences. AMS has more than 10,000 professional members from more than 100 countries and over 135 corporate and institutional members representing 40 countries.

The Education Program is the initiative of the American Meteorological Society fostering the teaching of the atmospheric and related oceanic and hydrologic sciences at the precollege level and in community college, college and university programs. It is a unique partnership between scientists and educators at all levels with the ultimate goals of (1) attracting young people to further studies in science, mathematics and technology, and (2) promoting public scientific literacy. This is done via the development and dissemination of scientifically authentic, up-to-date, and instructionally sound learning and resource materials for teachers and students.

Online Weather Studies, a new component of the AMS education initiative, is an introductory undergraduate meteorology course offered partially via the Internet in partnership with college and university faculty. **Online Weather Studies** provides students with a comprehensive study of the principles of meteorology while simultaneously providing classroom and laboratory applications focused on current weather situations. It provides real experiences demonstrating the value of computers and electronic access to time-sensitive data and information.

Developmental work for **Online Weather Studies** was supported by the Division of Undergraduate Education of the National Science Foundation under Grant No. DUE - 9752416.

This project was supported, in part,
by the
National Science Foundation
Opinions expressed are those of the authors and
not necessarily those of the Foundation

Online Weather Studies / Joseph M. Moran. — 2nd edition
ISBN: 1-878220-50-0

Copyright © 2002 by the American Meteorological Society

Published by the American Meteorological Society
45 Beacon Street, Boston, MA 02108

PREFACE

Welcome to *Online Weather Studies*! You are about to embark on an exciting study of the science of the atmosphere by following weather as it happens, in real-time via the Internet. The purpose of this book is to provide you with background information on the properties of the atmosphere, the interactions between the atmosphere and the other components of the Earth system, and the principles that govern weather and climate. This information will assist you as you complete companion investigations whose components are available in the *Online Weather Study Guide* and via the *Online Weather Studies Homepage*. This inquiry-based approach is designed to promote critical thinking as you analyze and interpret real-time weather information.

This book consists of 12 main chapters plus 3 optional chapters. Each of the first 12 chapters corresponds to one week of the *Online Weather Studies* course. Chapters are organized so that concepts build logically one upon the other so that Earth's atmosphere emerges as an interactive system subject to physical laws. Twice-weekly investigations, partially delivered via the Internet and focused on actual atmospheric conditions, are tied directly to each chapter. Topics covered include sources of weather information (Chapter 1), composition and structure of the atmosphere (Chapter 2), the planet's radiation balance and controls of temperature (Chapters 3 and 4), air pressure (Chapter 5), humidity, clouds and precipitation (Chapter 6 and 7), forces governing atmospheric circulation (Chapter 8), and weather systems (Chapters 9, 10, 11, and 12).

Each of the first 12 chapters opens with a *Case-in-Point*, an authentic, relevant, and real-life event or issue that highlights or applies one or more of the main concepts covered in the chapter. In essence, the Case-in-Point previews the chapter and is intended to engage reader interest early on. Chapter 7 (Clouds, Precipitation, and Weather Radar), for example, opens with a discussion of the effects of jet aircraft contrails on cloud cover between major cities. The Case-in-Point is followed by a sample *Driving Question*, a broad-based query that links chapter concepts and provides a central focus for that week's study. Chapter content is science-rich and informs additional driving questions. Each chapter closes with a list of *Basic Understandings*. One or more *Essays* at the end of Chapters 1-12 address in some depth specific topics that complement or supplement a concept covered by the narrative. Examples include: *The Atmosphere of Mars*, *Hazards of Solar Ultraviolet Radiation*, and *Rainmaking*. All bold-faced terms are defined in the *Glossary* at the back of the book.

In this second edition, we have updated the science, included more Essays, added more photographs and line drawings, and used full color throughout. Three optional chapters are included so that *Online Weather Studies* will mesh with the various calendars of colleges and universities nationwide. These chapters cover weather analysis and forecasting (Chapter A), atmospheric optics (Chapter B), and climate and climate change (Chapter C). Each chapter has accompanying investigations in the *Online Weather Study Guide*. Appendixes cover unit conversions, milestones in the history of atmospheric science, and climate classification.

Online Weather Studies is pedagogically guided by a teaching approach (*Project-Based Science*) that seeks to engage learners in exploring their world by investigating meaningful questions. The course incorporates driving questions, investigations, collaboration, technology, and artifacts. *Online Weather Studies* offers one driving question per chapter but each chapter plus investigations explore and inspire many other driving questions. Each investigation has printed and electronic components that make use of meteorological and climatological data available on the Internet. Investigations engage the student in observation, prediction, data analysis, inference, and critical thinking (processes of science). The course presents opportunities for students to collaborate with their teacher and fellow students as together they negotiate understanding. Application of information-age technology helps the student's ability to retrieve and analyze real-world data and share interpretations. Throughout the course, students assemble learning materials (artifacts) for assessment purposes.

Development work for *Online Weather Studies* was originally supported by the National Science Foundation under Grant No. DUE-9752416. This edition of the book was prepared for use in course updating and for course implementation initiatives funded by the National Science Foundation through its Opportunities for Enhancing Diversity in the Geoscience (OEDG) Program, Grant No. GEO-0119740, and its Course, Curriculum and Laboratory Improvement – National Dissemination (CCLI-ND) Program, Grant No. DUE-0126032.

Online Weather Studies is the product of collaboration among many individuals having considerable teaching experience in the atmospheric sciences. This book is primarily the work of Joseph M. Moran of the American Meteorological Society's Education Program and Professor Emeritus of Earth Science at the University of Wisconsin-Green Bay. Development of this second edition benefited greatly from suggestions and critical reviews provided by Ira W. Geer, Bernard A. Blair, and Elizabeth W. Mills of the AMS Education Program, Robert S. Weinbeck of SUNY College at Brockport and the AMS Education Program, Edward J. Hopkins of the University of Wisconsin-Madison, James A. Brey of the University of Wisconsin-Fox Valley, William Porter of Elizabeth City State University (NC), David R. Smith of the U.S. Naval Academy, and Jeff Clark of Lawrence University. For numerous comments and suggestions, a special thanks is extended to all past instructors of *Online Weather Studies* and their students.

Norman J. Frisch of Brockport, NY did an excellent job of turning line drawings into final art. Unless otherwise indicated, Joseph M. Moran supplied photographs. Bernard Blair of the AMS Education Program met the numerous technical challenges in turning the original manuscript into this book with his usual skill, attention to detail, dedication, and enthusiasm. J. Randy McGinnis of the University of Maryland, College Park, provided valuable advice on project planning, learning strategies, and evaluation.

A special note concerns the use of units in *Online Weather Studies* learning materials, including this book. Generally the International System of Units (abbreviated SI, for Systéme Internationale d'Unitès) is employed with equivalent English or other units following in parentheses. Exceptions are units used by convention or convenience in meteorology or the user community (e.g., knots, calories, millibars). Also, the equivalence between units is given in context; that is, where general estimates are used, approximate values are shown in all units. Conversion factors are given in Appendix I.

Ira W. Geer
AMS Education Program

BRIEF CONTENTS

CONTENTS

CHAPTER 4   HEAT, TEMPERATURE, AND ATMOSPHERIC CIRCULATION   69

CHAPTER 5   AIR PRESSURE   93

CHAPTER 6   HUMIDITY, SATURATION, AND STABILITY   109

CHAPTER 7   CLOUDS, PRECIPITATION, AND WEATHER RADAR 139

CHAPTER 8   WIND AND WEATHER   169

CHAPTER 11 THUNDERSTORMS AND TORNADOES   249

CHAPTER 12 TROPICAL WEATHER SYSTEMS   279

CHAPTER A   WEATHER ANALYSIS AND FORECASTING   301

CHAPTER B   ATMOSPHERIC OPTICS   319

CHAPTER C   CLIMATE AND CLIMATE CHANGE   329

CHAPTER 1

MONITORING WEATHER

A generation goes, and a generation comes. But the earth remains forever.
The sun rises and the sun goes down, and hastens to the place where it rises.
The wind blows to the south, and goes round to the north; round and round goes the wind, and on its circuits the wind returns.

ECCLESIASTES 1:4-6

Case-in-Point

On 12–14 March 1888, a powerful coastal storm accompanied by heavy snow and strong winds blasted the eastern seaboard from Washington, DC northward into southern New England. The so-called *Blizzard of '88* brought virtually all activities to a standstill. Particularly hard hit was New York City where the three-day snowfall total was 53 cm (21 in.) and 65 km (40 mi) per hr winds whipped snow into drifts 4.5 to 6.0 m (15 to 20 ft) deep. Much greater snowfalls (generally 100 to 125 cm or 40 to 50 in.) and deeper snowdrifts (to 12 m or 40 ft) were reported over southeastern New York, western Connecticut and western Massachusetts. An estimated 300 people (200 in New York City) lost their lives to exposure, accidents, or overexertion; almost 200 ships were sunk or damaged with nearly 100 lives lost; hundreds of trains stalled in deep snow drifts, marooning passengers for days; and telephone and telegraph lines were severed.

The *Blizzard of '88* was a total surprise. The U.S. Army Signal Corps had operated the nation's weather service for more than 17 years, and observers at some 154 weather stations nationwide telegraphed observations three times daily to headquarters in Washington, DC. *Indications* (later called weather forecasts) were issued based on surface weather observations alone. The technology of the day and the limited understanding of the workings of the atmosphere put meteorologists at a distinct disadvantage in forecasting storms, especially storms that tracked offshore and parallel to the coast (so-called *nor'easters*). There were no means of monitoring the upper atmosphere, no satellites, no radar, no wireless communications with ships at sea, no computer models. This was before radio, television, and the Internet so that the public received most if not all their weather information from newspapers.

More than a century later, in December 2000, another powerful coastal storm threatened to paralyze the major metropolitan areas from Washington, DC northeastward to Boston with heavy snow and strong and gusty winds. This time, weather forecasters were much better prepared and there would be no surprises. Satellites continually monitored the storm's developing cloud shield and tracked its movements; radar located heavy snow bands sweeping onshore; a flood of weather observational data from the surface and upper atmosphere fed into sophisticated computer models running on supercomputers. Well in advance of the storm's arrival, the National Weather Service issued winter storm watches and warnings that were rapidly communicated via the electronic media to the public and public service agencies. Some school districts announced closings more than 12 hrs before the first snowflake was expected, and people stocked up on food, videos, and other supplies in the event that they became marooned at home.

But in spite of the great technological advances in weather observation, forecasting, and communications, the much anticipated snowstorm spared the major metropolitan areas. Washington, DC, for example, reported only a few snowflakes. What happened? Post-event analysis by meteorologists revealed that the storm failed to intensify where expected over the ocean because sea-surface temperatures were too low. One of the computer models they relied on for forecasting the development of the storm uses satellite-derived sea-surface temperatures. But persistent cloudiness off the East Coast in the weeks before the storm prevented updating of sea-surface temperature values. Using temperatures that were too high, the model overestimated the amount of ocean-water evaporation, key to the intensification of the storm. Meteorologists learned from this experience and that new knowledge is now incorporated in computerized storm-prediction models, thereby benefiting future forecasts. The December 2000 storm reminds us that while weather analysis and forecasting has greatly benefited from new knowledge and technological advances of the twentieth century, we still have more to learn about the workings of Earth's atmosphere.

Driving Question:

What are some basic concepts regarding the atmosphere and weather?

We are about to embark on a systematic study of the continually changing atmospheric environment. We will observe weather as it happens while becoming familiar with some of the basic scientific principles and understandings that govern the atmospheric environment. As our study progresses, we will combine our observational experiences with what we have learned in exploring a wide variety of atmospheric phenomena ranging from cold waves to hurricanes. Long before the end of the course we will understand that the behavior of the atmosphere is not arbitrary; its workings are explainable in scientific terms. At the same time, we are likely to still have questions about past, present, and future weather and climate.

If you are counting on studying storms, tornadoes, fronts, hailstorms, and the like right away, you will not be disappointed. Significant weather events will be observed and discussed as they happen via information delivered by the Internet, and they will be explored in greater detail after relevant scientific foundations have been established. You can begin some activities now—today, in fact—that will involve you in weather events, and enrich and enliven your study of the atmosphere. Our objective in this first chapter is to introduce and describe some of the tools and basic understandings that will guide your investigation.

Sources of Weather Information

Everyone has considerable experience with (and understanding of) the subject matter of this course. After all, each of us has been living with weather all our lives. No matter where we live or what our occupations are, we are well aware of the far-reaching influence of weather. To a large extent, weather dictates how we dress, how we drive, and even our choice of recreational activities. Before setting off in the morning, most of us check the latest weather forecast on radio or TV and glance out the window to look at the sky or read the thermometer. Every day, we acquire information on weather through the media, our senses, and perhaps our own weather instruments. And from that information, we derive a basic understanding of how the atmosphere functions.

Until now, for most of us, keeping track of the weather has been a casual part of daily life. From now on, or at least for the duration of this course, weather observation will be a more formal and regular activity. As a key part of this course, you will be accessing current weather data via the Internet several times a week. In addition, we suggest that you tune to a televised weathercast at least once a day. Weathercasts are routine segments of the local morning, noon, and evening news reports. If you have cable television, you may choose to watch *The Weather Channel*, a 24-hr-a-day telecast devoted exclusively to weather reports and forecasts.

If television is not available, weather information may be obtained by reading local or national newspapers or by listening to radio broadcasts. Most newspapers include a weather column or page featuring maps and statistical summaries. Radio stations provide the latest local weather conditions and forecasts, but often do not include a summary of weather conditions across the nation unless some newsworthy event has occurred such as a hurricane or tornado outbreak.

Another valuable source of local weather information is via broadcasts of the **NOAA Weather Radio**. As a public service, the National Oceanic and Atmospheric Administration (NOAA), the parent organization of the National Weather Service (NWS) operates low-power, VHF high band FM radio transmitters that broadcast continuous weather information (e.g., regional conditions, local forecasts, marine warnings) directly from NWS forecast offices 24 hrs a day. A series of messages is repeated every 4 to 6 minutes and some messages are updated hourly. Regular reports are interrupted with watches, warnings, and advisories when weather-related hazards threaten. A special *weather radio* is required to receive NOAA transmissions because the seven broadcast frequencies (from 162.40 to 162.55 MHz) are outside the range of standard AM/FM radios (Figure 1.1). Some weather radios are designed to sound a tone alarm or switch on automatically when NWS forecasters issue a weather watch or warning or other emergency information. The latest generation weather radio is equipped with the Specific Area Message Encoding (SAME) feature that sounds an alarm only if a weather watch or warning is issued specifically for a county programmed (selected)

FIGURE 1.1
At the push of a button, this weather radio issues National Weather Service weather reports and forecasts.

by the user. Depending on terrain, the maximum range of NOAA weather radio broadcasts is about 65 km (40 mi). As of this writing, over 600 transmitters are operating in 50 states, Puerto Rico, the U.S. Virgin Islands, and the U.S. Pacific Territories. Expansion of this service is expected to eventually bring 95% of the U.S. population within range of NOAA weather radio broadcasts. In many communities, NOAA weather radio broadcasts are also available on cable TV and broadcast television's secondary audio programming channels.

Another valuable source of weather information is the Internet. The World Wide Web provides real-time access to weather maps, satellite images, national weather radar summaries, and weather forecasts, plus updates on environmental issues, such as trends in global climate, stratospheric ozone, and air quality. NWS forecast offices maintain web sites that provide links to a variety of meteorological, climatological, and hydrological information. This course's homepage brings you a host of real-time weather data via the Internet.

Weather Systems and Weather Maps

Weather information received via Internet, television, or newspapers may include: (1) national and regional weather maps; (2) satellite or radar images (or video loops) depicting large-scale cloud, precipitation and atmospheric circulation patterns; (3) data on current and past (24-hr) weather conditions; and (4) weather forecasts for the short-term (24 to 48 hrs) and long-term (up to 5 days or longer). So that your weather watching is more meaningful and useful, the remainder of this chapter is devoted to a description of what to watch for, beginning with features plotted on the national weather map.

Temperature, dewpoint, wind, and air pressure are among the many atmospheric variables that are routinely measured at weather stations and plotted on weather maps. In order to represent the state of the atmosphere at a particular time, weather observations are taken simultaneously around the world. (For information on time keeping see the Essay accompanying this chapter.) Special symbols are used on national weather maps to plot the location of the principal weather-makers, that is, pressure systems and fronts (Figure 1.2).

Pressure systems are of two types, *highs* (or anticyclones) and *lows* (or cyclones). The *high* and *low* designations refer to air pressure. We can think of **air**

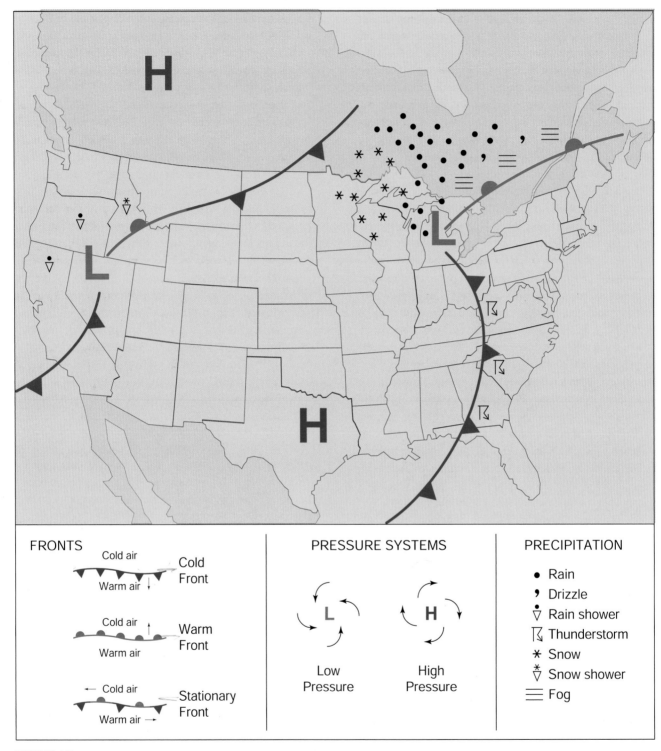

FIGURE 1.2
On a weather map, special symbols represent the state of the atmosphere over a broad geographical area at a specific time.

pressure as the weight per unit area of a column of air that stretches from the Earth's surface (or any altitude within the atmosphere) to the top of the atmosphere. At any specified time, air pressure at the Earth's surface varies from one place to another across the continent. On a weather map, *H* or *HIGH* symbolizes regions where the air pressure is relatively high compared to surrounding areas, and *L* or *LOW* symbolizes regions where the air pressure is relatively low compared to surrounding areas.

As you examine weather maps, note the following about pressure systems:

1. Usually highs are accompanied by fair weather and hence are described as *fair-weather systems*. Highs that originate in northwestern Canada bring cold, dry weather in winter and cool, dry weather in summer to much of the coterminous United States. Highs that develop further south bring hot, dry weather in summer and mild, dry weather in winter.

2. Viewed from above, surface winds in a high-pressure system blow in a clockwise (in the Northern Hemisphere) and outward spiral as shown in Figure 1.3A. Calm conditions or light winds are typical over a broad area about the center of a high.

3. Most lows produce cloudy, rainy or snowy weather and are often described as *stormy-weather systems*. An exception may be lows that develop over arid or semiarid terrain, especially in summer. In such areas, intense solar heating of the ground raises the air temperature and lowers the air pressure, producing a low that remains stationary over the hot ground and is not accompanied by stormy weather.

4. Viewed from above, surface winds in a low-pressure system blow in a counterclockwise (in the Northern Hemisphere) and inward spiral as shown in Figure 1.3B.

5. Both highs and lows move with the prevailing wind several kilometers above the surface, generally eastward across North America, and as they do, the weather changes. Highs follow lows and lows follow highs. As a general rule, highs track toward the east and southeast whereas lows track toward the east and northeast. An important exception is tropical low-pressure systems (e.g., hurricanes) that often move from east to west over the tropical Atlantic and Pacific before turning north and then eastward in midlatitudes.

6. Lows that track across the northern United States or southern Canada are more distant from sources of moisture and usually produce less rain- or snowfall than lows that track further south (such as lows that travel out of eastern Colorado and move along the Gulf Coast or up the eastern seaboard).

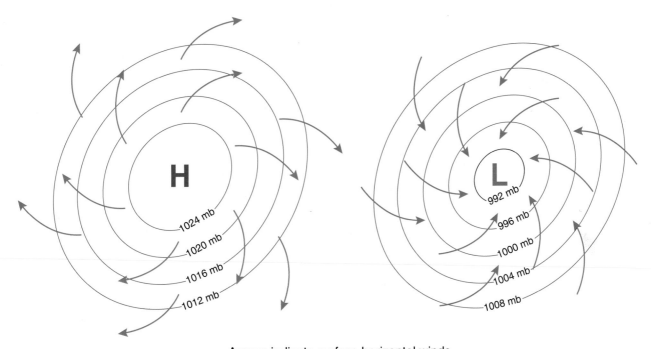

Arrows indicate surface horizontal winds

A   B

FIGURE 1.3
Viewed from above in the Northern Hemisphere, surface winds blow (A) clockwise and outward in a high pressure system, and (B) counterclockwise and inward in a low pressure system. Blue lines are isobars, passing through places having the same air pressure in millibars (mb).

7. Weather to the left side (west and north) of a storm's track (path) tends to be relatively cold, whereas weather to the right (east and south) of a storm's track tends to be relatively warm. For this reason, winter snows are most likely to the west and north of the path of a low-pressure system.

Air masses and fronts are also important weather-makers. An **air mass** is a huge volume of air covering hundreds of thousands of square kilometers that is horizontally relatively uniform in temperature and humidity. The specific characteristics of an air mass depend on the type of surface over which the air mass forms (its *source region*) and travels. Cold air masses form at polar latitudes over surfaces that are often snow or ice covered, whereas warm air masses form in the tropics where the Earth's surface is relatively warm year-round. Humid air masses form over moist maritime surfaces (e.g., Pacific Ocean, Gulf of Mexico), and dry air masses develop over dry continental surfaces (e.g., desert Southwest, northwestern Canada). The four basic types of air masses are cold and dry, cold and humid, warm and dry, and warm and humid.

A **front** is a narrow zone of transition between air masses that differ in temperature, humidity, or both. Fronts form where contrasting air masses meet, and the associated air movements often give rise to cloudiness and precipitation. The most common fronts are stationary, cold, and warm; weather map symbols for all three are shown in Figure 1.2. As the name implies, a *stationary front* is just that, stationary (or nearly so). On both sides of a stationary front, winds blow roughly parallel to the front but in opposite directions. A shift in wind direction may cause a portion of a stationary front to advance northward (becoming a warm front) or southward (becoming a cold front). At the same pressure, warm air is less dense than cold air so that a warm air mass advances by gliding up and over a retreating cold air mass. The cold air forms a wedge under the warm air and the leading edge of warm air at the Earth's surface is plotted on a weather map as a *warm front* (Figure 1.4A). On the other hand, cold air advances by sliding under and pushing up the less dense warm air and the leading edge of cold air at the Earth's surface is

FIGURE 1.4
The two most common types of fronts are (A) a warm front that marks the boundary between the advancing relatively warm (less dense) air and retreating cold (more dense) air, and (B) a cold front that marks the boundary between advancing cold air and retreating warm air. Both diagrams are vertical cross-sections with the vertical scale greatly exaggerated.

plotted on a weather map as a *cold front* (Figure 1.4B). Consequently, a warm front slopes more gently with altitude than does a cold front.

As you examine surface weather maps, note the following about air masses and fronts:

1. In response to regular seasonal changes in the duration and intensity of sunlight, polar air masses are much colder in winter and milder in summer. By contrast, in the tropics, sunlight is nearly uniform in duration and intensity throughout the year so that tropical air masses exhibit less seasonal variation in temperature.

2. An air mass modifies (becomes warmer, colder, wetter, drier) as it moves away from its source region with the degree of modification dependent on the properties of the surface over which the air mass travels. For example, a cold air mass warms more if it travels over ground that is bare rather than snow-covered.

3. Fronts are three-dimensional and the map symbol for a front is plotted where the front intersects Earth's surface.

4. Most cloudiness and precipitation associated with a warm front occur over a broad band, often hundreds of kilometers wide, in advance of where the front intersects Earth's surface. Precipitation ahead of a warm front generally is light to moderate in intensity and may persist at a particular location from 12 to 24 hrs or longer.

5. Most cloudiness and precipitation associated with a cold front occur as a narrow band along or just ahead of where the front intersects Earth's surface. Although precipitation often is showery and may last from a few minutes to a few hours, it can be very heavy.

6. Wind directions are different on the two sides of a front.

7. Some fronts are marked by neither cloudiness nor precipitation. Passage of the front is accompanied by a shift in wind direction and a change in air temperature and/or humidity.

8. In summer, air temperatures can be nearly the same ahead of and behind a cold front. In that case, the air masses on opposite sides of the front differ primarily in humidity; that is, the air mass ahead of the advancing front is more humid (and therefore less dense) and the air mass behind the front is drier (denser). With passage of the front, refreshingly drier air replaces uncomfortably humid air.

9. Cold and warm fronts are plotted on a weather map

as heavy lines that are often anchored at the center of a low-pressure system. The counterclockwise and inward circulation about a low brings contrasting air masses together to form fronts.

10. A low-pressure system may develop along a stationary front and travel rapidly like a large ripple from west to east along the front.

11. Thunderstorms and associated severe weather (e.g., tornadoes, hail) most often develop to the south and southeast of a low-pressure system in the warm, humid air mass that is located between the cold front and the warm front.

As you monitor national and regional weather maps, also watch for the following:

1. Cool sea breezes or lake breezes push inland perhaps 10 to 50 km (6 to 30 mi) and lower summer afternoon temperatures in coastal areas.

2. In late fall and throughout much of the winter, heavy lake-effect snows fall in narrow bands on the downwind (eastern and southern) shores of the Great Lakes and Great Salt Lake.

3. Severe thunderstorms and tornadoes are most common in spring across the central United States, especially from east Texas northward to Nebraska and from Iowa eastward to central Indiana.

4. Thunderstorms are relatively rare along the Pacific coast and on the Hawaiian Islands and most frequent in Florida and on the western High Plains.

5. Tropical storms and hurricanes occasionally impact the Atlantic and Gulf coasts, primarily from August through October.

Describing the State of the Atmosphere

Internet, television, and newspaper weather reports usually include statistical summaries of present and past weather conditions. It is useful to briefly comment on the meaning of the most common weather parameters:

1. *Maximum temperature.* The highest air temperature recorded over a 24-hr period, usually between midnight of one day and midnight of the next day. Typically, but not always, the day's maximum temperature occurs in the early to mid-afternoon. In the United States, surface air temperatures are reported in degrees Fahrenheit (°F) and in most other countries in degrees Celsius (°C).

2. *Minimum temperature.* The lowest air temperature recorded over a 24-hr period, usually between midnight of one day and midnight of the next day. Typically, but not always, the minimum temperature occurs around sunrise.

3. *Dewpoint* (or *frost point*). The temperature to which air must be cooled at constant pressure for dew (or frost) to begin forming on relatively cold surfaces.

4. *Relative humidity.* A measure of the actual concentration of the water vapor component of air compared to the concentration the air would have if saturated with water vapor. Relative humidity is always expressed as a percentage. Because the saturation concentration varies with air temperature so too does the relative humidity. On most days, the relative humidity is highest during the coldest time of day (around sunrise) and lowest during the warmest time of day (early to mid-afternoon).

5. *Precipitation amounts.* Rainfall or melted snowfall over a 24-hr period, from midnight of one day to midnight of the next, usually measured to one-hundredth of an inch in the United States and in millimeters elsewhere. On average, 10 inches of snow melt down to 1 inch of water.

6. *Air pressure.* The weight of a column of air over a unit area of the Earth's surface. With a mercury barometer, the traditional instrument for measuring air pressure, the pressure exerted by the atmosphere supports a column of mercury to a certain height and the mercury column fluctuates up and down as the air pressure rises and falls. This is the reason air pressure is commonly reported in units of length, that is, inches or millimeters of mercury. In the United States, meteorologists express air pressure in millibars (mb), a unit of pressure. The average air pressure at sea-level is 1013.25 mb, corresponding to 29.92 in. (760 mm) of mercury. Falling air pressure often signals an approaching low-pressure system and a turn to stormy weather. Rising air pressure, on the other hand, indicates an approaching high-pressure system and clearing skies or continued fair weather.

7. *Wind direction* and *wind speed.* Wind direction is the compass direction *from which* the wind blows (Figure 1.5). A southeast wind blows from the southeast toward the northwest and a west wind blows from the west toward the east. As a general rule, a wind shift from east to northeast to north is accompanied by falling air temperatures. On the other hand, a wind shift from east to southeast to

FIGURE 1.5
Wind vane atop the Smithsonian Building in Washington, DC. The arrow of the wind vane points in the direction from which the wind is blowing, as confirmed by the flags stretched out in the downwind direction. [Photo by R. S. Weinbeck]

south usually brings warmer weather. Over a broad area about the center of a high-pressure system calm air or light winds prevail. Wind speed tends to increase as a cold front approaches and winds are particularly strong and gusty in the vicinity of thunderstorms.

8. *Sky cover.* Based on the fraction of the sky that is cloud covered, the sky is described as clear (no clouds), a few clouds (1/8 to 2/8 cloud cover), scattered clouds (3/8 to 4/8), broken clouds (5/8 to 7/8), and overcast (completely cloud-covered). All other factors being equal, nights are coldest when the sky is clear and the air is relatively dry. An overcast sky elevates the day's minimum temperature and lowers the day's maximum temperature.

9. *Weather watch.* Issued by the National Weather Service when hazardous weather (e.g., tornadoes, heavy snowfall) is considered possible based on current or anticipated atmospheric conditions.

10. *Weather warning.* Issued by the National Weather Service when hazardous weather is imminent or actually taking place.

Weather Satellite Imagery

Satellite images and video loops (composed of successive images) are routine components of many televised and Internet-delivered weather reports. Some newspaper weather pages also feature satellite images. Sensors on weather satellites orbiting Earth provide a unique and valuable perspective on the state of the

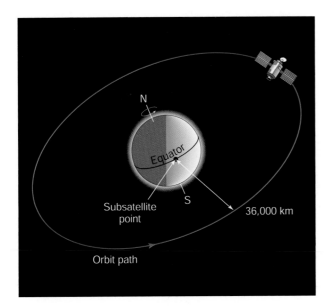

FIGURE 1.6
The orbit of a geostationary weather satellite.

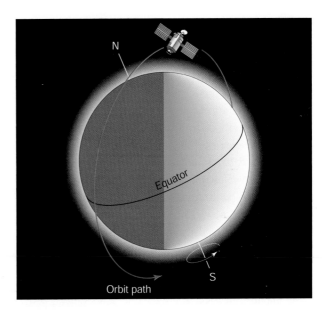

FIGURE 1.7
The path of a polar-orbiting weather satellite.

atmosphere and enable meteorologists to remotely measure temperature and humidity and to locate and track weather systems.

Weather satellite images most familiar to us are those obtained by sensors aboard a **geostationary satellite** that orbits Earth about 36,000 km (22,300 mi) above the planet's surface. The satellite orbits the planet at a rate that matches Earth's rotation and in the same eastward direction so that it is always positioned over the same spot on the Earth's surface and has the same field of view (Figure 1.6). The *subsatellite point*, the location on Earth's surface directly below the satellite, is on the equator. Two geostationary satellites, one at 75 degrees W longitude and the other at 135 degrees W longitude, provide a complete view of much of North America and adjacent portions of the Pacific and Atlantic Oceans to latitudes of about 60 degrees. Considerable distortion sets in near the edge of the field of view so that polar-orbiting satellites complement geostationary satellites in monitoring the planet.

A **polar-orbiting satellite** travels in a relatively low (800 to 1000 km) nearly north–south orbit passing over polar areas (Figure 1.7). The satellite's orbit defines a plane in space while the planet continually rotates on its axis through the plane of the satellite's orbit. With each orbit, points on Earth's surface (except near the poles) move eastward so that onboard sensors sweep out successive overlapping north–south strips. Sun-synchronous polar-orbiting weather satellites pass over essentially the same area twice every 24 hrs.

Sensors aboard a weather satellite receive two types of signals from Earth: reflected sunlight and emitted infrared radiation (IR). Sunlight reflected by Earth's surface and atmosphere produces images that are essentially black and white photographs of the planet. Highly reflective surfaces such as cloud tops and snow-covered ground appear bright white whereas less reflective surfaces such as evergreen forests and the ocean appear much darker. Cloud patterns on a **visible satellite image** are of particular interest to meteorologists (Figure 1.8). From analysis of cloud patterns displayed

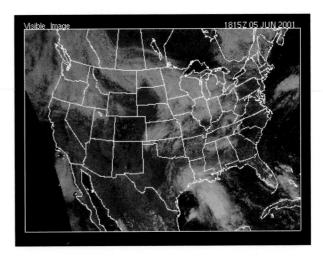

FIGURE 1.8
A sample visible satellite image from a geostationary satellite.

on the image, they can identify not only a specific type of weather system (such as a hurricane), but also the stage of its life cycle and its direction of movement.

A second type of sensor aboard a weather satellite detects infrared radiation (IR). IR is an invisible form of radiation that is emitted by all objects continually, both day and night. Hence, an **infrared satellite image** of the planet can provide useful information at any time whereas visible weather satellite images are of value only during daylight hours. (Because of around-the-clock availability, infrared satellite images are usually shown on television.) IR signals are routinely calibrated to give the temperature of objects in the sensor's field of view. This is possible because the intensity of IR emitted by an object depends on its surface temperature; that is, relatively warm objects emit more intense IR than do relatively cold objects. IR-derived temperature measurements are calibrated on a gray scale such that the brightest white indicates the lowest temperature and dark gray indicates the highest temperature. Alternately, a color scale is used so that, for example, reds and oranges represent high temperatures and blues and violets signal low temperatures.

The temperature dependency of IR emission makes it possible, for example, to distinguish low clouds from high clouds on an IR satellite image. In the part of the atmosphere where most clouds occur (the lowest 10 km or 6 mi or so), air temperature drops with increasing altitude. High clouds are colder than low clouds and emit less intense IR radiation. In the sample IR satellite image in Figure 1.9, high clouds appear bright white whereas low clouds are darker.

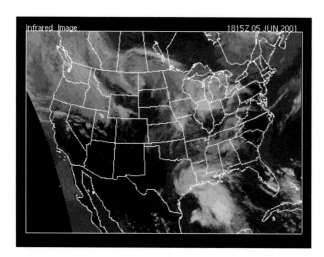

FIGURE 1.9
A sample infrared satellite image.

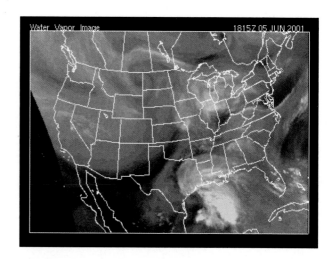

FIGURE 1.10
A sample water vapor satellite image.

Water vapor satellite imagery is a valuable tool in weather analysis and forecasting, enabling meteorologists to track the movement of plumes of moisture over distances of thousands of kilometers. Water vapor is an invisible gas and does not appear on visible or conventional infrared satellite images. But water vapor efficiently absorbs and emits certain wavelength bands of IR so that sensors aboard weather satellites that are sensitive to these wavelength bands can detect water vapor. Water vapor imagery displays the water vapor concentration between altitudes of about 5000 m and 12,000 m (16,000 ft and 40,000 ft) on a gray scale (Figure 1.10). At one extreme, black indicates little or no water vapor whereas at the other extreme, milky white indicates a relatively high concentration of water vapor. Upper-level clouds appear as bright white blotches on water vapor images.

Weather Radar

Weather radar complements satellite surveillance of the atmosphere by locating and tracking the movement of areas of precipitation and monitoring the circulation within small-scale weather systems such as thunderstorms. Weather radar continually emits pulses of microwave energy that are reflected by raindrops, snowflakes, or hailstones. The reflected signal (*radar echo*) is displayed as blotches on a television-type screen and superimposed on a map of the region surrounding the radar. The heavier the precipitation, the more intense is

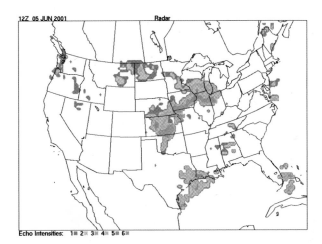

FIGURE 1.11
A sample composite national radar image.

FIGURE 1.12
These high, thin cirrus clouds appear fibrous because they are composed of mostly ice crystals.

the echo. Echo intensity is calibrated on a color scale so that light green indicates light precipitation whereas dark red signals heavy precipitation. From analysis of radar echoes, meteorologists can determine the intensity of thunderstorms, track the movement of areas of precipitation, and predict when precipitation is likely to begin or end at a particular place. Composite maps of radar echoes from a number of weather radars around the nation are useful in following the progress of large-scale weather systems (Figure 1.11). It is also useful to overlay composite radar echoes on satellite images of clouds to determine the most active portion of a weather system.

Weather radar also monitors movement of raindrops, snowflakes, and hailstones within a storm system. Using the same operating principle as the device that measures the speed of a pitched baseball (the *Doppler effect*), weather radar detects the circulation within a storm system. Early identification of tornado development is a potentially life-saving application of Doppler weather radar. With advance warning of a tornado before a funnel cloud touches the ground, the public has time to seek shelter or move out of its path.

Sky Watching

At this beginning stage in our study of the atmosphere and weather, it is also a good idea to develop the habit of observing the sky, watching for changes in clouds and cloud cover. Sky watching makes us more aware of the dynamic nature of the atmosphere and may reveal clues

to future weather. Here are some things to watch for:

1. Clouds are aggregates of tiny water droplets, ice crystals, or some combination of both. Ice-crystal clouds occur at high altitudes where air temperatures are relatively low and they have a fibrous or wispy appearance (Figure 1.12). Water-droplet clouds occur at lower altitudes where temperatures are higher and their edges are more sharply defined (Figure 1.13).

2. A cloud that is very near or actually in contact with Earth's surface is *fog*. By convention, fog is a suspension of tiny water droplets or ice crystals that reduces visibility to less than 1.0 km (5/8 of a mile).

3. Some clouds form horizontal layers (*stratiform clouds*) whereas others are puffy (*cumuliform clouds*). Stratiform clouds develop where air ascends gently over a broad region whereas

FIGURE 1.13
These relatively low clouds are composed of water droplets and have sharply defined edges.

FIGURE 1.14
Fair-weather cumulus clouds.

FIGURE 1.16
Clouds at different altitudes may move in different directions indicating a change in wind direction with altitude.

cumuliform clouds are produced by more vigorous ascent of air over a much smaller area. Often stratiform clouds develop ahead of a warm front and cumuliform clouds, especially those having great vertical development, form along or just ahead of a cold front.

4. Arrival of high, thin clouds in the western sky is often the first sign of an approaching warm front. In time, clouds gradually lower and thicken so that eventually they block out the sun during the day or the moon at night.

5. The day may begin clear but after several hours of bright sunshine, small white clouds appear, resembling puffs of cotton floating in the sky (Figure 1.14). These are fair-weather *cumulus clouds* that usually vaporize rapidly near sunset.

6. During certain atmospheric conditions, cumulus clouds build vertically and merge laterally, eventually forming a thunderstorm cloud, called a *cumulonimbus cloud* (Figure 1.15). Intense cumulonimbus clouds can produce severe weather including frequent lightning, torrential rains, hail, strong and gusty winds, and even tornadoes.

7. Clouds at different altitudes sometimes move horizontally in different directions (Figure 1.16). Because clouds move with the wind, this observation indicates that the horizontal wind shifts direction with increasing altitude.

Conclusions

We can learn much about the atmosphere and weather by keeping track of local, regional, and national weather patterns via the Internet, television, radio, and newspapers. Weather maps, satellite images, and radar displays are particularly valuable in following the development and movement of weather systems. In addition, we are well advised to develop the habit of watching the sky for changing conditions and to monitor weather instruments if they are available. In this way, we are able to get involved with the subject matter of this course from the beginning and what we learn becomes

FIGURE 1.15
Cumulonimbus (thunderstorm) cloud.

more meaningful and practical. Our study of the atmosphere, weather, and climate begins in the next chapter with an examination of the evolution and fundamental properties of the atmosphere.

Basic Understandings

- From the experiences of everyday life, all of us have derived a basic understanding of weather and the atmosphere.
- Sources of weather information include televised news programs, *The Weather Channel*, newspaper weather pages or columns, standard AM/FM radio, the NOAA weather radio, and the Internet.
- On a weather map, H or HIGH symbolizes a locale where air pressure is relatively high compared to the surrounding area. Viewed from above in the Northern Hemisphere, surface winds blow clockwise and outward about the center of a high. In general, a high is a fair-weather system.
- On a weather map, L or LOW symbolizes a locale where air pressure is relatively low compared to the surrounding area. Viewed from above in the Northern Hemisphere, surface winds blow counterclockwise and inward about the center of a low. In general, a low is a stormy weather system.
- An air mass is a huge volume of air covering hundreds of thousands of square kilometers that is relatively uniform horizontally in temperature and humidity. The temperature and humidity of an air mass depend on the properties of its source region and the nature of the surface over which the air mass travels.
- A front is a narrow zone of transition between air masses that contrast in temperature and/or humidity. They form where air masses meet and associated air movements often give rise to cloudiness and precipitation. The most common fronts are stationary, warm, and cold.
- Sensors onboard weather satellites orbiting the planet provide a unique and valuable perspective on the state of the atmosphere and enable meteorologists to remotely measure temperature and humidity and track weather systems. These sensors measure two types of radiation: sunlight that is reflected or scattered by Earth and infrared radiation that is emitted by Earth. Weather satellites are in either geostationary or polar orbits and generate data that are processed into visible and infrared images.
- Weather radar continually emits pulses of microwave radiation that are reflected by raindrops, snowflakes, or hailstones. The reflected signal, known as a radar echo, appears as blotches on a television-type screen. From analysis of radar echoes, meteorologists determine the intensity and track the movement of areas of precipitation. Using the Doppler effect, weather radar can also monitor the circulation of air within a storm system and provide advance warning of severe weather.
- Observing the development, type, and movement of clouds may provide clues as to future weather.

ESSAY: Time Keeping

Weather observations are made simultaneously at weather stations around the world. Simultaneous observations are necessary if the state of the atmosphere is to be portrayed accurately on a weather map at a specific time. The weather maps that you will use in this course are given in *Z time*. The meaning of *Z time* and the conventional basis for time keeping are subjects of this Essay.

For centuries, humans have kept track of their activities by the daily motions of the sun. Local noon was a convenient reference, marking the time when the sun was highest in the sky. However, locations even a few tens of kilometers to the east or west might have different *sun times*. After the Civil War, with advances in transportation and communication made possible by railroads and telegraphy, travel east or west meant that a person's local time kept changing. To eliminate confusion arising from the great number of locally observed times, the railroads argued for the simplified standardized time keeping scheme we use today. Civil time zones were initially instituted in the United States and Canada in November 1883 to standardize time keeping. The concept of international time zones was officially adopted one year later at the International Meridian Conference in Washington, DC.

Time zones were established on the basis of longitude, which is measured as so many degrees east and west of the *prime meridian*, that is, zero degrees longitude. Because some of the best early astronomical determinations of time had been made at The Old Royal Observatory in Greenwich, England, the meridian of longitude passing through the observatory was designated the *prime meridian*.

For more than 50 years, Greenwich Mean Time (GMT) was used for essentially all meteorological reports. GMT is based upon the daily rotation of Earth with respect to a "mean sun." Often the single letter Z (phonetically pronounced "zulu") designated the time within the Greenwich time zone (centered on the prime meridian). Today, the preferred time system is the more precise Coordinated Universal Time or Universel Temps Coordinné (UTC), based on an atomic clock and reckoned according to the stars. For all practical purposes, GMT and UTC are equivalent.

Earth makes one complete rotation (360 degrees) on its axis with respect to the sun once every 24 hrs. Hence, Earth rotates through 15 degrees of longitude every hour (360 degrees divided by 24 hrs equals 15 degrees per hr). Ideally, there should be 24 civil time zones each having a width of 15 degrees of longitude. The central meridian of each time zone is defined as a longitude that is evenly divisible by 15. For example, the Central Time Zone's central meridian is 90 degrees W longitude. Ninety degrees divided by 15 equals 6 so that Central Standard Time (CST) is six hrs different from the time at Greenwich. Earth rotates from west to east so that Greenwich time is ahead of CST by 6 hrs. If it is noon at Greenwich, it is 7 a.m. EST, 6 a.m. CST, 5 a.m. Mountain Standard Time (MST), and 4 a.m. Pacific Standard Time (PST).

Boundaries between some time zones have been adjusted to accommodate political boundaries of various nations. In a few cases, nations adhere to a local civil time that may differ by one half hour from the central meridian. To reduce confusion, time is expressed according to a 24-hr clock so that 7:45 a.m. is 0745 and 2:20 p.m. is 1420. While most of the United States observes Daylight Savings Time from April through October, UTC is fixed and does not shift to a summer schedule. Hence, you will have to adjust the time by one hour where and when Daylight Savings Time is in use. For example, during summer, residents of the U.S. Eastern Time zone lag Greenwich time by 4 instead of 5 hrs. Hence, 0800 Eastern Daylight Time (EDT) equals 0700 Eastern Standard Time (EST) equals 1200 Z.

By international agreement, surface weather observations are taken at least four times per 24 hrs, that is, at 0000 Z, 0600 Z, 1200 Z, and 1800 Z, with upper-air (radiosonde) measurements made at 0000 Z and 1200 Z. In the United States, surface observations are taken hourly (at the top of the hour), composite radar charts are also issued hourly at 35 minutes past the hour, and fronts are analyzed on weather maps every 3 hrs beginning at 0000 Z.

CHAPTER 2

ATMOSPHERE: ORIGIN, COMPOSITION, AND STRUCTURE

We have to understand that the world can only be grasped by action, not by contemplation. The hand is more important than the eye…The hand is the cutting edge of the mind.

JACOB BRONOWSKI

Case-in-Point

For many years, scientists suspected that significant amounts of red iron- and clay-rich soils on islands throughout the Caribbean and in Bermuda owed their composition partially to dust transported by northeast trade winds from the arid lands of North Africa. They also proposed that wind-borne North African dust supplies nutrients to the Amazon rain forest. Remote sensing by satellite now confirms transport of dust from North Africa to the western Atlantic (Figure 2.1). In recent years, scientists have found evidence that North African dust adversely affects air quality over the southeastern United States, may contribute to the formation of red tides in the Gulf of Mexico, and may threaten coral reefs in the Caribbean.

As weather disturbances sweep across North Africa, their associated strong winds pick up dust particles from the dry topsoil and carry them to altitudes of 3,000 m (10,000 ft) or so. Northeast trade winds then transport plumes of the smallest dust particles across the Atlantic and over the Caribbean, Central America, and the southeastern United States (Florida receives more than half of all North African dust delivered to the United

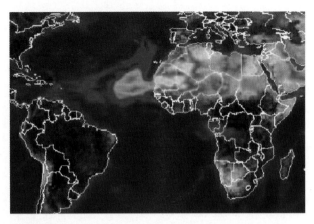

FIGURE 2.1
This satellite image taken by NASA's Total Ozone Mapping Spectrometer (TOMS) instrument shows dust coming off land sources in North Africa and moving westward across the Atlantic as a dust plume on 17 June 1999. [Source: NASA.]

States). This transoceanic journey takes about 1 to 2 weeks and dust plumes are observed primarily from June to October, peaking in July. Persistent drought in sub-Saharan Africa (from the mid-1960s through the early 1990s) may explain a recent increase in the volume of dust transported to the western Atlantic.

North African dust contributes to a reddish haze and colorful sunsets over much of the southeastern United States. The dust particles also can harbor in their microscopic cracks and crevasses bacteria and fungi, potential disease-causing organisms, that may pose a health risk for people suffering from respiratory illnesses or having weakened immune systems. Even without these pathogens, the dust is known to trigger allergic and respiratory reactions. Furthermore, the North African dust arrives over the southeastern United States at the same time of year when photochemical air pollutants are at their highest levels, further exacerbating regional air quality problems.

In the summer of 2001, scientists reported that the iron in North African dust particles fertilizes the Gulf of Mexico waters, increasing the probability of blooms of toxic algae, commonly known as red tides. Red tides have been implicated in the die-off of great numbers of fish, shellfish, marine mammals, and birds as well as respiratory problems and skin irritations in humans. Enhanced levels of iron enable specialized bacteria to convert nitrogen in the water to a form that can be used by other marine life, triggering an explosive growth in the populations of toxic algae. North African dust may also be harming coral reefs in the Caribbean through nutrient enrichment that spurs the growth of populations of algae and phytoplankton that colonize the same environment as coral and interfere with its growth. Furthermore, North African dust may also contain a soil fungus that attacks coral reefs.

The story of North African dust transport illustrates how our investigation of the atmosphere is important in developing a better understanding not only of weather and climate but also of environmental issues that involve the atmosphere and atmospheric processes. Hence, our study will focus on the atmosphere's interactions with the other components of the Earth system, including land, ocean, and living organisms. The significance of these interactions will become evident as we develop basic understandings in atmospheric science.

Driving Question:

What is the composition and structure of the atmosphere?

Almost everyone seems to be interested in the weather, probably because it affects virtually every aspect of daily life—our clothing, the price of oranges and coffee in the grocery store, and sometimes even the outcome of a football game. Tranquil, pleasant weather allows us to enjoy a variety of outdoor activities (Figure 2.2). A turn

FIGURE 2.2
The variability of weather makes possible a wide variety of outdoor recreational activities.

FIGURE 2.3
Heavy snowfall can disrupt motor vehicle traffic and cause considerable inconvenience.

to stormy weather can bring mixed blessings: heavy rains wash out picnics but also benefit crops wilting under the searing summer sun. Occasionally, the weather is severe, and the impact may range from mere inconvenience to a disaster that is costly in human lives and property. Thick fog causes flight delays and cancellations, a night of subfreezing temperatures takes its toll on Florida citrus, and heavy snowfall snarls commuter traffic (Figure 2.3). But these impacts are minor when compared to the death, injury, and property damage that a tornado or hurricane can cause.

Regardless of where we live, each of us is well aware from personal experience that weather is variable. This variability prompted Mark Twain, not one to shy from exaggeration, to quip of spring weather in New England: "In the spring I have counted one hundred and thirty-six different kinds of weather inside of four-and-twenty hours."[*] Of course, weather is not equally variable everywhere. For example, the temperature contrast between winter and summer is much more pronounced in the Dakotas, where summers are very warm and winters are bitter cold, than in south Florida, where the weather is usually warm year-round.

This chapter covers the evolution of the atmosphere, the composition of air, how meteorologists monitor surface and upper-air properties of the atmosphere, the temperature profile of the atmosphere, and the special electromagnetic characteristics of the upper atmosphere. We begin by defining weather and climate.

---

[*] "Address of Mr. Samuel L. Clemens: The Weather in New England," *New England Society in the City of New York*, *Annual Report*, 1876, p. 59.

Atmosphere, Weather, and Climate

Earth's **atmosphere** encircles the globe as a relatively thin envelope of gases and tiny, suspended particles. Compared to the planet's diameter, the atmosphere is like the thin skin of an apple. About half of the atmosphere's mass is concentrated within about 5500 m (18,000 ft) of Earth's surface and 99% of its mass is below an altitude of 32 km (20 mi).[**] Yet, this thin atmospheric skin is essential for life and the orderly functioning of physical and biological processes on Earth. The atmosphere shields organisms from exposure to hazardous levels of ultraviolet radiation; it contains the gases necessary for life-sustaining processes of cellular respiration and photosynthesis; and it supplies the water required by all forms of life.

The atmosphere is the site of weather and climate. **Weather** is defined as the state of the atmosphere at some place and time, described in terms of such variables as temperature, humidity, cloudiness, precipitation, and wind speed and direction. A place and time must be specified when describing weather because the atmosphere is dynamic and its state changes from one place to another and with time. At the same hour, the weather may be cold and snowy in New York City, warm and humid in Miami, and hot and dry in Phoenix. *If you don't like the weather, wait a minute* is an old saying that is not far from the truth in many areas of the nation. From personal experience, we know that tomorrow's weather may differ markedly from today's weather. **Meteorology** is the study of the atmosphere and the processes that cause weather.

Climate is popularly defined as weather conditions at some locality averaged over a specified time period. By international convention, average values of climatic elements are computed over a 30-year period beginning with the first year of a decade. At the close of a decade, the averaging period is moved forward ten years. For the first decade of the twenty-first century, 1971–2000 is the official averaging period. Thirty-year average monthly and annual temperatures and precipitation totals are commonly used to describe the climate of some locality. Other useful climatic parameters include average seasonal snowfall, length of growing season, percent of possible sunshine, and number of days with dense fog. Climate is the ultimate environmental control in that it governs, for example,

---

[**] For unit conversions, see Appendix I.

what crops can be cultivated, the fresh water supply, and the average heating and cooling requirements for homes.

In addition to average values of weather elements, climate encompasses extremes in weather (e.g., highest and lowest temperature, greatest 24-hr snowfall, most intense rainfall). Tabulation of extreme values usually covers the entire period of record (or at least for the period when observations were made at the same location). Specifying extremes in weather provides information on the variability of climate at a particular place and gives a more complete and useful description of climate. Farmers, for example, are interested in knowing not only the average summer rainfall, but also the frequency of extremely wet or dry summers. **Climatology** is the study of climate, its controls, and spatial and temporal variability.

Evolution of the Atmosphere

Earth's atmosphere is the product of a lengthy evolutionary process that began at the planet's birth about 4.6 billion years ago. Astronomers scanning the solar system and geologists analyzing evidence obtained from meteorites, rocks, and fossils have given us a reasonable, albeit as yet incomplete, account of the origins of the atmosphere.

PRIMEVAL PHASE

Earth, the sun, and the entire solar system are believed to have evolved from an immense cloud of dust and gases within the Milky Way galaxy. Earth's mass grew by accretion as the planet swept up cosmic dust in its path and its surface was bombarded by meteorites. At the very beginning, Earth's atmosphere was probably mostly hydrogen and helium plus some hydrogen compounds including methane (CH_4) and ammonia (NH_3). This earliest atmosphere eventually escaped to space. In time, as Earth congealed, volcanic activity began spewing huge quantities of lava, ash, and gases. By about 4.4 billion years ago, the planet's gravitational field was strong enough to retain a thin gaseous envelope, Earth's primeval atmosphere.

The principal source of atmospheric gases was **outgassing**, the release of gases from rock through volcanic eruptions and the impact of meteorites on the rocky surface of the planet. Perhaps as much as 85% of all outgassing took place within a million or so years of the planet's formation, but outgassing has persisted at a

slower rate throughout the planet's existence. Outgassing produced a primeval atmosphere that was mostly carbon dioxide (CO_2), with some nitrogen (N_2) and water vapor (H_2O), and trace amounts of methane, ammonia, sulfur dioxide (SO_2), and hydrochloric acid (HCl). Radioactive decay of an isotope of potassium (potassium-40) in the planet's bedrock added argon (Ar), an inert (chemically non-reactive) gas, to the evolving atmosphere. Dissociation of water vapor into its constituents, hydrogen and oxygen, by solar ultraviolet radiation contributed a small amount of free oxygen to the primeval atmosphere. (The lighter hydrogen, having high molecular speed, escaped to space.) Also, some oxygen was combined with other elements in various chemical compounds such as carbon dioxide.

Scientists suggest that between 4.5 and 2.5 billion years ago, the sun was about 30% fainter than it is today. This did not mean a cooler planet, however, because of the abundance of carbon dioxide. Earth's CO_2-rich atmosphere was perhaps 10 to 20 times denser than the present atmosphere. Carbon dioxide slows the escape of Earth's heat to space, so that average surface temperatures may have been as high as 85 °C to 110 °C (185 °F to 230 °F).

By perhaps 4 billion years ago, cooling of the planet caused atmospheric water vapor to condense into clouds, and torrential rains gave rise to the ocean that eventually covered about 95% of the planet's surface. These rains also were key to a substantial decline in the concentration of atmospheric CO_2. Carbon dioxide dissolves in rainwater producing weak carbonic acid that reacts chemically with bedrock. The net effect of this large-scale geochemical process was increasing amounts of carbon chemically locked in rocks and minerals and less and less CO_2 in the atmosphere.

Living organisms also played an important role in the evolving atmosphere, primarily through **photosynthesis**, the process whereby plants use sunlight, water, and carbon dioxide to manufacture their food. A byproduct of this process is oxygen (O_2). Although vegetation is a *sink* for carbon dioxide, photosynthesis likely played a secondary role to the geochemical processes just described in removing carbon dioxide from the atmosphere. Photosynthesis was much more important in increasing the amount of free oxygen in the atmosphere and dates back at least 2.5 billion years when the first primitive forms of life (blue-green algae) appeared in ancient seas. During the subsequent 500 million years, most oxygen generated via photosynthesis combined with ocean sediments and very little entered the atmosphere. By about 2 billion years ago, however, oxidation of ocean sediments tapered off and photosynthetically generated oxygen began cycling into the atmosphere. With the concurrent decline in CO_2 due to geochemical processes, within 500 million years, oxygen became the second most abundant atmospheric gas after nitrogen.

Nitrogen (N_2), a product of outgassing, became the most abundant atmospheric gas because it is relatively inert chemically and has molecular speeds too slow to readily escape to space. Furthermore, nitrogen is not very soluble in water as compared to other atmospheric gases such as oxygen and carbon dioxide. All these factors greatly limit the rate at which nitrogen is cycled out of the atmosphere. While nitrogen continues to be generated as a minor component of volcanic eruptions, today the principal source of free nitrogen entering the atmosphere is the denitrification process involved in bacterial decay of the remains of plants and animals. But this is countered by an amount taken out of air by biological fixation (i.e., direct nitrogen uptake by leguminous plants such as clover and soybeans) and atmospheric fixation (i.e., the process whereby the high temperatures associated with lightning causes nitrogen to combine with oxygen to form nitrates).

With oxygen emerging as a major component of Earth's atmosphere, the *ozone shield* formed. In the portion of the upper atmosphere known as the stratosphere, incoming solar ultraviolet (UV) radiation powers reactions that convert oxygen to ozone (O_3) and ozone to oxygen (Chapter 3). Absorption of UV radiation in these reactions prevents potentially lethal intensities of UV radiation from reaching Earth's surface. Formation of the stratospheric ozone shield made it possible for terrestrial forms of life to evolve.

Although carbon dioxide has been a *minor* component of the atmosphere for at least 3.5 billion years, from time to time in the geologic past its concentration fluctuated with important implications for climate. All other factors being equal, more CO_2 in the atmosphere means higher temperatures at the Earth's surface. Geologic evidence points to a burst of volcanic activity on the Pacific Ocean floor about 100 to 120 million years ago. Some of the CO_2 released during that activity eventually found its way to the atmosphere and raised the global mean temperature by as much as 10 Celsius degrees (18 Fahrenheit degrees). During the Pleistocene Ice Age (1.7 million to 10,500 years ago), atmospheric carbon dioxide levels fluctuated, decreasing during episodes of glacial expansion and increasing

during episodes of glacial recession (although it is not clear whether variations in atmospheric CO_2 were the cause or effect of global-scale climate variations).

MODERN PHASE

Ultimately, these gradual evolutionary processes produced the modern atmosphere. The lower atmosphere continually circulates so that the principal atmospheric gases are well mixed and occur almost everywhere in about the same relative proportions up to an altitude of about 80 km (50 mi). That portion of the atmosphere is called the **homosphere**. Above 80 km, gases are stratified such that concentrations of the heavier gases decrease more rapidly with altitude than do concentrations of the lighter gases. The region of the atmosphere above 80 km is known as the **heterosphere**.

Nitrogen and oxygen are the chief gases of the homosphere. Not counting water vapor (which has a highly variable concentration), nitrogen (N_2) occupies 78.08% by volume of the homosphere, and oxygen (O_2) is 20.95% by volume. Henry Cavendish discovered this basic gaseous composition of air in 1781. The next most abundant gases are argon (0.93%) and carbon dioxide (0.0369%). As shown in Table 2.1, the atmosphere also has small quantities of helium (He), methane (CH_4), hydrogen (H), ozone (O_3), and many other gases. Unlike

TABLE 2.1
Gases Composing Dry Air in the Lower Atmosphere (below 80 km)

Gas	% by Volume	Parts per Million
Nitrogen (N_2)	78.08	780,840.0
Oxygen (O_2)	20.95	209,460.0
Argon (Ar)	0.93	9,340.0
Carbon dioxide (CO_2)	0.03694	369.4
Neon (Ne)	0.0018	18.0
Helium (He)	0.00052	5.2
Methane (CH_4)	0.00014	1.4
Krypton (Kr)	0.00010	1.0
Nitrous oxide (N_2O)	0.00005	0.5
Hydrogen (H)	0.00005	0.5
Xenon (Xe)	0.000009	0.09
Ozone (O_3)	0.000007	0.07

the atmosphere's principal gases, the percent volume of some of these trace gases varies with time and location within the homosphere.

In the heterosphere, above about 150 km (95 mi), oxygen is the chief atmospheric gas but occurs primarily in the atomic (O) rather than diatomic (O_2) form. Solar ultraviolet radiation dissociates O_2 into its constituent atoms. Two oxygen atoms can recombine into a molecule only by colliding with another atom or molecule, but the number of molecules per unit volume (the *number density*) at these high altitudes is so low that such collisions are infrequent. At lower altitudes the intensity of incoming solar ultraviolet radiation is less, thereby reducing the rate of dissociation of O_2. Also, with more molecules per unit volume at lower altitudes, molecular collisions are more frequent. Below about 100 km (60 mi), oxygen atoms recombine at a faster rate than oxygen molecules dissociate, so oxygen occurs mostly as O_2.

Earth's nitrogen/oxygen-dominated atmosphere contrasts strikingly with the carbon-dioxide-rich atmospheres of neighboring planets Venus and Mars. The atmosphere of Venus is almost 100 times denser than Earth's atmosphere and features an average surface temperature of about 460 °C (860 °F). The Martian atmosphere, on the other hand, is much thinner than Earth's atmosphere and has an average surface temperature ranging from about −60 °C (−76 °F) at the equator to as low as −123 °C (−189 °F) at the poles. These temperature differences exist in spite of the likelihood that all three planets began with chemically similar atmospheres. The atmospheres of Earth, Mars, and Venus evidently followed different evolutionary paths; for more on the Martian atmosphere, see this chapter's first Essay.

In addition to gases, Earth's atmosphere contains minute liquid and solid particles, collectively called **aerosols**. A flashlight beam in a darkened room reveals an abundance of suspended dust particles. Most aerosols individually are too small to be visible but in aggregates, such as the water droplets and ice crystals that compose clouds, they may be visible. Most aerosols occur in the lower atmosphere near their sources on Earth's surface and are derived from wind erosion of soil, ocean spray, forest fires, volcanic eruptions, and industrial and agricultural activities. Also, some aerosols, such as meteoric dust, enter the atmosphere from above.

It may be tempting to dismiss as unimportant those substances that make up only a small fraction of the atmosphere, but the significance of an atmospheric gas or aerosol is not necessarily related to its relative abundance. Water vapor, carbon dioxide, and ozone (O_3) occur in minute concentrations, yet they are essential for life. By volume, no more than about 4% of the lowest kilometer of the atmosphere is water vapor even in the warm, humid air over tropical seas and rainforests. Without water vapor, however, there would be no clouds, and no rain or snow to replenish soil moisture, rivers, lakes, and seas. Although comprising only 0.0369% of the homosphere, carbon dioxide is essential for photosynthesis. Furthermore, water vapor and carbon dioxide absorb and emit infrared radiation, elevating the temperature of the lower atmosphere to a range that makes life possible on Earth. Although the volume percentage of ozone is minute, the formation and dissociation of this essential gas shields organisms, including ourselves, from exposure to potentially lethal intensities of ultraviolet radiation from the sun.

The aerosol concentration of the atmosphere is also relatively small, yet these suspended particles participate in important processes. As demonstrated in Chapter 7, some aerosols function as nuclei for the development of clouds. Furthermore, certain aerosols such as sulfurous particles can influence air temperature by interacting with incoming solar radiation (Chapter C).

AIR POLLUTION

Human activity also plays a role in the composition of air primarily by contributing air pollutants. An **air pollutant** is a gas or aerosol that occurs at a concentration that threatens the wellbeing of living organisms (especially humans) or disrupts the orderly functioning of the environment. Many of these substances occur naturally in the atmosphere. Sulfur dioxide (SO_2) and carbon monoxide (CO), for example, are normal minor gaseous components of the atmosphere that become pollutants when their concentrations approach or exceed the tolerance limits of organisms. At sufficiently high concentrations, sulfur dioxide irritates the throat and lungs and impairs the respiratory system's defenses against foreign particles and bacteria. Carbon monoxide is a colorless, odorless, and tasteless asphyxiating agent (reducing the blood's oxygen-carrying ability). Certain air pollutants, however, do not occur naturally in the atmosphere, and some of these are hazardous to human health even at very low concentrations. An example is benzene, which is known to cause cancer.

A substance that is harmful immediately upon emission into the atmosphere is designated a *primary air pollutant*. Carbon monoxide in automobile exhaust is an example. In addition, within the atmosphere, chemical reactions involving primary air pollutants, both gases and aerosols, produce *secondary air pollutants*. An example is *photochemical smog*, a mixture of aerosols, ozone, nitrogen oxides, and hydrocarbons, generated by the action of sunlight on motor vehicle exhaust, some industrial emissions, and volatile organic compounds emitted by certain vegetation.

Air pollutants are products of both natural events and human activities. Natural sources of air pollutants include forest fires, dispersal of pollen, wind erosion of soil, decay of dead plants and animals, and volcanic eruptions. The single most important human-related source of air pollutants is the internal combustion engine that propels most motor vehicles. Many industrial sources also contribute to air quality problems. Unless emissions are controlled, pulp and paper mills, zinc and lead smelters, oil refineries, and chemical plants can be major sources of air pollutants. Additional pollutants come from fuel combustion for space heating and generation of electricity, refuse burning, and various agricultural activities such as crop dusting.

The U.S. Environmental Protection Agency (EPA) has set ambient air quality standards for six criteria air pollutants: carbon monoxide, lead, oxides of nitrogen, ozone, particulate matter, and sulfur dioxide. *Ambient air* is the outside air that we breathe. Standards for the six criteria air pollutants are based on scientific studies that demonstrate that adverse effects are likely only after concentrations exceed a specific *threshold* value (hence, the *criteria* designation). Standards are of two types, primary and secondary. *Primary air quality standards* are defined as the maximum exposure levels that can be tolerated by humans without ill effects. *Secondary air quality standards* are defined as the maximum levels that are allowable to minimize the impact on crops, visibility, personal comfort, and climate. Emission standards for stationary sources (e.g., coal-fired electric power plants) and mobile sources (e.g., motor vehicles) are set to ensure, theoretically at least, that once pollutants are emitted to the atmosphere, ambient air quality standards are not exceeded.

A geographical region where the ambient air meets or does better than primary air quality standards is designated an *attainment area*. Compliance is established by direct sampling of air over specified time periods. A region where the ambient air does not meet the primary standard for one or more criteria air pollutants is designated a *nonattainment area* for those

pollutants. In that case, remedial action (i.e., reduction of emissions) is required.

Investigating the Atmosphere

Our understanding of the atmosphere, weather, and climate is the culmination of centuries of painstaking inquiry by scientists from many disciplines. Physicists, chemists, astronomers, and others have applied basic scientific principles in unlocking the mysteries of the atmosphere. The roots of modern meteorology, in fact, go back to about 340 B.C. and Aristotle's *Meteorologica*, the first treatise on atmospheric science.[***] Although much progress has been made over the years, many questions remain regarding the workings of the atmosphere, weather, and climate. Today's atmospheric scientists (meteorologists and climatologists) continue the efforts of their predecessors, and although armed with more sophisticated tools such as satellites, radar, and supercomputers, they still rely on the scientific method of inquiry.

THE SCIENTIFIC METHOD

The **scientific method** is a systematic form of inquiry involving observation, interpretation, speculation, and reasoning. The Antarctic ozone hole illustrates how scientists apply the scientific method. In 1985, the British Antarctic Survey team reported a drastic decline in the amount of stratospheric ozone over Antarctica during the Southern Hemisphere spring (mainly in September and October). The region of depleted ozone covered an area almost as large as North America and was later dubbed the *Antarctic Ozone Hole*. In checking records, scientists found evidence from ground-based instruments of massive ozone depletion during each of the eight prior Antarctic springs.

At first, scientists dismissed the Antarctic ozone hole as a product of instrument error—not unexpected considering the brutal environmental conditions in Antarctica. Other scientists argued that the Antarctic ozone hole was real and the product of natural seasonal changes in the polar atmospheric circulation. A third hypothesis attributed the Antarctic ozone hole to pollution. In an effort to solve the mystery, an intensive field study was launched during the Antarctic spring of

---

[***] For a timeline of historical events in meteorology, see Appendix II.

1987 involving monitoring of the Antarctic stratosphere by satellites, specially outfitted aircraft, and instrumented balloon probes. That field investigation not only confirmed the existence of the Antarctic ozone hole, but also detected exceptionally high levels of chlorine monoxide (ClO), a gas known from laboratory studies to be a product of chemical reactions that destroy ozone. For many scientists, discovery of ClO in the Antarctic stratosphere was the *smoking gun* linking ozone depletion to a group of chemicals known as chlorofluorocarbons (CFCs). We have more to say about threats to stratospheric ozone in Chapter 3.

From our Antarctic ozone example, it is evident that the scientific method involves a sequence of steps in which scientists (1) identify specific questions related to the problem at hand; (2) propose an answer to one of these questions in the form of an educated guess; (3) state the educated guess in such a way that it can be tested, that is, formulate a **hypothesis**; (4) predict the outcome of the test if the hypothesis were correct; (5) test the hypothesis by checking to see if the prediction is correct; and (6) reject or revise the hypothesis if the prediction is wrong. A hypothesis that has stood the test of time and is generally accepted by the scientific community is a **scientific theory**.

In our Antarctic ozone illustration we divided the scientific method into a sequence of steps but in actual practice, scientists do not always follow this scheme cookbook style, and discrete steps are often combined into a single avenue of inquiry. Furthermore, the scientific method is not a formula for creativity because it does not provide the key idea, the hunch, or the educated guess that spontaneously springs to mind and forms the basis of a hypothesis. Rather, it is a technique for assessing the validity of a creative idea regardless of how it originates.

As in the Antarctic ozone hole example, a hypothesis is a tool that suggests new field studies, experiments or observations, or opens new avenues of inquiry. Hence, even a rejected hypothesis may be fruitful. Above all, scientists must keep in mind that a hypothesis is merely a working assumption that may be accepted, modified, or rejected. They must be objective in evaluating a hypothesis and not allow personal biases or expectations to cloud that evaluation. In fact, scientists search for observations or information that could disprove their hypothesis. If they find such evidence, they revise their hypothesis to account for disparate observations or information. Inquiry, creative

thinking, and imagination are stifled when a hypothesis is considered immutable.

A new hypothesis (or an old, resurrected one) is sometimes hotly debated within the scientific community. History attests to the natural human resistance to accepting new ideas that threaten to displace long-held notions. Also, disagreements among scientists on a particularly controversial issue sometimes receive considerable media attention, which may confuse the general public. For some time now, atmospheric scientists have been engaged in a sometimes contentious public debate as to if, when, and to what extent burning coal and oil for power might impact global climate. In some cases, public reaction may be summed up as follows: "Well, if the so-called experts can't agree among themselves, who am I to believe? Is there really a problem after all?" Debate and disagreement among scientists are essential steps in the process of reaching understanding; they generate useful suggestions, stimulate new thinking, and uncover errors. In fact, debate and skepticism buffer the scientific community from too hastily accepting new ideas. If a hypothesis survives the scrutiny and skepticism of scientists and the public, it is probably correct.

ATMOSPHERIC MODELS

In applying the scientific method, scientists often find that models aid their inquiry and this is certainly the case in the atmospheric sciences. Because we use models throughout this course, we consider at the outset the general objectives and limitations of scientific models, especially as they apply to meteorology and climatology.

A **scientific model** is an approximate representation or simulation of a real system. A *system* is an entity having components that function and interact with one another in an orderly and predictable manner that can be described by fundamental physical principles (or natural laws). The **Earth–atmosphere system**, for example, encompasses the Earth's surface (i.e., continents, ocean, ice sheets) plus atmosphere. A model includes only what are considered to be the essential variables or characteristics of a system. For example, to learn how to improve the fuel efficiency of automobiles, engineers might examine a scale-model automobile in a wind tunnel. An automobile can be designed to minimize its air resistance, thereby increasing its miles per gallon. The shape of the model automobile is the critical variable and is the focus of study. Other variables, such as the color of the model automobile or whether it is equipped with a miniature CD-player, are irrelevant to the experiment and are ignored. In constructing a model, trial and error often determine which variables are or are not essential.

Because models are not cluttered with extraneous and distracting details, they may provide important insights as to how things interact, or they may facilitate critical thinking about complex phenomena. Models also are used to predict how a real system might respond to internal or external forcing (i.e., environmental change). Depending on their particular function, scientific models are classified as conceptual, graphical, physical, or numerical.

A *conceptual model* is an abstract idea that represents some fundamental law or relationship. The *geostrophic wind* (described in Chapter 8) is a conceptual model that relates the interaction of certain forces operating in the atmosphere to straight, horizontal winds that blow at altitudes of 1000 m (3300 ft) or higher. A *graphical model* compiles and displays data in a format that readily conveys meaning. For example, a weather map (Figure 1.2) integrates weather observations taken simultaneously at hundreds of locations into a coherent representation of the state of the atmosphere at a specific time.

A *physical model* is a miniaturized version of some system. Almost half a century ago, scientists studied the patterns exhibited by a fluid in a flat rotating pan in experiments designed to simulate the planetary scale atmospheric circulation. More recently, Purdue University scientists simulated a tornadic circulation in a specially designed *Tornado Vortex Chamber*, measuring 5 m wide, 5 m deep, and 8 m high (Figure 2.4). Precision weather instruments monitored air motions within small tornado-like vortices (swirls) that developed. Using this model, scientists improved their understanding of the internal characteristics of real tornadoes.

Today, atmospheric scientists usually depend on numerical models rather than physical models for investigating the atmosphere, weather, and climate. A *numerical model* consists of one or more mathematical equations that describe the relationship among variables of a system. Numerical models of the Earth–atmosphere system are usually programmed in a computer that can accommodate enormous quantities of data and perform calculations with lightning speed. Variables in the numerical model (such as temperature or cloud cover) are

A

B

FIGURE 2.4
(A) Exterior view of Purdue University's Tornado Vortex Chamber, used to simulate the circulation within tornadoes (B). [Courtesy of John T. Snow, University of Oklahoma, and C.R. Church, Miami University of Ohio.]

manipulated, individually or in groups, to assess their influence on the behavior of the system.

Numerical models have been used to forecast weather since the 1950s. More recently, other types of numerical models have been used to predict the potential impact on the global climate of rising levels of atmospheric carbon dioxide and other infrared-absorbing gases. As discussed in more detail in Chapters 3 and C, the atmospheric carbon dioxide concentration has been rising for more than a century, primarily as a byproduct of fossil fuel (coal, oil, and natural gas) burning. Higher air temperatures may be the consequence because, as noted earlier, carbon dioxide slows the escape of Earth's heat to space. Atmospheric scientists employ numerical *global climate models* in experiments to compute the magnitude of warming that might accompany a continued increase in atmospheric carbon dioxide concentration. Essentially they follow three steps: (1) Design a global climate model that incorporates the various controls of climate and accurately depicts the long-term average pattern of air temperature worldwide. (2) Holding all other climate-control variables constant, elevate the carbon dioxide concentration (typically, it is doubled), and the global climate model computes a new worldwide average air temperature pattern. (3) Subtract the present temperature pattern from the temperature pattern predicted for the CO_2-enriched atmosphere to obtain the net temperature change.

All models *simulate* reality and as such are subject to error. A weather map portrays the state of the atmosphere at a particular time based on weather observations at discrete locations (weather stations), which may be hundreds of kilometers apart. Weather satellite imagery provides a much more continuous field of view but the spatial resolution of satellite sensors is still limited. A potential difficulty with numerical models concerns the accuracy of their component equations. Equations only approximate the way a system really works in nature, and may not adequately account for all relevant variables. This is one reason why the skill of long-range weather forecasting, based on numerical models of the Earth–atmosphere system, declines in accuracy as the forecast period lengthens (Chapter A).

Monitoring the Atmosphere

Much of what we know about Earth's atmosphere, weather, and climate is derived from direct (*in situ*) and

remote sensing of the Earth–atmosphere system. Since invention of the first weather instruments in the seventeenth century, monitoring of the atmosphere has undergone considerable refinement. Denser monitoring networks, more sophisticated instruments and communications systems, and better trained observers have produced an increasingly detailed, reliable, and representative record of weather and climate. Monitoring has progressed on two fronts: surface and upper-air observations.

SURFACE OBSERVATIONS

Early weather observers used primitive instruments or made qualitative assessments of the weather, jotting down observations in journals or diaries. The first systematic weather observations in North America were made in 1644–45 at Old Swedes Fort (now Wilmington, DE). The observer was John Campanius Holm, chaplain of the Swedish military expedition. Other temperature records began in Philadelphia in 1731; in Charleston, SC in 1738; and in Cambridge, MA in 1753. The New Haven, CT temperature record began in 1781 and continues uninterrupted today. Thomas Jefferson (third President of the United States) and the Reverend James Madison (president of the College of William and Mary) are credited with the nation's first simultaneous weather observations. Over a six-week period in 1778, they recorded temperature, air pressure, and wind at Monticello and Williamsburg, VA.

On 2 May 1814, James Tilton, M.D., U.S. Surgeon General, issued an order that in retrospect was the first step in the eventual establishment of a national network of weather-observing stations. Tilton directed the Army Medical Corps to begin a diary of weather conditions at army posts, with responsibility for observations falling to the post's chief medical officer. Tilton's objective was to assess the relationship between weather and the health of the troops, for it was a popular notion at the time that weather and its seasonal changes were important factors in the onset of disease. (Even well into the twentieth century, more troops died from disease than combat.) He also wanted to learn more about the climate of the then sparsely populated interior of the continent.

The ongoing War of 1812 prevented immediate compliance with Tilton's order. In 1818, Joseph Lovell, M.D., succeeded Tilton as Surgeon General and issued formal instructions for taking weather observations. Reports soon began trickling into the Army Medical Department and by 1838, 16 army posts had compiled at least 10 complete (although not always successive) years of weather records. By the close of the U.S. Civil War, weather records had been tabulated for varying periods at 143 army posts. In 1826, Lovell began compiling, summarizing and publishing the data and for this reason Lovell, rather than Tilton, is sometimes credited with founding the federal government's system of weather observation.

In the mid-1800s, Joseph Henry, first secretary of the Smithsonian Institution in Washington, DC, established a national network of volunteer observers who mailed in monthly weather reports. The number of citizen observers (consisting of farmers, educators, public servants) peaked at nearly 600 just prior to the U.S. Civil War. Henry understood the value of rapid communication of weather data and was quick to realize the potential of the newly invented telegraph in achieving this goal. In 1849, Henry persuaded the heads of several telegraph companies to direct their telegraphers in major cities to make and transmit weather observations at the opening of business each day free of charge. Henry supplied thermometers and barometers (for measuring air pressure). Using simultaneous weather observations, in 1850 Henry prepared the first current national weather map and later regularly displayed the daily weather map for public viewing in the Great Hall of the Smithsonian building. By 1860, some 42 telegraph stations, mostly east of the Mississippi River, were participating in the Smithsonian network.

In the 1860s, growing public alarm over the great loss of life and property from shipwrecks caused by surprise storms sweeping across the Great Lakes spurred the federal government to take a greater role in weather observation and forecasting. In 1868 and 1869 alone, storms sunk or damaged 3,078 vessels with the loss of 530 lives. In response, the naturalist Increase A. Lapham of Milwaukee, WI urged his congressman, Halbert E. Paine, to convince Congress to establish a telegraph-based storm-warning system for the Great Lakes. Lapham was inspired by the success of Henry's Smithsonian network and another telegraph-based network operated by Cleveland Abbe in the Midwest. President Ulysses S. Grant signed the resolution into law on 9 February 1870 and the network, initially composed of 24 stations, began operating on 1 November 1870 under the auspices of the U.S. Army Signal Corps.

Although the storm-warning network was originally authorized for the Great Lakes, in 1872 Congress appropriated funds to expand the network to the entire nation. The new weather-observing network soon encompassed stations formerly operated by the Surgeon General, Smithsonian Institution, the U.S. Army Corps of Engineers, and Cleveland Abbe. The number of Signal Corps stations regularly reporting daily weather observations by telegraph reached 110 by 1880.

On 1 July 1891, the nation's weather network was transferred from military to civilian hands in a new Weather Bureau within the U.S. Department of Agriculture, with a special mandate to provide weather and climate guidance for farmers. Later, aviation's growing need for weather information spurred the shift of the Weather Bureau to the Commerce Department on 1 July 1940. Many cities saw their Weather Bureau offices moved from a downtown location to an airport usually in a rural area well outside the city (Figure 2.5). In 1965, the Weather Bureau was reorganized as the National Weather Service (NWS) within the Environmental Science Services Administration (ESSA), which became the National Oceanic and Atmospheric Administration (NOAA) in 1971.

During the early and mid-1990s, facilities of the National Weather Service underwent extensive modernization with the goal of upgrading the quality and reliability of weather observations and forecasts. As part of this $4.5 billion project, new or restructured NWS Forecast Offices were established at 115 locations across the nation (Figure 2.6). Some 1700 automated weather stations, many of them at airports, replaced the old

FIGURE 2.5
U.S. Weather Bureau meteorologist Hilda Goodrich climbs a 12-m (40-ft) tower on the roof of a six-story building in Green Bay, WI to repair wind equipment. Ms. Goodrich was reported to be the first woman in charge of a U.S Weather Bureau office (1943–44). At the time, the city's Weather Bureau office was located downtown, but was later relocated to the airport 15 km southwest of downtown.

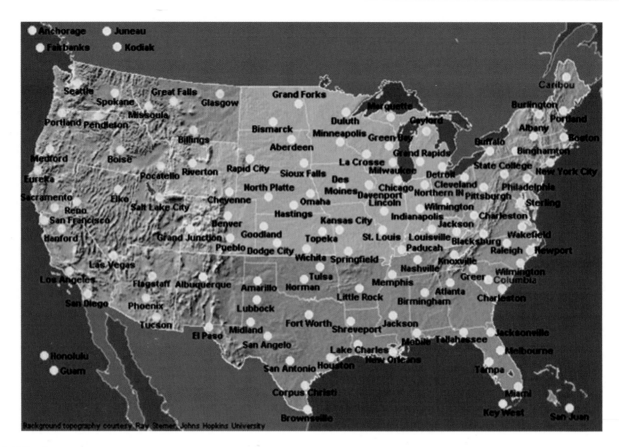

FIGURE 2.6
Locations of National Weather Service Weather Forecast Offices.

manual system of hourly observations. This **Automated Surface Observing System (ASOS)** consists of electronic sensors, computers, and fully automated communications ports (Figure 2.7). Twenty-four hours a day, ASOS feeds data to NWS Weather Forecast Offices and airport control towers. In addition, Doppler weather radar was installed at 113 sites nationwide.

Besides the numerous weather stations that provide observational data for weather forecasting and aviation, another 8000 cooperative weather stations are scattered across the nation (Figure 2.8). These stations (rooted in the old Army Medical Corps and Smithsonian networks) are cooperative in that volunteers supply their time and labor to monitor instruments and the National Weather Service provides instruments and data management. The principal function of member stations of the **NWS Cooperative Observer Network** is to record daily precipitation and maximum/minimum temperatures for hydrologic, agricultural, and climatic purposes.

FIGURE 2.7
Sensors in the Automated Surface Observing System (ASOS) of the National Weather Service measure cloud height, visibility, precipitation, temperature, dewpoint, and wind speed, and identify freezing rain and snow.

FIGURE 2.8
This NWS cooperative observer station at Mount Mary College, Milwaukee, WI is equipped with a rain gauge and thermometers housed in an instrument shelter.

UPPER-AIR OBSERVATIONS

Kites were used in early investigations of the upper atmosphere. In July 1749 in Glasgow, Scotland, Alexander Wilson attached several thermometers to six paper kites and flew them in tandem. Wilson designed the apparatus so that the thermometers would fall to the ground unbroken at predetermined intervals. In this way, he was the first to obtain a free-air temperature profile of the lower atmosphere (up to an altitude of perhaps 60 m or 200 ft).

Benjamin Franklin, an inventive genius, is credited with performing an experiment using a kite to demonstrate the electrical nature of lightning. On 10 June 1752, during a developing thunderstorm, Franklin touched a metal key that he had attached to the end of a kite string. Fortunately, lightning did not strike the kite because a single bolt could have killed him. Franklin detected static electricity that had built up along the kite string, an observation that inspired him the next year to extol the effectiveness of lightning rods in his *Poor Richard's Almanac.*

On 27 August 1804, French scientists J. L. Gay Lussac and Jean Biot ushered in the age of manned balloon exploration of the atmosphere. They took air samples, measured temperature and humidity, and on one ascent reached an altitude of 7000 m (23,000 ft). In 1862, the British scientist James Glaisher and his fellow aeronaut Henry Coxwell took weather instruments aloft in a series of balloon ascents over Wolverhampton, England. They nearly perished from severe cold and oxygen deprivation when they set a manned balloon altitude record of 9000 m (29,500 ft).

Through the early part of the twentieth century, weather instruments borne by kites, aircraft, and balloons provided data chiefly on the lowest 5000 m (16,000 ft) of the atmosphere. On 4 August 1894 at Harvard's Blue Hill Observatory near Boston, MA, kites were used for

the first time to carry aloft a self-recording thermometer (*thermograph*). The instrument provided an air temperature profile to an altitude of 427 m (1400 ft) above the ground. From 1907 to 1933, the U.S. Weather Bureau operated a network of weather kite stations at several locations, mostly in the central part of the nation. Box kites were equipped with a recording instrument (called a *meteorgraph*) that profiled air pressure, temperature, humidity, and wind speed up to a maximum altitude of about 3,000 m (10,000 ft). The longest operating of those stations, at Ellendale, ND was closed in 1933. Among reasons cited for discontinuing the kite network was the relatively low maximum altitude reached by kites. For several years afterward, the U.S. Weather Bureau relied mostly on regular 5 a.m. (EST) aircraft observations originating at as many as 30 locations to probe the atmosphere to altitudes up to 4900 m (16,000 ft), but this practice proved too hazardous and costly.

A leap forward in monitoring higher altitudes came in the late 1920s with invention of the first radiosonde. A **radiosonde** is a small instrument package equipped with a radio transmitter that is carried aloft by a helium- (or hydrogen-) filled balloon. This device transmits altitude readings, called a **sounding**, of temperature, air pressure, and dewpoint (a measure of humidity) to a ground station. Radiosonde data are received immediately; no recovery of a recording instrument is needed. The first official Weather Bureau radiosonde was launched at East Boston, MA in 1937. By World War II, meteorologists were tracking radiosonde movements from ground stations using radio direction-finding antennas, thereby monitoring variations in wind direction and speed with altitude. A radiosonde used in this way is called a **rawinsonde**.

Today, radiosondes are launched simultaneously at 12-hr intervals (at 0000Z and 1200Z) from hundreds of ground stations around the world (Figure 2.9). The balloon bursts at an altitude of about 30,000 m (100,000 ft), and the instrument package descends to the surface under a parachute. In the United States, about 20% of radiosondes are recovered, refurbished, and reused. Each radiosonde contains a prepaid mailbag and instructions to its finder for returning the instrument to the National Weather Service.

A **dropwindsonde** is similar to a rawinsonde except that instead of being launched by a balloon from a surface station, it is dropped from an aircraft. The instrument package descends on a parachute at about 18 km (11 mi) per hr and along the way radios data back to the aircraft every few seconds. The dropwindsonde was

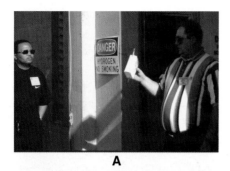

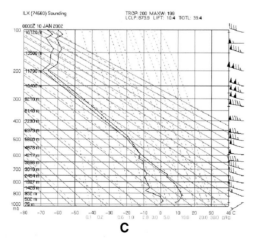

FIGURE 2.9
A radiosonde (A) is a small instrument package equipped with a radio transmitter. Borne by a hydrogen or helium-filled balloon (B), a radiosonde measures vertical profiles of air temperature, pressure, and dewpoint (C) up to an altitude of about 30,000 m (100,000 ft). Tracking horizontal movements of the radiosonde gives a profile of wind speed and direction (symbols on right).

developed at the National Center for Atmospheric Research (NCAR) in Boulder, CO to obtain soundings over the ocean where conventional rawinsonde stations are few and far between. Dropwindsondes provide vertical profiles of air temperature, pressure, dewpoint, and wind.

Robert H. Goddard is credited with conducting the first rocket probe of the atmosphere in 1929. The payload of Goddard's primitive rocket included a thermometer and a barometer. World War II spurred advances in rocketry so that by the late 1940s, instrumented rockets were investigating the composition of the middle and upper atmosphere. In March 1947, a vertically fired V2 rocket took the first successful photographs of Earth's cloud cover from altitudes of 110 to 165 km (70 to 100 mi). This and subsequent rocket probes of the atmosphere convinced scientists of the value of cloud pattern photography in monitoring weather systems and inspired the first serious proposals for orbiting a weather satellite.

In the mid- to late 1950s, the United States' fledgling space program was directed at developing a launch vehicle (rocket) capable of putting a satellite in orbit. With the former Soviet Union's successful orbiting of *Sputnik I* on 4 October 1957, the age of remote sensing by satellite had begun. On 1 April 1960 the United States orbited the world's first weather satellite, TIROS-1 (Television and Infrared Observation Satellite). Since then, a series of weather satellites equipped with increasingly sophisticated sensors have been orbited.

Weather satellites are invaluable tools in weather observation and storm surveillance. These *eyes in the sky* offer distinct advantages over the network of surface weather stations by providing a broad and nearly continuous field of view. Surface weather stations are discrete and often widely spaced data sources, and weather observations are sparse or absent over vast areas of Earth's surface, especially the ocean. Five geostationary satellites provide nearly complete coverage of the globe poleward to about 60 degrees latitude. Weather satellites carry sensors capable of monitoring cloud patterns, temperature, water vapor concentration, upper-air winds, rainfall, and the life cycles of severe storms. By the early 1990s, vertical profiling of the atmosphere's temperature and humidity by satellite became routine practice.

A variety of technologies provides detailed information on the state of the atmosphere. The remainder of this chapter considers two important properties of the atmosphere: its average vertical temperature profile and electromagnetic characteristics of the upper atmosphere.

Temperature Profile of the Atmosphere

For convenience of study, the atmosphere is subdivided into concentric layers based upon the vertical profile of the average air temperature, as shown in Figure 2.10. Almost all weather occurs within the lowest layer, the **troposphere**, which extends from Earth's surface to an average altitude ranging from about 6 km (3.7 mi) at the poles to about 20 km (12 mi) at the equator. Normally, but not always, the temperature within the troposphere falls with increasing altitude. Hence, the air temperature is usually lower on mountaintops than in surrounding

FIGURE 2.11
Within the troposphere, average air temperature falls with increasing altitude so that it is generally colder on mountain peaks than in lowlands.

lowlands (Figure 2.11). On average, within the troposphere, the air temperature drops 6.5 Celsius degrees for every 1000 m increase in altitude (3.5 Fahrenheit degrees per 1000 ft). The upper boundary of the troposphere, called the **tropopause**, is a transition zone between the troposphere and the next higher layer, the stratosphere.

The **stratosphere** extends from the tropopause up to about 50 km (31 mi). On average, the temperature does not change with increasing altitude through the lower portion of the stratosphere. A constant temperature condition is described as *isothermal*. Above about 20 km (12 mi), the temperature rises with increasing altitude up to the top of the stratosphere, the **stratopause**. At the stratopause, the temperature is not much lower than it is at sea level. The stratosphere is ideal for jet aircraft travel because it is above the weather and offers excellent visibility and generally smooth flying conditions. But because little air is exchanged between the troposphere and the stratosphere, pollutants that reach the lower stratosphere may persist there for lengthy periods. Gases and aerosols thrown into the stratosphere during violent volcanic eruptions, for example, can reside there for many months to years and perhaps trigger short-term changes in climate (Chapter C). Other pollutants threaten the protective ozone layer within the stratosphere.

The stratopause is the transition zone between the stratosphere and the next higher layer, the **mesosphere**. Within this layer, the temperature once again falls with increasing altitude. The mesosphere

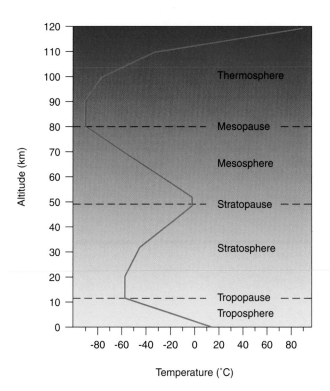

FIGURE 2.10
The variation in average air temperature with altitude within the atmosphere. Based on this vertical temperature profile, the atmosphere is subdivided vertically into the troposphere, stratosphere, mesosphere, and thermosphere.

extends up to the **mesopause**, which is about 80 km (50 mi) above Earth's surface and features the lowest average temperature in the atmosphere (about −95 °C or −139 °F). Above this layer is the **thermosphere**, where temperatures at first are isothermal and then rise rapidly with increasing altitude. Within the thermosphere, air temperature is more variable with time than in any other region of the atmosphere.

The Ionosphere and the Aurora

The **ionosphere**, located primarily within the thermosphere, from a base of 70 to 80 km (43 to 50 mi) to an indefinite altitude, features a relatively high concentration of ions and electrons. An **ion** is an atomic-scale particle that carries an electrical charge. High-energy solar radiation entering the upper atmosphere strips electrons from oxygen and nitrogen atoms and molecules, converting them to positively charged ions. The highest concentration of ions is in the lower portion of the thermosphere.

Although conditions in the upper atmosphere apparently have no influence on day-to-day weather in the troposphere, the ionosphere reflects radio waves and is important for long-distance radio transmission. Radio signals travel in straight lines and bounce back and forth between Earth's surface and the ionosphere. By repeated reflections, a radio signal may travel completely around the globe. For more on this phenomenon, see the Essay.

The ionosphere is also the site of the spectacular **aurora** (Latin for "dawn")—aurora borealis (northern lights) in the Northern Hemisphere and aurora australis (southern lights) in the Southern Hemisphere. Auroras often appear in the night sky as overlapping curtains of greenish-white light, occasionally fringed with pink (Figure 2.12). The bottom of the curtains is at an altitude of about 100 km (62 mi) and the top is at 400 km (250 mi) or higher.

An aurora is triggered by the **solar wind**, a stream of super-hot electrically charged subatomic particles (protons and electrons) that continually emanates from the sun and travels into space at speeds of 400 to 500 km (250 to 300 mi) per second. Like a rock in a mountain stream, Earth's magnetic field deflects the solar wind and, in so doing, is deformed into a teardrop-shaped cavity surrounding the planet known as Earth's **magnetosphere** (Figure 2.13). The solar wind compresses the front end of the magnetosphere and elongates the back end. A complex interaction between the solar wind and the magnetosphere generates beams of electrons that collide with atoms and molecules within the ionosphere. Collisions rip apart molecules, excite atoms, and increase ion and electron densities. As atoms shift down from their excited (energized) state and as ions combine with free electrons, they emit radiation, part

FIGURE 2.12
Aurora borealis (northern lights) nearly directly overhead, photographed at Menasha, WI at 9:30 p.m. CST on 5 November 2001. The curving arcs of orange light are reflections from a light at a nearby parking lot. [Courtesy of John Beaver, University of Wisconsin, Fox Valley.]

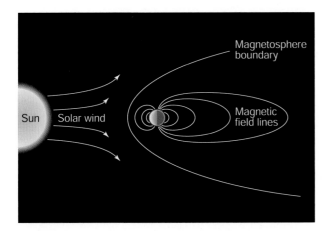

FIGURE 2.13
Earth's magnetic field deflects the solar wind and is thereby
deformed into the teardrop-shaped magnetosphere that
surrounds the planet.

of which is visible as the aurora. Excited nitrogen
molecules emit pinkish or magenta light, whereas excited
oxygen atoms emit greenish light.

The aurora is usually visible only at higher
latitudes. Earth's magnetic field channels some solar
wind particles into two doughnut-shaped belts that are
centered on the planet's north and south geomagnetic
poles. These belts of more or less continuous auroral
activity, known as *auroral ovals*, are located between 20
and 30 degrees of latitude from the geomagnetic poles
and bulge equatorward on the dark (night) side of the
planet. The Northern Hemisphere auroral oval is
centered on the northwest tip of Greenland at latitude
78.5 degrees N and longitude 69 degrees W.

Auroral activity varies directly with the sun's
activity. When the sun is quiet, the auroral oval shrinks,
but when the sun is active, the auroral oval expands
toward the equator, and the aurora may be visible across
southern Canada and the northern United States or,
rarely, further south. Solar activity follows a roughly 11-
year cycle, with the most recent solar maximum in 2000
(Chapter C). In 1989, during a particularly active solar
phase, the aurora was visible as far south as Mexico's
Yucatán Peninsula. Gigantic explosions, called *solar
flares*, characterize an active sun. A solar flare is a brief
event (lasting perhaps an hour) that produces a shock
wave that propagates rapidly (500 to 1000 km per
second) through the solar wind. Collision of the shock
wave with the magnetosphere causes the auroral oval to
expand equatorward. During some active solar episodes,
the phenomenon that produces auroras also disrupts the

operation of electric power grids, telecommunications
systems, and satellites.

Conclusions

In this chapter we distinguished between weather and
climate, and covered the origin, evolution, composition,
and structure of the atmosphere. We emphasized the
importance of minor gases in the functioning of the
atmosphere, and surveyed some of the various
technologies used to monitor the atmosphere. We also
saw how the vertical profile of average air temperature is
the basis for subdividing the atmosphere into four layers.
Because the primary focus of this course is weather and
climate, we will be concerned primarily with atmospheric
processes operating within the troposphere.

Our next major objective is to examine the
driving force behind weather. To do so, we require an
understanding of energy input and energy conversions
within the Earth–atmosphere system. In the next chapter
we learn how the sun supplies the energy that drives the
atmosphere's circulation. As we will see, circulation of
the atmosphere ultimately is responsible for variations in
weather from one place to another and with time.

Basic Understandings

- Earth's atmosphere encircles the planet as a thin
  envelope of gases and suspended solid and liquid
  particles (aerosols).
- Weather is defined as the state of the atmosphere at a
  specified place and time and described in terms of
  such variables as temperature, humidity, cloudiness,
  precipitation, and wind speed and direction. Climate
  encompasses average weather plus extremes in
  weather at some location over a specified time
  period.
- The modern atmosphere is the product of a lengthy
  evolutionary process that began about 4.6 billion
  years ago. Outgassing (the release of gases from
  rock through volcanic eruptions and meteorite
  impacts) played a key role in this evolution by contributing water vapor, carbon dioxide, and nitrogen.
  Photosynthesis is the principal source of free oxygen
  in the atmosphere.
- Within the homosphere, the lowest 80 km (50 mi) of
  the atmosphere, the principal atmospheric gases,

nitrogen (N_2) and oxygen (O_2), occur everywhere in the same relative proportion (about 4 to 1). In addition to gases, Earth's atmosphere contains minute solid and liquid particles, collectively called aerosols.

- The significance of an atmospheric gas or aerosol is not necessarily related to its relative concentration. Water vapor, carbon dioxide, and ozone are *minor* in concentration but extremely important in the life-sustaining roles they play on Earth.

- An air pollutant is a gas or aerosol occurring in concentrations that adversely affect the wellbeing of organisms (especially humans) or disrupt the orderly functioning of the environment. Human activities and natural processes are sources of air pollutants.

- The scientific method is a systematic form of inquiry that requires the formulation and testing of hypotheses. A hypothesis is a working assumption that may be accepted, modified, or rejected. Debate and disagreement among scientists on controversial issues are usual and important elements of the scientific method.

- A scientific model is an approximate representation of a real system and may be conceptual, graphical, physical, or numerical. Examples of models commonly used in meteorology and climatology include weather maps and global climate models.

- The Army Medical Corps operated the nation's first weather observation network in the 1800s. Invention of the telegraph made possible rapid communication of weather data and the first near real-time weather maps.

- In the 1860s, growing public concern over the great loss of life and property from surprise storms sweeping the Great Lakes spurred the federal government to create a telegraph-linked national weather observation network in 1870 under the auspices of the U.S. Army Signal Corps, forerunner of today's National Weather Service.

- Through the years, tools for investigating the upper atmosphere progressed from instrumented kites and manned balloons to rockets, radiosondes, and satellites.

- A radiosonde consists of an instrument package and radio transmitter carried aloft by balloon to altitudes up to 30,000 m (100,000 ft); it provides vertical profiles (called *soundings*) of air temperature, pressure, and dewpoint (a measure of humidity).

- The atmosphere is subdivided into four concentric layers (troposphere, stratosphere, mesosphere, and thermosphere) based on the average vertical temperature profile. Almost all weather takes place in the troposphere, the lowest subdivision of the atmosphere.

- The ionosphere is situated primarily within the thermosphere and consists of a relatively high concentration of charged particles (ions and electrons).

- An aurora (northern or southern lights) is a spectacular display of curtains of color in the night sky of higher latitudes. It develops in the ionosphere when the solar wind interacts with Earth's magnetic field and varies with solar activity.

ESSAY: The Atmosphere of Mars

In 1976, sensors aboard NASA's Viking spacecraft confirmed much earlier speculation that the Martian atmosphere differs considerably from Earth's atmosphere. The Martian atmosphere is 95% carbon dioxide, 2% to 3% nitrogen, 1% to 2% argon, and 0.1% to 0.4% oxygen. Also, the Martian atmosphere is much thinner; its surface pressure is only 0.6% of Earth's average sea-level air pressure. In the beginning, more than 4 billion years ago, however, the atmospheres of both planets probably were quite similar, but for several reasons, followed distinctly different evolutionary paths.

Outgassing was responsible for the primeval atmospheres of Earth and Mars. Gases were released from ancient planetary rock through volcanic eruptions and the impact of meteorites on the rocky surfaces of both planets. The gases were probably the same because the source rocks on both planets were chemically the same. Through outgassing, the primeval atmosphere of both planets was mostly carbon dioxide along with some nitrogen and water vapor, the principal gaseous emissions of volcanoes, both ancient and modern.

As noted elsewhere in this chapter, geochemical processes plus photosynthesis gradually altered Earth's primeval atmosphere so that eventually nitrogen and oxygen became the principal gases and carbon dioxide was reduced to minor status. From the beginning, however, carbon dioxide has remained the chief gaseous constituent of the Martian atmosphere, although through the eons its density declined significantly with major implications for the planet's climate.

Contrasts in the volcanic histories of Earth and Mars may help to explain why the atmospheres of the two planets evolved differently. On Mars, the bulk of volcanic activity apparently took place during the planet's first 2 billion years, whereas on Earth, volcanism has been more or less continuous throughout the planet's history. The decline in volcanism on Mars cut the supply of nitrogen, and much of the original nitrogen escaped Mars's relatively weak gravitational field. Gravity, the force that holds an atmosphere to a planet, is about 38% weaker on Mars than on Earth because Mars is less massive than Earth (11% of Earth's mass).

The decline in Martian volcanism also meant less CO_2 released to the planet's atmosphere. At the same time, other geological processes removed carbon dioxide and reduced the density of the Martian atmosphere. In the beginning, the CO_2-rich Martian atmosphere produced a surface pressure three times that of Earth's modern atmosphere. Enormous amounts of CO_2 escaped to space, some adhered to the fine sediment that blankets the planet's surface, and some was locked up in carbonate rocks. Carbon dioxide slows the loss of the planet's heat to space so that as CO_2 thinned, more heat escaped and temperatures on the surface of the planet fell. Today, the mean temperature on the Martian surface ranges from about −60 °C (−76 °F) at the equator to as low as −123 °C (−189 °F) at the poles.

Evidence that Mars was warmer and wetter several billion years ago supports the view that Mars's original CO_2-rich atmosphere was denser than it is now. Mariner satellite missions of the 1960s and Viking spacecraft missions of the 1970s photographed glacial features, eroded craters, and valleys on the Martian surface that presumably were cut by running water. These findings were confirmed 21 years later by NASA's Pathfinder mission to Mars. On 4 July 1997, the Pathfinder spacecraft landed on an ancient floodplain that had been scoured by a catastrophic flood an estimated 1.8 to 3.5 billion years ago. Also, Pathfinder's tiny rover, *Sojourner*, photographed pebbles and rocks on the Martian surface having physical properties indicative of running water. In 2000, Mars Observer images revealed sedimentary rock strata (likely aqueous in origin) in impact craters near the Martian equator.

In 1994, S.W. Squyres of Cornell University and J.F. Kasting of Pennsylvania State University questioned the assumption that in the beginning Mars was much warmer and wetter than it is today. They argued that Mars was considerably colder than has been previously assumed. As pointed out elsewhere in this chapter, at that time the sun was about 30% dimmer than it is today. Mars's greater distance from the sun compared to Earth would only exacerbate the cooling and there wasn't enough CO_2 in the Martian atmosphere to offset the cooling. Furthermore, Squyres and Kasting proposed that it was cold enough to form clouds of CO_2 ice crystals that reflected away sunlight, further contributing to cooling. How do they explain running water on Mars?

According to Squyres and Kasting, one possibility is that the water flowing in the valleys did not originate as runoff from precipitation. Rather, water seeped out of the ground where it was kept in the liquid state by the slowly cooling but still very hot interior of Mars. Explaining why the water did not freeze while flowing on the surface is more of a challenge. One possibility is that the Martian atmosphere contained some methane and/or ammonia, gases that have the same warming effect as CO_2. Together, these gases may have kept the surface just warm enough for running water to remain liquid. Today, water in unknown quantities likely remains on Mars mixed with solid carbon dioxide (dry ice) in the polar ice caps and as patches of permafrost (permanently frozen ground). In winter, perhaps 20% to 30% of the Martian atmosphere freezes out as dry ice. Severe cold means no life, no photosynthesis, and hence, only a trace of free oxygen and no ozone in the Martian atmosphere.

In summary, the difference in evolutionary paths taken by the atmospheres of Earth and Mars may be due largely to contrasts in the volcanic history and physical characteristics of the two planets. Earth has been more volcanically active (greater outgassing) and is more massive (stronger gravitational field) than Mars.

ESSAY: Radio Transmission and the Ionosphere

Reception of distant radio signals (waves) at night is not at all unusual. Late-night radio listeners in Illinois can, for example, routinely pick up WBZ, a Boston radio station (1030 on the AM dial) even though the station's transmitter is more than 1500 km (930 mi) away.

Arrival of distant radio waves at night and their subsequent disappearance during sunlit hours is due to interactions of those waves with the ionosphere. A radio wave is a form of *electromagnetic radiation* that travels in straight paths in all directions away from its source transmitter (Chapter 3). Earth is essentially a sphere so its surface gradually curves under and away from direct radio waves. A *quiet zone*, where direct radio waves are not received, begins about 160 km (100 mi) from a transmitter tower of average height. At night, beyond the distant edge of the quiet zone, reception resumes because radio waves are reflected back to Earth's surface from the upper ionosphere.

Recall from elsewhere in this chapter that the ionosphere is a region of ions and free electrons. Highly energetic solar radiation splits molecular nitrogen (N_2) and molecular oxygen (O_2) into atoms, positively charged ions, and free electrons. The production rate of ions and electrons depends on two factors, both of which vary with altitude: (1) the number density of atoms and molecules available for ionization, which decreases rapidly with altitude, and (2) the intensity of solar radiation, which increases with altitude. Combined, these two factors maximize the concentration of ions and free electrons in the ionosphere.

By convention, the ionosphere is subdivided vertically into several layers. From lowest to highest, layers are designated D (upper mesosphere below 100 km), E (100 to 120 km), F_1 (150 to 190 km), and F_2 (above 200 km). The original basis for this subdivision was the belief that each layer is a distinct zone of maximum electron density. Measurements by rockets and satellites, however, show that the ionosphere is not made up of discrete layers; rather, electron density increases nearly continuously with altitude to a maximum at an average altitude close to 300 km (186 mi). The D, E, F_1, and F_2 labels in this discussion refer to regions within the ionosphere.

Radio waves that enter the ionosphere interact with free electrons in the D, E, and F regions and are either absorbed or reflected back toward Earth's surface. At night, in the absence of ionizing radiation, the D region virtually disappears as ions and electrons recombine into neutral particles. The recombination rate depends on air density; that is, the higher the density, the greater the likelihood of collision of particles and capture of electrons by positive ions. In the E, F_1, and F_2 regions, air is so rarefied that collisions are infrequent, and although the E region weakens, these regions persist through the night. Radio waves that reach the F region are reflected back toward Earth's surface. (Exceptions are radio waves that enter the F region at nearly a right angle; these waves pass on into space.) At night, because of F-region reflection, radio waves propagate many hundreds of kilometers from their point of origin.

With the return of the sun's ionizing radiation during the day, the D region redevelops. Most radio waves that reach the D region are absorbed rather than reflected. The F region reflects waves that penetrate the D region back to the D region where they are absorbed. Consequently, during sunlit hours, radio-wave propagation is not aided by F-region reflection. In summary, radio waves travel greater distances at night because the upper ionosphere reflects them back to Earth's surface.

Ionizing radiation from the sun generates the ionosphere so that any solar activity that disturbs the flow of this radiation may affect ion density and consequently, radio communication on Earth. *Sudden ionospheric disturbances (SIDs)*, typically lasting 15 to 30 minutes, are caused by bursts of ultraviolet radiation from the sun. Ionization temporarily increases, D-region absorption strengthens, and radio transmissions fade. The same solar activity responsible for the aurora also increases ionization and causes radio fadeout.

CHAPTER 3

SOLAR AND TERRESTRIAL RADIATION

> Splendid with splendor hid you come,
>   from your Arab abode,
> a fiery topaz smothered in the hand of
>   a great prince who rode
> before you, Sun—whom you outran,
> piercing his caravan.
>
> MARIANNE MOORE, "Sun"

Case-in-Point

Ancient peoples were well aware of the sun's annual cycle and the progression of the seasons. Knowing when seasons begin and end was critical to the timing of planting and harvesting of crops. In monsoon climates, the timing of the rainy and dry seasons is closely tied to the solar cycle. While these people did not possess a calendar in the modern sense, their knowledge of the path of the sun through the sky and other regular astronomical events inspired them to construct astronomical calculators in the form of elaborate megaliths.

Probably the best known of the ancient astronomical calculators is Stonehenge in Southern England, the earliest portion of which dates to about 2950 B.C. As early as the eighteenth century, scientists noticed that the horseshoe arrangement of great stones opened in the direction of sunrise on the summer solstice. Stones also point in the direction of the mid–winter sunset. In more recent years, scientists discovered that the arrangement of stones and other features at Stonehenge could be used to predict solar and lunar eclipses. Native Americans living at Cahokia (just east of modern St. Louis, MO) erected similar solar calendars consisting of wooden posts arranged in circles. So-called Woodhenge calendars date from the period 900 to 1100 A.D.

Predating Stonehenge by some two thousand years, the oldest astronomical calculator discovered so far consists of megaliths and a stone circle, located near Nabta in the Nubian Desert of southern Egypt. A global positioning satellite confirms the stone's alignment with the position of the rising sun on the summer solstice (as it would have been 6000 years ago). For the people living there then this was a very significant date because monsoon rains typically begin shortly after the summer solstice.

Driving Question:

How does energy flow into and out of the Earth–atmosphere system?

The sun drives the atmosphere; that is, the sun is the source of energy that drives the circulation of the atmosphere and powers winds and storms. (*Energy* is defined as the capacity for doing work and occurs in many different forms such as radiation and heat.) The circulation of the atmosphere ultimately is responsible for weather and its temporal and spatial variability.

The sun ceaselessly emits energy to space in the form of electromagnetic radiation. A very small portion of that energy is intercepted by the Earth–atmosphere system and converted to other forms of energy including, for example, heat and the kinetic energy of the winds. Energy cannot be created nor destroyed although it can be converted from one form to another. This is the **law of energy conservation** (also known as the *first law of thermodynamics*).

In this chapter, we examine the basic properties of electromagnetic radiation, laws governing electromagnetic radiation, how solar radiation interacts with the components of the Earth–atmosphere system, and the conversion of solar radiation to heat. The Earth–atmosphere system responds to solar heating by emitting infrared radiation. Some of this infrared radiation is absorbed by certain atmospheric gases and radiated to Earth's surface elevating the temperature of the lower atmosphere to levels that make life possible (the so-called *greenhouse effect*). We begin with the nature of electromagnetic radiation and some of the properties of its various forms.

Electromagnetic Spectrum

Planet Earth is bathed continuously in **electromagnetic radiation**, so named because this form of energy has both electrical and magnetic properties. All objects emit electromagnetic radiation. Forms of electromagnetic radiation include radio waves, microwaves, infrared radiation, visible light, ultraviolet radiation, X-rays, and gamma radiation. Together, they make up the **electromagnetic spectrum**, illustrated in Figure 3.1.

Electromagnetic radiation travels as waves, described in terms of wavelength or frequency. **Wavelength** is the distance between successive wave crests (or equivalently, wave troughs), as shown in Figure 3.2. **Wave frequency** is defined as the number of crests (or troughs) that pass a given point in a specified period of time, usually 1 second. Passage of one complete wave is called a *cycle*, and a frequency of 1 cycle per second equals 1.0 hertz (Hz). Wave frequency is inversely proportional to wavelength; that is, the higher the frequency, the shorter is the wavelength. Radio waves have frequencies in the millions of hertz and wavelengths up to hundreds of kilometers. At the other end of the electromagnetic spectrum, by contrast, gamma rays have frequencies as high as 10^{24} (a trillion trillion) Hz and wavelengths as short as 10^{-14} (a hundred trillionth) m.

Electromagnetic waves travel through space and may pass through gases, liquids, and solids. In a vacuum, all electromagnetic waves travel at their maximum possible speed, that is, 300,000 km (186,000 mi) per second. All forms of electromagnetic radiation slow down when passing through materials, their speed varying with wavelength and type of material. As electromagnetic radiation passes from one medium into another, it may be reflected or refracted (that is, bent) at the interface. This happens, for example, when solar radiation strikes the ocean surface at an oblique angle—some radiation is reflected into the atmosphere and some bends (is refracted) as it penetrates the water. Electromagnetic radiation may also be absorbed, that is, converted to heat.

Although the electromagnetic spectrum is continuous, different names are assigned to different segments because we detect, measure, generate, and use those segments in different ways. Furthermore, the

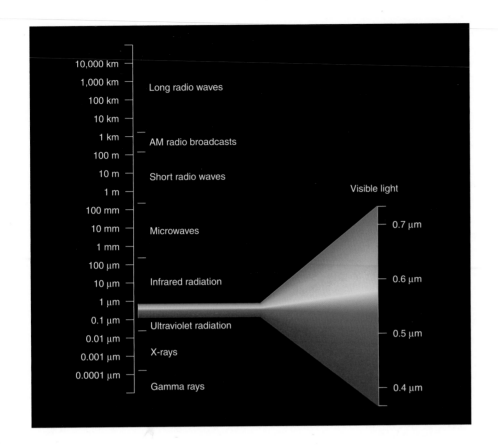

FIGURE 3.1
The electromagnetic spectrum consists of many forms of radiation distinguished by wavelength in micrometers (µm).

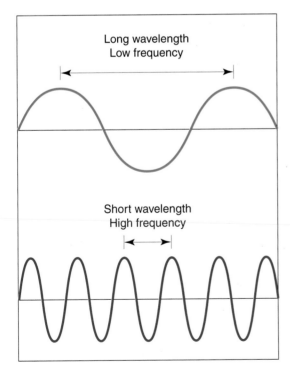

FIGURE 3.2
Wavelength is the distance between two successive crests or, equivalently, between two successive troughs. Wavelength is inversely related to wave frequency.

various types of electromagnetic radiation do not begin or end at precise points along the spectrum. For example, red light shades into invisible infrared radiation (infrared, meaning *below red*). At the other end of the visible portion of the electromagnetic spectrum, violet light shades into invisible ultraviolet radiation (ultraviolet, meaning *beyond violet*).

Beyond visible light on the electromagnetic spectrum and in order of increasing energy level, increasing frequency, and decreasing wavelength, are **ultraviolet radiation (UV)**, **X-rays**, and **gamma radiation**. All three types of radiation occur naturally and all can be produced artificially. All have medical uses: ultraviolet radiation is a potent germicide; X-rays are used as a powerful diagnostic tool; and both X-rays and gamma radiation are used to treat cancer patients. These three highly energetic types of radiation are dangerous as well as useful. Ultraviolet radiation can cause irreparable damage to the light-sensitive cells of the eye. Staring at the sun (for instance, during a partial solar eclipse) can permanently blind a person unless a filter is used to block out ultraviolet radiation. Also, overexposure to UV, X-rays, or gamma radiation can cause sterilization, cancer, mutations, or damage to a fetus. Fortunately, Earth's atmosphere blocks out most incoming ultraviolet radiation and virtually all X-rays and gamma radiation. Without this protective atmospheric shield, all life on Earth would perish.

At lower frequencies and longer wavelengths, UV radiation shades into **visible radiation**, that is, radiation that is perceptible by the human eye. Wavelengths of visible light range from about 0.40 micrometer at the violet end to approximately 0.70 micrometer at the red end. (One *micrometer* is a millionth of a meter, about one-tenth the thickness of a human hair.) Visible light is essential for many activities of plants and animals. In plants, light provides the energy needed for photosynthesis; it also coordinates the opening of buds and flowers in spring and the dropping of leaves in autumn. For animals, light regulates the timing of reproduction, hibernation, and migration and makes vision possible.

Between red light and microwave radiation is **infrared radiation (IR)**. IR is not visible, but we can feel the heat it generates when it is intense, as it is, for example, when emitted by a hot stove. Actually, every known object, including you and this book, emit small amounts of infrared radiation. And, as we will see later in this chapter, absorption and emission of IR by certain atmospheric gases is responsible for significant warming of the lower atmosphere (the *greenhouse effect*).

At longer wavelengths is the **microwave** portion of the electromagnetic spectrum, which includes wavelengths ranging from about 0.1 to 1000 millimeters. Some microwave frequencies are used for radio communication, in microwave ovens, and in weather radar. At the low energy, low frequency, long wavelength end of the electromagnetic spectrum are **radio waves**. Wavelengths range from a fraction of a centimeter up to hundreds of kilometers, and frequencies can extend to a billion Hz. FM (frequency modulation) radio waves, for example, span 88 million to 108 million Hz; hence, the familiar 88 and 108 at opposite ends of the FM radio dial.

Radiation Laws

Several physical laws describe the properties of electromagnetic radiation emitted by a perfect radiator, a so-called blackbody. By definition, a **blackbody** at a constant temperature absorbs all radiation that is incident on it and emits all the radiant energy it absorbs. A blackbody is both a perfect absorber and perfect emitter of radiation. The wavelengths of the emitted radiation are related to the temperature of the blackbody. Surfaces of real objects may approximate blackbodies for certain wavelengths of radiation but not for others. Freshly fallen snow, for example, is very nearly a blackbody for infrared radiation but not for visible light. (Note that *blackbody* does not refer to color.) Although neither the sun nor Earth is a blackbody, their absorption and emission of radiation is sufficiently close to that of a blackbody that we can apply blackbody radiation laws to them with some very useful results. Here we apply two blackbody radiation laws: Wien's displacement law and the Stefan–Boltzmann law.

All known objects emit and absorb all forms of electromagnetic radiation. The wavelength of most intense radiation (λ_{max}) emitted by a blackbody is inversely proportional to the absolute temperature (T) of the object. That is,

$$\lambda_{max} = C/T$$

where C, a constant of proportionality, has the value of 2897 if λ_{max} is expressed in micrometers, and T is in kelvins. *Absolute temperature* is the number of kelvins above absolute zero (-273.15 °C or -459.67 °F). This is a statement of **Wien's displacement law**.

According to Wien's displacement law, hot objects (such as the sun) emit radiation that peaks at relatively short wavelengths, whereas cold objects (such as the Earth–atmosphere system) emit peak radiation at longer wavelengths. The top burners of an electric range provide an illustration of Wien's displacement law. After switching on a burner, its metal coils warm and we readily feel the heat. We are actually feeling invisible infrared radiation. As the coil temperature continues to rise, the coils emit more intense infrared radiation with peak emission at shorter wavelengths. That is, as the coil temperature rises, the cell's peak radiation is *displaced* from the infrared toward the near-infrared portion of the electromagnetic spectrum.

Radiation emitted by the sun is similar to that emitted by a blackbody at temperature of about 6000 °C (11,000 °F). Figure 3.3 shows the flux (rate of energy transfer per unit time per unit area) of solar radiation received at the top of the atmosphere as a function of wavelength. The sun emits a band of radiation (at wavelengths mostly between 0.25 and 2.5 micrometers) that is most intense at a wavelength of about 0.5 micrometer (in the green of visible light). The flux of radiation emitted by Earth's surface is similar to that

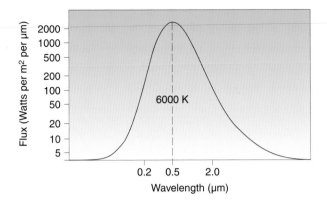

FIGURE 3.3
The flux of solar radiation incident at the top of the atmosphere as a function of wavelength. Radiation emitted by the sun is similar to that emitted by a blackbody at a temperature of about 6000 °C (6000 K).

emitted by a blackbody at a temperature of about 15 °C (59 °F) peaking in the infrared (Figure 3.4). Earth's surface emits a broad band of infrared radiation (at wavelengths mostly between 4 and 24 micrometers) with peak intensity at a wavelength of about 10 micrometers.

The area under the blackbody curves in Figures 3.3 (for the sun) and 3.4 (for Earth's surface) represent the total radiation energy emitted per unit time per unit surface area at all wavelengths. The vertical scales in the two figures are not the same because the sun emits immensely more total radiational energy than does Earth's surface. According to the **Stefan–Boltzmann law**, the total energy flux emitted by a blackbody across all wavelengths (E) is proportional to the fourth power of the absolute temperature (T^4) of the object; that is,

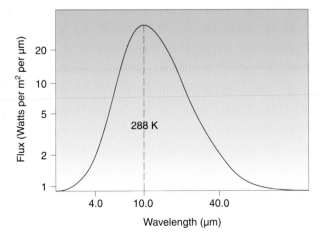

FIGURE 3.4
Flux of radiation as a function of wavelength emitted by a blackbody radiating at about the same average temperature as Earth's surface, that is, 15 °C (288 K).

$$E \sim T^4.$$

This relationship implies that a small change in the temperature of a blackbody results in a much greater change in the total amount of radiational energy emitted by the blackbody. The sun radiates at a much higher temperature than does Earth's surface, so that the Stefan–Boltzmann law predicts that the sun's energy output per square meter is about 190,000 times that of the Earth–atmosphere system.

As solar radiation travels away from the sun and spreads outward into space, its intensity (energy per unit time per unit area) diminishes rapidly, as the inverse square of the distance traveled. According to the *inverse square law*, doubling the distance traversed by radiation reduces its intensity to $(\frac{1}{2})^2$ or 1/4 of its initial value. Earth orbits the sun at an *ideal* distance in that it receives an intensity of solar radiation that favors a temperature range that supports life. Venus is closer to the sun, receives much more intense radiation, and is too hot for life. Mars is farther away from the sun, receives less intense radiation, and is too cold for life.

The total energy (in the form of solar radiation) absorbed by planet Earth is equal to the total energy (in the form of infrared radiation) emitted by the Earth–atmosphere system back to space. This balance between energy input and energy output for the Earth–atmosphere system is known as **global radiative equilibrium** and is an example of the law of energy conservation. We examine this important concept in much more detail in the next chapter.

Input of Solar Radiation

The sun, the closest star to Earth, is a huge gaseous body composed almost entirely of hydrogen (about 80% by mass) and helium with internal temperatures that may exceed 20 million °C. The ultimate source of solar energy is a continuous nuclear fusion reaction in the sun's interior. Simply put, in this reaction, four hydrogen nuclei (protons) fuse to form one helium nucleus (alpha particle). However, the mass of the four hydrogen nuclei is about 0.7% greater than the mass of one helium nucleus. This excess mass is converted to energy as described by Albert Einstein's equation

$$E = mc^2$$

where mass, m, is related to energy, E, and c is the speed of light (300,000 km per second). Note that c^2 is such a huge number that even a very small mass is converted to an enormous quantity of energy. Some of the energy produced by nuclear fusion in the sun is used to bind the helium nucleus together. The rest of the energy is radiated and convected to the sun's surface, and then radiated off to space.

At radiating temperatures near 6000 °C (11,000 °F), the visible surface of the sun, known as the **photosphere**, is much cooler than the sun's interior. A network of huge, irregularly shaped convective cells, called *granules*, gives the photosphere its honeycomb appearance. A typical granule is about 1000 km (600 mi) across, although some so-called *supergranules* may be 30,000 to 50,000 km (18,500 to 31,000 mi) in diameter. Most granules have a life expectancy of only a few minutes and consist of a broad central area of rising hot gas surrounded by a thin layer of cooler gas sinking back toward the center of the sun. This zone of convective activity encompasses the outer 200,000 km (125,000 mi) of the sun.

Relatively dark, cool areas, called *sunspots*, dot the surface of the photosphere. Sunspot temperatures may be 400 to 1800 Celsius degrees (720 to 3240 Fahrenheit degrees) lower than the photosphere's average temperature. Bright areas, known as *faculae*, usually occur near sunspots. Changes in the number of sunspots and faculae accompany changes in solar energy output and may influence Earth's climate (Chapter C).

Outward from the photosphere is the **chromosphere**, consisting of ions of hydrogen and helium at 4000 °C to 40,000 °C (7200 °F to 72,000 °F). Beyond this zone is the outermost portion of the sun's atmosphere, the **solar corona**, a region of extremely hot (1 to 4 million °C) and highly rarefied ionized gases (predominantly hydrogen and helium) that extends millions of kilometers into space. The solar wind originates in the corona and solar flares that erupt from the photosphere into the corona intensify the solar wind (Chapter 2). In its orbit about the sun, planet Earth intercepts only about one two-billionth of the enormous quantity of energy continually radiated by the sun to space.

SOLAR ALTITUDE

Our experience in midlatitudes tells us that the intensity of solar radiation striking Earth's surface varies significantly with the seasons. At noon, the summer sun is higher in the sky than the winter sun, and solar rays

striking Earth's surface are more concentrated in summer than in winter. Even over the course of a single day, regular changes occur in incoming solar radiation striking the Earth's surface; that is, the solar beam is more concentrated at noon than at sunrise or sunset. Evidently, the angle of the sun above the horizon, called the **solar altitude**, influences the intensity of solar radiation received at the Earth's surface. When and where the sun is directly overhead, the solar altitude has its maximum value of 90 degrees and solar rays are most concentrated (Figure 3.5). As the sun moves lower in the sky (that is, as the solar altitude declines), solar radiation spreads over an increasing area of Earth's surface and thus becomes less and less intense.

Solar altitude always varies with latitude. Earth is so far away from the sun (a mean distance of about 150 million km or 93 million mi) that solar radiation reaches the planet as essentially parallel beams of uniform intensity. But the nearly spherical Earth presents a

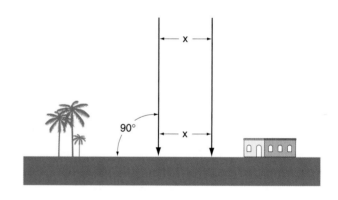

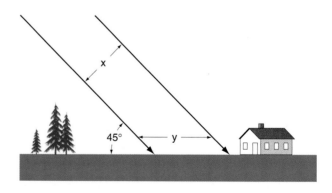

FIGURE 3.5
The intensity of solar radiation striking Earth's surface per unit area varies with the solar altitude. (A) Incident solar radiation is most intense when the sun is directly overhead (solar altitude of 90 degrees). With decreasing solar altitude, solar radiation received at Earth's surface spreads over an increasing area (y is greater than x) so that radiation is less concentrated (less radiational energy per unit area).

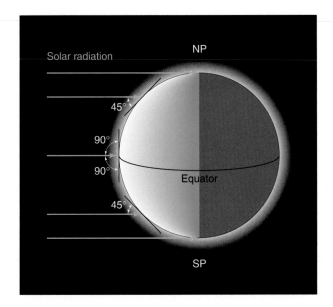

FIGURE 3.6
On any day of the year, the noon solar altitude always varies with latitude because Earth presents a curved surface to the incoming solar beam. In this example for the equinoxes, the solar altitude is 90 degrees at the equator and decreases with latitude (toward the poles). Hence, solar radiation striking horizontal surfaces per unit area is most intense at the equator and least intense at the poles.

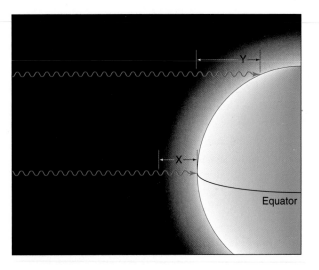

FIGURE 3.7
Solar radiation's path through the atmosphere lengthens with decreasing solar altitude, that is, as the sun moves lower in the sky. *X* is the path length at high solar altitude and *Y* is the path length at low solar altitude. Path Y may be more than 30 times longer than X.

curved surface to incoming solar radiation so that the noon solar altitude always varies with latitude (Figure 3.6). The intensity of solar radiation actually striking Earth's surface peaks at the latitude where the noon solar altitude is 90 degrees and decreases with distance north and south of that latitude.

Solar altitude also influences the interaction between solar radiation and the atmosphere. Decreasing solar altitude lengthens the path of the sun's rays through the atmosphere (Figure 3.7). As the path lengthens, solar radiation interacts more with clouds and atmospheric gases and aerosols causing its intensity to diminish. Even if skies are cloud-free, the longer the path of solar radiation through the atmosphere, the less intense is the radiation striking Earth's surface. The nature of the interactions between solar radiation and the atmosphere (absorption, reflection, and scattering) is discussed later in this chapter.

While solar altitude influences the intensity of solar radiation striking Earth's surface per unit area, the length of daylight affects the total amount of radiational energy that is received each day. For example, in summer the altitude of the noon sun is lower at high latitudes than at middle latitudes. Nonetheless, the

greater length of daylight at high latitudes may translate into more total radiation striking Earth's surface during a 24-hour day (Figure 3.8). Variations in both solar altitude and length of daylight accompany the annual march of the seasons. Before examining these relationships, first consider the fundamental motions of

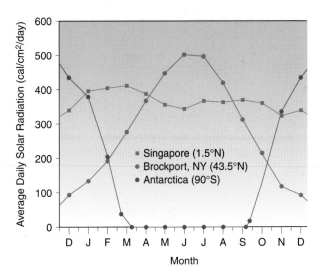

FIGURE 3.8
Average daily solar radiation (in calories per cm² per day) by month at Singapore, Brockport, NY, and Antarctica. During part of the year, the polar (Antarctic) and mid-latitude (Brockport) locations receive more daily radiation than the near-equator location (Singapore).

Earth in space: rotation of the planet on its spin axis and the planet's orbit about the sun.

EARTH'S MOTIONS IN SPACE AND THE SEASONS

Rotation of Earth on its axis accounts for day and night. Once every 24 hours, the nearly spherical Earth completes one rotation with respect to the sun and at any instant, half the planet is illuminated by visible solar radiation (daylight) and the other half is in darkness (night).

In one year, which is actually 365.2422 days, Earth makes one complete revolution about the sun in a slightly elliptical orbit (Figure 3.9). Earth's orbital eccentricity, that is, its departure from a circular orbit, is so slight that the Earth-to-sun distance varies by only about 3.3% through the year. Earth is closest to the sun (147 million km or 91 million mi) on about 3 January and farthest from the sun (152 million km or 94 million mi) on 4 July. These are the current dates of **perihelion** and **aphelion**, respectively. In the Northern Hemisphere, Earth is closest to the sun in winter and farthest from the sun in summer. And because of the spreading of radiation described by the *inverse square law*, Earth receives about 6.7% more solar radiation at perihelion than at aphelion. The eccentricity of Earth's orbit about the sun cannot explain the seasons. What does account for seasons?

The answer is found in the tilt of Earth's spin axis. Earth's equatorial plane is tilted 23 degrees 27 minutes to the plane defined by the planet's annual orbit about the sun (Figure 3.10). Equivalently, Earth's spin axis is tilted 23 degrees 27 minutes from a perpendicular to Earth's orbital plane. During its annual revolution about the sun, Earth's spin axis always remains in the same orientation relative to the stars with the North Pole aimed toward *Polaris*, the North Star, while its orientation to the sun changes continually. Accompanying these changes are regular variations in solar altitude and length of daylight, which in turn affect the intensity and total amount of solar radiation received at Earth's surface. If Earth's rotational axis were perpendicular to its orbital plane (no tilt), Earth's axis would always have the same orientation to the sun. Without an axial tilt, only the changes in the Earth–sun distance between aphelion and perihelion would produce a seasonal contrast and it would be slight.

How does Earth's orientation to the sun change over the course of a year? Viewed from Earth's surface, the latitude where the sun's rays are most intense (solar altitude of 90 degrees) shifts from 23 degrees 27 minutes south of the equator to 23 degrees 27 minutes north of the equator, and then back to 23 degrees 27 minutes S. On about 21 March and again on about 23 September, the sun's noon position is directly over the equator. Day and night are approximately equal in length (12 hours) everywhere, except at the poles (Figure 3.11). For this reason, these dates are the **equinoxes** (from the Latin for "equal nights").

Following the equinoxes, the sun continues its apparent journey toward its maximum poleward

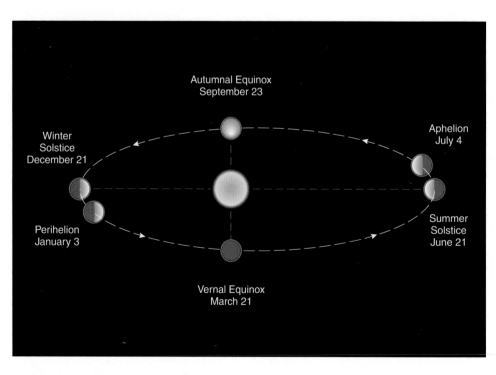

Autumnal Equinox
September 23

Aphelion
July 4

Winter
Solstice
December 21

Perihelion
January 3

Summer
Solstice
June 21

Vernal Equinox
March 21

FIGURE 3.9
Earth's orbit is an ellipse with the sun located at one focus. Earth is closest to the sun at *perihelion* (about 3 January) and farthest from the sun at *aphelion* (about 4 July). Note that the eccentricity of Earth's orbit is greatly exaggerated in this drawing.

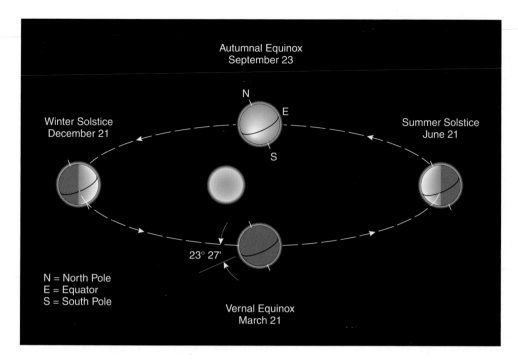

FIGURE 3.10
The seasons change because Earth's equatorial plane is inclined (at 23 degrees, 27 minutes) to its orbital plane. The seasons given are for the Northern Hemisphere. Note that the eccentricity of Earth's orbit is greatly exaggerated in this drawing.

locations. On 21 June, the sun's noon rays are vertical at 23 degrees 27 minutes N, the latitude circle known as the **Tropic of Cancer**. As shown in Figure 3.12, daylight is continuous north of the **Arctic Circle** (66 degrees 33 minutes N) and absent south of the **Antarctic Circle** (66 degrees 33 minutes S). Elsewhere, days are longer than nights in the Northern Hemisphere, where it is the first day of summer, and days are shorter than nights in the Southern Hemisphere, where it is the first day of winter. Thus 21 June is a **solstice** date—the summer solstice in the Northern Hemisphere and winter solstice in the Southern Hemisphere. *Solstice* is from the Latin, *solstitium*, referring to the sun standing still.

On 21 December, the noon sun is directly over 23 degrees 27 minutes S, the latitude circle known as the **Tropic of Capricorn**, and the situation is reversed (Figure 3.13). Daylight is continuous south of the Antarctic Circle and absent north

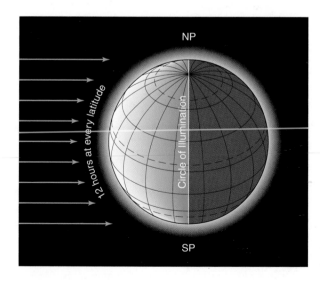

FIGURE 3.11
At the autumnal and spring equinoxes, the noon solar altitude is greatest (90 degrees) over the equator and day and night are about equal in length except at the poles.

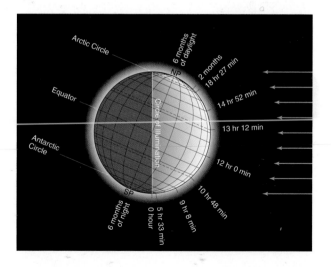

FIGURE 3.12
On the Northern Hemisphere summer solstice (21 June), the noon solar altitude is greatest (90 degrees) at 23 degrees 27 minutes N, and days are longer than nights everywhere north of the equator. Duration of daylight is given for every 20 degrees of latitude.

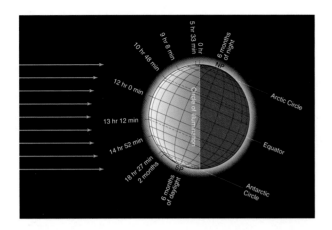

FIGURE 3.13
On the Northern Hemisphere winter solstice (21 December), the noon solar altitude is greatest (90 degrees) at 23 degrees 27 minutes S, and days are shorter than nights everywhere north of the equator. Duration of daylight is given for every 20 degrees of latitude.

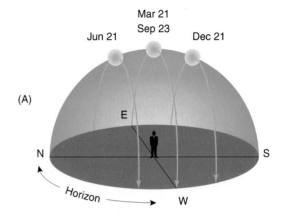

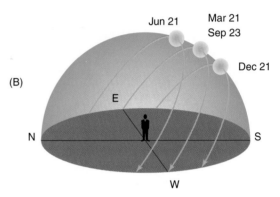

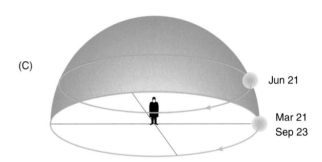

FIGURE 3.14
Path of the sun through the sky on the solstices and equinoxes at (A) the equator, (B) midlatitudes of the Northern Hemisphere, and (C) the North Pole.

of the Arctic Circle. Elsewhere, nights are longer than days in the Northern Hemisphere, where it is the first day of winter, and days are longer than nights in the Southern Hemisphere, where it is the first day of summer. Thus 21 December is the date of the winter solstice in the Northern Hemisphere and the summer solstice in the Southern Hemisphere.

As Earth's orientation to the sun changes through the course of a year, so too does the path of the sun through the local sky. Figure 3.14 portrays the path of the sun through the sky from sunrise to sunset for solstices and equinoxes at the equator, a midlatitude Northern Hemisphere location, and the North Pole. The high point of each path is the sun's position at local noon. At the midlatitude location and North Pole, the altitude of the noon sun is greatest on the summer solstice but at the equator, the noon solar altitude is maximum on the equinoxes (when the sun is directly overhead). Note that on 21 June at the North Pole, the sun circles the sky at a constant solar altitude (about 23.5 degrees) and neither rises nor sets.

Solar radiation incident on horizontal Earth surfaces is at maximum intensity where the noon sun is directly overhead. North and south of that latitude, the intensity of solar radiation diminishes because the solar altitude decreases. At the equinoxes, solar rays are most intense at the equator at noon and decrease with latitude toward the poles. At the Northern Hemisphere summer solstice, solar rays are most intense along the Tropic of Cancer and decrease to zero at the Antarctic Circle. On

that day, the noon sun has the same altitude at 47 degrees N as at the equator. At the Northern Hemisphere winter solstice, solar rays are most intense along the Tropic of Capricorn and decrease to zero at the Arctic Circle.

As noted earlier, days and nights are approximately equal in length (12 hours) everywhere (except at the poles) on only two days of the year, the spring and autumnal equinoxes. The length of daylight on the equinoxes is not precisely 12 hours because of optical effects of the atmosphere on the solar beam.

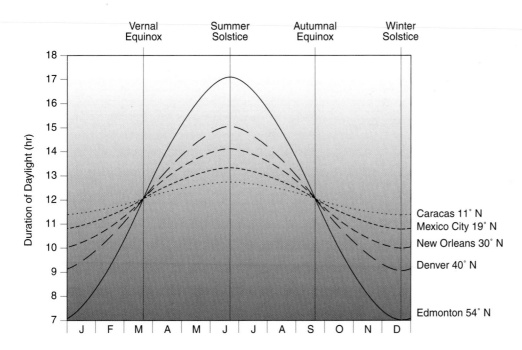

FIGURE 3.15
Variation in the length of daylight through the year increases with increasing latitude.

Times of sunrise and sunset refer to the hour and minute when the outer edge of the sun (not its center) is on the horizon. The atmosphere refracts (bends) the solar beam so that the sun appears to be higher than it actually is, lengthening the period of daylight. For example, at Washington, DC, the day length is 12 hours and 6 minutes on 23 September 2002, the autumnal equinox.

Between the March and September equinoxes, days are longer than nights in the Northern Hemisphere and days are shorter than nights in the Southern Hemisphere. Between the September and March equinoxes, days are shorter than nights in the Northern Hemisphere and days are longer than nights in the Southern Hemisphere. Furthermore, the seasonal (winter-to-summer) contrast in length of daylight increases with increasing latitude (Figure 3.15). At the equator, the period of daylight is essentially the same (slightly more than 12 hours) year round. At 60 degrees N, the length of daylight varies from 5 hours 53 minutes on 22 December to 18 hours 52 minutes on 21 June.

THE SOLAR CONSTANT

For convenience of study, the solar energy input into the Earth–atmosphere system is often expressed as the solar constant. The **solar constant** is defined as the rate at which solar radiation falls on a surface located at the outer edge of the atmosphere and oriented perpendicular to the incoming solar beam when Earth is at its mean distance from the sun. The *constant* designation is actually misleading because solar energy output fluctuates by a very small fraction of a percent over a year and exhibits long-term variations (Chapter C). The solar constant averages about 1.97 calories per square centimeter (cal/cm^2) per minute, or 1368 watts per square meter (W/m^2).

The rate of total solar energy input for the planet varies through the course of a year from a maximum when Earth is closest to the sun (perihelion) to a minimum when Earth is farthest from the sun (aphelion). At perihelion, Earth is about 3.3% closer to the sun than at aphelion. Applying the *inverse square law*, the planet intercepts about 6.7% more radiation at perihelion (2.04 cal/cm^2 per minute or 1417 W per m^2) than at aphelion (1.91 cal/cm^2 per minute or 1326 W per m^2).

The perihelion-to-aphelion contrast in solar energy input coupled with the seasonal variation in radiation has implications for global climate. The Southern Hemisphere receives more radiation during summer and less radiation during winter than does the Northern Hemisphere. Consequently, all other factors being equal, we might expect a greater winter-to-summer temperature contrast in the Southern Hemisphere. However, in the Southern Hemisphere the larger percentage of ocean surface area coupled with the relatively great thermal inertia of ocean water moderates seasonal temperature differences and largely offsets the greater seasonal contrast in incoming solar radiation (Chapter 4).

Solar Radiation and the Atmosphere

Solar radiation interacts with gases and aerosols as it travels through the atmosphere. These interactions consist of reflection, scattering, and absorption. Solar radiation that is not reflected or scattered back to space, or absorbed by gases and aerosols, reaches Earth's surface, where further interactions take place. Within the atmosphere, the percentage of solar radiation that is absorbed (*absorptivity*) plus the percentage reflected or scattered (*albedo*) plus the percentage transmitted to Earth's surface (*transmissivity*) must equal 100%. This relationship is another example of the law of energy conservation.

Reflection takes place at the interface between two different media, such as air and cloud, when some of the radiation striking that interface is redirected. At such an interface (Figure 3.16), the angle of incident radiation (*i*) equals the angle of reflected radiation (*r*); this is known as the **law of reflection**. Solar radiation is reflected without being converted to heat. The fraction of incident radiation that is reflected by a surface (or interface) is the **albedo** of that surface, that is,

albedo = [(reflected radiation)/(incident radiation)]

where albedo is expressed either as a percentage or a

FIGURE 3.16
Reflection of a solar ray by the top of a cloud illustrates the law of reflection whereby the angle of incident radiation (*i*) equals the angle of reflected radiation (*r*).

FIGURE 3.17
Cloud tops strongly reflect visible solar radiation and appear very bright as in this view from an airplane window.

fraction. Surfaces having a high albedo reflect a relatively large fraction of incident solar radiation and appear light in color. Surfaces having a relatively low albedo reflect a relatively small fraction of incident solar radiation and appear dark in color.

Within the atmosphere, the tops of clouds are the most important reflectors of solar radiation (Figure 3.17). Cloud top albedo depends primarily on cloud thickness and varies from under 40% for thin clouds (less than 50 m, or 165 ft, thick) to 80% or more for thick clouds (more than 5000 m, or 16,500 ft, thick). For this reason, during daytime, a high thin veil of cirrus clouds appears much brighter than the underside of a much thicker thunderstorm cloud. The average albedo for all cloud types and thickness is about 55%, and at any point in time, clouds cover about 60% of the planet. All other factors being constant, solar radiation reaching Earth's surface is more intense and daytime surface temperatures are higher when the sky is clear rather than cloudy.

With **scattering**, a particle disperses solar radiation in all directions—up, down, and sideways. (Actually, reflection is a special case of scattering.) Within the atmosphere, both gas molecules and aerosols (including the tiny water droplets and ice crystals composing clouds) scatter solar radiation but with some important differences. Scattering by molecules is wavelength dependent and the preferential scattering of blue-violet light by oxygen and nitrogen molecules is the principal reason for the color of the daytime sky (Chapter B). On the other hand, the tiny water droplets and ice crystals that compose clouds scatter visible solar radiation equally at all wavelengths so that clouds appear white.

Reflection and scattering within the atmosphere only alter the direction of incoming solar radiation. **Absorption**, however, is an energy conversion process whereby some of the radiation striking the surface of an object is converted to heat energy. Oxygen, ozone, water vapor, and various aerosols (including cloud particles) absorb a portion of the incoming solar radiation. Absorption by atmospheric gases varies by wavelength; that is, a specific gas absorbs strongly in some wavelengths and weakly or not at all in other wavelengths. As a consequence of absorption, radiation in that wavelength band is removed.

Within the stratosphere, oxygen (O_2) and ozone (O_3) strongly absorb solar ultraviolet radiation at wavelengths shorter than 0.3 micrometer. Oxygen absorbs UV at very short wavelengths (less than 0.2 micrometer), and ozone absorbs longer UV (0.22 to 0.29 micrometer). The net effect of this absorption is twofold: (1) a significant reduction in the intensity of UV that reaches Earth's surface and (2) a marked warming of the upper stratosphere (Figure 2.10). The clear atmosphere is essentially transparent to solar radiation in the wavelength range between about 0.3 and 0.8 micrometer (mostly visible radiation). Water vapor absorbs solar infrared radiation in certain wavelength bands greater than 0.8 micrometer. Clouds are relatively poor absorbers of solar radiation, typically absorbing less than 10% of the solar radiation that strikes the cloud top, although exceptionally thick clouds such as thunderclouds (cumulonimbus) absorb somewhat more.

The Stratospheric Ozone Shield

Ozone (O_3) is a relatively unstable molecule made up of three atoms of oxygen. It occurs naturally in the atmosphere mostly at altitudes below 48 km (30 mi). Depending on where it occurs in the atmosphere, ozone has either a positive or negative impact on life. On the negative side, ozone near the Earth's surface is a serious air pollutant and one of the chief constituents of photochemical smog (Chapter 2). On the positive side, ozone in the stratosphere shields organisms at the Earth's surface from exposure to potentially lethal intensities of solar ultraviolet radiation. Without this so-called **stratospheric ozone shield**, life as we know it could not exist on Earth.

Within the stratosphere, two sets of competing chemical reactions, both powered by solar ultraviolet radiation, continually generate and destroy ozone (Figure 3.18). During ozone production, UV strikes an oxygen molecule (O_2) causing it to split into two free oxygen atoms (O). Free oxygen atoms then collide with molecules of oxygen to form ozone molecules (O_3). At the same time, ozone is destroyed. Ozone absorbs ultraviolet radiation, splitting the molecule into one free oxygen atom (O) and one molecule of oxygen (O_2). The free oxygen atom then collides with an ozone molecule to form two molecules of oxygen.

The net effect of these opposing sets of chemical reactions is a minute reservoir of ozone that peaks at only about 10 parts per million (ppm) in the middle stratosphere. Ultraviolet radiation (at different wavelengths) powers both sets of chemical reactions so that much, but not all, UV radiation is prevented from reaching Earth's surface. Prolonged exposure of the skin to the UV that does penetrate the ozone shield can cause serious health problems. The most dangerous portion of UV radiation that reaches Earth's surface, designated UVB, spans the wavelength band from 0.28 to 0.32 micrometer. For more on the hazards of overexposure to solar UV radiation, see the Essay.

One of today's major environmental concerns involves a group of chemicals known as CFCs (for *chlorofluorocarbons*), which threatens the stratospheric ozone shield. First synthesized in 1928, CFCs were widely used as chilling (heat-transfer) agents in refrigerators and air conditioners, for cleaning electronic circuit boards, and in the manufacture of foams used for insulation. F.S. Rowland and M.J. Molina of the University of California at Irvine first warned of the threat of CFC to the stratospheric ozone shield in 1974. Use of CFCs as propellants in common household aerosol sprays such as deodorants, hairsprays, and furniture polish was banned in the United States, Canada, Norway, and Sweden in 1979. Because of the acknowledged threat to the stratospheric ozone shield, by international agreement, worldwide production and use of CFCs for any purpose was phased out beginning in 1996. For their pioneering studies of the depletion of stratospheric ozone, Rowland, Molina (now at the Massachusetts Institute of Technology), and P.J. Crutzen (of the Max Planck Institute for Chemistry, Germany) were awarded the 1995 Nobel Prize in chemistry.

Certain CFCs are inert (chemically nonreactive) in the troposphere, where they have accumulated for decades. Atmospheric circulation trans-

OZONE PRODUCTION

High energy ultraviolet radiation
strikes an oxygen molecule...

...and causes it to split into
two free oxygen atoms.

The free oxygen atoms collide
with molecules of oxygen...

To form ozone molecules.

OZONE DESTRUCTION

Ozone absorbs a range of
ultraviolet radiation...

...splitting the molecule into
one free oxygen atom and
one molecule of ordinary oxygen.

The free oxygen atom then can
collide with an ozone molecule...

To form two molecules of oxygen.

FIGURE 3.18
Within the stratosphere, two sets of competing chemical reactions continually produce and destroy ozone (O_3). [Adapted from "Ozone: What is it and why do we care about it?" *NASA Facts*, NASA Goddard Space Flight Center, Greenbelt, MD, 1993.]

ports CFCs into the stratosphere where, at altitudes above about 25 km (15 mi), intense UV radiation breaks down CFCs, releasing chlorine (Cl), a gas that readily reacts with and destroys ozone. Products of this reaction are chlorine monoxide (ClO) and molecular oxygen (O_2). Chlorine (Cl) acts as a catalyst in chemical reactions that convert ozone to oxygen. In this way, each chlorine atom destroys perhaps tens of thousands of ozone molecules.

A thinner ozone shield would likely mean more intense UV radiation received at Earth's surface and, for humans, a greater risk of skin cancer, cataracts of the eye, and immune deficiencies. As a general rule, every 1%

decline in stratospheric ozone concentration translates into a 2% increase in the intensity of UV that passes through the ozone shield. Various studies suggest that a 2.5% thinning of the ozone shield could boost the rate of human skin cancer by 10%. The increase in UVB that actually reaches Earth's surface hinges on the cloudiness and dustiness of the atmosphere.

As noted in Chapter 2, the first sign of a thinning of the stratospheric ozone shield came from Antarctica. For about six weeks during the Southern Hemisphere spring (mainly in September and October), the ozone layer in the Antarctic stratosphere (mostly at altitudes between 14 and 19 km, or 9 and 12 mi) thins drastically (Figure 3.19A). Antarctic stratospheric ozone recovers during November. Satellite measurements indicate that the *Antarctic ozone hole* has steadily deepened since the late 1970s (Figure 3.19B). The 2000 ozone hole was the largest on record, covering an area larger than Antarctica and exposing southern portions of Argentina and Chile to elevated levels of UV.

Research conducted during the National Ozone Expeditions to the U.S. McMurdo Station in 1986–87 plus NASA aircraft flights into the Antarctic stratosphere in 1987 led to the discovery of relatively high concentrations of chlorine monoxide (ClO) in the Antarctic stratosphere. This discovery established a convincing link between ozone depletion and CFCs.

What causes the Antarctic ozone hole and why does it fill in by November? During the long, dark Antarctic winter, extreme radiational cooling causes temperatures in the stratosphere to plunge below -85 °C (-121 °F). At such frigid temperatures, what little water vapor that exists in the stratosphere forms clouds composed of ice crystals. Those ice crystals are key to ozone depletion by providing surfaces on which chlorine compounds that are inert toward ozone are converted to active forms that destroy ozone. Once the sun reappears in spring, solar radiation supplies the energy that causes active forms of chlorine to begin destroying ozone.

Ozone depletion takes place while the Antarctic atmosphere is essentially cut off from the rest of the planetary-scale atmospheric circulation by the **circumpolar vortex**, a belt of strong winds that encircles the outer margin of the Antarctic continent. A month or so into spring, however, the circumpolar vortex begins to weaken and allows warmer ozone-rich air from lower latitudes to invade the Antarctic stratosphere. Ice crystal clouds vaporize, and the stratospheric ozone concentration returns to normal levels; that is, the Antarctic ozone hole fills in.

Scientists investigating stratospheric chemistry in the Arctic in early 1989 discovered ozone-destroying chlorine compounds and a slight thinning of ozone. An Arctic ozone hole comparable in magnitude to the Antarctic ozone hole is unlikely for two reasons. For one, in winter the Arctic stratosphere averages about 10 Celsius degrees (18 Fahrenheit degrees) warmer than the Antarctic stratosphere, making formation of stratospheric ice crystal clouds improbable. Secondly, the circumpolar vortex that surrounds the Arctic weakens earlier than its Antarctic counterpart. An exceptionally cold Arctic winter coupled with an unusually persistent circumpolar

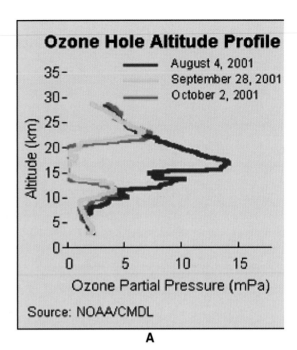

FIGURE 3.19
The Antarctic ozone hole involves a widespread depletion of stratospheric ozone that develops over Antarctica during the Southern Hemisphere spring. (A) Concentration of stratospheric ozone over Antarctica expressed as partial pressure (in mPa or millipascals) as a function of altitude for 4 August 2001 (green), 28 September 2001 (yellow), and 2 October 2001 (red). (B) Minimum values of stratospheric ozone over Antarctica during the period from 5 September to 25 October from 1979 to 2001. Ozone concentration is expressed in Dobson units where one *Dobson unit* corresponds to one-hundredth of a millimeter and is the depth of ozone produced if all the ozone in a column of the atmosphere were brought to sea level temperature and pressure. [Source: NOAA and NASA Goddard Space Flight Center.]

vortex could, however, translate into considerable ozone depletion in the Arctic, a region not very far from human population centers.

The Antarctic ozone hole and its probable link to CFCs spurred questions regarding trends in stratospheric ozone elsewhere around the globe. In the spring of 1992, Richard Stolarski of NASA's Goddard Space Flight Center reported a downward trend in stratospheric ozone at mid-latitude ground stations for all seasons since 1970. Shorter duration satellite measurements (about 12 years) revealed a negative ozone trend in both hemispheres except near the equator, where no significant change was detected. At mid–latitudes, the greatest rate of depletion occurred in late winter and early spring and peaked at slightly more than 6% per decade. The first well-documented reports of a significant increase in ultraviolet radiation striking Earth's surface came from Toronto, Ontario. James Kerr and Thomas McElroy of Canada's Atmospheric Environment Service (AES) found that summer levels of UVB increased about 7% annually from 1989 to 1993. In the same period, winter levels of UVB climbed more than 5% per year.

The phase out of CFCs and other ozone-destroying chemicals has slowed the release of chlorine into the atmosphere. However, CFCs have a long residence time so that the concentration of chlorine in the stratosphere is not expected to begin to decline until about 2010 and complete recovery of the Antarctic ozone layer will not occur until at least 2050. Meanwhile, scientists continue to monitor stratospheric ozone levels worldwide.

Solar Radiation and Earth's Surface

Incoming solar radiation has both direct and diffuse components. Solar radiation that passes directly through the atmosphere to Earth's surface is the direct component whereas solar radiation that is scattered and/or reflected to Earth's surface is the diffuse component. Direct plus diffuse solar radiation that strikes Earth's surface is either reflected or absorbed depending on the surface albedo. The fraction that is not reflected is absorbed (that is, converted to heat) and the fraction that is not absorbed is reflected.

As noted earlier, light surfaces are more reflective of solar radiation than are dark surfaces. Skiers and snow boarders who have been sunburned on the slopes on a sunny day are well aware of the high reflectivity of a snow cover. The albedo of fresh-fallen snow typically ranges between 75% and 95%; that is, 75% to 95% of the solar radiation striking a fresh snow cover is reflected, and the rest (5% to 25%) is absorbed (converted to heat). A fresh snow cover is highly reflective because the snow surface consists of a multitude of randomly oriented crystals each having many reflecting surfaces. As snow ages, the surface albedo declines as snow crystals convert to spherical ice particles having fewer reflecting surfaces. The albedo of old snow typically ranges between 40% and 60%.

At the other extreme, the albedo of a dark surface, such as a black topped road or a spruce forest, may be as low as 5%. From these values, we can understand why light-colored clothing is usually a more comfortable choice than dark-colored clothing during sunny, hot weather. Albedos of some common surfaces are listed in Table 3.1.

TABLE 3.1

Average Albedo (Reflectivity) of Some Common Surface Types for Visible Solar Radiation

Surface	Albedo (% reflected)
Deciduous forest	15-18
Coniferous forest	9-15
Tropical rainforest	7-15
Tundra	15-35
Grasslands	18-25
Desert	25-30
Sand	30-35
Soil	5-30
Green crops	15-25
Sea ice	30-40
Fresh snow	75-95
Old snow	40-60
Glacial ice	20-40
Water body (high solar altitude)	3-10
Water body (low solar altitude)	10-100
Asphalt road	5-10
Urban area	14-18
Cumulonimbus cloud	90
Stratocumulus cloud	60
Cirrus cloud	40-50

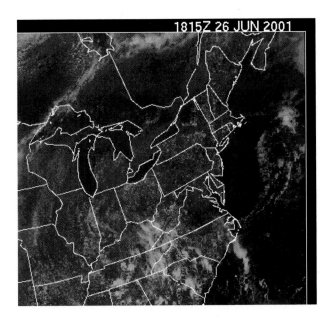

FIGURE 3.20
In this visible satellite image of 26 June 2001 at 1815 UTC, water surfaces appear black because of their low albedo. Note the appearance of Lakes Michigan, Huron, Erie and Ontario, and the near-shore ocean from North Carolina northward to Long Island, NY. Clusters of fair-weather cumulus clouds are visible from Wisconsin and Illinois eastward to the Atlantic coast. Bright white blotches over portions of Kentucky, Tennessee, and western Virginia and North Carolina are thicker convective clouds (including some thunderstorm clouds).

In the visible satellite image in Figure 3.20, the ocean surface appears dark because of its low albedo and strong absorption of solar radiation. The albedo of the ocean (or lake) surface varies with the angle of the sun above the horizon (*solar altitude*). Under clear skies, the albedo of a flat, tranquil water surface decreases with increasing solar altitude (Figure 3.21). The albedo approaches a mirror-like 100% near sunrise and sunset (when the solar altitude is near 0 degrees) but declines sharply as the solar altitude approaches 20 degrees. With overcast skies, only diffuse solar radiation strikes the water surface and the albedo varies little with solar altitude and is uniformly less than 10%. On a global basis, the albedo of the ocean surface averages only about 8%; that is, the ocean absorbs 92% of incident solar radiation.

Solar radiation is selectively absorbed by wavelength as it passes through ocean or lake waters. Absorption increases with increasing wavelength so that within clear, clean water, red light is completely absorbed within about 15 m (50 ft) of the surface, whereas green and blue-violet light may penetrate to depths approaching 250 m (800 ft). More green and blue light is scattered to our eyes, explaining the blue/green

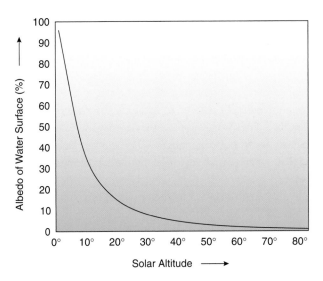

FIGURE 3.21
Variation in albedo of a flat and undisturbed water surface under clear skies with changes in solar altitude. A wave-covered water surface has a slightly higher albedo at high solar altitudes and a slightly lower albedo at low solar altitudes.

color of the open ocean. Suspended sediment significantly boosts absorption so that sunlight often is completely absorbed at shallower depths. In fact, most near-shore waters are so turbid (cloudy) that little if any sunlight reaches much below 10 m (35 ft). Suspended particles preferentially scatter yellow and green light giving these waters their characteristic color.

Significant changes in surface albedo occur seasonally and affect the fraction of incident solar radiation that is converted to heat. In autumn in forested areas, loss of leaves from deciduous trees raises the surface albedo. At middle and high altitudes, significant increases in surface albedo accompany the winter freeze-over of lakes, the formation of sea ice, and development of a thick winter snow cover.

Global Solar Radiation Budget

Measurements by sensors aboard satellites indicate that the Earth–atmosphere system reflects or scatters back to space on average about 31% of the solar radiation intercepted by the planet. This is Earth's **planetary albedo**. By contrast, the moon's albedo is only about 7%, primarily because of the absence of clouds in the highly rarefied lunar atmosphere. Hence, Earth viewed from the moon (by American astronauts) is more than four times brighter than the moon viewed from Earth on a clear night.

The atmosphere (i.e., gases, aerosols, clouds) absorbs only about 20% of the total solar radiation intercepted by the Earth–atmosphere system. In other words, the atmosphere is relatively transparent to solar radiation. The remaining 49% of solar radiation (31% was reflected or scattered to space) is absorbed by Earth's surface, chiefly because of the low average albedo of ocean water covering about 71% of the globe. The global annual solar radiation budget is summarized in Table 3.2.

Earth's surface is the principal recipient of solar

TABLE 3.2
Earth's Solar Radiation Budget

Reflected by the Earth–atmosphere system	31%
Absorbed by the atmosphere	20%
Absorbed by Earth's surface	49%
Total	100%

heating, and heat is transferred from Earth's surface to the atmosphere, which eventually radiates this energy to space. The Earth's surface is thus the main source of heat for the atmosphere; that is, the atmosphere is heated from below. This is evident in the vertical temperature profile of the troposphere. Normally, air is warmest close to the Earth's surface, and air temperature drops with increasing altitude in the troposphere, that is, away from the main source of heat (Figure 2.10).

Infrared Response and the Greenhouse Effect

If solar radiation were continually absorbed by the Earth–atmosphere system without any compensating flow of heat out of the system, Earth's surface temperature would rise steadily. Eventually, life would be extinguished and the ocean would boil away. Actually, the global air temperature changes little from one year to the next. Global radiative equilibrium keeps the planet's temperature in check; that is, emission of heat to space in the form of infrared radiation balances solar radiational heating of the Earth–atmosphere system. Although solar radiation is supplied only to the illuminated portion of the planet, infrared radiation is emitted to space

ceaselessly, day and night, by the entire Earth–atmosphere system. This explains why nights are usually colder than days and why air temperatures usually drop throughout the night.

While the clear atmosphere is relatively transparent to solar radiation, certain gases in the atmosphere impede the escape of infrared radiation to space thereby elevating the temperature of the lower atmosphere. This important climate control is the so-called *greenhouse effect*.

GREENHOUSE WARMING

The **greenhouse effect** refers to the heating of Earth's surface and lower atmosphere caused by strong absorption and emission of infrared radiation by certain atmospheric gases, known as *greenhouse gases*. Solar radiation and terrestrial infrared radiation peak in different portions of the electromagnetic spectrum, their properties differ, and they interact differently with the atmosphere. As noted earlier, the atmosphere absorbs only about 20% of the solar radiation intercepted by the planet. The atmosphere absorbs a greater percentage of the infrared radiation emitted by Earth's surface and the atmosphere, in turn, radiates some IR to space and some to Earth's surface. Hence, Earth's surface is heated by absorption of both solar radiation and infrared radiation.

The similarity in radiational properties between infrared-absorbing atmospheric gases and the glass panes of a greenhouse is the origin of the term *greenhouse effect*. Window glass, like Earth's atmosphere, is relatively transparent to visible solar radiation but strongly absorbs infrared radiation. A greenhouse, where plants are grown, takes advantage of the radiational properties of glass and is constructed almost entirely of glass panes (Figure 3.22). Sunlight readily penetrates greenhouse glass and is absorbed (converted to heat) within the greenhouse. Objects in the greenhouse emit infrared radiation that is strongly absorbed by the glass panes. Glass, in turn, emits IR to the atmosphere and the greenhouse interior, thereby raising the temperature within the greenhouse.

Although the greenhouse analogy is widely used, absorption and emission of IR radiation by glass is only part of the reason why the interior of a greenhouse is relatively warm. A greenhouse is also a shelter from the wind and reduces heat loss to the external environment by conduction and convection (Chapter 4). As a rule, the thinner the greenhouse glass and the higher the external wind speed, the more important the shelter effect is compared to radiational effects. For this reason, some

FIGURE 3.22
The glass of a greenhouse behaves similarly to certain gases (e.g., water vapor, carbon dioxide) in the atmosphere that are transparent to sunlight but strongly absorb and emit infrared radiation.

atmospheric scientists argue that the greenhouse analogy is inappropriate and the phenomenon should be renamed the *atmospheric effect*. But use of the term *greenhouse effect* is so common among scientists, public policymakers, and the media that we continue use of the term in this book.

The greenhouse effect is responsible for considerable warming of Earth's surface and lower atmosphere. Viewed from space, the planet radiates at about −18 °C (0 °F), whereas the average temperature at the Earth's surface is about +15 °C (59 °F). The temperature difference is due to the greenhouse effect and amounts to

$$[15 \text{ °C} - (-18 \text{ °C})] = 33 \text{ Celsius degrees}$$
$$\text{or}$$
$$[59 \text{ °F} - 0 \text{ °F}] = 59 \text{ Fahrenheit degrees.}$$

Without the greenhouse effect, Earth would be too cold to support most forms of plant and animal life.

Water vapor is the principal greenhouse gas. Other greenhouse gases are carbon dioxide, ozone, methane (CH_4), and nitrous oxide (N_2O). The percentage of infrared radiation absorbed by these gases varies with wavelength (Figure 3.23). Significantly, the percentage absorbed by all atmospheric gases is very low at wavelengths near the planet's peak infrared intensity (about 10 micrometers). Through these so-called **atmospheric windows**, wavelength bands in which little or no radiation is absorbed, most heat from Earth's surface escapes to space as infrared radiation.

The warming effect of atmospheric water vapor is evident even at the local or regional scale. Consider an example. Locations in the American desert southwest and along the Gulf Coast are at about the same latitude and receive essentially the same input of solar radiation on a clear summer day. In both places, summer afternoon high temperatures commonly top 32 °C (90 °F). At night, however, air temperatures often differ markedly. Air is relatively dry (low humidity) in the Southwest so that terrestrial infrared radiation readily escapes to space and air temperatures near Earth's surface may drop well under 15 °C (59 °F) by sunrise. People who hike or camp in the desert are aware of the dramatic fluctuations in temperature between day and night. Infrared radiation does not escape to space as readily through the Gulf Coast atmosphere where the air is more humid. Water vapor strongly absorbs outgoing IR and emits IR to Earth's surface so that early morning low temperatures may dip no lower than the 20s Celsius (70s Fahrenheit). The smaller diurnal temperature contrast along the Gulf Coast is due to more water vapor and a stronger greenhouse effect.

Clouds are composed of IR-absorbing water droplets and/or ice crystals and also contribute to the greenhouse effect. All other factors being equal, nights

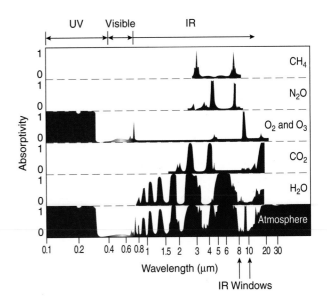

FIGURE 3.23
Absorption of radiation by selected gaseous components of the atmosphere as a function of wavelength. *Absorptivity* is the fraction of radiation absorbed and ranges from 0 to 1 (0% to 100% absorption). Absorptivity is very low or near zero in *atmospheric windows*. Note the infrared windows near 8 and 10 micrometers.

usually are warmer when the sky is overcast than when the sky is clear. Even high, thin cirrus clouds through which the moon is visible can reduce the temperature drop at Earth's surface by several Celsius degrees. Clouds thus affect climate in two opposing ways: By absorbing and radiating IR, clouds warm Earth's surface and by reflecting solar radiation, they cool Earth's surface. On a global scale, which one of these two opposing effects is more important? Analysis of satellite measurements of incoming and outgoing radiation indicates that clouds have a net cooling effect on global climate. That is, all other factors being equal, a more extensive cloud cover would cool the planet and less extensive cloud cover would warm the planet.

GREENHOUSE GASES AND GLOBAL CLIMATE CHANGE

Many atmospheric scientists as well as public policymakers are concerned about the possible impact on global climate of the steadily rising concentrations of atmospheric carbon dioxide (CO_2) and other infrared absorbing gases. Higher levels of these gases appear likely to *enhance* the natural greenhouse effect and could contribute to warming on a global scale.

Systematic monitoring of atmospheric carbon dioxide began in 1957 at NOAA's Mauna Loa Observatory in Hawaii under the direction of Charles D. Keeling of Scripps Institution of Oceanography. The observatory is situated on the slope of a volcano 3400 m (11,200 ft) above sea level in the middle of the Pacific Ocean—sufficiently distant from major industrial sources of air pollution that carbon dioxide levels are considered representative of at least the Northern Hemisphere. Also since 1957, atmospheric CO_2 has been monitored at the South Pole station of the U.S. Antarctic Program and that record closely parallels the one at Mauna Loa. The Mauna Loa record shows a sustained increase in average annual atmospheric carbon dioxide concentration from about 315.98 ppmv (parts per million by volume) in 1959 to 369.40 ppmv in 2000 (Figure 3.24). Superimposed on this upward trend is an annual carbon dioxide cycle caused by seasonal changes in Northern Hemisphere vegetation: Carbon dioxide level falls during the growing season, reaching a minimum in September, and rises in winter, reaching a maximum in April, May, and June.

The upward trend in atmospheric carbon dioxide was underway long before Keeling's monitoring and appears likely to continue well into the future. Human contributions to the buildup of atmospheric CO_2

began roughly three centuries ago with land clearing for agriculture and settlement. Land clearing contributes CO_2 to the atmosphere via burning, decay of wood residue, and reduced photosynthetic removal of carbon dioxide from the atmosphere. By the middle of the 19th century, growing dependency on coal burning associated with the beginnings of the Industrial Revolution caused a more rapid rise in CO_2 concentration. Carbon dioxide is a byproduct of the burning of coal and other fossil fuels. The concentration of atmospheric CO_2 has increased by about 31% since 1750. The present atmospheric CO_2 concentration has not been exceeded over at least the past 420,000 years and the current rate of increase is unprecedented in the past 20,000 years. Fossil fuel combustion accounted for roughly 75% of the increase in atmospheric carbon dioxide over the past two decades while deforestation (and other land clearing) is likely responsible for the balance. If the growth in fossil fuel combustion continues, the atmospheric carbon dioxide concentration could top 550 ppmv (double the pre-industrial level) by the end of the 21st century.

Furthermore, rising levels of other infrared-absorbing gases (e.g., methane and nitrous oxide) could also enhance the natural greenhouse effect. Methane's concentration in the atmosphere has increased by 1060 ppb (parts per billion) (151%) since 1750. About half of all methane emissions are anthropogenic. Methane is a product of organic decay in the absence of oxygen and its concentration may be rising because of more rice cultivation, cattle, landfills, and/or termites, all sources of

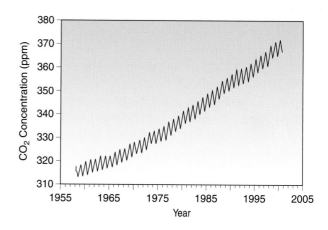

FIGURE 3.24
Concentration of atmospheric carbon dioxide (CO_2) as measured at the Mauna Loa Observatory in Hawaii. Annual cycles in photosynthesis and cellular respiration account for seasonal variations in CO_2. [Source: C.D. Keeling *et al.*, Scripps Institution of Oceanography, University of California, La Jolla, CA.]

methane. The atmospheric concentration of nitrous oxide (N_2O) has increased by 46 ppb (17%) since 1750. The upward trend in nitrous oxide is likely linked to industrial air pollution, cattle feedlots, and agricultural soils. Although occurring in extremely low concentrations, methane and nitrous oxide are very efficient absorbers of infrared radiation (absorbing in the *atmospheric windows* mentioned earlier) and potentially could also contribute to global warming.

As noted in Chapter 2, atmospheric scientists use numerical global climate models to predict the magnitude of global warming that could accompany the upswing in atmospheric CO_2. Based on the most recent runs of global climate models, in 2000 the *Intergovernment Panel on Climate Change (IPCC)*, a standing international committee sponsored by the United Nations, revised upward its earlier projections of the magnitude of global warming that could attend an enhanced greenhouse effect. Various models project that the globally averaged surface temperature will increase by 1.4 to 5.8 Celsius degrees (2.5 to 10.4 Fahrenheit degrees) over the period 1990 to 2100. As discussed in more detail in Chapter C, climate change is geographically non-uniform so that this projected warming is not necessarily representative of what might happen everywhere. In any event, enhancement of the natural greenhouse effect could cause a climate change that would be greater in magnitude than any prior climate change of the past 10,000 years.

What are the potential societal impacts of an enhanced greenhouse effect and global warming? According to estimates by the IPCC, North American climatic zones could shift northward by as much as 550 km (350 mi) and would likely affect all sectors of society. In some areas, heat and moisture stress would cut crop yields, and traditional farming practices would have to change. Across the North American grain belt, for example, higher temperatures and more frequent drought during the growing season might necessitate a switch from corn to wheat and require more irrigation.

Global warming may also cause sea level to rise in response to melting of land-based polar ice sheets and mountain glaciers, and expansion of seawater. (Melting of floating ice does not raise sea level.) Past large-scale climate changes apparently were amplified at higher latitudes, prompting speculation on the fate of the Greenland and West Antarctic ice sheets with global warming. Climate models predict that local warming over Greenland may be up to three times the global average. Tide gauging stations document a sea level rise

of 10 to 20 cm (4 to 8 in.) over the past century and global climate models predict that this rise will accelerate. According to IPCC estimates, melting glaciers combined with thermal expansion of ocean water could raise global mean sea level by 9 to 88 cm (4 to 35 in.) over the period 1990 to 2100. This rise in sea level would accelerate coastal erosion and inundate islands and low-lying coastal plains, some of which are densely populated.

On the positive side, global warming would lengthen the growing season at high latitudes. Also, more atmospheric carbon dioxide is known to spur plant growth. Warmer winters would also lengthen the navigation season on lakes, rivers, and harbors where seasonal ice cover is a problem. For most Canadians and Americans living in the northern states, energy savings due to less space heating in winter would more than compensate for the greater energy requirements anticipated with more air conditioning in summer.

Is an enhanced greenhouse effect underway? Some atmospheric scientists point to recent trends in climate that are consistent with changes predicted by global climate models. In Spring 2000, the National Climatic Data Center (NCDC) reported that global warming accelerated during the final quarter of the 20th century and that the seven warmest years in the instrument-based record occurred in the 1990s. During 1997 and 1998, record high global mean surface temperatures were set over 16 consecutive months. In early 2001, the IPCC reported that the global mean annual surface temperature was about 0.6 Celsius degree (1.1 Fahrenheit degrees) higher at the end of the 20th century than at the close of the 19th century (Figure 3.25).

The IPCC also reports that according to satellite measurements, the extent of snow cover has decreased about 10% since the late 1960s. Ground-based observations show a reduction of about two weeks in the annual duration of lake- and river-ice cover at mid and high latitudes of the Northern Hemisphere during the 20th century. Many mountain glaciers are retreating in non-polar regions. Also, in Northern Hemisphere spring and summer, sea-ice extent has decreased by about 10% to 15% since the 1950s.

But before residents of low-lying coastal areas head for high ground to avoid rising ocean water, some words of caution are warranted. Even though some recent climatic trends and environmental changes are consistent with changes expected to accompany greenhouse warming, there is as of yet little evidence of a direct cause-effect relationship. Climate controls other

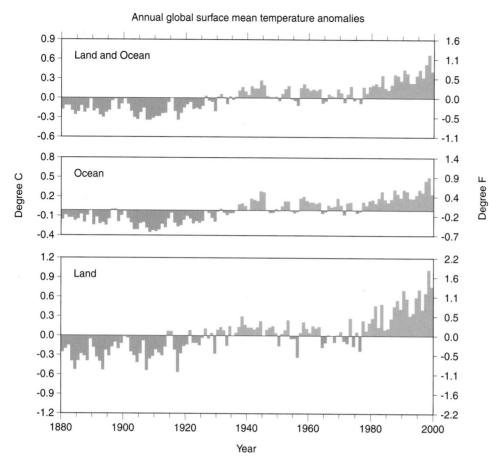

Annual global surface mean temperature anomalies

FIGURE 3.25
Anomalies in annual mean land-, sea-surface and global (land and ocean)temperatures. Anomalies are differences in degrees Celsius from the long-term average (1961–1990 mean). [From NOAA, National Climatic Data Center, Asheville, NC]

cycling of sulfurous particles out of the stratosphere, warming resumed in 1994.

Some scientists question the integrity of the global mean temperature record that shows a gradual warming trend between the 1880s and 1990s. Questions center on the possible influence of (1) changes in instrument technology during the record period, (2) warming caused by urbanization (Chapter 4), and (3) gaps in the spatial coverage of the networks of weather stations, especially over the ocean. Criticism is also directed at global climate models. For one, models vary in sophistication and in their ability to simulate the climatic influence of ocean currents, cloud cover, and atmospheric water vapor. Furthermore, the current generation of climate models does not do well in forecasting regional responses to hemispheric- or global-scale climate change.

In spite of scientific uncertainties, many experts argue that society has so much at stake that action should be taken immediately to head off possible enhanced greenhouse warming. They call for (1) sharp reductions in coal and oil consumption, (2) greater reliance on non-fossil fuel energy sources (see this chapter's Essay on solar power), (3) higher energy efficiencies (e.g., more vehicle miles per gallon), and (4) massive reforestation and a halt to deforestation. They hasten to point out that such actions are advisable even if greenhouse warming does not happen because they will help alleviate other serious environmental problems. For example, reducing reliance upon fossil fuels will also cut air pollution. Coal and oil burning produces not only CO_2, but also sulfur dioxide (SO_2) and particulate matter that are serious air pollutants. We have more on climate and climate change

than (or in addition to) greenhouse gases may well be responsible for recent trends. Furthermore, the atmosphere is a complex and highly interactive system, so that in the future other processes may compensate for an enhanced greenhouse effect. For example, higher temperatures in polar latitudes may mean more snowfall and eventually more, not less, glacial ice.

The global climatic impact of the eruption of Mount Pinatubo in the Philippines underscores the dependency of climate on many variables. The June 1991 eruption, the most violent of the 20th century, injected huge quantities of sulfur dioxide (SO_2) into the stratosphere. SO_2 converted to tiny sulfuric acid droplets and sulfate particles that absorbed some solar radiation and scattered other solar radiation back to space. Less solar radiation reaching Earth's surface likely contributed to a temporary downturn in global mean temperature that persisted through 1992 and 1993. With the eventual in Chapter C.

Monitoring Radiation

The **pyranometer** is the standard instrument for measuring the intensity of solar radiation striking a horizontal surface. The instrument consists of a sensor enclosed in a transparent hemisphere that transmits total (direct plus diffuse) solar short-wave (less than 3.0 micrometers) radiation (Figure 3.26). In one design, the sensor is a disk consisting of alternating black and white wedge-shaped segments forming a star-like pattern. Black wedges are highly absorptive whereas white wedges are highly reflective. Differences in absorptivity and albedo mean that the temperatures of the black and white portions of the sensor respond differently to the same intensity of solar radiation. The temperature contrast between the black and white segments is calibrated in terms of radiation flux (e.g., watts per m^2). A pyranometer may be linked electronically to a pen recorder or computer that traces a continuous record of incident solar radiation. Special care must be taken in mounting and maintaining a pyranometer. The instrument should be situated where it will not be affected by shadows, any nearby highly reflective surface, or other sources of radiation. The special enclosing bulb must also be kept clean and dry.

As discussed in Chapter 1, sensors onboard weather satellites measure two types of radiation emanating from the Earth–atmosphere system: reflected (or scattered) visible solar radiation and emitted infrared radiation. The first is processed into visible satellite images and the second is processed into infrared satellite images. An *infrared radiometer* is an instrument that measures the intensity of infrared radiation emitted by the surface of some object such as clouds or the ocean. The temperature dependency of IR emission means that infrared sensors can be calibrated to remotely sense the temperature of the Earth's surface or cloud tops.

Recall from earlier in this chapter, for a *blackbody* (perfect radiator), the wavelength of maximum radiation emission varies inversely with the radiation temperature (*Wien's displacement law*). Hence, a sensor that measures the intensity of radiation emitted by an object can determine the radiation temperature of that object. For a blackbody, the actual temperature equals the radiation temperature. For all other objects, the actual temperature is higher than the radiation temperature, the difference depending on the object's **emissivity**, a measure of how closely an object approximates a blackbody. The emissivity of a blackbody is 1.0 (or 100%) and less than 1.0 for all other objects. As a general rule, every 1% change in emissivity is associated with a change in the object's surface temperature of slightly less than 1.0 Celsius degree.

With adjustments made for differences in emissivity of objects within its field of view, an IR sensor (radiometer) aboard a weather satellite measures the temperature of land and sea surfaces and the tops of clouds. Based on measured differences in IR emission, sensors can distinguish low clouds, which are relatively warm, from high clouds, which are relatively cold. These measurements are calibrated on a gray scale (in which low clouds appear dark gray and high clouds appear bright white) or a color scale. One advantage of IR satellite images over visible satellite images is their availability both day and night.

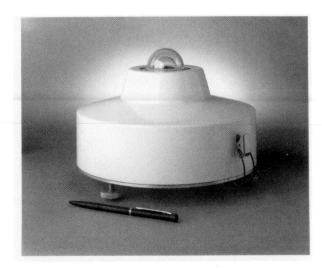

FIGURE 3.26
Solar pyranometer, the standard instrument for monitoring solar radiation in the wavelength band from 0.3 to 3.0 micrometers.

Conclusions

Earth intercepts only a tiny fraction of the total radiational energy output of the sun. Movements of the planet in space (rotation on its spin axis and revolution about the sun) distribute this energy unequally within the Earth–atmosphere system. As solar radiation passes through the atmosphere, a portion is scattered and re-flected back to space and a portion is absorbed (that is, converted to heat). Overall, however, the atmosphere is relatively transparent to solar radiation and that which reaches Earth's surface is either reflected or absorbed. The Earth–atmosphere system responds to solar heating by emitting infrared radiation to space. Greenhouse gases and clouds absorb and emit IR to Earth's surface, thereby significantly elevating the average temperature of the lower atmosphere (the greenhouse effect).

Net incoming solar radiation is balanced by the infrared radiation emitted to space by the Earth–atmosphere system. Absorption of solar radiation causes warming and emission of infrared radiation to space causes cooling. Within the Earth–atmosphere system, however, the rates of radiational heating and radiational cooling are not the same everywhere. In the next chapter, we examine the reasons for these energy imbalances and their implications for atmospheric circulation and weather. But first we need to distinguish between heat and temperature and describe heat transfer processes.

Basic Understandings

- All objects emit and absorb energy in the form of electromagnetic radiation. The many forms of electromagnetic radiation make up the elec-tromagnetic spectrum and are distinguished on the basis of wavelength, frequency, and energy level.
- Earth and sun closely approximate perfect radiators (*blackbodies*) so that blackbody radiation laws may be applied to them with useful results.
- *Wien's displacement law* predicts that the wavelength of most intense radiation is inversely proportional to the absolute temperature of a radiating object. Hence, solar radiation peaks in the visible, and Earth–atmosphere (terrestrial) radiation peaks in the infrared (IR).
- According to the *Stefan–Boltzmann law*, the total energy radiated by an object at all wavelengths is directly proportional to the fourth power of the object's absolute temperature. Hence, the much hotter sun emits considerably more radiational energy per unit area than does the Earth–atmosphere system.
- The total energy (in the form of solar radiation) that is absorbed by the Earth–atmosphere system equals the total energy (in the form of infrared radiation) emitted by the Earth–atmosphere system to space. This balance of energy input and output is known as *global radiative equilibrium* and is the prevailing condition on planet Earth.
- Solar altitude, the angle of the sun above the horizon, influences the intensity of solar radiation that strikes Earth's surface. All other factors being equal, as the solar altitude increases, the intensity of solar radiation received at Earth's surface (energy per unit area) also increases.
- As a consequence of Earth's elliptical orbit, nearly spherical shape, and tilted spin axis, solar radiation is distributed unevenly over Earth's surface and changes through the course of a year.
- In the summer hemisphere, solar altitudes are higher, daylight is longer, there is more solar radiation, and air temperatures are higher. In the winter hemisphere, solar altitudes are lower, daylight is shorter, there is less solar radiation, and air temperatures are lower.
- Solar radiation that is not absorbed by the atmosphere (converted to heat) or reflected or scattered to space reaches Earth's surface.
- Absorption of ultraviolet radiation during the natural formation and destruction of ozone within the stratosphere shields organisms from exposure to potentially lethal levels of UV.
- Solar radiation that reaches Earth's surface is either reflected or absorbed depending on the surface albedo. Surfaces that appear light-colored have a relatively high albedo for visible radiation whereas surfaces that appear dark-colored have a relatively low albedo.
- The atmosphere is heated from below; that is, on average heat flows from Earth's surface to the overlying air. This is evident in the average temperature profile of the troposphere; that is, air temperature drops with increasing altitude.
- Water vapor, carbon dioxide, and several other atmospheric trace gases absorb and emit infrared radiation to Earth's surface, thereby significantly elevating the average temperature of the lower atmosphere. Clouds also contribute to this so-called

greenhouse effect.   Water vapor is the dominant greenhouse gas.

- Rising levels of atmospheric CO_2 and several other IR-absorbing gases may enhance the natural greenhouse effect and cause global climate change. Combustion of fossil fuels and, to a lesser extent, deforestation are responsible for an upward trend in atmospheric carbon dioxide.

- The pyranometer is the standard instrument for monitoring the total flux of solar radiation whereas the infrared radiometer selectively monitors IR.

ESSAY: Hazards of Solar Ultraviolet Radiation

Most of us welcome sunny weather. Bright, sunny skies not only make possible a variety of outdoors activities, but they also seem to energize and cheer most people. During warm weather, many people spend as much time as possible in the sun, some of them sunbathing for hours to develop a dark tan (see Figure). However, there is overwhelming evidence that too much sun can cause premature aging of the skin and serious health problems, including skin cancer and cataracts of the eye. The culprit is overexposure to the ultraviolet portion of solar radiation.

FIGURE
Sunbathers run the risk of excessive exposure to solar ultraviolet (UV) radiation that can cause serious health problems.

Just as visible light is made up of all the colors of the rainbow, ultraviolet radiation is also divided into segments: UVA, UVB, and UVC. UVC rays, the most energetic of the three, readily kill exposed cells. Fortunately for life on Earth, the stratospheric ozone shield prevents UVC from reaching Earth's surface. Some UVB penetrates the ozone shield and strikes Earth's surface. Although less energetic than UVC, UVB rays are sufficiently intense to damage cells and cause sunburn, skin cancer, and other health problems. UVA, the weakest of the three forms of solar ultraviolet radiation, poses a much smaller health risk than UVB but does accelerate the natural aging of the skin and exacerbates the damaging effects of UVB.

Some consequences of overexposure to solar ultraviolet radiation appear within hours (e.g., sunburn) but others typically take many years to show up (e.g., skin cancer and cataracts). Probably the most noticeable impact of long-term overexposure to the sun is premature aging of the skin, called *photoaging*. UV rays that penetrate the skin damage the structural proteins (known as collagen and elastin) that compose the elastic fibers responsible for the skin's strength and resilience. As damage progresses, the skin develops a leathery texture, wrinkles, and irregular pigmentation. People who have experienced long-term exposure to the sun, perhaps because of their occupation or recreational interests, may exhibit symptoms of photoaging much earlier (prior to age 30) than people who minimize their exposure (after age 40).

Ultraviolet radiation contributes to certain types of cataracts, a major cause of vision impairment worldwide. A *cataract* is a cloudy (opaque) region that develops in the lens of the eye, the part of the eye responsible for focusing light and producing sharp, clear images. If untreated, cataracts can cause partial or total blindness. Research suggests that 20% of all cataracts are related to UV exposure.

Most worrisome is the role of ultraviolet radiation in the development of skin cancer, the most common form of cancer in the United States. Researchers continue to accumulate evidence that links overexposure to UVB radiation to the development of the three most common forms of skin cancer: basal cell carcinoma, squamous cell carcinoma, and malignant melanoma. In each case, UVB damages the DNA of the respective cells, resulting in uncontrolled growth and subsequent formation of tumors. The American Cancer Society estimates that more than 500,000 new cases of skin cancer (responsible for more than 9000 deaths) are diagnosed each year in the United States. About 1 in 7 Americans will contract this disease during their lifetime and perhaps 90% of these cases will be due to overexposure to the sun's UV rays.

How can people protect themselves from the health hazards of solar ultraviolet radiation? They can begin by monitoring their level of exposure. The National Weather Service, Environmental Protection Agency, and the Centers for Disease Control and Prevention developed a special index that alerts the public to the risk of exposure to solar ultraviolet radiation. Issued once daily by NOAA's Climate Prediction Center for 58 cities across the nation, the *UV Index* predicts the relative risk of sunburn at solar noon. The index is based on day of the year, station elevation and latitude, total ozone in the atmosphere above the location (measured by satellite sensors), and anticipated cloud cover. The index is designed so that the higher the rating, the greater the expected exposure to UV and the quicker a person is likely to sunburn. As shown in the Table, the *UV Index* ranges from 0 to 15 and, for an expected UV level, specifies the number of *minutes to burn* depending on the sensitivity of a person's skin. A sample UV-Index forecast is shown on the map.

TABLE
UV-Index and Minutes to Burn

Exposure category	Index value	Minutes to burn for most susceptible skin type	Minutes to burn for least susceptible skin type
Minimal	0–2	30.0	>120
Low	3	20.0	90
	4	15.0	75
Moderate	5	12.0	60
	6	10.0	50
High	7	8.5	40
	8	7.5	35
	9	7.0	33
Very High	10	6.0	30
	11	5.5	27
	12	5.0	25
	13	<5.0	23
	14	4.0	21
	15	<4.0	20

The safe level of exposure to UVB depends upon skin phototype. People who never tan and always burn (most-sensitive phototype) can tolerate considerably less exposure to UV than people who always tan and rarely burn (least-sensitive phototype). For a moderate UV Index value of 6, the skin of a person with the most-sensitive phototype begins to burn after only 10 minutes of exposure to the sun; the time to burn for a person having the least-sensitive phototype is 5 times longer. The UV Index is intended only as a general guide. Because skin phototype varies considerably within human populations, some people will find that they require more protection from the sun than indicated by the UV Index.

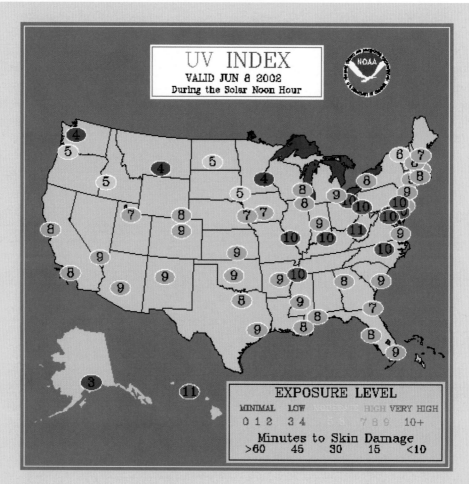

FIGURE
A sample UV Index forecast [NOAA, National Climate Prediction Center].

If possible, people should avoid the sun when its rays are most intense, that is, between about 10 a.m. and 2 p.m. However, this general rule fails to account for the width of time zones (15 degrees of longitude typically), daylight savings time, and seasonal changes in solar altitude. Leith Holloway of the Geophysical Fluid Dynamics Laboratory in Princeton, NJ, developed a more reliable and simple guide for UV exposure. Holloway found that during a clear midsummer day, as solar altitude increases in the morning, the intensity of UV striking Earth's surface increases slowly until the sun is about 45 degrees above the horizon. At higher solar altitudes, the intensity of UV radiation striking Earth's surface increases rapidly as the solar beam passes through less and less atmosphere. In fact, total UV exposure at solar altitudes greater than 45 degrees is about 5 times greater than the total exposure during the remaining daylight hours. How then does a person determine if the solar angle is greater than or less then 45 degrees?

Shadows on a horizontal surface are shorter than the objects casting them when the solar angle is greater than 45 degrees (and vice versa). This is the basis for Holloway's *shadow rule*. If a person's shadow is shorter than he or she is tall, then the person should avoid the sun or apply a sunscreen to exposed skin; if a person's shadow is longer than she or he is tall, then the health risk is much less. Simply stated, "Short shadow? Seek shade!"

Solar altitudes are greatest in summer (May through August in the Northern Hemisphere) and in tropical latitudes so that UV intensity is also greatest at such times and locations. Unfortunately, summer is the season and the tropics the place where people tend to spend more time outdoors and wear less protective clothing. Most skin cancers appear on exposed skin surfaces such as on the forearms, face, and neck. Hence, wearing protective clothing (e.g., wide-brimmed hats, long-sleeved shirts, long pants, tightly woven fabrics) is a wise strategy. Sunglasses that are UV-rated worn along with a wide brimmed hat provide adequate eye protection.

Sunscreens have ingredients that selectively absorb UV wavelengths and should be applied to all exposed skin approximately 20 minutes before venturing outdoors. Be sure to read the label; medical experts strongly recommend a broad-spectrum sunscreen that provides protection against both UVB and UVA. Also be aware of the degree of protection offered, that is, the *sun protection factor* (*SPF*), a measure of the time that the skin can be safely exposed to the sun when wearing a particular sunscreen versus wearing no sunscreen. For example, if a person normally begins to turn pink after 10 minutes in the sun, applying a product with SPF 15 provides about 2.5 hours of protection (10 minutes × 15 = 150 minutes). The higher the SPF, the longer the protection lasts. Health professionals recommend a minimum SPF of 15, but opting for a higher SPF is a good idea if a person is fair-skinned, spends a lot of time outdoors, or lives at a high elevation. Be sure to apply sunscreen liberally to all exposed skin to ensure optimal protection. Perspiration and water remove sunscreens (even so-called waterproof sunscreens) so they should be reapplied every two hours, even on cloudy days.

ESSAY: Solar Power

Solar power is a renewable energy source that will last as long as the sun, billions of years. Solar power is an environmentally attractive alternative to conventional energy sources, especially coal, oil, and nuclear fuels. Furthermore, the cost of solar power is becoming more competitive with the price of conventional fuels in a number of applications and environments.

If all the solar radiation that is transmitted through the day-lit atmosphere were distributed uniformly over Earth's surface, about 180 watts would illuminate every square meter. (By comparison, the *solar constant* is 1368 watts per m^2.) This amounts to about 4 kilowatt-hours per square meter per day, sufficient to meet nearly half the heating and cooling requirements of a typical American home. The total solar energy that falls yearly on America's land surface is about 600 times greater than the nation's total annual energy demand.

All technologies based on the collection of solar radiation are limited by the variability of the energy source. There is no solar radiation at night and its intensity varies with time of day, day of the year, latitude and cloud cover. Seasonal fluctuations in incident solar radiation are especially pronounced at middle and high latitudes. For example, in relatively sunny eastern Washington, average annual solar radiation striking Earth's surface is 194 watts per m^2, but monthly mean values range from a low of 50 to a high of 343 watts per m^2, a sevenfold difference through the course of the year. The American desert southwest is the most promising area for solar power because of minimal cloud cover and less seasonal variability in solar radiation. Fluctuations in consumer demand for energy further complicate matters so that matching energy supply with demand requires use of energy storage systems. Solar collectors may produce more heat during the day than can be used at that time so that excess heat is usually stored in insulated water tanks or compartments filled with rocks. At middle and high latitudes, a conventional heating system is usually needed as a backup to solar collectors, particularly during extended spells of cloudy or very cold weather.

A traditional *solar collector* is basically a framed panel of glass that is designed to tap solar power. Sunlight passes through two layers of glass before being absorbed by a blackened (low albedo) metal plate. Heat is then conducted from the absorbing plate to either air or a liquid, which is conveyed by a fan or pump to wherever the heat is needed. Solar collectors typically capture 30% to 50% of the solar energy that strikes them and are most commonly used for space or water heating in small buildings such as homes, schools, and apartment houses. Solar collectors do not produce high enough temperature to turn water into steam, so these devices are not suitable for most industrial purposes.

To partially compensate for the variability of incoming solar radiation due to changes in solar altitude, some solar collectors are tilted. The advantage of tilted collectors over ones that are horizontal depends on the average cloud cover and latitude of the site. An optimal situation occurs in winter at a midlatitude locality favored by clear skies. There, tilted solar collectors can double the amount of solar radiation that is absorbed. Some solar collectors are designed to track the sun so that they are always oriented perpendicular to the solar beam (see Figure). However, the cost of tracking devices usually exceeds the value of the energy obtained.

FIGURE
A panel of interconnected photovoltaic cells, computer-controlled and designed to follow the sun and always present a horizontal surface to the incoming solar beam.

Scientists and engineers are developing ways to efficiently convert solar energy to electricity on a large scale. In one conversion method, a *power tower system*, an array of computer-controlled flat mirrors, called *heliostats*, track the sun and focus its radiation on a single heat collection point at the top of a tower. Concentrated sunlight in these systems can produce temperatures up to 480 °C (900 °F), high enough to convert water to high-pressure steam for driving turbines that generate electricity. Other designs utilize a large parabolic dish or trough as a collection device.

An alternative to solar-driven turbines for generating electricity is the *photovoltaic cell* (*PV cell*), also known as a *solar cell*. Solar cells convert solar radiation directly into electricity and routinely power many devices, including handheld calculators, parking lot lights, and space vehicles. Solar cells use sunlight to create a *voltage*, that is, a difference in electrical potential, in a *diode*. When sunlight strikes the diode, an electric current flows through the circuit to which it is connected. Only special materials develop the necessary voltage to produce a direct current when they are exposed to sunlight. These materials, called *semiconductors*, are composed of highly purified silicon to which tiny amounts of certain impurities have been added. Semiconductors are manufactured in the form of tiny wafers or sheets. Electricity travels through metal contact wires on the front and back of the wafer. Groups of wafers are wired together to form photovoltaic modules, and these are interconnected to form a photovoltaic panel. About 40 wafers must be linked to match the power output of a single automobile battery.

One serious drawback of today's solar cells is low efficiency. In this context, *efficiency* is defined as the percentage of solar radiation striking a photovoltaic cell that is converted to electrical energy. The efficiency of a handheld solar-powered calculator, for example, is only 3% or less. Some mass-produced photovoltaic panels have conversion efficiencies of 10% to 12%. In 1988, researchers at Sandia National Laboratories attained an all-time high efficiency rating of 31% by stacking two solar cells (a gallium arsenide cell on top and a silicon cell on the bottom). Among the many factors that contribute to low solar-cell efficiency are cell reflectivity, conversion of radiation to heat, and the sensitivity of cells to only a portion of the solar spectrum.

The main obstacle to greater use of solar cells for generating electricity is cost, although future prospects are encouraging. The cost of solar cell generated electricity has dropped substantially since 1980. Future cost reductions are unlikely to be achieved by boosting the efficiency of traditional solar cells because, as a general rule, production costs soar with increasing solar cell efficiency. A more promising alternative is further development of thin-film solar cells, which are less efficient but also much less expensive to manufacture than are traditional solar cells. A *thin-film solar cell* consists of a film of silicon or other light-sensitive substance deposited on a base material, whereas traditional solar cells consist of individual crystals.

Commercial-scale testing of the world's largest array of thin-film photovoltaic modules took place at Davis, CA, in late 1992. By early the following year, the array was feeding electricity to a grid operated by the Pacific Gas and Electric Company. The photovoltaic system supplies up to 480 kilowatts, sufficient for the electrical needs of about 125 households.

About 50 megawatts of photovoltaic capacity are installed annually around the world. In the future, multi-megawatt solar-cell power plants are expected to routinely feed electricity into regional grids at costs that are competitive with conventional power plants. This outlook is based on current technological trends in developing more efficient thin-film solar cells (up to almost 18% efficient in 1996), plus declining manufacturing costs made possible by mass production and economy of scale. In addition, environmental concerns and the inevitable decline in supplies of fossil fuels should spur greater demand for solar power in the future.

CHAPTER 4

HEAT, TEMPERATURE, AND ATMOSPHERIC CIRCULATION

The mere formulation of a problem is far more essential than its solution, which may be merely a matter of mathematical or experimental skill. To raise new questions, new possibilities, to regard old problems from a new angle requires creative imagination and marks real advances in science.

UNKNOWN

Case-in-Point

Most of us would hardly notice a half-degree change in air temperature. But for the operators of a Midwestern natural gas utility, a half-degree change in average winter temperature can mean a significant change in consumer energy demand for space heating that translates into a major fluctuation in revenue stream for the utility.

Energy utilities are not the only weather-sensitive businesses. An estimated 70% of all U.S. businesses are sensitive to temperature and other weather variables, representing at least $1 trillion of the nation's economic activity. Besides utilities, weather-sensitive businesses include transportation, agriculture, retail, tourism, recreation, and insurance. For example, a mild winter may mean fewer weather delays for aircraft and cross-country truckers, less risk of freezes in the Florida citrus groves, fewer northerners vacationing in southern Arizona, and reduced consumer demand for snow shovels, skis, and winter coats.

By purchasing insurance, businesses can protect themselves against losses caused by high risk, low probability weather extremes (e.g., hurricanes, floods). How can businesses protect themselves against decreased revenues arising from low risk, high probability weather events (e.g., mild winter, exceptionally cool summer)? These businesses can minimize their losses by purchasing *weather derivatives*, a risk-management contract that protects a weather-sensitive business against weather-related losses. Consider an example.

An electrical utility has decreased revenues during a mild winter and wants to hedge against future mild winters so it purchases a weather derivative. The derivative buyer (DB) and derivative seller (DS) agree that the contract will be based on heating degree-days. As described in detail in this chapter's first Essay, heating degree days are a measure of household energy consumption for space heating. Heating degree-days are computed for days when the mean daily outdoor temperature is less than 65 °F. For example, if the mean temperature is 35 °F, there are 65 − 35 = 30 heating degree-days for the day. Cumulative totals of degree-days are maintained throughout the heating season.

Prior to the contract period, DB pays DS a premium to assume DB's risk. Meteorologists working for DS prepare a probabilistic forecast of seasonal heating degree-days for DB's location based on long-range weather outlooks (issued by the National Weather Service). DB and DS agree on a threshold value for the number of heating degree-days accumulated during the period of the contract (i.e., the upcoming winter heating season). If the actual number of heating degree-days is less than the threshold value, DS pays DB an amount based on the difference between the threshold and actual values and the effect of that difference on consumer demand for electricity. If the actual number of heating degree-days is greater than or equal to the threshold value, no payment is made to DB, and the premium becomes profit for DS.

Weather derivatives are a relatively new strategy for business to deal with the economic impacts of weather events. The first contract was written in 1997 and by mid-2000 about 4000 contracts had been negotiated and the weather derivatives market had grown to more than $8 billion.

Driving Question:

What is the consequence of heat transfer in the Earth-atmosphere system?

Temperature is one of the most important and common variables we use to describe the state of the atmosphere; it is a usual element in weather observation and forecasting. From everyday experience, we know that air temperature varies with time: from one season to another, between day and night, and even from one hour to the next. Air temperature also varies from one place to another: highlands and higher latitudes are usually colder than lowlands and lower latitudes.

We also know that temperature and heat are closely related concepts. Heating a pan of soup on the stove causes a rise in the temperature of the soup whereas dropping an ice cube into a beverage causes a drop in the temperature of the beverage. Granted that the two concepts are related, what is the precise distinction between heat and temperature? This is one of the questions we consider in this chapter. We also compare temperature scales, describe how temperature is measured, explain how heat is transferred in response to a temperature gradient, and describe how the temperature of a substance responds to an addition or loss of heat.

This chapter then examines heat imbalances that develop within the Earth-atmosphere system and the various processes whereby heat is transferred from Earth's surface to the troposphere and from the tropics to middle and high latitudes. All this enables us to better understand atmospheric circulation, energy conversions within the Earth-atmosphere system, and the controls of air temperature.

Distinguishing Temperature and Heat

All matter is composed of a multitude of minute particles (atoms or molecules) that are in continual random motion. The energy represented by this motion is called kinetic molecular energy or just *kinetic energy*, the energy of motion. In any substance, atoms or molecules actually move about with a range of kinetic energy. **Temperature** is directly proportional to the *average* kinetic energy of atoms or molecules composing a substance. At the same temperature, a drop of water has the same average kinetic molecular energy as a bathtub of water.

Internal energy encompasses all the energy in a substance, that is, the kinetic energy of atoms and molecules plus the potential energy arising from forces between atoms or molecules. If two objects are at different temperatures (different average kinetic molecular energy) and are brought into contact, energy will be transferred between the two objects. **Heat** is the name given to this energy in transfer. Heat transferred from one object reduces the internal energy of that object whereas heat absorbed by another object increases its internal energy.

Differences in temperature rather than differences in the amount of internal energy govern the direction of heat transfer. Heat energy is always transferred from an object with a higher temperature to an object with a lower temperature. Heat is not necessarily transferred from an object having greater internal energy to an object having less internal energy. A hot marble (at 40 °C) is dropped into 5 liters of cold water (at 5 °C). The water has much more internal energy than the marble, but heat is transferred from the warmer marble to the cooler water until both are at the same temperature.

The following illustrates the distinction between temperature and heat. A cup of water at 60 °C (140 °F) is much hotter than a bathtub of water at 30 °C (86 °F); that is, the average kinetic energy of water molecules is greater at 60 °C than at 30 °C. Although lower in temperature, the greater volume of water in the bathtub means that it contains more total kinetic molecular energy than does the cup of water. If in both cases, the water is warmer than its environment, energy (i.e., heat) is transferred from water to its surroundings. But much more heat energy must be transferred from the bathtub water than from the cup of water in order for both to cool to the same temperature.

From the above discussion, we might assume that any substance, including air, always changes temperature whenever it gains or loses heat. However, this is not necessarily the case. Water is a component of air and occurs in all three phases (ice crystals, droplets, and vapor), and, as we will see later in this chapter, heat must be added from or is released to the environment for water to change phase. Furthermore, air is a compressible mixture of gases; that is, an air sample can change volume. As we will discuss in Chapter 6, heat energy is required for the work of expansion or compression of air. Hence, as a sample of air gains or loses heat, that heat

may be involved in some combination of temperature change, phase change of water, or volume change.

TEMPERATURE SCALES AND HEAT UNITS

For most scientific purposes, temperature is measured on the Celsius scale. First proposed by Swedish astronomer Anders Celsius in 1742, the Celsius temperature scale has the numerical convenience of a 100-degree interval between the melting point of ice and the boiling point of pure water at sea level. The United States is virtually the only nation that still uses the numerically more cumbersome Fahrenheit scale for everyday temperature measurements, including weather reports. A German physicist, Gabriel Daniel Fahrenheit, introduced this scale in 1714. If a thermometer graduated in both scales is immersed in a glass containing a mixture of ice and water, the temperature will be 0 degrees on the Celsius scale (0 °C) and 32 degrees on the Fahrenheit scale (32 °F). In boiling water at sea level, the readings will be 100 °C and 212 °F. The average kinetic molecular energy decreases with falling temperature. There is, theoretically at least, a temperature at which all molecular motion ceases, called **absolute zero**, corresponding to -273.15 °C (-459.67 °F). Actually, some atomic-level activity likely occurs at absolute zero, but an object at that temperature emits no electromagnetic radiation.

On the Kelvin scale, temperature is the number of *kelvins* above absolute zero; hence, the Kelvin scale is a more direct measure of average kinetic molecular activity than either the Celsius or Fahrenheit temperature scales. The Kelvin scale is named for Lord Kelvin (William Thomson) who developed it in the mid 19th century. Whereas units of temperature are expressed as *degrees Celsius* (°C) on the Celsius scale and *degrees Fahrenheit* (°F) on the Fahrenheit scale, temperature is expressed simply in *kelvins* (K) on the Kelvin scale. Nothing can be colder than absolute zero so the Kelvin scale has no negative values and a 1-kelvin increment corresponds precisely to a 1-degree increment on the Celsius scale. The three temperature scales are contrasted in Figure 4.1, and conversion formulas are presented in Table 4.1.

Although temperature is a convenient way to describe the degree of hotness or coldness, we can quantify heat energy directly. Until recently, meteorologists commonly measured heat energy in units called calories, where one **calorie (cal)** is defined as the quantity of heat needed to raise the temperature of 1 gram of water 1 Celsius degree (technically, from 14.5 °C

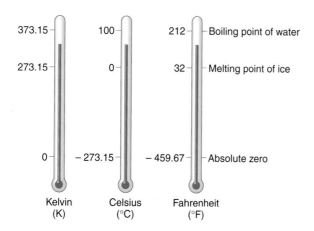

FIGURE 4.1
A comparison of three temperature scales.

TABLE 4.1
Temperature Conversion Formulas

°F = 9/5 °C + 32°
°C = 5/9 (°F – 32°)
K = 5/9 (°F + 459.67)
K = °C + 273.15

to 15.5 °C). (The term *calorie* has two definitions, which can cause confusion. The *calorie* unit is also used to measure the energy content of food and is actually 1000 heat calories or 1.0 kilocalorie. Here we refer to the "small" calorie.) Today, the more usual unit for energy of any form, including heat, is the *joule (J)*. One calorie equals 4.1868 J. In the English system, heat is quantified as British thermal units. One **British thermal unit (Btu)** is defined as the amount of heat required to raise the temperature of 1 pound (lb) of water 1 Fahrenheit degree (technically, from 62 °F to 63 °F). One Btu is equivalent to 252 cal and to 1055 J.

MEASURING AIR TEMPERATURE

A **thermometer** is the usual instrument for monitoring variations in air temperature. Galileo Galilei is credited with its invention in 1592. (Actually, Galileo invented a *thermoscope*, an instrument in which a fluid such as colored water responds to a gain or loss of heat but has no scale.) Perhaps the most common type of thermometer consists of a liquid-in-glass tube attached to a graduated scale (Figure 4.2). Typically, the liquid is

either mercury (which freezes at -39 °C or -38 °F) or alcohol (which freezes at -117 °C or -179 °F). Both the glass and the liquid (mercury or alcohol) expand when heated but the liquid much more so than the glass. As air warms, heat is transferred to the thermometer, and the expanding liquid rises in the glass tube. As air cools, heat is transferred from the thermometer, and the liquid contracts, dropping in the tube.

A second type of thermometer uses a bimetallic sensing element to take advantage of the expansion and contraction that accompany the heating and cooling of metals. A bimetallic sensing element consists of strips of two different metals welded together back to back. The two metals have different rates of thermal expansion; that is, one metal expands more than the other in response to the same heating. Because the two metals are bonded together, heating causes the bimetallic strip to bend; the greater the heating, the greater the bending. For example, the rate of thermal expansion of brass is about twice that of iron so that a bimetallic strip composed of those two metals will bend in the direction of the iron when heated. A series of gears or levers translates the response of the bimetallic strip to a pointer and a dial calibrated to read in °C or °F. Alternatively, this device may be rigged to a pen and a clock-driven drum to produce a continuous trace of temperature with time; this instrument is called a **thermograph**.

Another common type of thermometer employs an electrical conductor whose resistance changes as the temperature fluctuates, permitting a calibration between electrical resistance and temperature. Figure 4.3 shows the change of temperature during passage of a cold front as determined by an electronic thermometer. A radiosonde is equipped with this type of thermometer. Digital read-out thermometers widely available in a range of consumer products are also of this type. Some electronic thermometers are designed to give remote

FIGURE 4.2
Liquid-in-glass thermometer.

Tuesday, April 10, 10:35:04, 2001

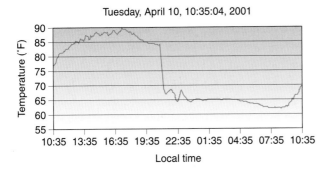

FIGURE 4.3
Temperature changes over a 24-hour period during a cold front passage.

temperature readings by mounting the sensor at the end of a long cable joined to the instrument. This system has replaced standard liquid-in-glass thermometers at National Weather Service facilities nationwide. Also, as described in Chapters 1 and 3, satellite sensors monitor surface temperatures remotely by measuring the intensity of emitted infrared radiation (Chapter 3).

Regardless of the type of thermometer used, two important properties of the instrument are accuracy and response time. For most weather and climate purposes, a thermometer that is accurate to within 0.3 Celsius degree (0.5 Fahrenheit degree) will suffice. Response time refers to the instrument's capability of resolving changes in temperature. Electrical resistance thermometers have rapid response times, liquid-in-glass somewhat less, whereas bimetallic thermometers tend to be more sluggish.

For representative measurements of air temperature, ideally a thermometer should be adequately ventilated and shielded from precipitation, direct sunlight, and the night sky. Enclosing temperature sensors (and other weather instruments) in a white, louvered wooden shelter had been standard practice for official temperature measurements. The sensor for the new National Weather Service electronic thermometer is mounted inside a ventilated shield (Figure 4.4) made of white plastic and the digital read-out box is indoors. So that temperature readings are comparable, an instrument shelter should be located in an open grassy area well away from trees, buildings, or other obstacles, and at a standard height above the ground. As a general rule, the shelter should be no closer than four times the height of the nearest obstacle. Where a shelter is not available, mounting a thermometer outside a window on the shady north side of a building is usually sufficient for general purposes.

FIGURE 4.4
Enclosure for the National Weather Service electronic temperature sensor.

In addition to using standard thermometers, air temperature can sometimes be determined in unconventional ways. One interesting and surprisingly accurate method is to count cricket chirps. Crickets are cold-blooded organisms so their activity (including frequency of chirping) depends on air temperature. For air temperatures above about 12 °C (54 °F), the number of cricket chirps heard in an 8-second period plus 4 approximates the air temperature in °C.

Heat Transfer

In large part due to imbalances in radiational heating and radiational cooling within the Earth-atmosphere system, air temperature varies from one place to another. A change in temperature with distance is known as a **temperature gradient**. A familiar temperature gradient prevails between the hot equator and the cold poles (a horizontal temperature gradient). Another is the vertical temperature gradient between the relatively mild Earth's surface and the relatively cold tropopause.

Heat flows in response to a temperature gradient, according to the **second law of thermodynamics**. Simply put, this law states that all systems tend toward a state of disorder. You probably have personal experience with some implications of the second law. If you avoid cleaning up your home or room, for example, it rapidly becomes more and more disorganized. The presence of a gradient of any kind within a system signals order within that system. Hence, as a system tends toward disorder, gradients decrease. The second law predicts that where a temperature gradient exists, heat is transferred in a direction so as to eliminate the gradient; that is, heat flows from locations of higher temperature toward locations of lower temperature. In addition, the greater the temperature difference (i.e., the steeper the temperature gradient), the more rapid is the rate of heat flow.

Within the Earth-atmosphere system, heat is transferred via radiation, conduction and convection, and phase changes of water (i.e., *latent heating*).

RADIATION

Radiation is both a form of energy and a means of energy transfer (Chapter 3). Electromagnetic radiation can be considered as a spectrum of waves traveling at the speed of light. But unlike conduction and convection, radiation can travel through a vacuum; it requires no intervening physical medium. Although not actually a vacuum, interplanetary space is so highly rarefied that conduction and convection play essentially no role in transporting heat from the sun to Earth (or any other planet). Rather, radiation is the principal means whereby the Earth-atmosphere system gains heat from the sun. Radiation is also the principal means whereby heat escapes from the planet to space.

As we saw in Chapter 3, absorption of radiation involves a conversion of electromagnetic energy to heat. In contrast, emission of electromagnetic energy is a loss of heat from the radiating object to the environment. All objects both absorb and emit electromagnetic radiation. If an object absorbs radiation at a greater rate than it emits radiation, the temperature of the object will rise. This type of imbalance in the flux of radiation is known as **radiational heating**. If an object emits radiation at a greater rate than it absorbs radiation, the temperature of the object will fall. This type of imbalance in the flux of radiation is called **radiational cooling**. At radiative equilibrium, when absorption and emission of radiation are equal, the object's temperature is constant.

Radiative equilibrium does not necessarily mean that the temperature stays constant among all components of a system. The entire Earth-atmosphere system is in radiative equilibrium with its surroundings. Nonetheless, heat may be redistributed among the various components of the Earth-atmosphere system (for example, among the ocean, land, and glaciers) so that air temperature at a specified location may undergo significant short- and long-term variations. Hence, global radiative equilibrium does not preclude changes in weather and climate.

CONDUCTION AND CONVECTION

Conduction occurs within a substance or between substances that are in direct physical contact. **Conduction** (of heat) refers to the transfer of kinetic energy of atoms or molecules via collisions between neighboring atoms or molecules. This is why a metal spoon heats up when placed in a steaming cup of coffee. As the more energetic molecules of the hot coffee collide with the less energetic atoms of the cooler spoon, some kinetic energy is transferred to the atoms of the spoon. These atoms then transmit some of their heat energy, via collisions, to neighboring atoms, so that heat is conducted up the handle of the spoon and eventually the handle becomes hot to the touch.

Some substances are better conductors of heat than others. *Heat conductivity* is defined as the ratio of the rate of heat transport across an area to a temperature gradient. Hence, in response to a specified temperature gradient, substances having higher heat conductivities have greater rates of heat transport. As a rule, solids are better conductors of heat than are liquids, and liquids are better heat conductors than gases. At one extreme, metals are excellent conductors of heat and, at the other extreme, still air is a very poor conductor of heat. Heat conductivities of some common substances are listed in Table 4.2.

Differences in heat conductivity can be the reason one object feels colder than another, even though both objects have the same temperature. For example, at

TABLE 4.2
Heat Conductivity of Some Familiar Substances[a]

Copper	0.92
Aluminum	0.50
Iron	0.16
Ice (at 0 °C)	0.0054
Limestone	0.0048
Concrete	0.0022
Water (at 10 °C)	0.0014
Dry sand	0.0013
Air (at 20 °C)	0.000061
Air (at 0 °C)	0.000058

[a]*Heat conductivity* is defined as the quantity of heat (in calories) that would flow through a unit area of a substance (cm^2) in one second in response to a temperature gradient of one Celsius degree per centimeter. Hence, heat conductivity has units of calories per cm^2 per sec per C° per cm.

the same relatively low temperature, a metallic object feels colder than a wooden object. The heat conductivity of metals is much greater than that of wood so that when you grasp the two objects, your hand more rapidly conducts heat to the metallic object than to the wooden object. Consequently, you have the sensation that the metal is colder than the wood.

The relatively low heat conductivity of air makes it a good heat insulator. Heat conductivity is lower for still air than for air in motion so to take maximum advantage of air as a heat insulator, air must be confined. For example, when a fiberglass blanket is used as attic insulation, it is primarily the motionless air trapped between individual fiberglass fibers that inhibits heat loss. In time, the fibers settle, air is excluded, and the blanket loses much of its insulating value.

The heat conductivity of a fresh snow cover is low because of the air trapped between individual snowflakes. A thick snow cover (20 to 30 cm or 8 to 12 in.) can thus inhibit or prevent freezing of the underlying soil, even though the temperature of the overlying air drops well below freezing. In time, however, like the fiberglass, a snow cover loses some of its insulating property as the snow settles. As snow settles, air escapes, snow density increases, and heat conductivity increases as shown in Figure 4.5.

Heat is conducted from sun-warmed ground to

cooler overlying air, but because air is a poor conductor of heat, conduction is significant only in a very thin layer of air that is in immediate contact with Earth's surface. Much more important than conduction in transporting heat vertically within the troposphere is convection. **Convection** is the transport of heat within a fluid via motions of the fluid itself. Although conduction takes place in solids, liquids, or gases, convection generally occurs only in liquids or gases. (An exception is the convection currents that likely occur in the Earth's interior under conditions of tremendous confining pressure.)

Convection in the atmosphere is the consequence of differences in air density. At the same pressure, cold air is denser than warm air. As heat is conducted from the relatively warm ground to cooler overlying air, that air is heated and becomes less dense. Cool, denser air from above sinks and forces the warmer, less dense air at the ground to rise. (This is similar to what happens when cold tap water flows into a tub of hot water; that is, the cold denser water sinks and forces the hot, less dense water to rise.) Ascending warm air expands and cools and eventually sinks back to the ground. Meanwhile, air now in contact with the warm ground is heated and rises—displaced by the sinking cooler denser air. In this way, as illustrated in Figure 4.6, convection transports heat vertically from Earth's surface thousands of meters into the troposphere.

We can readily observe convection currents in a pan of water on a hot stove. By adding a drop or two of food coloring to the water, we actually see the circulating water redistributing heat that is conducted from the bottom of the pan into the water. In this example as well as in the troposphere, conduction and convection work together.

PHASE CHANGES OF WATER

Another important heat transfer process involves phase changes of water. Water occurs in the Earth-atmosphere system in all three phases, as solid (ice or snow), liquid, and vapor and is continually changing phase. Depending on the specific phase change, water either absorbs heat from its environment or releases heat to its environment. The quantity of heat that is involved in phase changes of water is known as **latent heat**. Heat is required for phase changes because of differences in molecular activity represented by the three physical phases of water.

In the solid phase (ice), water molecules are relatively inactive and vibrate about fixed locations.

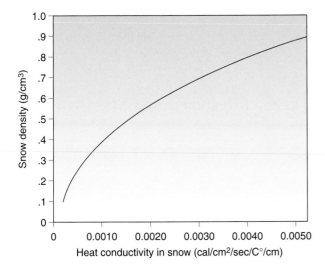

FIGURE 4.5
A thick layer of fresh snow is a good heat insulator primarily because of air trapped between the individual snowflakes. (Air is a poor conductor of heat.) But in time, the snow settles, air escapes, snow density increases, and the snow cover's insulating property diminishes.

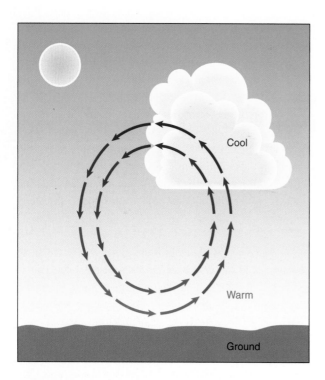

FIGURE 4.6
Convection currents transport heat from Earth's surface into the troposphere.

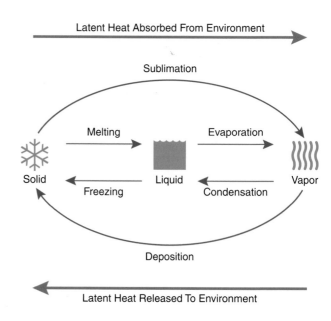

FIGURE 4.7
When water changes phase, heat is either absorbed from or released to the environment.

Hence, at subfreezing temperatures, an ice cube or any other piece of ice maintains its shape. In the liquid phase, molecules move about with greater freedom, so that liquid water takes the shape of its container. In the vapor phase, water molecules exhibit maximum activity and diffuse readily throughout the entire volume of its container. A change in phase is thus linked to a change in level of molecular activity, which is brought about by either an addition or loss of heat (Figure 4.7). Heat is absorbed from the environment during *melting* (phase change from solid to liquid), *evaporation* (phase change from liquid to vapor), and *sublimation* (phase change directly from solid to vapor). Heat is released to the environment during *freezing* (phase change from liquid to solid), *condensation* (phase change from vapor to liquid), and *deposition* (phase change directly from vapor to solid). During any phase change, heat is exchanged between water and its environment. Although the temperature of the environment changes in response, the temperature of the water undergoing the phase change remains constant until the phase change is complete. That is, the available heat (latent heat) is involved exclusively in changing the phase of water and not in changing its temperature.

Latent heating refers to the transport of heat from one location to another as a consequence of changes in the phase of water. For example, the latent heat that is supplied to change liquid water to vapor (evaporation) in one place is released to the environment when that water vapor condenses at some other place. Prior to condensation, winds can transport water vapor to distances of thousands of kilometers and to altitudes of thousands of meters so that latent heating can involve long-range transport of heat. We have more to say about latent heating later in this chapter.

Thermal Response and Specific Heat

Whether by radiation, conduction, convection, or phase changes of water, transfer of heat from one place to another within the Earth-atmosphere system is accompanied by changes in temperature. A heat gain causes a rise in temperature whereas a heat loss causes a drop in temperature.

The temperature change associated with an input (or output) of a specified quantity of heat varies from one substance to another. The amount of heat that will raise the temperature of 1 gram of a substance by 1 Celsius degree is defined as the **specific heat** of that substance. Joseph Black, a Scottish chemist, first proposed the concept of specific heat in 1760. The

TABLE 4.3
Specific Heat of Some Familiar Substances[a]

Water	1.000
Wet mud	0.600
Ice (at 0 °C)	0.478
Wood	0.420
Aluminum	0.214
Brick	0.200
Granite	0.192
Sand	0.188
Dry air[b]	0.171
Copper	0.093
Silver	0.056
Gold	0.031

[a]Calories per gram per Celsius degree.

[b]At constant volume.

FIGURE 4.8
The contrast in specific heat is one reason why the sand is hotter than the water.

specific heat of all substances is measured relative to that of liquid water, which is defined as 1 calorie per gram per Celsius degree (at 15 °C). The specific heat of ice is about 0.5 calorie per gram per Celsius degree (near 0 °C). Specific heats of some familiar substances are listed in Table 4.3. The variation in specific heat from one substance to another implies that different materials have different capacities for storing internal energy.

Upon absorbing the same quantity of heat energy, a substance with a high specific heat experiences a smaller rise in temperature (warms less) than a substance having a low specific heat. Water has the greatest specific heat of any naturally occurring substance. From Table 4.3, water's specific heat is about five times that of dry sand. One calorie of heat will raise the temperature of 1 gram of water 1 Celsius degree, whereas 1 calorie of heat will raise the temperature of 1 gram of dry sand by 5 Celsius degrees. This contrast in specific heat helps explain why at the beach in summer the sand feels considerably hotter to bare feet than the water (Figure 4.8). This also largely explains why wet sand feels cooler than dry sand.

Water's exceptional capacity to store heat has important implications for weather and climate. A large body of water (such as the ocean or Great Lakes) can significantly influence the climate of downwind localities. The most persistent influence is on air temperature. Compared to an adjacent landmass, a body of water does not warm as much during the day (or in summer) and does not cool as much at night (or in winter). In other words, a body of water exhibits a greater resistance to temperature change, called **thermal inertia**, than does a landmass. Whereas the difference in specific heat between land and water is the major reason for this contrast in thermal inertia, differences in heat transport between land and water also contribute. Sunlight penetrates water to some depth and is absorbed (converted to heat) through a significant volume of water. But sunlight cannot penetrate the opaque land surface and is absorbed only at the surface. Furthermore, circulation of ocean and lake waters transports heat through great volumes of water, whereas heat is conducted only very slowly into soil. The input (or output) of equal amounts of heat energy causes a land surface to warm (or cool) more than the equivalent surface area of a body of water.

Air temperature is regulated to a considerable extent by the temperature of the surface over which the air resides or travels. Air over a large body of water tends to take on similar temperature characteristics as the surface water. Places immediately downwind of the ocean experience much less contrast between average winter and summer temperature and the climate is described as *maritime*. Places at the same latitude but well inland experience a much greater contrast between winter and summer temperature and the climate is described as *continental*. That is, at the same latitude, summers are cooler and winters are milder in maritime climates than in continental climates.

Consider an example of the temperature contrast between continental and maritime climates. The latitude of Stevens Point, WI (44.5 degrees N) is about the same

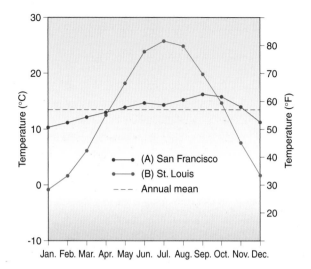

FIGURE 4.9
Variation in the march of monthly mean temperature for (A) maritime San Francisco, CA and (B) continental St. Louis, MO. The two cities are located at about the same latitude.

as that of Newport, OR (44.6 degrees N) so that the seasonal variation in the amount of solar radiation striking Earth's atmosphere is about the same at both places. Stevens Point is located far from the moderating influence of the ocean and its climate is continental. The average summer (June, July, and August) temperature at Stevens Point is 19.9 °C (67.9 °F) and its average winter (December, January, and February) temperature is −8.2 °C (17.2 °F), giving an average summer-to-winter

seasonal temperature contrast of 28.1 Celsius degrees (50.7 Fahrenheit degrees). Newport is located on the Oregon coast, immediately downwind of the Pacific Ocean; its climate is maritime. The average summer temperature at Newport is 13.6 °C (56.5 °F) and the average winter temperature is 7.0 °C (44.6 °F), giving an average seasonal temperature contrast of only 6.6 Celsius degrees (11.9 Fahrenheit degrees). Another example of the contrast in temperature between maritime and continental localities is shown in Figure 4.9.

Heat Imbalance: Atmosphere versus Earth's Surface

Weather is not a capricious act of nature. Ultimately, weather is a response to unequal rates of radiational heating and radiational cooling within the Earth-atmosphere system. Imbalances in rates of radiational heating and cooling from one place to another within the Earth-atmosphere system produce temperature gradients. In response to temperature gradients, the atmosphere circulates and thereby redistributes heat.

Figure 4.10 shows how solar radiation intercepted by planet Earth interacts with the atmosphere and Earth's surface. Numbers are global and annual averages. For every 100 units of solar radiation that enters the upper atmosphere, the Earth-atmosphere

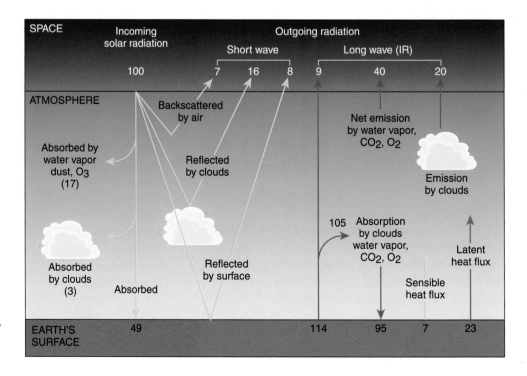

FIGURE 4.10
Globally and annually averaged disposition of 100 units of solar radiation entering the atmosphere. Solar radiation fluxes are depicted at the left, infrared radiation fluxes in the middle, and latent and sensible heat flux at the right.

system reflects or scatters 31 units (31%) to space, the atmosphere absorbs 20 units (20%), and Earth's surface (principally the ocean) absorbs 49 units (49%). In response to radiational heating, Earth's surface emits 114 units of infrared radiation. Atmospheric gases and clouds absorb 105 units of terrestrial infrared radiation and emit 95 units to Earth's surface (the *greenhouse effect*). A total of 69 units of IR radiation is emitted out the top of the atmosphere and to space, equal to the amount of solar radiation absorbed by the Earth-atmosphere system.

The global annual distribution of incoming solar radiation and outgoing infrared radiation implies net warming of Earth's surface and net cooling of the atmosphere (Table 4.4). At Earth's surface, absorption of solar radiation is greater than emission of terrestrial infrared radiation. In the atmosphere, on the other hand, emission of terrestrial infrared radiation to space is greater than absorption of solar radiation. That is, on a global average annual basis, Earth's surface undergoes net radiational heating while the atmosphere undergoes net radiational cooling.

TABLE 4.4
Global Radiation Balance

Solar radiation intercepted by Earth	100 units
Solar radiation budget	
Scattered and reflected to space (7 + 16 + 8)	31
Absorbed by the atmosphere (17 + 3)	20
Absorbed at the Earth's surface	49
Total	100 units
Radiation budget at the Earth's surface	
Infrared cooling (95 – 114)	-19
Solar heating	+49
Net heating	+30 units
Radiation budget of the atmosphere	
Infrared cooling (- 40 – 20 + 105 – 95)	-50
Solar heating	+20
Net cooling	-30 units
Heat transfer: Earth's surface to atmosphere	
Sensible heating (conduction/convection)	7
Latent heating (phase changes of water)	23
Net transfer	30 units

The atmosphere is not actually cooling relative to Earth's surface. In response to the radiationally induced temperature gradient between Earth's surface and atmosphere, heat is transferred from Earth's surface to the atmosphere. A combination of latent heating and sensible heating (conduction and convection) is responsible for this transfer of heat. As shown in Figure 4.10, on a global annual average basis, 30 units of heat energy are transferred from Earth's surface to atmosphere: 23 units (about 77% of the total) by latent heating and 7 units (about 23%) by sensible heating.

LATENT HEATING

As Earth's surface absorbs radiation (both solar and infrared), some of the heat energy is used to vaporize water from the ocean, lakes, rivers, soil, and vegetation. The latent heat required for vaporization is supplied at the Earth's surface, and that same heat is subsequently released to the atmosphere during cloud development. Within the troposphere, clouds form as some of the water vapor condenses into liquid water droplets or deposits as ice crystals. During cloud formation, water changes phase and latent heat is released to the atmosphere. Through latent heating, then, heat is transferred from Earth's surface into the troposphere. In fact, latent heating is more important than either radiational cooling or sensible heating in cooling Earth's surface (Figure 4.11).

Unusually large quantities of heat are required to bring about phase changes of water as compared to phase changes of other naturally occurring substances. Consider, for example, the quantity of heat involved in changing the phase of a one-gram ice cube as it is heated from an initial temperature of -20 °C (-4 °F) to a final temperature of 100 °C (212 °F) (Figure 4.12). The specific heat of ice is about 0.5 cal per gram per Celsius degree, which means that 0.5 cal of heat must be supplied for every 1 Celsius degree rise in temperature. Hence, warming our ice cube from -20 °C to 0 °C (-4 °F to 32 °F) requires an input of 10 cal of heat. Once the freezing (or melting) point is reached, an additional 80 cal of heat, called **latent heat of melting**, must be supplied per gram to break the forces that bind water molecules in the ice phase. The temperature of the water and ice mixture remains at 0 °C until all the ice melts.

The specific heat of liquid water is 1 cal per gram per Celsius degree. Once the ice cube melts, 1 cal is needed for every 1 Celsius degree rise in water temperature. Evaporation can occur at any temperature and requires the addition of much more heat than does a

FIGURE 4.11
Earth's surface is cooled via (A) vaporization of water, (B) net emission of infrared radiation, and (C) conduction plus convection. Numbers are global annual averages.

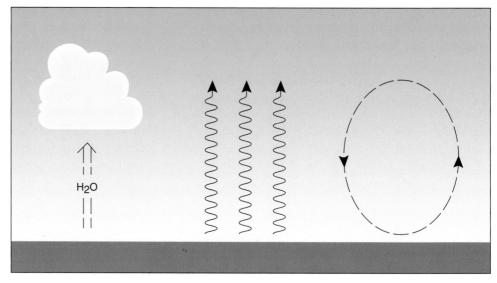

A. Latent Heating
23 Units
(46.9%)

B. Net IR Radiation
19 Units
(38.8%)

C. Sensible Heating
7 Units
(14.3%)

phase change from ice to liquid water. The **latent heat of vaporization** varies from about 600 cal per gram at 0 °C (32 °F) to 540 cal per gram at 100 °C (212 °F). The water boils at a temperature of 100 °C (212 °F). For our

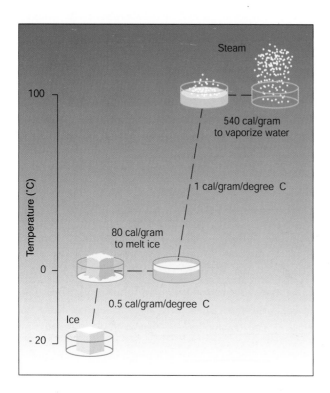

FIGURE 4.12
Heating a 1-gram ice cube causes a rise in temperature plus phase changes.

1-gram ice cube to vaporize directly without melting first (*sublimation*), the latent heat of melting plus the latent heat of vaporization must be supplied to the ice cube. This amounts to 680 cal per gram at 0 °C.

If the process just described is reversed, that is, if the water vapor is cooled until it becomes liquid (*condensation*) and then ice (*freezing*), the water temperature drops and phase changes take place as equivalent amounts of latent heat are released to the environment. When water vapor becomes liquid, latent heat of vaporization is released to the environment, and when water freezes, **latent heat of fusion** is released. If water vapor changes to ice without first becoming liquid (*deposition*), the latent heats of vaporization plus fusion are released to the environment.

SENSIBLE HEATING

Heat transfer via conduction and convection can be monitored (*sensed*) by temperature changes; hence, **sensible heating** encompasses both of these processes. Heat is conducted from the relatively warm surface of the Earth to the cooler overlying air. Heating reduces the density of that air, which is forced to rise by cooler denser air replacing it at the surface (Figure 4.6). In this way, convection transports heat from Earth's surface into the troposphere. Because air is a relatively poor conductor of heat, convection is much more important than conduction as a heat transport mechanism within the troposphere.

Often sensible heating combines with latent

FIGURE 4.13
Cumulus clouds form in the ascending branch of a convective circulation.

heating to channel heat from Earth's surface into the troposphere. Updrafts (the ascending branch) of vapor-laden air in convection currents often produce **cumulus clouds**, which resemble puffs of cotton floating in the sky (Figure 4.13). These clouds are sometimes referred to as *fair-weather* cumulus because they seldom produce rain or snow. On the other hand, if atmospheric conditions are favorable (described in Chapter 11), convection currents can surge to great altitudes, and cumulus clouds merge and billow upward to form towering **cumulonimbus clouds**, also known as thunderstorm clouds (Figure 4.14). In retrospect, two important heat transfer processes (a combination of latent heating and sensible heating) took place last summer when that thunderstorm washed out your ball game or sent you scurrying for shelter at the beach.

FIGURE 4.14
A cumulonimbus (thunderstorm) cloud forms when atmospheric conditions favor deep convection.

At times, heat transport is directed from the troposphere to the Earth's surface, the reverse of the global average annual situation. This reversal in direction of heat transport occurs, for example, when mild winds blow over cold, snow-covered ground or when warm air moves over a relatively cool ocean surface. Heat transport from the atmosphere to Earth's surface is the usual situation at night (especially when skies are clear) when radiational cooling causes Earth's surface to become colder than the overlying air.

As noted in Chapter 2, nearly all weather is confined to the troposphere, implying that heat transport by sensible and latent heating operates primarily within the lower atmosphere. Radiational processes determine heat and temperature distributions above the troposphere.

BOWEN RATIO

The **Bowen ratio** is a useful parameter that describes how the heat energy received at the Earth surface by absorption of solar and infrared radiation is partitioned between sensible heating and latent heating. That is,

Bowen ratio = [(sensible heating)/(latent heating)]

At the global scale,

Bowen ratio = [(7 units)/(23 units)] = 0.3

As shown in Table 4.5, the Bowen ratio varies from one locality to another depending on the amount of

TABLE 4.5	
Bowen Ratio[a] of Various Geographical Areas	
All Oceans	0.11
Atlantic Ocean	0.11
Pacific Ocean	0.10
Indian Ocean	0.09
All Land	0.96
North America	0.74
South America	0.56
Europe	0.62
Asia	1.14
Africa	1.61
Australia	2.18
Globe	0.30

[a]Ratio of sensible heating to latent heating.

surface moisture. The more moist the surface, the less important is sensible heating and the more important is latent heating. The Bowen ratio ranges from about 0.1 (one-tenth as much sensible as latent heating) for the ocean to about 5.0 (five times as much sensible as latent heating) in deserts. Ocean waters cover much of Earth's surface so it is not surprising that the global Bowen ratio is relatively low (0.3).

Heat Imbalance: Tropics versus Middle and High Latitudes

On a global scale, imbalances in radiational heating and radiational cooling occur not only vertically, between Earth's surface and the atmosphere, but also horizontally, with latitude. Because the planet is nearly a sphere, parallel beams of incoming solar radiation strike lower latitudes more directly than higher latitudes. (That is, solar altitudes are higher at lower latitudes and lower at higher latitudes.) At higher latitudes, solar radiation spreads over a greater area and is less intense per unit surface area than at lower latitudes (Chapter 3).

Emission of infrared radiation (IR) by the Earth-atmosphere system also varies with latitude but less so than solar radiation. IR emission declines with increasing latitude in response to the drop in average temperature with latitude. (Recall from Chapter 3 that

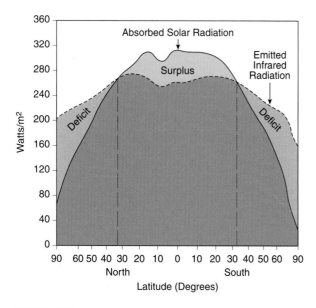

FIGURE 4.15
Variation by latitude of absorbed solar radiation and outgoing infrared radiation based on satellite measurements. [From NOAA/NESDIS]

radiation emission is temperature dependent.) Consequently, over the course of a year at higher latitudes, the rate of infrared cooling to space exceeds the rate of warming caused by absorption of solar radiation. At lower latitudes the reverse is true, that is, over the course of a year, the rate of solar radiational heating is greater than the rate of infrared radiational cooling (Figure 4.15). Averaged over all latitudes, incoming energy (absorbed solar radiation) must equal outgoing energy (infrared radiation emitted to space). That is, the areas under the two curves in Figure 4.15 are equal. This *global radiative equilibrium* is an illustration of the law of energy conservation.

Satellite measurements indicate that the division between regions of net radiational cooling and regions of net radiational warming is close to the 35-degree latitude circle in both hemispheres. By implication, latitudes poleward of about 35 degrees N and 35 degrees S experience net cooling over the course of a year, whereas tropical latitudes are sites of net warming. In fact, lower latitudes do not become progressively warmer relative to higher latitudes because heat is transported poleward, from the tropics into middle and high latitudes. **Poleward heat transport** is brought about by (1) air mass exchange, (2) storms, and (3) ocean circulation. According to recent research conducted at the National Center for Atmospheric Research (NCAR) at Boulder, CO, atmospheric processes account for 78% of total poleward heat transport in the Northern Hemisphere and 92% in the Southern Hemisphere.

Recall that an **air mass** is a huge volume of air covering thousands of square kilometers that is relatively uniform horizontally in temperature and humidity (water vapor concentration). The properties of an air mass largely depend on the characteristics of the surface over which the air mass resides or travels. Air masses that form at high latitudes over cold, often snow- or ice-covered surfaces are relatively cold. Those air masses that form at low latitudes are relatively warm. Air masses that develop over the ocean are humid and those that form over land are dry. Hence, there are four basic types of air masses: cold and humid, cold and dry, warm and humid, and warm and dry. Warm air masses that form in lower latitudes flow poleward and are replaced by cold air masses that flow toward the equator from source regions at high latitudes. In this exchange of air masses, sensible heat is transported poleward.

Release of latent heat in storms is associated with air mass exchange in the poleward transport of heat. At low latitudes, water that evaporates from the warm ocean surface is drawn into the circulation of a developing cyclone (low pressure system). As the cyclone travels into higher latitudes, some of that water vapor condenses into clouds, thereby releasing latent heat to the troposphere. Latent heat of vaporization acquired at low latitudes is thereby delivered to middle and high latitudes. This mechanism of poleward heat transport is readily apparent in water vapor imagery (Chapters 1 and 6).

The ocean's contribution to poleward heat transport is primarily via wind-driven surface currents and huge conveyor-belt-like currents that traverse the length of the ocean basins. Surface water that is cooler than the overlying air is a *heat sink* for the atmosphere; that is, heat is conducted from air to sea. And surface water that is warmer than the overlying air is a *heat source* for the atmosphere; that is, heat is conducted from sea to air. Cold surface currents, such as the California Current, flow from high to low latitudes, absorbing heat from the relatively warm tropical troposphere. Warm surface currents such as the Gulf Stream flow from the tropics into middle latitudes, supplying heat to the cooler mid-latitude troposphere (Figure 4.16).

According to W. S. Broecker of Columbia University's Lamont-Doherty Earth Observatory, in the Atlantic conveyor-belt system, warm surface waters flow northward from the Tropics to near Greenland. Chilling of waters in northern latitudes raises its density so that it

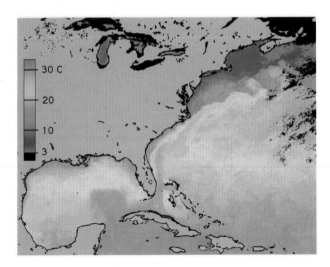

FIGURE 4.16
A composite infrared satellite image showing sea-surface temperatures in °C. The warm Gulf Stream is clearly discernible as a ribbon of 30 °C water flowing along the East Coast from Florida north to off the coast of New Jersey. This composite is based on NOAA geostationary satellite monitoring from 14 to 18 January 1999. [From NOAA]

sinks and forms a bottom current that flows into the South Atlantic as far south as Antarctica. In that region, the bottom current is warmer (and less dense) than the surface waters so that the current wells up to the surface, is chilled, and sinks again. Branches of that cold bottom current then spread northward into the Atlantic, Indian, and Pacific Oceans. The importance of oceanic conveyor belts in poleward heat transport is illustrated by the temperature contrast between surface and deep currents in the far North Atlantic. Surface waters approaching Greenland are some 8 Celsius degrees (14 Fahrenheit degrees) warmer than the bottom current that flows southward. The drop in water temperature is a measure of the magnitude of heat transport from the ocean surface to the overlying troposphere.

Why Weather?

As we have seen, imbalances in rates of radiational heating and radiational cooling give rise to temperature gradients between (1) Earth's surface and troposphere and (2) low and high latitudes. In response, heat is transported within the Earth-atmosphere system via conduction, convection, cloud development, air mass exchange, and storms. That is, the atmosphere circulates, bringing about changes in the state of the atmosphere (weather). A cause-and-effect chain thus operates in the Earth-atmosphere system, starting with the sun as the prime energy source and resulting in weather.

We have also seen that within the Earth-atmosphere system, some solar radiation is absorbed, that is, converted to heat, and eventually all of this heat is emitted to space as infrared radiation. Some solar energy is also converted to **kinetic energy**, the energy of motion, in the circulation of the atmosphere. Kinetic energy is manifested in winds, in convection currents, and in the north-south exchange of air masses. Circulation (weather) systems do not last indefinitely, however. The kinetic energy of atmospheric circulation ultimately is dissipated as frictional heat as winds blow against Earth's surface. This heat, in turn, is emitted to space as infrared radiation. In summary, the sun drives the atmosphere: imbalances in solar heating spur atmospheric circulation, which redistributes heat. Hence, solar energy is the ultimate source of kinetic energy in weather systems.

The rate of heat redistribution within the Earth-atmosphere system varies seasonally so that atmospheric circulation and weather also change through the year. When steep temperature gradients prevail across North America, the weather tends to be more dynamic. Storm systems are large and intense, winds are strong, and the weather is changeable. Such weather is typical of winter, when it is not unusual for daily temperatures in the southern United States to be more than 30 Celsius degrees (54 Fahrenheit degrees) higher than temperatures across southern Canada. When air temperature varies little across the continent, as in summer, the weather tends to be more tranquil, and large-scale weather systems are generally weak and ill-defined. Nevertheless, summer weather is sometimes very active. Intense heating of the ground by the summer sun coupled with lifting processes often triggers deep convection and development of thunderstorms. Some of these weather systems spawn destructive hail, strong and gusty winds, and heavy rains. However, these systems are usually shorter lived and more localized than winter storms.

Variation of Air Temperature

Air temperature is variable, fluctuating from hour to hour, from one day to the next, with the seasons, and from one place to another. Our discussion of the basic reasons for atmospheric circulation and weather provides some insight as to why air temperature is variable. The radiation balance plus movements of air masses regulate air temperature locally. Although these two factors actually work in concert, for purpose of study, we initially consider them separately.

RADIATIONAL CONTROLS

Many factors govern the local radiation balance and air temperature. These factors include the following: (1) time of day and day of the year, which determine the solar altitude and the intensity and duration of solar radiation striking Earth's surface; (2) cloud cover, because cloudiness affects the flux of both incoming solar and outgoing terrestrial radiation; and (3) surface characteristics, which determine the albedo and the percentage of absorbed radiation (heat) used for sensible heating and latent heating. Hence, air temperature is generally higher in July than in January (in the Northern Hemisphere), during the day than at night, under clear rather than overcast daytime skies, when the ground is bare instead of snow-covered, and when the ground is dry rather than wet.

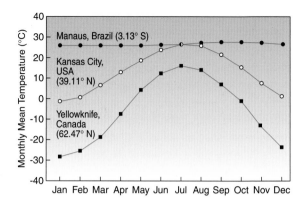

FIGURE 4.17
March of monthly mean temperature at Manaus, Brazil, Kansas City, MO, and Yellowknife, Canada.

The annual temperature cycle (also called the *march of mean monthly temperature*) reflects the systematic variation in incoming solar radiation over the course of a year (Figure 4.17). In the latitude belt between the Tropics of Cancer and Capricorn, solar radiation varies little through the course of a year so that average monthly air temperatures exhibit minimal seasonal contrast. In fact, in the tropics, the temperature difference between night and day often is greater than the winter-to-summer temperature contrast. At middle latitudes, solar radiation features a pronounced annual maximum and minimum. At high latitudes, poleward of the Arctic and Antarctic circles, the seasonal difference in solar radiation is extreme, varying from zero in fall and winter to a maximum in spring and summer. This marked periodicity of solar radiation outside of the tropics accounts for the distinct winter-to-summer temperature contrasts observed in middle and high latitudes.

At middle and high latitudes, the march of mean monthly temperature lags behind the monthly variation in solar radiation so that the warmest and coldest months of the year typically do not coincide with the times of maximum and minimum solar radiation, respectively. The troposphere's temperature profile takes time to adjust to seasonal changes in solar energy input. Typically, the warmest portion of the year is about a month after the summer solstice, and the coldest part of the year usually occurs about a month after the winter solstice. In the United States, the temperature cycle lags the solar cycle by an average of 27 days. However, in coastal localities with a strong maritime influence (e.g., Florida, the shoreline of New England, and coastal California), the average lag time is up to 36 days. In addition, as we saw earlier in this chapter, the maritime influence reduces the amplitude of the annual march of mean monthly temperature; that is, the winter-to-summer temperature contrast is less in maritime climates.

Through the course of a 24-hour day, surface air temperature responds to regular variations in solar and terrestrial infrared radiation. Typically, the day's lowest (minimum) temperature occurs shortly after sunrise. The day's highest temperature is usually recorded in early or mid-afternoon, even though solar radiation peaks around noon. Air temperature depends on the balance between incoming solar radiation and outgoing terrestrial radiation. Beginning shortly after sunrise, solar radiation exceeds emission of terrestrial IR and the air temperature rises. By early to mid-afternoon, solar radiation balances terrestrial radiation and air temperature reaches a maximum. Subsequently, terrestrial radiation exceeds solar radiation and the temperature falls. Overnight, only outgoing terrestrial radiation is occurring, further cooling the surface and air above.

The diurnal lag between solar radiation and air temperature explains why in summer the greatest risk of sunburn is around noon (the time of peak solar altitude) and not during the warmest time of day. Incoming solar ultraviolet radiation, the cause of sunburn, is most intense near noon, but the air temperature doesn't reach a maximum until several hours later.

As a further illustration of local radiational controls, consider the influence of ground characteristics on air temperature. All other factors being equal, in response to the same intensity of solar radiation striking Earth's surface, air over a dry surface warms more than air over a moist surface. When the surface is dry, absorbed radiation is used primarily for sensible heating of the air (conduction and convection of heat from the surface into the overlying air). Hence, the air temperature is higher (and so is the Bowen ratio). On the other hand, when the surface is moist, much of the absorbed radiation is used to evaporate water, the air temperature is lower (and so is the Bowen ratio).

Dry soil (and a relatively high Bowen ratio) helps explain why unusually high air temperatures often accompany *drought*, a lengthy period of moisture deficit. Soils dry out, crops wither and die, and lakes and other reservoirs shrink. Because less surface moisture is available for vaporization, more of the available heat is channeled into raising the air temperature through conduction and convection. Consider as an example the severe drought that gripped a ten-state area of the

southeastern United States between December 1985 and July 1986. In most places, rainfall was less than 70% of the long-term average, and in the hardest hit areas, portions of the Carolinas, it was less than 40%. By July, many weather stations in the drought-stricken region were setting new high temperature records. Columbia, SC, Savannah, GA, and Raleigh-Durham, NC reported the warmest July on record. Also contributing to record heat was more intense solar radiation reaching the ground, a consequence of less than the usual daytime cloud cover. The same association between exceptionally dry surface conditions and unusually high air temperatures was observed during the severe drought that afflicted the Midwest and Great Plains during the summer of 1988. At many long-term weather stations, the summer of 1988 was one of the driest and hottest on record.

Relative lack of moisture in cities is one of the reasons why the average annual surface temperature is slightly higher in a city than in the surrounding countryside. A city is an island of warmth surrounded by cooler air—an **urban heat island**. Snow melts faster and flowers bloom earlier in a city. City sewer systems efficiently carry off most runoff from rain and snowmelt whereas the countryside typically has considerable standing water (e.g., lakes, rivers, moist soils) and much more vegetative cover for transpiration (emission of water vapor to the air by plants). In a city more of the available heat from absorbed radiation is used to raise the temperature of surfaces (sensible heating) and less for evaporation of water (latent heating). In the moister countryside, less of the heat from absorption of radiation is used for sensible heating and more for latent heating. On average, the Bowen ration is about four times greater in a city than the countryside.

Other factors that contribute to an urban heat island include the greater concentration of heat sources in a city (e.g., motor vehicles, space heaters, air conditioners). On a cold winter day in New York City, heat from urban sources may approach 100 W per m^2, about 7% of the solar constant. City surfaces generally have a lower albedo than the vegetative cover of rural areas. A city's canyon-like terrain of narrow streets and tall buildings causes multiple reflections of sunlight increasing the amount of solar radiation that is absorbed. Urban building materials (e.g., concrete, asphalt, and brick) conduct heat more readily than the soil and vegetation of rural areas so that release of heat to the urban atmosphere from buildings and streets partially counters radiational cooling especially at night.

An urban heat island is best developed at night when the air is calm and the sky is clear. Under those conditions, the nighttime temperature contrast between a city and its surroundings can be as great as 10 Celsius degrees (18 Fahrenheit degrees). Strong winds mix the city and country air, greatly diminishing the temperature contrast.

Snow has a relatively high albedo and substantially reduces the amount of solar radiation that is absorbed at the surface and converted to heat. Furthermore, snow-covered ground reduces sensible heating of the overlying air because some of the available heat is used to vaporize or melt snow. Consequently, a snow cover lowers the day's maximum air temperature. Because snow is also an excellent emitter of infrared radiation, nocturnal radiational cooling is extreme where the ground is snow-covered, especially when skies are clear (minimum greenhouse effect). Cooling near the Earth's surface is further enhanced if winds are very light or the air is calm. Light winds or calm conditions reduce vertical mixing of air. On such nights, the air temperature near the surface may be 10 Celsius degrees (18 Fahrenheit degrees) or more lower than if the ground were bare of snow. By reducing both the maximum and minimum daily air temperature, a snow cover significantly lowers the 24-hr average temperature.

COLD AND WARM AIR ADVECTION

Air mass advection refers to the movement of an air mass from one locality to another. With advection, one air mass replaces another air mass having different temperature (and/or humidity) characteristics. **Cold air advection** occurs when the wind transports colder air into a previously warmer area. On a weather map cold air advection is indicated by winds blowing across regional isotherms from a colder area to a warmer area (arrow A in Figure 4.18). Cold air advection occurs behind a cold front (Chapter 10). **Warm air advection** takes place when the wind blows across regional isotherms from a warmer area to a colder area (arrow B in Figure 4.18). Warm air advection occurs behind a warm front and ahead of a cold front (Chapter 10). *Isotherms* are lines drawn on a map through localities having the same air temperature and usually must be interpolated between reporting weather stations. Recall from earlier in this chapter that air mass exchange is a major contributor to poleward heat transport.

The significance of air mass advection for local temperature variations depends on (1) the initial temperature of the new air mass, and (2) the degree of

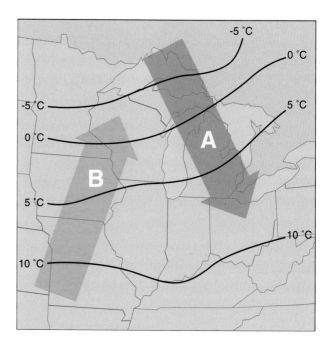

FIGURE 4.18
Cold air advection occurs when (A) horizontal winds blow across regional isotherms from colder areas toward warmer areas, and warm air advection occurs when (B) horizontal winds blow across regional isotherms from warmer toward colder areas. Solid lines are isotherms in °C.

modification the air mass undergoes as it travels over Earth's surface. For example, a surge of bitterly cold arctic air loses much of its punch when it travels over ground that is not snow covered, because the arctic air is warmed from below by sensible heating (conduction and convection). In contrast, modification of an arctic air mass by sensible heating is minimized when the air mass travels over a cold, snow-covered surface.

So far, we have been describing how horizontal movement of air (advection) might influence air temperature at some locality. However, as we saw in our discussion of convection currents, air also moves vertically. As air ascends and descends, its temperature changes; air cools as it rises and warms as it descends. We discuss the reasons for these temperature changes in Chapter 6.

Although we have considered local radiational controls and air mass advection separately, the two actually regulate air temperature together. Sometimes air mass advection compensates for, or even overwhelms, local radiational influences on air temperature. As noted earlier, in response to the local radiation balance, the air temperature usually climbs from a minimum near sunrise to a maximum in early or midafternoon. This typical pattern can change, however, if an influx of cold air occurs at the same time. Depending on how cold the incoming air is, air temperatures may climb more slowly than usual, remain steady, or even fall during daylight hours. If cold air advection is extreme, air temperatures may drop precipitously throughout the day, in spite of bright, sunny skies. In another example, air temperatures may climb through the evening hours as a consequence of strong warm air advection, so the day's high temperature occurs at night.

Conclusions

Heat and temperature are distinct yet closely related quantities. Unequal rates of radiational heating and cooling give rise to temperature gradients within the Earth-atmosphere system.   In response to these temperature gradients, heat is transferred from warmer to colder localities via radiation, conduction and convection, and latent heating. The temperature response of a substance to a gain or loss of heat depends primarily on its specific heat. Chiefly because of contrasts in specific heat, the temperature of a body of water is much less variable than that of land surfaces. This contrast in thermal inertia influences the temperature of the overlying air so that regions downwind of the ocean (or large lakes) exhibit less temperature contrast between summer and winter than locations far from large water bodies.

Imbalances in radiational heating and radiational cooling within the Earth-atmosphere system are ultimately responsible for the circulation of the atmosphere. These imbalances occur both vertically (between Earth's surface and the atmosphere) and horizontally (between tropical and higher latitudes). Through atmospheric circulation, heat is redistributed within the Earth-atmosphere system. This enables us to understand how air temperature is regulated by a combination of local radiational controls and air mass advection.

Another important consequence of atmospheric circulation is the formation of clouds that can produce rain, snow, and other forms of precipitation. Before examining cloud and precipitation forming processes in detail, we must first consider another important variable of the atmosphere, air pressure. We do this in the next chapter.

Basic Understandings

- Temperature is directly proportional to the average kinetic energy of the atoms or molecules composing a substance. Heat is the name we give to energy transferred from a warmer object to a colder object.

- As a sample of air gains or loses heat, that heat may be used for some combination of changes in temperature, phase of water, or volume of the air sample.

- The Fahrenheit temperature scale is still commonly used in the United States for meteorological and other purposes. The Celsius temperature scale is more convenient in that a 100-degree interval separates the freezing and boiling points of pure water at sea level. The Kelvin scale is based on absolute zero and is a more direct measure of average kinetic molecular activity than either the Fahrenheit or Celsius scale. An object at absolute zero (0 kelvin) emits no electromagnetic radiation. Heat energy is quantified as calories, joules, or British thermal units.

- Common types of thermometers are liquid-in-glass, bimetallic, and electronic. Thermometers used for weather observations should be mounted where they are sheltered from precipitation and direct sunshine.

- In response to a temperature gradient, heat always flows from locations of higher temperature toward locations of lower temperature. This is a consequence of the second law of thermodynamics. Heat transfer occurs via radiation, conduction and convection, and phase changes of water (latent heating).

- All objects absorb and emit radiation. If absorption exceeds emission (radiational heating), the temperature of an object rises, and if emission exceeds absorption (radiational cooling), the temperature of an object falls. At radiative equilibrium, absorption balances emission and the temperature of the object is constant.

- As a rule, solids (especially metals) are better conductors of heat than are liquids, and liquids are better conductors than gases. Motionless air is a very poor conductor of heat. Convection is the transport of heat within a fluid via motion of the fluid itself and is much more important than conduction in transporting heat within the troposphere.

- When water changes phase, heat is either absorbed

from or released to the environment. Latent heat is absorbed during melting, evaporation, and sublimation. Latent heat is released during freezing, condensation, and deposition.

- The temperature response to an input or output of heat differs from one substance to another depending primarily on the specific heat of each substance. Water bodies, such as the ocean or large lakes, exhibit less temperature variability from day to night and from summer to winter than do landmasses because the specific heat of water is higher than that of land, solar radiation readily penetrates water, and water circulates. The winter-to-summer temperature contrast is greater in continental climates than in maritime climates.

- On a global average annual basis, radiational cooling is greater than radiational heating of the atmosphere. On the other hand, radiational heating is greater than radiational cooling of Earth's surface. In response, heat is transported from the warmer Earth's surface to the cooler troposphere via latent heating (vaporization of water followed by cloud development) and sensible heating (conduction plus convection).

- The ratio of sensible heating to latent heating, the Bowen ratio, depends on the amount of moisture at the Earth's surface. The Bowen ratio is relatively high for dry surfaces and relatively low for wet surfaces. The global Bowen ratio is 0.30.

- Poleward of about 35 degrees latitude, over the course of a year, the rate of cooling due to infrared emission to space is greater than the rate of warming due to absorption of incoming solar radiation. In tropical latitudes, on the other hand, the rate of warming due to absorption of solar radiation is greater than the rate of cooling due to emission of infrared radiation. Poleward heat transport within the Earth system is the consequence.

- Poleward heat transport is brought about chiefly by atmospheric processes (north-south exchange of air masses and release of latent heat in storm systems) and, to a lesser extent, by the flow of warm and cold ocean currents.

- The sun drives the circulation of the atmosphere by causing heat imbalances within the Earth-atmosphere system. Some solar energy is converted to the kinetic energy of atmospheric circulation, which is ultimately dissipated as frictional heat that is radiated to space.

- Radiation balance plus air mass advection govern variations in local air temperature. Radiation balance varies with time of day, day of the year, cloud cover, and characteristics of the Earth's surface. Air mass advection occurs when winds blow across regional isotherms. In some cases, cold or warm air advection compensates for or even overwhelms the effects of the local radiation balance on air temperature.

ESSAY: Heating and Cooling Degree-Days

Television and newspaper weather summaries routinely report heating or cooling degree-day totals in addition to daily maximum and minimum temperatures.   Heating and cooling degree-days are indicators of household energy consumption for space heating and cooling, respectively.

In the United States, *heating degree-days* are based on the Fahrenheit temperature scale and are computed only for days when the mean outdoor air temperature is lower than 65 °F (18 °C).   Heating engineers who formulated this index early in this century found that when the mean outdoor temperature drops below 65 °F, space heating is required in most buildings to maintain an average indoor air temperature of 70 °F (21 °C).   The mean daily temperature is the simple arithmetic average of the 24-hr maximum and minimum air temperatures.   Each degree of mean temperature below 65 °F is counted as *one heating degree-day*.   Subtracting the mean daily temperature from 65 °F yields the number of heating degree-days for that day.   Suppose, for example, that this morning's low temperature was 36 °F, and this afternoon's high temperature was 52 °F.   Today's mean temperature would then be 44 °F, for a total of 21 heating degree-days (65 - 44 = 21).   It is usual to keep a running total of heating degree-days, that is, to add degree-days for successive days through the heating season (actually from 1 July of one year through 30 June of the next).

Fuel distributors and power companies closely monitor heating degree-days.   Fuel oil dealers base fuel use rates on cumulative degree-days and schedule deliveries accordingly.   Natural gas and electrical utilities anticipate power demands on the basis of degree-day totals, and implement priority use policies on the same basis when capacity fails to keep pace with demand.

The map is a plot of the average annual heating degree-day totals over the United States.   Outside of mountainous areas, regions of equal heating degree-day totals tend to parallel latitude circles with degree-day totals increasing poleward.   As an example, the average annual space-heating requirement in Chicago (6100 heating degree-days) is about four times that of New Orleans (1500 heating degree-days).   If per unit fuel costs are the same in both cities, then, in an average winter, Chicago homeowners can expect to pay four times as much for space heating as homeowners in New Orleans.   This assumes that buildings in the two cities are comparable in structure and insulation.

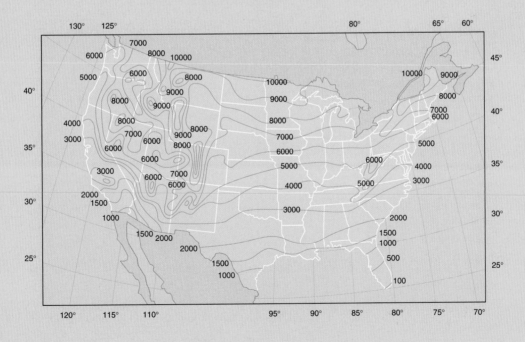

FIGURE
Average annual heating degree-day totals over the lower 48 states using a base of 65 °F.

Cooling degree-days are computed only for days when the mean outdoor air temperature is higher than 65 °F (although higher base temperatures are sometimes used). Supplemental air conditioning may be needed on such days. Again, a cumulative total is maintained through the cooling season (1 January through 31 December). Across the United States, average annual cooling degree-day totals range from less than 500 in the northern tier states and along much of the Pacific coast to more than 4000 in South Texas, South Florida, and the Desert Southwest.

Indexes of heating and cooling requirements are based on outdoor air temperatures and do not take into account other weather elements, such as air circulation and humidity, which also influence human comfort and demands for space heating and cooling. Heating and cooling degree-days are therefore only approximations of residential fuel demands for heating and cooling.

ESSAY: Windchill

At low air temperature, the wind increases human discomfort outdoors and heightens the danger of *frostbite*, the freezing of body tissue, or *hypothermia,* a potentially lethal condition brought on by a drop in the temperature of the body's vital organs (e.g., heart, lungs). Air in motion is more effective than still air in transporting heat away from the body. To account for the effect of wind on human comfort and wellbeing, weather reports during winter in northern and mountainous regions include the *windchill equivalent temperature (WET)* along with the actual air temperature.

A very thin layer of motionless air (thickness measured in millimeters) next to the skin helps insulate the body from heat loss to the environment. Within this so-called *boundary layer*, heat loss is through the very slow process of conduction. (Recall that calm air is a poor conductor of heat.) The boundary layer is thickest and offers maximum insulation when the wind speed is under 0.5 mi per hr (0.2 m per sec). Because water vapor molecules diffuse slowly through the skin's boundary layer, the rate of evaporative cooling is also relatively slow. Hence, even at very low air temperatures, people who are appropriately clothed are comfortable as long as winds are light. As winds strengthen, the boundary layer becomes thinner and the rate of heat and water vapor transport from the skin surface increases. Thinning of the boundary layer (and accelerated heat loss) is most pronounced at low to moderate wind speeds. As wind speed increases above about 35 mi (55 km) per hour, incremental heat loss becomes relatively small as the boundary layer approaches its minimum thickness.

The *windchill equivalent temperature (WET)*, also called the *windchill index*, is a measure of the rate of body heat loss due to a combination of wind and low air temperature. The original index was based upon research conducted in Antarctica by R.A. Siple and C.F. Passel during the winter of 1941. Their original objective was to measure the time required for water to freeze under various weather conditions. Only later did the idea arise to develop a windchill index based upon this research. Their experiments using a small water-filled non-insulated plastic cylinder provide, at best, a rough approximation of the thermal response of a human body dressed appropriately for cold and windy weather. The windchill index also failed to account for the appreciable warming that occurs when the body absorbs solar radiation—Siple and Passel made their measurements in darkness.

In the Winter of 2001-02, the National Weather Service (NWS) and the Meteorological Services of Canada (MSC) implemented a major revision of the windchill index. The new index incorporates recent advances in science, technology, and computer modeling to provide a more accurate, understandable, and useful formula for representing the combined hazard of wind and low air temperatures. The new windchill index uses wind speed adjusted for the average height (1.5 m or 5 ft) of the human body's face instead of standard anemometer height (10 m or 33 ft), applies modern heat transfer theory, and assumes the worst case scenario for solar radiation (clear night sky). The formula is

$$\text{WET (°F)} = 35.74 + 0.6215T - 35.75 (V^{0.16}) + 0.4275T (V^{0.16})$$

Where T is the temperature in °F and V is the wind speed in miles per hour. This formula is the basis for the Windchill Table. (For computing windchill, calm is defined as 3 mi per hr or less.)

Suppose, for example, the air temperature is 35 °F and the air is calm; then the windchill equivalent temperature (35 °F) is the same as the actual air temperature. If the wind strengthens to 25 mi per hr with no change in actual air temperature, the WET drops to 23 °F. Contrary to popular opinion, this does not mean that the temperature of exposed skin actually drops to 23 °F. Skin temperature can drop no lower than the actual air temperature, which in this example is 35 °F. When skin temperature equals the air temperature, the temperature gradient between the skin surface and adjacent air is zero and no heat is exchanged between the skin and the air. Even though the WET may fall below the freezing point, skin will not freeze if the actual air temperature is above the freezing point.

On a cold and windy day, an exposed body part loses heat to the environment at the same rate that it would if the air were calm and the actual air temperature equaled the windchill equivalent temperature. In fact, many combinations of air temperature and wind speed produce about the same rate of heat loss. For instance, heat loss is essentially the same for air temperature/wind speed combinations of 15 °F/5 mi per hr and 25 °F/35 mi per hr. In both cases, the windchill equivalent temperature is 7 °F.

New Wind Chill Chart
Wind (mph)

Temperature (°F)

Calm	5	10	15	20	25	30	35	40	45	50	55	60
40	36	34	32	30	29	28	28	27	26	26	25	25
35	31	27	25	24	23	22	21	20	19	19	18	17
30	25	21	19	17	16	15	14	13	12	12	11	10
25	19	15	13	11	9	8	7	6	5	4	4	3
20	13	9	6	4	3	1	0	-1	-2	-3	-3	-4
15	7	3	0	-2	-4	-5	-7	-8	-9	-10	-11	-11
10	1	-4	-7	-9	-11	-12	-14	-15	-16	-17	-18	-19
5	-5	-10	-13	-15	-17	-19	-21	-22	-23	-24	-25	-26
0	-11	-16	-19	-22	-24	-26	-27	-29	-30	-31	-32	-33
-5	-16	-22	-26	-29	-31	-33	-34	-36	-37	-38	-39	-40
-10	-22	-28	-32	-35	-37	-39	-41	-43	-44	-45	-46	-48
-15	-28	-35	-39	-42	-44	-46	-48	-50	-51	-52	-54	-55
-20	-34	-41	-45	-48	-51	-53	-55	-57	-58	-60	-61	-62
-25	-40	-47	-51	-55	-58	-60	-62	-64	-65	-67	-68	-69
-30	-46	-53	-58	-61	-64	-67	-69	-71	-72	-74	-75	-76
-35	-52	-59	-64	-68	-71	-73	-76	-78	-79	-81	-82	-84
-40	-57	-66	-71	-74	-78	-80	-82	-84	-86	-88	-89	-91
-45	-63	-72	-77	-81	-84	-87	-89	-91	-93	-95	-97	-98

Frostbite Times

■ 30 Minutes   ■ 10 Minutes   ■ 5 Minutes

$$\text{Wind Chill } (^{\circ}F) = 35.74 + 0.6215T - 35.75(V^{0.16}) + 0.4275T(V^{0.16})$$

Where, T = Air Temperature (°F)
V = Wind Speed (mph)

FIGURE

New windchill table introduced by the National Weather Service in the Winter of 2001-02.

During cold and blustery weather, if the human body cannot supply heat to the skin at a rate sufficient to compensate for heat loss, skin temperature declines. Initially a person has a sense of discomfort. If the rate of heat loss from exposed skin causes skin temperature to fall to subfreezing levels, then frostbite may ensue. Body parts that are usually exposed and have a relatively high surface-to-volume ratio, such as the ears, nose, and fingers, are particularly susceptible to frostbite. Frostbite may occur in 15 minutes or less at windchill values of –18 °F (–28 °C) or lower. In extreme circumstances, the rate of heat loss from the body may be great enough to cause life-threatening hypothermia.

CHAPTER 5

AIR PRESSURE

"Yes, what a climb that was!
I was scared to death, I can tell you.
Sixteen hundred meters—that is over five
thousand feet, as I reckon it..."

And Hans Castrop took in a deep,
experimental breath of the strange air. It was
fresh, and that was all. It had no perfume, no
content, no humidity; it breathed in easily, and
held for him no association.

THOMAS MANN
The Magic Mountain

Case-in-Point

At 8850 m (29,035 ft) above sea level, the summit of Mount Everest is the highest mountain peak in the world. This massive glacier-encrusted mountain is located at the edge of the Tibetan Plateau on the border between Tibet and Nepal in the central Himalayas. In 1856, it was named for Sir George Everest, surveyor general of India. For most Nepali people, the mountain is *Sagarmatha* ("forehead in the sky") while the Sherpa people of northern Nepal call it *Chomolungma* ("Goddess Mother of the World"). Local peoples revered the mountain as sacred and did not attempt to scale it prior to the 20th century. For years, restrictions on foreign visitors, brutally harsh weather, the threat of massive snow and ice avalanches on the mountain, and the thin air at higher elevations were insurmountable barriers for climbers bent on conquering the peak. The summit of Mount Everest was not reached until 28 May 1953 when scaled by Sir Edmund Hillary of New Zealand and Sherpa Tenzing

Norgay of Nepal. Since then, more than 4000 people have attempted to climb to the summit of Mount Everest; 660 were successful and more than 140 died in the effort.

The latitude of Mount Everest (28 degrees N) is about the same as Tampa, FL but the rapid decline of air temperature with elevation means that the upper reaches of the mountain are perpetually at subfreezing temperatures. Estimated January mean temperature at the summit of Mount Everest is -36 °C (-33 °F). In July, the warmest month, the mean temperature is about -19 °C (-2 °F). From June through September, thick clouds shroud the mountain and heavy snows are fed by moist winds blowing inland from off the Indian Ocean, the wet monsoon circulation. From November through February, the jet stream dips down from the north, replacing the monsoon flow, and bringing to the mountain strong winds frequently blowing in excess of hurricane force. The combination of high winds and low temperatures make frostbite and hypothermia a constant threat for climbers.

Air also becomes thinner with increasing elevation; that is, the number of air molecules per unit volume decreases. Although the ratio of nitrogen (N_2) to oxygen (O_2) remains constant at about 4 to 1, the decline in the amount of oxygen (coupled with the extreme weather) can quickly lead to life-threatening conditions. The decline in air density with altitude is accompanied by a decrease in air pressure (the weight per unit area of a column of air). At the summit of Mount Everest, the pressure exerted by oxygen is only about one-third of its sea level value. Without a supplemental oxygen supply, people cannot survive for very long at the summit. Climbers gradually ascend to higher altitudes taking several weeks for their bodies to adjust to thin air. But even with acclimatization (discussed in this chapter's first Essay), the summit of Mount Everest is so high and the air so thin that until recently experts considered survival near or at the summit to be impossible without a supplemental oxygen supply. The experts were proved wrong in 1978 when the Austrians Reinhold Messner and Peter Habeler ascended to the summit without a supplemental oxygen supply and survived. In spite of this remarkable accomplishment, the extreme environment at the summit of Mount Everest remains very close to the physiological limits of human life.

Driving Question:

What is the significance of horizontal and vertical variations in air pressure?

In describing the state of the atmosphere, television and radio weathercasts usually cite the latest air pressure reading along with air temperature, relative humidity, and other weather elements. Although we are physically aware of changes in temperature and humidity, we do not sense changes in air pressure as readily. If we follow air pressure reports over a period of time, however, we quickly learn that important changes in weather accompany relatively small variations in air pressure.

In this chapter, we examine the properties of air pressure, how air pressure is measured, and the reasons for spatial and temporal variations in air pressure. In later chapters, we describe how this variability of air pressure contributes to the circulation of the atmosphere and weather.

Defining Air Pressure

Air exerts a force on the surfaces of all objects that it contacts. Air pressure is a measure of that force per unit surface area. (A *force* is a push or pull on an object and is computed as mass times acceleration.) Molecules composing air are always in rapid, random motion, and each molecule exerts a force as it collides with others on the surface of a solid (e.g., the ground) or liquid (e.g., the ocean). In a millionth of a second, billions upon billions of gas molecules bombard every square centimeter of Earth's surface. The total air pressure is the cumulative force of a multitude of molecules colliding with a unit surface area of any object in contact with air.

The pressure produced by the gas molecules composing air depends on (1) the mass of the molecules, and (2) the kinetic molecular activity. In the larger sense, we can think of **air pressure** at a given location on the Earth's surface as the weight per unit area of the column of air above that location. The pressure at any point within the atmosphere is equal to the weight per unit area of the atmosphere above that point. Weight is the force exerted by gravity on a mass, that is,

$$\text{weight} = (\text{mass}) \times (\text{acceleration of gravity})$$

The average air pressure at sea level is about 1.0 kg per square centimeter (14.7 lb per square inch). This is the same pressure as produced by a column of water about 10 m (33 ft) high. Hence, the total weight of the atmosphere on the roof of a typical three-bedroom ranch-style house at sea level is about 2.1 million kg (4.6 million lb), equivalent to the combined weight of 1500 full-sized autos. Why doesn't the roof collapse? It doesn't because air pressure at any point is the same in all directions (up, down, and sideways); that is, air is pushing up from under the roof with the same pressure. Air pressure within the house exactly counterbalances air pressure outside the house so that the *net* pressure acting on the roof is zero. This pressure balance (or equilibrium) is the prevailing condition in the atmosphere.

Air Pressure Measurement

A **barometer** is the instrument used to measure air pressure and monitor its changes. The basic types are mercury and aneroid barometers.

760 mm mercury = 29.92 in Mercury
= 1013.25 mB

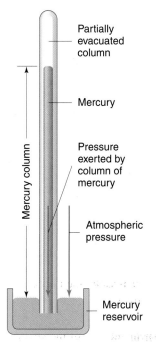

FIGURE 5.1
Schematic drawing of a mercury barometer.

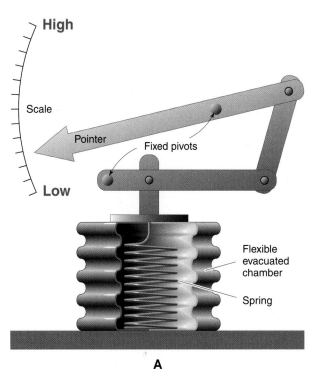

A

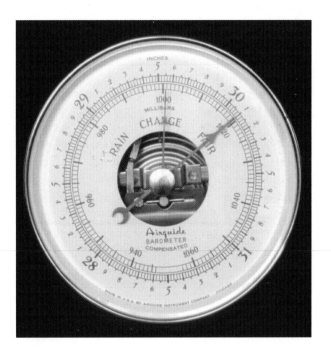

B

FIGURE 5.2
An aneroid barometer, a portable instrument used to monitor air pressure. internal view (A) and external view (B). [Adapted from Snow, et al., "Basic Meteorological Observations for Schools: Atmospheric Pressure." *Bulletin of the American Meteorological Society* 73(1992):785]

The more accurate, though cumbersome, of the two is the **mercury barometer**, invented in 1643 by Evangelista Torricelli, an Italian mathematician and student of Galileo. The instrument consists of a glass tube a little less than 1.0 m (39 in.) long, sealed at one end, open at the other end, and filled with mercury which is about 13 times denser than liquid water. The open end of the tube is inverted into a small open container of mercury, as shown in Figure 5.1. Mercury settles down the tube (and into the container) until the pressure of the mercury column exactly balances the pressure of the atmosphere acting on the surface of the mercury in the container.

The average air pressure at sea level will support the mercury column in the tube to a height of 760 mm (29.92 in.). When air pressure changes, however, the height of the mercury column changes. Falling air pressure allows the mercury column to drop, whereas increasing air pressure forces the mercury column to rise with the height of the mercury column directly proportional to air pressure. This is the origin of the common practice of expressing air pressure in units of length. Air pressure readings by a mercury barometer require adjustments for (1) the expansion and contraction of mercury that accompanies changes in temperature, and (2) the slight variation of gravity with latitude and altitude. By convention, readings are adjusted to standard conditions of 0 °C (32 °F) and 45 degrees latitude at sea level.

An **aneroid** (nonliquid) **barometer** is less precise but more portable than a mercury barometer. It consists of a flexible chamber from which much of the air has been evacuated (Figure 5.2A). A spring prevents the chamber from collapsing. As air pressure changes, the chamber flexes, compressing when air pressure rises and expanding when air pressure drops. A series of gears and

FIGURE 5.3
A barograph provides a continuous trace of air pressure variations with time.

levers transmits these movements to a pointer on a dial, which is calibrated to read in equivalent millimeters (or inches) of mercury, or to read directly in units of air pressure (Figure 5.2B). The latest aneroid barometers provide direct digital readouts (and even have been incorporated into wristwatches). Also, many new aneroid barometers are *piezoelectric*, that is, they depend on effect of air pressure on a crystalline substance.

Indoor air pressure quickly adjusts to changes in outdoor air pressure so that a barometer is usually mounted indoors for convenience. The instrument should be anchored to a sturdy wall not in direct sunlight. Mercury barometers must be vertical and most aneroid barometers are designed to be read in a vertical position. In taking a reading from a dial-type aneroid barometer, it is a good idea to first gently tap the barometer because friction in the mechanism may cause the pointer to stick.

Aneroid barometers intended for home use typically have dials with legends, such as *fair, changeable,* and *stormy*, corresponding to certain ranges of air pressure. These designations should not be taken literally because a given air pressure reading does not always correspond to a specific type of weather. Much more useful than these legends for local weather forecasting is **air pressure tendency**, that is, the change in air pressure with time. Rising air pressure usually means continued fair or clearing weather, whereas falling air pressure generally signals the approach of stormy weather. For determining pressure tendency, some aneroid barometers are equipped with a second pointer

that serves as a reference marker. By turning the knob on the barometer face, the user sets the second pointer to correspond to the current air pressure reading. At a later time (perhaps in 2 or 3 hrs or less during changeable weather), the user can observe the new pressure reading and compare it with the earlier set reading to determine the air pressure tendency. An aneroid barometer may be linked to a pen that records on a clock-driven drum chart. This instrument, called a **barograph**, provides a continuous trace of air pressure with time, making it easier to determine pressure tendency (Figure 5.3).

Air Pressure Units

On television and radio weathercasts, air pressure readings are usually reported in units of length (millimeters or inches of mercury). It is scientifically more appropriate to express air pressure in units of pressure. Physicists use the *pascal (Pa)* as the metric unit of pressure and have determined that the average air pressure at sea level is 101,325 Pa, 1013.25 hectopascals (hPa), or 101.325 kilopascals (kPa). U.S. meteorologists, on the other hand, traditionally designate air pressure in *millibar (mb)* units, where 1 mb equals 1 hPa or 100 Pa. In turn, 1 mb is the equivalent of 0.02953 in. of mercury (Table 5.1).

The usual worldwide range in sea-level air pressure is roughly 970 to 1040 mb (28.64 to 30.71 in. of mercury). The lowest sea-level air pressure ever recorded was 870 mb (25.69 in. of mercury), measured on 12 October 1979 in the eye of Typhoon Tip over the Pacific Ocean northwest of Guam. The air pressure no doubt is lower than 870 mb in some tornadoes, but such low readings have never been confirmed because

TABLE 5.1
Conversion Factors for Units of Air Pressure

1 bar = 1000 millibars (mb)
1 mb = 0.02953 in. of mercury
1 inch of mercury = 33.8639 mb
1 kilopascal (kPa) = 1000 pascals (Pa)
1 hectopascal (hPa) = 100 Pa
1 mb = 1 hPa
1 inch of mercury = 33.8639 hPa
1 inch of mercury = 25.4 mm of mercury

tornadic winds are so strong that they destroy barometers. The highest sea-level air pressure ever recorded was 1083.8 mb (32.01 in. of mercury) at Agata, Siberia on 31 December 1968 and was produced by an extremely cold, dense air mass.

Variation in Air Pressure with Altitude

We know from pumping up a bicycle tire that air is compressible; that is, its volume and density are variable. The pull of gravity compresses the atmosphere so that the maximum **air density** (mass of molecules per unit volume) is at the Earth's surface. In other words, the atmosphere's gas molecules are most closely spaced at the Earth's surface, and the spacing between molecules increases with increasing altitude. The number of gas molecules per unit volume, air's **number density**, thus decreases with altitude. *Thinning* of air is so rapid in the lower atmosphere that at an altitude of 16 km (10 mi), air density is only about 14% of its average sea-level value.

Thinning of air with altitude is accompanied by a decline in air pressure (Figure 5.4). Blaise Pascal of France was the first to propose that air pressure decreases with increasing altitude. Poor health prevented Pascal from verifying his hypothesis by actual field measurements, so he asked his brother-in-law, Florin Perier, for assistance. Perier agreed, and on 19 September 1648, Perier and some clerics and laymen from his hometown took barometer readings at the base and summit of Puy de Dôme (elevation, 1464 m or 4800 ft) in the Auvergene region of southern France. Those barometer readings confirmed Pascal's hypothesis.

Because air is compressible, the rate of air pressure drop with altitude is greatest in the lower tropos-phere and then becomes more gradual aloft. For example, air pressure decreases about 25% in the first 2500 m (8200 ft) of altitude, but a further ascent of 3000 m (9800 ft) is required for another 25% drop in air pressure.

Vertical profiles of average air pressure (Figure 5.4) and temperature (Figure 2.10) are based on the standard atmosphere, a model of the real atmosphere. The **standard atmosphere** is the state of the atmosphere averaged for all latitudes and seasons. It features a fixed sea-level air temperature (15 °C or 59 °F) and pressure (1013.25 mb) and fixed vertical profiles of air temperature and air pressure (Table 5.2). The actual

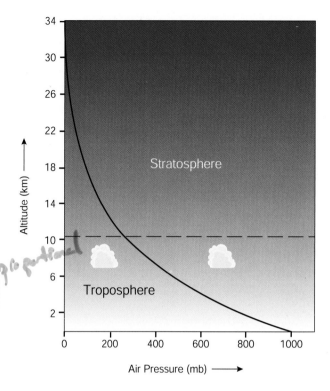

FIGURE 5.4
Average variation in air pressure in millibars (mb) with altitude (km). The average air pressure at sea level is 1013.25 mb.

altitude of a specific air pressure (e.g., 500 mb) varies with time and from one place to another. In fact, meteorologists routinely draw upper-air weather maps on which are plotted contours of elevation of an *isobaric surface*, that is, an imaginary surface where the air pressure is the same everywhere. For example, contour maps are constructed for the 200-mb, 500-mb, and 850-mb levels. Observational data for these maps are obtained primarily from radiosondes.

Although air pressure and density drop with increasing altitude, it is not possible to specify an altitude where Earth's atmosphere definitely ends. That is, no one altitude can be clearly identified as the beginning of interplanetary space. At best, we can describe the vertical extent of the atmosphere in terms of the relative distribution of its mass with altitude. Half the atmosphere's mass lies between Earth's surface and an average altitude of about 5500 m (18,000 ft). About 99% of the atmosphere's mass is below 32 km (20 mi). Above 80 km (50 mi), that is, above the homosphere, the relative proportions of atmospheric gases change markedly, and by about 1000 km (620 mi), the atmosphere merges with the highly rarefied interplanetary gases, mainly hydrogen and helium. Interestingly, if Earth's atmosphere had a

TABLE 5.2
The Standard Atmosphere

Altitude (km)	Temperature (°C)	Pressure (mb)	P/P_o*	Density (kg/m³)	D/D_o*
30.00	-46.6	11.97	0.01	0.02	0.02
25.00	-51.6	25.49	0.03	0.04	0.03
20.00	-56.5	55.29	0.05	0.09	0.07
19.00	-56.5	64.67	0.06	0.10	0.08
18.00	-56.5	75.65	0.07	0.12	0.09
17.00	-56.5	88.49	0.09	0.14	0.12
16.00	-56.5	103.52	0.10	0.17	0.14
15.00	-56.5	121.11	0.12	0.20	0.16
14.00	-56.5	141.70	0.14	0.23	0.19
13.00	-56.5	165.79	0.16	0.27	0.22
12.00	-56.5	193.99	0.19	0.31	0.25
11.00	-56.4	226.99	0.22	0.37	0.30
10.00	-49.9	264.99	0.26	0.41	0.34
9.50	-46.7	285.84	0.28	0.44	0.36
9.00	-43.4	308.00	0.30	0.47	0.38
8.50	-40.2	331.54	0.33	0.50	0.40
8.00	-36.9	356.51	0.35	0.53	0.43
7.50	-33.7	382.99	0.38	0.56	0.45
7.00	-30.5	411.05	0.41	0.59	0.48
6.50	-27.2	440.75	0.43	0.62	0.50
6.00	-23.9	472.17	0.47	0.66	0.54
5.50	-20.7	505.39	0.50	0.70	0.57
5.00	-17.5	540.48	0.53	0.74	0.60
4.50	-14.2	577.52	0.57	0.78	0.63
4.00	-11.0	616.60	0.61	0.82	0.67
3.50	-7.7	657.80	0.65	0.86	0.70
3.00	-4.5	701.21	0.69	0.91	0.74
2.50	-1.2	746.91	0.74	0.96	0.78
2.00	2.0	795.01	0.78	1.01	0.82
1.50	5.3	845.59	0.83	1.06	0.86
1.00	8.5	898.76	0.89	1.11	0.91
0.50	11.8	954.61	0.94	1.17	0.95
0.00	15.0	1013.25	1.00	1.23	1.00

*P/P_o = ratio of air pressure to its sea-level value; D/D_o = ratio of air density to its sea-level value.

uniform density and temperature throughout, it would have a well-defined top. Assuming a temperature equal to the average sea level value, the top of a uniform density atmosphere would be at an altitude of only 8 km (5 mi).

From a somewhat different perspective, at an altitude of only 32 km (20 mi), air pressure is less than 1% of its average sea-level value. The rapid vertical pressure drop in the lower reaches of the atmosphere means that appreciable changes in air pressure accompany even relatively minor changes in land elevation. For example, the average air pressure at Denver, the *mile-high city*, is about 83% of the average air pressure at Boston, located just above sea level. The expansion and thinning of air that accompany the fall in air pressure with altitude can cause discomfort and even serious illness for people who visit high altitudes. For more on this topic, refer to the Essay, Human Responses to Changes in Air Pressure.

Very low air density at high altitudes also has interesting implications for air temperature and heat transfer. In the thermosphere, the highest thermal subdivision of the atmosphere (Figure 2.10), individual atoms and molecules move about with an average kinetic activity indicative of very high temperatures. There are so few atoms and molecules per unit volume, however, that very little heat is transferred. In spite of temperatures that approach 1200 °C (2200 °F) in the thermosphere, heat is not readily conducted to cooler bodies. For example, satellites orbiting at these altitudes do not acquire such temperatures.

Because air pressure drops with altitude, an aneroid barometer can be calibrated to monitor altitude. Such an instrument is called an **altimeter**. For more on this, see the Essay,

Determining Altitude from Air Pressure.

Horizontal Variations in Air Pressure

Air pressure differs from one place to another, and variations are not always due to differences in the elevation of the land. In fact, meteorologists are more interested in air pressure variations that arise from factors other than land elevation. Hence, weather observers determine an equivalent sea-level air pressure value; that is, for stations located above sea level, they increase local air pressure readings to approximately what the air pressure would be if the station were actually located at sea level. The simplest method of adjustment is to assume an imaginary column of air extending from the station down to sea level and having the properties of the standard atmosphere. When this reduction to sea level is carried out everywhere, air pressure is observed to vary from one place to another (Figure 5.5) and fluctuate from day to

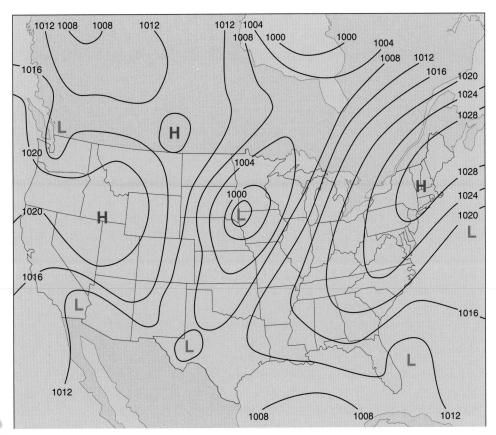

FIGURE 5.5
A surface weather map shows spatial variations in air pressure (reduced to sea level). Dark lines are isobars (in mb) passing through localities having the same air pressure. *L* is plotted where air pressure is relatively low and *H* is plotted where air pressure is relatively high.

Time

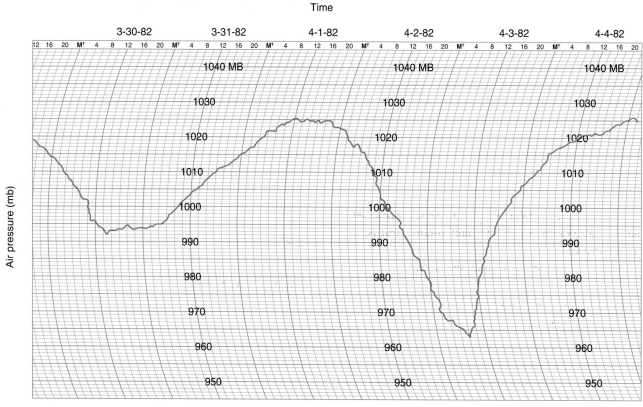

M = midnight on scale at top of record; time lines are in 2 hour increments

FIGURE 5.6
A trace from a baragraph showing the variation in air pressure in millibars reduced to sea level at Green Bay, Wisconsin, from 30 March through 4 April 1982. During the period from 2 to 3 April, Green Bay was under the influence of a very intense low pressure system. Note that significant changes in air pressure occur from day to day and even from one hour to the next.

day and even from hour to hour (Figure 5.6).

Air pressure readings, adjusted to sea level, range in value from one place to another, but the magnitude of this variation is much less than the rate at which air pressure drops with altitude. The air pressure change of about 3 mb observed in the lowest 30 m (100 ft) of the troposphere may not be equaled over a horizontal distance of 200 km (125 mi) at sea level. Nonetheless, these relatively small horizontal changes in air pressure can be accompanied by important changes in weather.

In middle latitudes, weather is dominated by a continuous procession of different air masses that bring about changes in air pressure and weather. Recall that an *air mass* is a huge volume of air that is relatively uniform horizontally in temperature and water vapor concentration. As air masses move from place to place, surface air pressures fall or rise, and the weather changes. As a general rule, weather becomes stormy when air

pressure falls and clears or remains fair when air pressure rises.

Why do some air masses exert greater pressure than other air masses? One reason is the contrast in air density that arises from differences in air temperature or water vapor concentration. As a rule, temperature has a much more important influence on air density and pressure than does water vapor concentration. Also, diverging or converging winds may bring about a change in air pressure.

INFLUENCE OF TEMPERATURE AND HUMIDITY

Recall from Chapter 4 that temperature is a measure of the average kinetic energy of individual molecules. Rising air temperature corresponds to an increase in average kinetic molecular activity. If air is heated within a closed container, such as a rigid metal can, we would expect the air pressure acting on the internal walls of the container to rise as the increasingly

Tendencies of warm/cold
high/low pressure at f

energetic molecules bombard the walls with greater force. The air density inside the container does not change because no air is added to or removed from the container and the air volume is constant. By contrast, except for Earth's surface, walls do not confine the atmosphere, so air is free to expand and contract. That is, within the atmosphere, air density is variable.

Within the atmosphere, when air is heated (e.g., by absorption of radiation, conduction, convection, or some phase changes of water), air density decreases because the greater activity of the heated molecules increases the spacing between neighboring molecules (Figure 5.7). As a column of air is heated, decreasing numbers of molecules exist per unit volume; that is, air density decreases. The lower total mass in the column will then exert less pressure on Earth's surface.

The greater density of cold air versus warm air affects the rate at which air pressure drops with increasing altitude. Air pressure drops more rapidly with

FIGURE 5.7
A hot air balloon ascends within the atmosphere because heated air within the balloon is less dense than the cooler air surrounding the balloon.

altitude within a column of cold (more dense) air than in a column of warm (less dense) air. Beginning with equal pressures at Earth's surface, equivalent pressure surfaces (e.g., the 500-mb level) occur at a lower altitude in a cold column of air than in a warm column of air. Within the troposphere, isobaric surfaces (surfaces of constant pressure) slope downward from the relatively warm tropics toward relatively cold high latitudes. As noted in this chapter's second Essay, the sloping of isobaric surfaces has important implications for altimetry.

Increasing humidity has the same influence on air density as rising air temperature. That is, the greater the concentration of water vapor, the less dense is the air. This statement is contrary to the popular perception that humid air is *heavier* than dry air. Although hot, muggy air may weigh heavily on a person's disposition, humid air is, in fact, less dense than dry air at the same temperature and total air pressure. Water vapor reduces the density of air because the molecular weight of water is less than the average molecular weight of dry air. The molecular weight of water is 18 atomic mass units whereas the mass-weighted mean molecular weight of dry air is about 29 atomic mass units. Also, the number of molecules in a given volume of a gas or mixture of gasses at a steady temperature and pressure must remain the same (*Avogadro's law*). This implies that when water molecules enter the atmosphere as a gas, they take the place of other gas molecules, principally nitrogen and oxygen. The molecular weight of water vapor (H_2O) is less than that of either N_2 or O_2. With all other factors equal, as the water vapor concentration in air increases (perhaps as the consequence of evaporation of water at Earth's surface), the net effect is for air density to decrease. For equal volumes at the same temperature, then, a column of humid air exerts less pressure than a column of relatively dry air.

Cold, dry air masses are denser and usually produce higher surface pressures than warm, humid air masses. Warm, dry air masses, in turn, often exert higher surface pressures than equally warm, but more humid, air masses. Hence, a change in surface air pressure usually accompanies the replacement of one air mass by another, that is, *cold* or *warm air advection*. *Air mass modification* (changes in air mass temperature and/or water vapor concentration) also can produce changes in surface air pressure. These modifications may occur when an air mass travels over different surface types (from cold snow cover to mild bare ground, for example) or, if the air mass is stationary, when the air is locally heated or cooled. In Chapter 4, we examined the regulation of air temperature by the local radiation balance plus air mass advection. From the above discussion, it is evident that local conditions and air mass advection can also influence surface air pressure.

INFLUENCE OF DIVERGING AND CONVERGING WINDS

In addition to variations in air temperature and (to a lesser extent) humidity, divergence or convergence of winds may also bring about changes in surface air pressure. Divergence or convergence are produced by a circulation pattern in which (1) horizontal winds blow toward or away from some location, or (2) wind speed changes in a downstream direction. We consider the first mechanism here and the second in Chapter 8. Suppose that at the Earth's surface, horizontal winds blow away from a column of air, as in Figure 5.8. This is an example of **diverging winds**. At the same time, horizontal winds aloft blow toward the air column; this is an example of **converging winds**. Within the air column, air descends from above and takes the place of air diverging at the surface. If more air diverges at the surface than converges aloft, then air density and surface air pressure decrease. On the other hand, if more air converges aloft than diverges at the surface, density of the air column and surface air pressure increase. In later chapters, we discuss in greater detail how divergence and

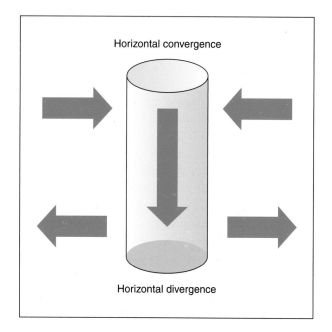

Horizontal convergence

Horizontal divergence

FIGURE 5.8
Winds diverging at the surface and converging aloft induce descending air motion.

convergence of air cause air pressure changes within weather systems.

Highs and Lows

We now have more insight into the meaning of those *H* (or *HIGH*) and *L* (or *LOW*) symbols on surface weather maps (Figure 5.5). To locate *HIGHs* and *LOWs*, a map is drawn that portrays the pattern of surface air pressure. First, simultaneous air pressure readings at all weather stations are adjusted to sea level, plotted on a weather map, and then isobars are drawn. An **isobar** is a line passing through locations having the same air pressure and usually requires interpolation between reporting weather stations. By U.S. convention, isobars are drawn at 4-mb intervals (e.g., 996 mb, 1000 mb, 1004 mb).

A *HIGH* or *H* symbol is used to designate places where sea-level air pressure is relatively high compared to the air pressure in surrounding areas. A *HIGH* is also known an *anticyclone*. A *LOW* or *L* symbol signifies regions where sea-level air pressure is relatively low compared to surrounding areas. A *LOW* is also known as a *cyclone*. For reasons presented in Chapter 8, a *HIGH* usually is a fair weather system, and a *LOW* is a stormy weather system. Viewed from above (in the Northern Hemisphere), surface winds blow clockwise and diverge outward about the center of a *HIGH* and counterclockwise and converge inward about the center of a *LOW*. Winds are strong where isobars are close together and weak where isobars are widely spaced. Isobars generally are more closely spaced near the center of a *LOW* than a *HIGH*.

The Gas Law

To this point, we have described the state of the atmosphere in terms of variations in temperature, pressure, and density. These important properties, collectively known as **variables of state**, change in magnitude from one place to another across Earth's surface, with altitude above Earth's surface, and with time. The three variables of state are also interrelated through the **gas law**. Although the gas law was derived for a single ideal gas, the law provides a reasonably accurate description of the behavior of air, which, as noted earlier, is a mixture of many different gases.

Simply put, the gas law states that the pressure exerted by air is directly proportional to the product of its density and temperature. Expressed as a word equation, the gas law becomes

air pressure = (constant) × (density) × (temperature)

The constant is an experimentally derived number that changes a proportional relationship to an equation. From the gas law equation, the following is evident:

1. The density of air within a rigid closed container remains constant because the volume of the container and the mass of its contents are fixed. A rise in temperature is accompanied by an increase in pressure exerted by the gas molecules on the interior walls of the container.

2. Consider a parcel of air (an arbitrary amount of air useful for visualizing atmospheric processes) that contains a fixed number of molecules. Its volume can change, but its mass remains the same. If the temperature is held constant, compressing the air parcel causes its density to increase because its volume decreases (while its mass remains constant). Hence, at a fixed temperature, pressure is directly proportional to density.

3. For the same air parcel at constant pressure, a rise in temperature is accompanied by a decrease in density. The density decreases because the more energetic molecules cause the parcel to expand, that is, its volume increases while its mass remains constant. Hence, at a fixed pressure, temperature is inversely proportional to density.

In the atmosphere, however, the situation is more complicated because all three variables of state may change simultaneously. Hence, as the air temperature rises (perhaps in response to radiational heating), air expands, air density decreases, and air pressure at the Earth's surface falls. On the other hand, in winter, it is usual for air temperature to drop (not rise) as the surface air pressure rises. The gas law is satisfied in the atmosphere because air density increases as the temperature drops.

Conclusions

We now have a working definition of air pressure, and we have examined the causes of spatial and temporal variations in air pressure within the Earth-atmosphere system. Air pressure drops rapidly with altitude in the lower troposphere and then more gradually aloft. At great altitudes, the atmosphere gradually merges with the gases of interplanetary space so that the atmosphere has no clearly defined top. Adjusting barometer readings to sea level eliminates the influence of weather station elevation on air pressure. Then, surface air pressure depends on air density, which in turn, is governed by air temperature and, to a lesser extent, by the concentration of water vapor in air. Diverging or converging winds may also affect air density and surface air pressure.

The range of air temperature and pressure in the Earth-atmosphere system allows water to occur in all three phases. In Chapter 4, we saw how phase changes of water help redistribute heat within the Earth-atmosphere system (latent heating). Changes in air temperature also trigger the phase changes of water that cause clouds to form or to dissipate. In the next chapter, we take a closer look at water within the atmosphere with a special emphasis on cloud-forming processes.

Basic Understandings

- The pressure (force per unit area) exerted by the atmosphere depends on the pull of gravity and the mass and kinetic energy of the gas molecules that compose air. We can think of air pressure as the weight of a column of air acting on a unit area of Earth's surface. The air pressure at any specified altitude is equal to the weight per unit area of the atmosphere above that altitude.

- At any specified point within the atmosphere, air pressure has the same magnitude in all directions.

- A barometer is the weather instrument that monitors changes in air pressure. A mercury barometer is more accurate but far less portable than an aneroid barometer. The aneroid barometer can be calibrated to measure altitude.

- Air pressure and air density decrease rapidly with increasing altitude in the lower atmosphere and then more gradually aloft. The atmosphere has no clearly defined upper boundary; rather, Earth's atmosphere gradually merges with the highly rarefied mostly hydrogen/helium atmosphere of interplanetary space. About 50% of the mass of the atmosphere occurs below an altitude of 5.5 km (3.5 mi), and 99% below an altitude of 32 km (20 mi).

- Barometer readings are adjusted to sea level in order

to compare all air pressure readings at the same elevation, that is, to remove the influence of station elevation.

- Within the atmosphere, at constant pressure, air density is inversely proportional to air temperature. Hence, all other factors being equal, cold air masses are denser and exert higher pressure at Earth's surface than do warm air masses.

- Within the atmosphere, air density is also inversely proportional to water vapor concentration. Hence, at equivalent temperatures, dry air masses are denser than humid air masses. As a rule, temperature has a much more significant influence on air density and air pressure than does humidity.

- Air pressure may fluctuate in response to divergence or convergence of air, which is produced by changes in wind speed or direction.

- Important changes in weather often accompany relatively small changes in air pressure at the Earth's surface. As a rule, high or rising pressure signals fair weather, whereas low or falling pressure means stormy weather.

- Variables of state of the atmosphere (temperature, pressure, and density) are related through the gas law.

ESSAY: Human Responses to Changes in Air Pressure

Visitors to mountain elevations above about 2500 m (8000 ft) may develop symptoms of *mountain sickness*, that is, headache, shortness of breath, fatigue, insomnia, and nausea. Such illness is not uncommon—perhaps one of every four people who ascend to high elevations experiences one or more of these distressing symptoms. Some people who spend more than 36 to 72 hrs at elevations greater than about 2800 m (9000 ft) develop *high-altitude pulmonary edema.* Symptoms of this potentially life-threatening condition are somewhat similar to those of pneumonia, that is, severe cough, shortness of breath, lethargy, mild fever, and a buildup of fluid (edema) in the lungs. Onset of high-altitude pulmonary edema requires prompt medical attention.

Breathing air that does not supply adequate oxygen (O_2) can bring on altitude sickness. Thinning of air that accompanies the fall in air pressure with altitude also means a rapid decrease in the *number density* of the atmosphere's principal gases; that is, a decline in the number of nitrogen (N_2) and oxygen (O_2) molecules per unit volume of air. At altitudes of 2450 to 2750 m (8000 to 10,000 ft), oxygen's number density is only about 71% to 74% of its sea level value. Less oxygen per unit volume of air may lead to *hypoxia,* a deficiency of oxygen in the body. Symptoms of mild hypoxia are commonly known as mountain sickness, whereas symptoms of more severe (acute) hypoxia constitute high altitude pulmonary edema.

As a person ascends to high altitudes, the body attempts to compensate for the decline in O_2 by altering breathing. The breathing rate accelerates and breathing becomes deeper so that more oxygen enters the lungs per unit time. Accelerated and deeper breathing provides no relief from hypoxia unless oxygen that is delivered to the lungs is quickly transported to body tissues. Hence, the heart rate and stroke volume (the quantity of blood pumped out of the heart during each contraction) also increase. Although the quantity of oxygen per unit volume of blood remains essentially constant, the heart propels more blood (and hence, more oxygen) per unit time to body tissue.

The easiest way to alleviate acute mountain sickness is to descend to and remain at a lower altitude, especially at night. The altitude at which a person sleeps is more important than the altitude reached during the day. If a person remains at a high altitude, emergency adjustments by the heart and the muscles that control breathing reduce the symptoms of hypoxia while other longer-term changes take place in the body. Symptoms of acute mountain sickness gradually subside after several days as additional hemoglobin (molecules that pick up oxygen from air in the lungs) and red blood cells (cells that transport hemoglobin molecules in the bloodstream) are manufactured in the bone marrow. Although increased cardiac output generally tapers off within a week or so, accelerated breathing persists as long as a person remains at high altitude. The body thus adjusts to low oxygen levels that otherwise could cause serious health problems; this is an example of *acclimatization.*

A person can hasten acclimatization to high altitudes. Although a person may be eager to put in a full day of skiing or backpacking upon arrival in the mountains, he or she is well advised to minimize physical activity the first day or two. Here the old adage *the impatient becomes the patient* applies. Overexertion often results in more severe and persistent symptoms. Because alcohol aggravates mountain sickness, its use should be avoided, particularly during the first 48 hrs.

Some people are better able to acclimatize than others because of differences in their genetic makeup and general health. Nonetheless, there is a limit to acclimatization. Long-term residence at high altitude cannot fully restore the body's capacity to perform work to the level that is possible at sea level. At altitudes above about 5200 m (17,000 ft), without supplementary oxygen, nothing can be done to prevent a continual decline in all bodily functions. Hence, as described in this chapter's Case-in-Point, the summit of Mount Everest, which at 8850 m (29,035 ft) is the world's highest peak, is very close to (if not beyond) the limits of human survival.

Symptoms of hypoxia become more severe as the air thins; hence, an aircraft cabin must be pressurized at altitudes above 4600 m (15,000 ft) unless a supplemental oxygen supply is available to the flight crew and passengers. In actual practice, commercial aircraft cabins are pressurized beginning at takeoff and remain pressurized throughout the flight. A cabin is typically pressurized to about 75% of average sea-level air pressure. Although the aircraft cabin is pressurized, people commonly feel the effects of changing air pressure in a rapidly ascending or descending aircraft by a popping sensation in their ears. Rapid ascent in an elevator or in a car on mountain roads often produces the same sensation. Earpopping is symptomatic of a natural process that helps to protect the eardrum from damage.

The eardrum separates the outer ear from the middle ear chamber. As an aircraft takes off and cabin pressure drops, air pressure in the outer ear declines. As air pressure in the outer ear changes, the eardrum is distorted unless a compensating pressure change takes place in the middle ear. If the pressure does not equalize between the outer and middle ear, the eardrum bulges outward. On the other hand, when an aircraft descends and cabin pressure increases, air pressure in the outer ear increases. Without a compensating pressure change in the middle ear, the eardrum bulges inward. In both cases, deformation of the eardrum not only causes physical discomfort, but a bulging eardrum does not vibrate efficiently and sounds are muffled. If the air pressure difference between the middle ear and outer ear continues to increase, the eardrum could rupture, perhaps causing permanent hearing loss.

Fortunately, the body has a natural mechanism that regulates air pressure in the middle ear. The *eustachian tube* connects the middle ear to the upper throat region which, in turn, leads to the outside via the oral and nasal cavities. Normally, the eustachian tube is closed where it enters the throat, but it opens if a sufficient air pressure difference develops between the middle ear and the throat. When the eustachian tube opens, air pressure in the middle ear quickly comes into equilibrium with the external air pressure and the eardrum pops back to its normal shape. Vibrations of the eardrum associated with its rapid change in shape are what we hear as *ear-popping*, the body's way of preventing a permanent hearing loss when a person experiences a rapid change in air pressure. To reduce discomfort, yawning or swallowing hastens the opening of the eustachian tube. For this reason, travelers in aircraft or on mountain roads are advised to chew gum because the swallowing that accompanies gum chewing helps to open the eustachian tube.

ESSAY: Determining Altitude from Air Pressure

Altimetry is the determination of altitude above mean sea level. A *pressure altimeter* is an aneroid barometer that is graduated in increments of altitude. The graduation (that is, the calibration of altitude against air pressure) is prescribed by the *standard atmosphere*, which is described elsewhere in this chapter. At any time and place, however, the real atmosphere usually differs from the standard atmosphere so that an altimeter typically does not give the true altitude. The *indicated altitude* (the altimeter reading) is the same as the *true altitude* only when air pressure and temperature match the standard atmosphere. Unless adjustments are made, the discrepancy between indicated and true altitudes can pose serious problems, especially for aircraft during the crucial takeoff and landing phases of flight.

Changes in surface air pressure en route are one cause of differences between indicated and true altitudes. For example, as an aircraft travels toward a destination reporting a lower surface air pressure than its departure point, the altimeter will read higher than the true altitude. Hence, aircraft altimeters are equipped with a movable scale that enables the pilot to adjust altimeter readings. The Federal Aviation Administration (FAA) requires all aircraft flying below 5500 m (18,000 ft) to calibrate altimeters to surface air pressure radioed from flight service stations en route. (Above 5500 m, aircraft fly along an isobaric surface with the altimeter zeroed at the standard sea-level pressure of 1013.25 mb.)

In-flight adjustments of altimeters to surface conditions, however, do not correct for pressure variations that arise en route principally from temperature variations within the air column beneath the aircraft. Cold air is denser than warm air so that air pressure drops more rapidly with altitude in cold air than in warm air. Hence, within a column of cold air, a given air pressure occurs at a lower altitude than does the same air pressure in a column of warm air (see Figure). This means, for example, that as an aircraft flies into a column of air that is warmer than specified by the standard atmosphere, the altitude indicated by the altimeter will be lower than the true altitude. Conversely, the altimeter aboard an aircraft flying into air colder than specified by the standard atmosphere will read too high.

The danger, of course, is that an erroneous altimeter reading may impede a pilot's ability to clear an obstacle such as a mountain peak. In practice, this hazard can be greatly reduced by an onboard computer that measures air temperature at flight level and makes appropriate adjustments to the altimeter reading. Note that this correction is based on the mean temperature of the air column so that the error is reduced but not eliminated.

In summary, differences between altimeter readings and true altitude arise from en route changes in surface air pressure and/or average temperature in the air column beneath the aircraft. Even with adjustments in altimeter readings, pilots are well advised to follow the adage *"cold or low, look out below."* Hence, pilots should always select a flight altitude that will allow for a margin of safety, especially when flying over mountainous terrain or during conditions of restricted visibility.

An alternative to an air pressure-based altimeter is a *radio altimeter*, which is flown aboard most commercial aircraft. This instrument emits radio waves from the plane to the ground, where the signal is reflected back to the aircraft. Altitude is calibrated in terms of the time elapsed between emission and reception of the radio signal. The longer the signal takes, the greater is the indicated altitude of the aircraft.

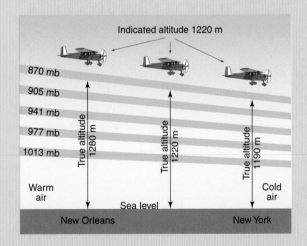

Figure
An aircraft pressure altimeter is initially calibrated based on the relationship between air pressure and altitude as specified by the standard atmosphere. But as the aircraft flies into colder or warmer air, the indicated altitude may differ from the true altitude because pressure drops more rapidly with altitude in cold air than in warm air.

CHAPTER 6

HUMIDITY, SATURATION, AND STABILITY

Imagination is more important than knowledge. For while knowledge defines all we currently know and understand, imagination points to all we might yet discover and create.

ALBERT EINSTEIN

Case-in-Point

Some tropical and subtropical forests are perpetually shrouded in clouds and mist. These so-called *cloud forests* play an important role in the local fresh water supply and are home to a wide variety of plants and animals, most of which have yet to be studied by scientists. However, pressures from land developers (e.g., for cattle grazing, logging, coca plantations) perhaps exacerbated by climate change make cloud forests one of the world's most threatened ecosystems.

Cloud forests are confined to mountainous areas of the tropics and subtropics generally at elevations from 2000 to 3500 m (6500 to 11,500 ft) above sea level.

According to the Tropical Montane Cloud Forest Initiative, there are 605 cloud forests in 41 nations. Most are in Central and South America from Honduras, Panama, and Costa Rica to as far south as northern Argentina. They are also found at lower elevations (as low as 500 m or 1600 ft) on humid portions of Caribbean and Hawaiian Islands.

A prevailing onshore and upslope flow of warm, humid air is responsible for the low clouds, fog, and mist that characterize cloud forests. The air is initially unsaturated but as it flows up the mountain slopes, it expands and cools. Expansional cooling raises its relative humidity to saturation (100%) and water vapor condenses into low clouds and fog. By trapping cloud droplets and promoting coalescence of tiny cloud droplets into larger drops, the tree canopy strips moisture from the windblown clouds and this water drips to the forest floor. The volume of water provided in this way is equivalent to 20% to 60% of the local rainfall. In this way, the cloud forest provides a steady source of water for mountain streams that in turn supply water to downstream cities and towns. Persistent cloud cover also shapes the cloud forest environment by reducing incoming solar radiation and suppressing evapotranspiration (the combination of direct evaporation and transpiration by plants.) Cloud forests provide unique habitats for lichens, tree firns, orchids, and large numbers of species of birds and mammals.

If global warming translates into higher sea surface temperatures (SSTs) in the tropics, the cloud forests could be impacted. Higher SSTs would have two opposing effects on the properties of the air flowing onshore and upslope. For one, that air would be warmer so that greater ascent (i.e., expansional cooling) would be required to produce clouds. Clouds would form at higher elevations and perhaps lift off mountaintops entirely. The original cloud forest would become drier, dramatically altering its habitats for plants and animals. However, higher SSTs would also increase the amount of evaporation of ocean water and result in a more humid onshore and upslope airflow. More humid air would condense into clouds at lower elevations. Using a numerical climate model to simulate the effects of higher SSTs on tropical airflow, Stephen Schneider and his colleagues at Stanford University recently concluded that higher air temperatures would be the more important factor in any displacement of cloud forests. Cloud forests appear to be particularly sensitive to climate variations and may be among the early indicators of the environmental effects of global-scale climate change.

Driving Question:

How is water cycled between Earth's surface and atmosphere?

Water occurs in all three phases in the atmosphere: as water vapor (an invisible gas) and as aggregates of tiny ice crystals and water droplets, visible as clouds. The total amount of water in the atmosphere is very small, and most of that is confined to the lower portion of the troposphere. If at any moment all water vapor were removed from the atmosphere as rain and distributed uniformly over the globe, this water would cover Earth's surface to a depth of only about 2.5 cm (1.0 in.). Water continually cycles into the atmosphere as vapor from reservoirs of water at the Earth's surface (e.g., ocean, lakes, soil, vegetation), and water continually leaves the atmosphere, returning to Earth's surface as rain, snow, and other forms of precipitation. On average, the residence time of a water molecule in the atmosphere is about 10 days. This cycling is an essential component of the global water cycle.

In this chapter, we consider how the global water cycle functions, particularly as it relates to the transfer of water between Earth's surface and atmosphere. We learn how to quantify the water vapor component of air, how air becomes saturated through uplift and expansional cooling, and how atmospheric stability influences ascent of air. All this is important because as air nears saturation, cloud development becomes more and more likely and clouds are required for formation of precipitation.

Global Water Cycle

We can reasonably assume that the total amount of water in the Earth-atmosphere system is neither increasing nor decreasing although natural processes continually generate and breakdown water. Water vapor accounts for perhaps half of all gases emitted during a volcanic eruption; at least some of this water was originally sequestered in magma and solid rock. Volcanic activity is more or less continuous on Earth and adds to the supply of water. Also, a minute amount of water is added to Earth by meteors and other extraterrestrial debris continually bombarding the upper atmosphere. At the same time, intense solar radiation entering the upper atmosphere, converts (*photodissociates*) a small amount of water vapor into its component hydrogen and oxygen atoms, which may escape to space. Also, water chemically reacts with other substances and is thereby locked up in various chemical compounds. Annually, additions of water during volcanic eruptions roughly equal losses of water through photodissociation of water vapor and chemical reactions. This balance of give and take has prevailed on Earth for perhaps millions of years.

The essentially fixed supply of water in the Earth-atmosphere system is distributed in its various phases among oceanic, terrestrial (land-based), atmospheric, and biospheric reservoirs (Table 6.1). The ocean is the largest of these reservoirs by far, accounting for 97.2% of all water on the planet; most of the rest is tied up as ice sheets up to 3 km (1.8 mi) thick that cover most of Antarctica and Greenland. Relatively small amounts of water occur in living organisms (plants and animals), rivers and lakes, and occupy the tiny pore spaces and fractures within soil, sediment, and bedrock (soil moisture and groundwater). Even smaller amounts of water occur in the atmosphere as clouds, precipitation, and invisible water vapor.

The ceaseless movement of water among the various reservoirs on a planetary scale is known as the **global water cycle** (Figure 6.1.) In brief, water vaporizes from the ocean and continents to the atmosphere where winds can transport water vapor thousands of kilometers. Clouds form and rain, snow and other forms of precipitation fall from clouds to Earth's surface, recharging the ocean and the terrestrial reservoirs of water. From terrestrial reservoirs, water

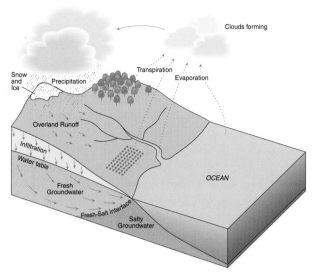

FIGURE 6.1
The global water cycle is a continuous flow of water and energy among oceanic, terrestrial, and atmospheric reservoirs.

flows to the ocean basins. The continuity of the global water cycle is captured in a verse from *Ecclesiastes 1:7*: "Every river flows into the sea, but the sea is not yet full. The waters return to where the rivers began and start all over again."

The sun drives the global water cycle. As we saw in Chapter 4, some of the radiation that strikes Earth's surface is absorbed, that is, converted to heat, and some of this heat is used to vaporize water. If water did not vaporize, there would be no clouds, no precipitation, and no global water cycle. Here we focus on a critical link in the water cycle—the link between the atmosphere and the oceanic and terrestrial reservoirs of water.

TRANSFER PROCESSES

As part of the global water cycle, water is transferred between Earth's surface and the atmosphere via phase changes (evaporation, condensation, transpiration, sublimation, and deposition) and precipitation. Within the usual range of air temperature and pressure on Earth, all three phases of water coexist naturally; that is, water vapor is in equilibrium with water's liquid and solid phases. At the interface between liquid water and air (e.g., lake or sea surface), water molecules continually change phase: some crossing the interface from water to air and others from air to water. If more water molecules enter the atmosphere as vapor than return as liquid, a net loss occurs in liquid water mass. This process is known as **evaporation**. On the other hand, if more water molecules return to the water

TABLE 6.1
Water Stored in Reservoirs of the Global Water Cycle

Reservoir	Percent of total water
Ocean	97.20
Ice sheets and glaciers	2.15
Groundwater	0.62
Lakes (freshwater)	0.009
Inland seas, saline lakes	0.008
Soil moisture	0.005
Atmosphere	0.001
Rivers and streams	0.0001

surface as liquid than enter the atmosphere as vapor, a net gain of liquid water mass results. This process is called **condensation**. Evaporation explains the drying of soil and disappearance of puddles following a rain shower. Water evaporates from the surface of the ocean, lakes, and rivers as well as from soil and the damp surfaces of plant leaves and stems. Evaporation of ocean water is the principal source of atmospheric water vapor. Condensation of atmospheric water vapor can be observed when small droplets form on the cold surface of a glass of ice tea on a humid summer day.

Transpiration is the process whereby water that is taken up by plant roots escapes as vapor through tiny pores on the surface of green leaves. On land during the growing season, transpiration is considerable and is often more important than direct evaporation of water in delivering water vapor to the atmosphere. For example, a single hectare (2.5 acres) of corn typically transpires about 34,000 liters (L) (8800 gal) of water per day. Annually, a mature oak tree may transpire more than 150,000 liters (40,000 gal) of water. Measurements of direct evaporation from Earth's surface plus transpiration are usually combined as **evapotranspiration**.

At the interface between ice and air (e.g., the surface of a snow cover), water molecules are also continually changing phase: directly from ice to vapor and directly from vapor to ice. If more water molecules enter the atmosphere as vapor than make the transition to ice, a net loss of ice mass occurs. **Sublimation** is the process whereby ice or snow becomes vapor without first becoming a liquid. Sublimation explains the gradual disappearance of snow and ice on sidewalks and roads even while the air temperature remains well below

freezing. On the other hand, if more atmospheric water molecules transition to ice than move from ice to vapor, a net gain of ice mass results. **Deposition** is the process whereby water vapor becomes ice without first becoming a liquid. During a cold winter night, the formation of frost on automobile windows is an example of deposition. Condensation or deposition operating within the atmosphere produces clouds.

Precipitation in the form of rain, drizzle, snow, ice pellets, and hail returns a major portion of atmospheric water from clouds to the Earth's surface, where most of it eventually vaporizes back into the atmosphere. Evaporation (or sublimation) followed by condensation (or deposition) purifies water. As water vaporizes from the Earth's surface, all suspended and dissolved substances like sea salts and other contaminants are left behind. Through this natural cleansing mechanism, salty ocean water is the source of much of what eventually falls as freshwater precipitation onto Earth's surface. Purification of water through phase changes is known as **distillation**.

GLOBAL WATER BUDGET

Comparing the movement of water into and out of terrestrial reservoirs with the movement of water into and out of the ocean is instructive. The balance sheet for the input and output of water to and from the various global reservoirs is called the **global water budget** (Table 6.2). Over the course of a year, the volume of precipitation (rain plus melted snow) that falls on the continents exceeds the total volume of water that vaporizes (via evaporation, transpiration, and sublimation) from the continents by about one-third.

TABLE 6.2
Global Water Budget

Source	Cubic meters per year	Gallons per year
Precipitation on the ocean	3.24×10^{14}	85.5×10^{15}
Evaporation from the ocean	-3.60×10^{14}	-95.2×10^{15}
Net loss from the ocean	-0.36×10^{14}	-9.7×10^{15}
Precipitation on land	0.98×10^{14}	26.1×10^{15}
Evapotranspiration from land	-0.62×10^{14}	-16.4×10^{15}
Net gain on land	0.36×10^{14}	9.7×10^{15}

Over the same period, the volume of precipitation falling on the ocean is less than the volume of water that evaporates from the ocean. Comparison of the evaporation and precipitation components of the global water budget indicates an annual net gain of water mass on the continents and an annual net loss of water mass from the ocean, with the excess on the continents about equal to the deficit from the ocean. But year after year, the continents are not getting any wetter and sea level is not falling appreciably because the excess water seeps and flows from the continents to the ocean, thereby completing the global water cycle.

Precipitation strikes the ground directly or it may be intercepted by vegetation and then evaporate or drip to the ground (Figure 6.2). Also, as described in this chapter's Case-in-Point, some trees collect moisture from drifting fog or low clouds and that water drips to the ground (known as *fog drip*). Once water reaches Earth's surface, it follows various pathways. Some water vaporizes back to the atmosphere and some either flows on the surface as rivers or streams (the *runoff component*) or percolates into the ground as soil moisture or groundwater (the *infiltration component*). Roughly one-

third of the annual precipitation that falls on the continents runs off to the ocean. The ratio of the portion of water that infiltrates the ground to the portion that runs off depends on rainfall intensity, vegetation, topography, and physical properties of the surface. Rain falling on frozen ground or city streets mostly runs off whereas rain falling on unfrozen sandy soil readily soaks into the ground.

Rivers and streams plus their tributaries drain a fixed geographical area known as a *drainage basin* (or *watershed*). The quantity and quality of water flowing in a river depends on the climate, vegetation, topography, geology, and land use in its drainage basin. For example, in places where the climate features distinct rainy and dry season, stream flow can vary considerably through the year. A drainage basin may also include lakes, wetlands, and glaciers, temporary impoundments of the runoff component of the global water cycle.

How Humid Is It?

"It's not the heat, it's the humidity." This popular statement attributes the discomfort we feel on a hot, muggy day to the water vapor content of the air. **Humidity** is a general term referring to any one of many different ways of describing the amount of water vapor in the air. Humidity is an important determinant of our physical comfort, as discussed in the Essay, "Humidity and Human Comfort." Experience tells us that humidity varies with the time of year, from one day to the next, within a single day, and from one place to another. In most places summer days feel more humid than a typical winter day. In cold regions, dry winter air also causes some discomfort. This section covers the various measures of humidity, including vapor pressure, mixing ratio, specific humidity, absolute humidity, relative humidity, dewpoint, and precipitable water.

VAPOR PRESSURE

When water enters the atmosphere as vapor, water molecules disperse in all directions and mix with the other gases composing air (mostly nitrogen and oxygen) and contribute to the total pressure exerted by the atmosphere. The amount of pressure produced by water vapor molecules is a measure of the humidity: the more water vapor, the greater the pressure exerted by water molecules. Water vapor's contribution to the total air pressure is described by Dalton's law. According to

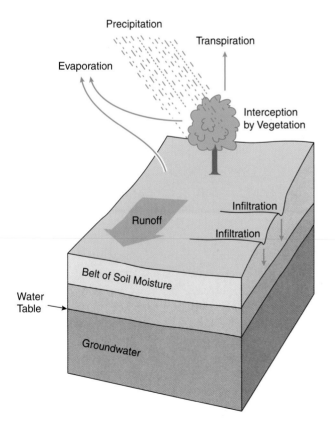

FIGURE 6.2
The various pathways taken by precipitation falling on land.

Dalton's law, the total pressure exerted by a mixture of gases equals the sum of the pressures produced by each constituent gas; that is, each gas species in the mixture acts independently of all the other molecules. Stated another way, each gas exerts a pressure as though it were the only gas present. The pressure exerted by water vapor alone is known as **vapor pressure**.

As the water vapor content of the atmosphere increases, its vapor pressure also increases. Water vapor is a highly variable component of air but composes at most no more than about 4% of the mass of the atmosphere's lowest kilometer—even in the sultry air over tropical ocean and rainforest. The total pressure exerted by all atmospheric gases at sea level is about 1000 mb. This means that the vapor pressure is very unlikely anywhere to exceed 40 mb (4% × 1000 mb = 40 mb) at sea level, and in most places the vapor pressure is considerably less than 40 mb.

MIXING RATIO, SPECIFIC HUMIDITY, AND ABSOLUTE HUMIDITY

Other ways of quantifying humidity are the mixing ratio, specific humidity, and absolute humidity. **Mixing ratio** is defined as the mass of water vapor per mass of the remaining dry air, usually expressed as so many grams of water vapor per kilogram of dry air. Typically the mixing ratio at sea level is less than 40 grams per kilogram. **Specific humidity** is defined as the ratio of the mass of water vapor (in grams) to the mass (in kilograms) of the air containing the water vapor (that is, the combined mass of dry air plus water vapor). Values of specific humidity and mixing ratio are so close that they usually can be considered equivalent.

Absolute humidity is defined as the mass of water vapor per unit volume of humid air, typically expressed as grams of water vapor per cubic meter of air. In other words, absolute humidity is the density of the water vapor component of air. Although conceptually simple, absolute humidity has limited application in meteorology because the volume of a parcel of air may change causing the absolute humidity to vary even though no water vapor is gained or lost by the parcel. For example, an air parcel's volume increases when it is heated or lifted in the atmosphere. Even though no water vapor enters or leaves the parcel, the absolute humidity decreases.

SATURATED AIR

Regardless of how humidity is quantified, at a specified temperature, the amount of water vapor in air has a practical upper limit. Air at its maximum humidity is described as *saturated* with respect to water vapor. Earlier in this chapter, we described a two-way exchange of water molecules at the interface between water and air (or between ice and air). Water molecules are in a continual state of flux between the liquid and vapor phases. During evaporation, more water molecules become vapor than return to the liquid phase, and during condensation, more water molecules return to the liquid phase than enter the vapor phase. Eventually, a dynamic equilibrium may develop such that the flux of water molecules is the same in both directions; that is, liquid water becomes vapor at the same rate that water vapor becomes liquid. At equilibrium, the air is saturated with water vapor and the vapor pressure is referred to as the **saturation vapor pressure**, the mixing ratio is the **saturation mixing ratio**, the specific humidity is the **saturation specific humidity**, and the absolute humidity is the **saturation absolute humidity**.

Changing the temperature disturbs this dynamic equilibrium at least temporarily. With a rise in water temperature, the average kinetic energy of individual water molecules increases and water molecules more readily escape the water surface as vapor. Initially, evaporation prevails. If the supply of water is sufficient and as long as water vapor is not continually carried away by winds, eventually a new dynamic equilibrium is established. That is, the flux of water molecules becoming liquid once again balances the flux of water molecules becoming vapor. This new equilibrium is achieved with more water vapor in the air at a higher temperature. Raising the temperature increases the saturation vapor pressure, saturation mixing ratio, saturation specific humidity, and saturation absolute humidity.

Conversely, with a drop in water temperature, the average kinetic energy of individual water molecules decreases and molecules less readily escape the water surface as vapor. Initially, condensation prevails but eventually a new equilibrium is established; that is, the flux of water molecules becoming vapor again balances the flux of water molecules becoming liquid. This new equilibrium is achieved with less water vapor in the air at a lower temperature. Dropping the temperature decreases the saturation vapor pressure, saturation mixing ratio, saturation specific humidity, and saturation absolute humidity.

Ultimately, the water vapor component of air depends on the rate of vaporization of water, which is regulated chiefly by temperature. The dependence of the

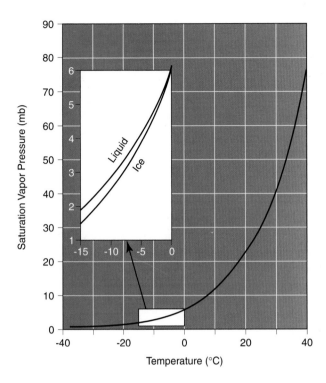

FIGURE 6.3
Variation in saturation vapor pressure with air temperature.

TABLE 6.3
Variation of Saturation Vapor Pressure with Temperature[a]

Temperature °C (°F)	Saturation Vapor Pressure (mb) Over water	Over ice
50 (122)	123.40	
45 (113)	95.86	
40 (104)	73.78	
35 (95)	56.24	
30 (86)	42.43	
25 (77)	31.67	
20 (68)	23.37	
15 (59)	17.04	
10 (50)	12.27	
5 (41)	8.72	
0 (32)	6.11	6.11
-5 (23)	4.21[a]	4.02[a]
-10 (14)	2.86	2.60
-15 (5)	1.91	1.65
-20 (-4)	1.25	1.03
-25 (-13)	0.80	0.63
-30 (-22)	0.51	0.38
-35 (-31)	0.31	0.22
-40 (-40)	0.19	0.13
-45 (-49)	0.11	0.07

[a]Note that for temperatures below freezing, two different values are given: one over supercooled water and the other over ice. Supercooled water remains liquid at subfreezing temperatures.

saturation vapor pressure on temperature is shown in Figure 6.3 and Table 6.3. The variation of the saturation mixing ratio with temperature is shown in Figure 6.4 and Table 6.4. Extensive laboratory work by 19th century steam engineers and scientists is the source of these data. For both measures of humidity, the relationship is non-linear but as a general rule, the saturation vapor pressure and saturation mixing ratio double in value for every 10 Celsius degree (20 Fahrenheit degree) rise in temperature. At 100 °C (212 °F) at sea level, the saturation vapor pressure is 1013.25 mb, the same as the standard sea level air pressure. Water boils when the saturation vapor pressure and ambient air pressure are equal.

The relationship between temperature and the saturation vapor pressure (or saturation mixing ratio) is popularly interpreted to mean that warm air can *hold* more water vapor than can cold air. That is, air is likened to a sponge that can soak up only so much water depending on the temperature. While intuitively appealing, this analogy can be misleading. Air does not literally hold water vapor like a sponge; rather, water vapor coexists with the other gases that form the mixture known as air. Recall from Dalton's law that each gas in a mixture of gases exerts a pressure as though it were the only gas present. Water vapor is just one of the many gases that compose air. To a large extent, temperature governs the rate of vaporization of water. From this perspective, the saturation vapor pressure (or saturation mixing ratio) is actually a measure of water's vaporization rate. At any specified temperature, the saturation vapor pressure (or saturation mixing ratio) would have the same value even if water vapor were the only gas in the atmosphere.

RELATIVE HUMIDITY

Relative humidity is the water vapor measure most often reported by television and radio weathercasters and is probably the most familiar for most of us. **Relative humidity** compares the actual amount of water vapor in the air with the amount of water vapor that would be present in the same air at saturation. Relative humidity (RH) is expressed as a percentage and can be computed from either the vapor pressure or mixing ratio.

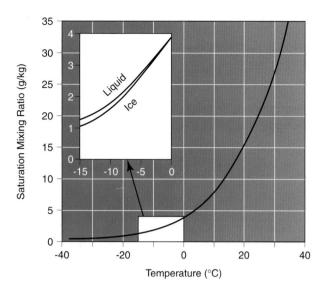

FIGURE 6.4
Variation in saturation mixing ratio with air temperature (at a pressure of 1000 mb).

TABLE 6.4
Variation of Saturation Mixing Ratio with Temperature (at a pressure of 1000 mb)[a]

Temperature °C (°F)	Saturation Mixing Ratio (g/kg)	
	Over water	Over ice
50 (122)	88.12	
45 (113)	66.33	
40 (104)	49.81	
35 (95)	37.25	
30 (86)	27.69	
25 (77)	20.44	
20 (68)	14.95	
15 (59)	10.83	
10 (50)	7.76	
5 (41)	5.50	
0 (32)	3.84	3.84
-5 (23)	2.64[a]	2.52[a]
-10 (14)	1.79	1.63
-15 (5)	1.20	1.03
-20 (-4)	0.78	0.65
-25 (-13)	0.50	0.40
-30 (-22)	0.32	0.24
-35 (-31)	0.20	0.14
-40 (-40)	0.12	0.08
-45 (-49)	0.07	0.05

[a]Note that for temperatures below freezing, two different values are given: one over supercooled water and the other over ice. Supercooled water remains liquid at subfreezing temperatures.

That is,

RH = [(vapor pressure)/(saturation vapor pressure)] × 100%

or

RH = [(mixing ratio)/(saturation mixing ratio)] × 100%

When the actual concentration of water vapor in air equals the water vapor concentration at saturation, the relative humidity is 100%; that is, the air is saturated with respect to water vapor.

Consider an example of how relative humidity is computed. Suppose that the air temperature is 10 °C (50 °F), and the vapor pressure is 6.1 mb. From Table 6.3, we determine that the saturation vapor pressure of air at 10 °C is 12.27 mb. Using the formula above, we compute the relative humidity to be 49.7%, that is,

RH = [(6.1 mb)/(12.27 mb)] × 100% = 49.7%

At constant temperature and pressure, the relative humidity varies directly with the vapor pressure (or mixing ratio); that is, the relative humidity increases as water vapor is added to air as long as the air temperature and pressure do not change. Because the saturation vapor pressure varies directly with temperature, the relative humidity varies inversely with temperature. If no water vapor is added to or removed from unsaturated air, the relative humidity increases as the temperature drops and decreases as the temperature rises. Consider a common example.

On a clear day with no wind, the air temperature usually rises from a minimum near sunrise to a maximum during early to mid-afternoon and then falls through the evening hours and overnight (Chapter 4). If the amount of water vapor in air remains constant throughout the day, then the relative humidity will vary inversely with air temperature. As shown in Figure 6.5, the relative humidity is highest when the air temperature is lowest and the relative humidity is lowest when the temperature is highest. After sunrise, as the air warms, the relative humidity drops, but not because the amount of water vapor is reduced. The relative humidity drops because the saturation vapor pressure increases as the air temperature rises.

In the previous example, the relative humidity responded only to variations in local air temperature; the air was calm so there was no cold or warm air advection. The situation is made more complicated by air mass advection, which can influence both local air temperature

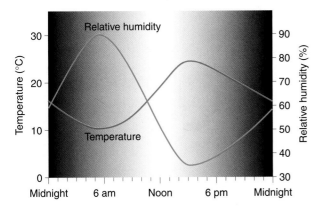

FIGURE 6.5
Variation in relative humidity and air temperature on a day when the air is calm and essentially no variation in vapor pressure. The relative humidity is highest when the temperature is lowest and the relative humidity is lowest when the temperature is highest.

and vapor pressure. For example, when a warm and humid air mass replaces a cool and dry air mass, both rising temperatures and increasing vapor pressure affect the relative humidity.

DEWPOINT

Dewpoint, often cited on television and radio weathercasts, is another useful measure of humidity. **Dewpoint** is the temperature to which air must be cooled at constant pressure to achieve saturation of air relative to liquid water. The higher the dewpoint, the greater is the concentration of water vapor in air. As unsaturated air is cooled at constant pressure, its relative humidity increases. When the relative humidity reaches 100%, the air is saturated and the air temperature is the same as the dewpoint. Conversely, the lower the relative humidity, the greater the difference between the actual air temperature and the dewpoint.

Dew consists of tiny droplets of water formed when water vapor condenses on a cold surface such as blades of grass on a clear, calm night. As discussed in Chapter 7, dew forms as a consequence of radiational cooling. (Dew that freezes after forming is known as *white dew* because it appears as white beads of ice.) Dew is not a form of precipitation because it does not fall from clouds to Earth's surface. For water vapor to condense as dew on the surface of an object, the temperature of that surface must cool *below* the dewpoint. When cooling at constant pressure produces saturation at an air temperature below freezing, water vapor deposits as **frost**. The air temperature at which frost forms is known

as the **frost point**. For water vapor to deposit on the surface of an object, the temperature of that surface must drop below the frost point.

Summer dewpoints are usually highest in the states bordering the Gulf of Mexico where the July mean dewpoint typically ranges between 21 °C and 24 °C (70 °F and 75 °F). During the oppressive heat waves that sometimes sweep over the continent east of the Rocky Mountains, the dewpoint may top the low 20s Celsius (low 70s Fahrenheit) even in the northern states. On 30 July 1999, during a particularly humid heat wave, the dewpoint at Milwaukee, WI tied a record high of 27.8 °C (82 °F). Summer dewpoints are lowest in the Rocky Mountain States and the American Southwest. From New Mexico northward into western Montana, the July mean dewpoint generally ranges between –1 °C and 7 °C (30 °F and 45 °F). The dewpoint (actually frost point) is exceptionally low in polar and arctic air masses that invade broad regions of the United States in winter.

PRECIPITABLE WATER

Another way of describing the amount of water vapor in the atmosphere is in terms of precipitable water. Unlike the other humidity measures that represent the amount of water vapor at a specific place (usually near the Earth's surface), **precipitable water** is the depth of water that would be produced if all the water vapor in a vertical column of air were condensed into liquid water. The air column is usually taken to extend from Earth's surface to the top of the troposphere, the portion of the atmosphere where most water vapor occurs. A reasonably good measure of precipitable water is obtained by radiosonde soundings.

Precipitable water is not always indicative of the amount of precipitation that might fall at a particular location. Numerous other factors, including advection of water vapor to or from the column along with recycling of water locally, ultimately determine the amount of precipitation.

Condensing all the water vapor in the atmosphere would produce a layer of water covering the entire Earth's surface to a depth of 2.5 cm (1.0 in.). Average precipitable water varies with latitude in response to the poleward decline in air temperature. This geographical distribution of precipitable water is usually evident on any given day as illustrated by Figure 6.6. Evaporation and precipitable water are less in cold regions. In fact, average annual precipitable water depths vary from more than 4.0 cm (1.6 in.) in the humid tropics to less than 0.5 cm (0.2 in.) in polar regions.

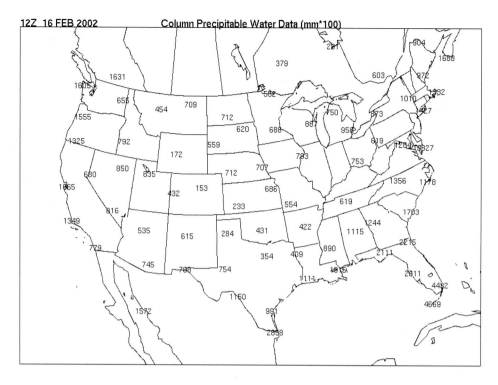

FIGURE 6.6
Precipitable water at various locations in the lower 48 states and southern Canada in units of millimeters × 100. Greatest values are in Florida and lowest values are in the far north.

Monitoring Water Vapor

A variety of techniques are used to monitor atmospheric water vapor both directly and remotely by satellite. For information on the measurement of evapotranspiration, see the Essay.

HUMIDITY INSTRUMENTS

Leonardo da Vinci, a creative genius of the 15th century, was probably the first to conceive of an instrument to gauge the water vapor content of air. His design was a simple balance. A small wad of dry cotton on one side of a scale exactly balanced a weight on the other side of the scale. As the cotton absorbed water vapor from the air, an imbalance developed and the amount of imbalance was a measure of humidity. Actual construction of this type of instrument is credited to Cardinal Nicholas of Cusa in about 1450.

The **hygrometer** is an instrument that measures the water vapor concentration of air; several different designs are available. The National Weather Service

most commonly employs a **dewpoint hygrometer**. With this instrument, air passes over the surface of a metallic mirror that is cooled electronically. An electronic sensor continually monitors the temperature of the mirror at the same time that an infrared beam is pointed at the mirror. When a thin condensation film forms on the mirror, the reflectivity of the infrared beam changes, and the mirror temperature is automatically recorded as the dewpoint. The mirror is then warmed electronically to evaporate the dew in preparation for the next measurement.

Some hygrometers take advantage of the sensitivity of organic materials to changes in humidity. One common design, called a **hair hygrometer**, uses human hair as the sensing element. Cells in the hair adsorb water and swell causing the hair to lengthen slightly as the relative humidity increases. Typically, hair changes length by about 2.5% over the full range of relative humidity from 0% to 100%. Usually, a sheaf of blond hair is linked mechanically to a pointer on a dial that is calibrated to read in percent relative humidity. A hair hygrometer may be designed to move a pen on a clock-driven drum. This instrument, called a **hygrograph**, gives a continuous record of fluctuations in relative humidity with time.

An **electronic hygrometer** is based on changes in the electrical resistance of certain chemicals as they adsorb water vapor from the air. The adsorbing element may be a thin carbon coating on a glass or plastic strip. The more humid the air, the more water adsorbed, and the lower is the resistance to an electric current passing through the sensing element. Variations in electrical resistance are calibrated in terms of percent relative humidity or dewpoint. An electronic hygrometer is flown aboard radiosondes (Chapter 2).

A psychrometer provides a less direct measure of relative humidity. A **psychrometer** consists of two identical liquid-in-glass thermometers mounted side by

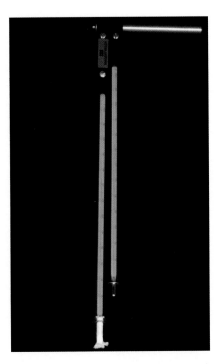

FIGURE 6.7
When whirled about, a sling psychrometer measures the actual air temperature (dry-bulb reading) and the wet-bulb temperature. The relative humidity is then determined from a psychrometric table (Table 6.5).

side with the bulb of one thermometer wrapped in a muslin wick. To take a reading, the wick-covered bulb is first soaked in water (preferably distilled) and then the instrument is ventilated. A *sling psychrometer* has an attached handle that allows it to be whirled about with ease (Figure 6.7), whereas a small fan is used to ventilate an *aspirated psychrometer*. The dry-bulb thermometer measures the actual air temperature. Water vaporizes from the muslin wick into the air streaming past the wet-bulb thermometer and evaporative cooling lowers the reading on this thermometer to the **wet-bulb temperature**. Ventilation continues until the reading on the wet-bulb thermometer steadies. The drier the air, the greater the evaporation, and the lower is the wet-bulb temperature compared to the dry-bulb temperature. The difference between the dry-bulb temperature and the wet-bulb temperature, known as the **wet-bulb depression**, is calibrated in terms of percent relative humidity on a psychrometric table (Table 6.5). This table was developed more than a century ago using a combination of theoretical considerations and extensive field experiments.

Consider an example of how relative humidity is determined using a psychrometer. Suppose that the air

temperature (dry-bulb reading) is 20 °C (68 °F) and the wet-bulb depression (dry-bulb reading minus wet-bulb reading) is 5 Celsius degrees (9 Fahrenheit degrees). From Table 6.5, the relative humidity is 58%. When there is no wet-bulb depression, the air is saturated; that is, the relative humidity is 100%.

A major drawback of a psychrometer is the difficulty of taking measurements at low and subfreezing temperatures. At low air temperatures, small differences in the wet-bulb depression correspond to greater increments in relative humidity than is the case at high temperatures. This situation requires a greater degree of precision in reading the instrument at low air temperatures. At freezing, ice forms on the wick inhibiting evaporative cooling. For this reason, an electronic or hair hygrometer is more desirable when the air temperature falls below 0 °C (32 °F). Furthermore, in very dry air the wick may dry out prior to reaching full depression value.

The dewpoint should not be confused with the wet-bulb temperature; they are not the same. The wet-bulb temperature is determined by inducing evaporative cooling. Adding water vapor to the air raises the temperature at which dew will form so that, except at saturation, the wet-bulb temperature is higher than the dewpoint. At saturation, the dewpoint, wet-bulb, and actual air temperatures are the same. The dewpoint also can be obtained from measurements of the dry-bulb temperature plus the wet-bulb depression (Table 6.6). For example, if the dry-bulb temperature is 20 °C (68 °F) and the wet-bulb depression is 5 Celsius degrees (9 Fahrenheit degrees), then the dewpoint is 11.6 °C (53 °F). In this case, the wet-bulb temperature is 15 °C (59 °F), some 3.4 Celsius degrees (6 Fahrenheit degrees) higher than the dewpoint.

WATER VAPOR SATELLITE IMAGERY

Water vapor satellite imagery is a valuable remote sensing product for tracking the broad scale distribution and movement of water vapor within the atmosphere (Figure 6.8). Water vapor is an invisible gas and does not appear on visible satellite images. But water vapor efficiently absorbs and emits certain wavelength ranges of infrared radiation and is readily detected by a special infrared sensor aboard weather satellites. The sensor that produces conventional infrared (thermal) satellite images is sensitive to infrared radiation emitted by the Earth-atmosphere system at wavelengths within atmospheric windows, that is, within wavelength bands from 10.5 to 12.6 micrometers. Recall from

TABLE 6.5
Psychrometric Table: Relative Humidity (Percent)

Dry-bulb Temp (°C)	Wet-bulb Depression (C°)														
	0.5	1.0	1.5	2.0	2.5	3.0	3.5	4.0	4.5	5.0	7.5	10.0	12.5	15.0	17.5
-10.0	85	69	54	39	24	10	—	—	—	—	—	—	—	—	—
-7.5	87	73	60	48	35	22	10	—	—	—	—	—	—	—	—
-5.0	88	77	66	54	43	32	21	11	1	—	—	—	—	—	—
-2.5	90	80	70	60	50	41	37	22	12	3	—	—	—	—	—
0.0	91	82	73	65	56	47	39	31	23	15	—	—	—	—	—
2.5	92	84	76	68	61	53	46	38	31	24	—	—	—	—	—
5.0	93	86	78	71	65	58	51	45	38	32	1	—	—	—	—
7.5	93	87	80	74	68	62	56	50	44	38	11	—	—	—	—
10.0	94	88	82	76	71	65	60	54	49	44	19	—	—	—	—
12.5	94	89	84	78	73	68	63	58	53	48	25	4	—	—	—
15.0	95	90	85	80	75	70	66	61	57	52	31	12	—	—	—
17.5	95	90	86	81	77	72	68	64	60	55	36	18	2	—	—
20.0	95	91	87	82	78	74	70	66	62	58	40	24	8	—	—
22.5	96	92	87	83	80	76	72	68	64	61	44	28	14	1	—
25.0	96	92	88	84	81	77	73	70	66	63	47	32	19	7	—
27.5	96	92	89	85	82	78	75	71	68	65	50	36	23	12	1
30.0	96	93	89	86	82	79	76	73	70	67	52	39	27	16	6
32.5	97	93	90	86	83	80	77	74	71	68	54	42	30	20	11
35.0	97	93	90	87	84	81	78	75	72	69	56	44	33	23	14
37.5	97	94	91	87	85	82	79	76	73	70	58	46	36	26	18
40.0	97	94	91	88	85	82	79	77	74	72	59	48	38	29	21

Chapter 3 that *atmospheric windows* refer to wavelengths of infrared radiation not absorbed by the atmosphere. On the other hand, the satellite water vapor sensor detects infrared radiation at a wavelength of 6.7 micrometers, which is strongly absorbed and emitted by water in the atmosphere.

Water vapor imagery displays only water vapor and clouds occurring at altitudes above about 3000 m (10,000 ft). Although this type of satellite imagery does not show water vapor (or clouds) at lower altitudes, it does detect otherwise transparent water vapor in the middle troposphere and can show moisture gathering in broad scale circulation patterns prior to the development of extensive cloudiness in an incipient storm system. Water vapor imagery utilizes a gray scale. At one extreme, black indicates little or no water vapor, whereas at the other extreme, milky white signals a relatively high concentration of water vapor. Clouds appear as bright white blotches.

Sequential water vapor images viewed in rapid succession to detect motion show water vapor being transported horizontally as huge swirling plumes, often originating over tropical seas. A typical **water vapor plume** can be several hundred kilometers wide and thousands of kilometers in length (Figure 6.9). Plumes supply moisture to hurricanes, thunderstorm clusters, and winter storms. Water vapor imagery also reveals something about the vertical (up and down) motion of air. Most water vapor originates at the surface of the Earth so that air is moving upward in humid regions (appearing milky white) and downward in dry regions (appearing black).

TABLE 6.6
Psychrometric Table: Dewpoint Temperature (°C)

Dry- bulb Temp (°C)	Wet-bulb Depression (C°)														
	0.5	1.0	1.5	2.0	2.5	3.0	3.5	4.0	4.5	5.0	7.5	10.0	12.5	15.0	17.5
-10.0	-12.1	-14.5	-17.5	-21.3	-26.6	-36.3	—	—	—	—	—	—	—	—	—
-7.5	-9.3	-11.4	-13.8	-16.7	-20.4	-25.5	-34.4	—	—	—	—	—	—	—	—
-5.0	-6.6	-8.4	-10.4	-12.8	-15.6	-19.0	-23.7	-31.3	-78.6	—	—	—	—	—	—
-2.5	-3.9	-5.5	-7.3	-9.2	-11.4	-14.1	-17.3	-21.5	-27.7	-41.3	—	—	—	—	—
0.0	-1.3	-2.7	-4.2	-5.9	-7.7	-9.8	-12.3	-15.2	-18.9	-23.9	—	—	—	—	—
2.5	1.3	0.1	-1.3	-2.7	-4.3	-6.1	-8.0	-10.3	-12.9	-16.1	—	—	—	—	—
5.0	3.9	2.8	1.6	0.3	-1.1	-2.6	-4.2	-6.1	-8.1	-10.4	-47.7	—	—	—	—
7.5	6.5	5.5	4.4	3.2	2.0	0.7	-0.8	-2.3	-4.0	-5.8	-21.6	—	—	—	—
10.0	9.1	8.1	7.1	6.0	4.9	3.8	2.5	1.2	-0.2	-1.8	-12.8	—	—	—	—
12.5	11.6	10.7	9.8	8.8	7.8	6.7	5.6	4.5	3.2	1.9	-6.8	-28.2	—	—	—
15.0	14.2	13.3	12.5	11.6	10.6	9.6	8.6	7.6	6.5	5.3	-1.9	-14.5	—	—	—
17.5	16.7	15.9	15.1	14.3	13.4	12.5	11.5	10.6	9.6	8.5	2.3	-7.0	-35.1	—	—
20.0	19.3	18.5	17.7	16.9	16.1	15.3	14.4	13.5	12.6	11.6	6.1	-1.4	-14.9	—	—
22.5	21.8	21.1	20.3	19.6	18.8	18.0	17.2	16.3	15.5	14.6	9.6	3.2	-6.3	-37.5	—
25.0	24.3	23.6	22.9	22.2	21.4	20.7	19.9	19.1	18.3	17.5	12.9	7.3	-0.2	-13.7	—
27.5	26.8	26.2	25.5	24.8	24.1	23.3	22.6	21.9	21.1	20.3	16.1	11.1	4.7	-4.7	-3.17
30.0	29.4	28.7	28.0	27.4	26.7	26.0	25.3	24.6	23.8	23.1	19.1	14.5	9.0	1.6	-11.1
32.5	31.9	31.2	30.6	29.9	29.3	28.6	27.9	27.2	26.5	25.8	22.1	17.8	12.8	6.6	-2.4
35.0	34.4	33.8	33.1	32.5	31.9	31.2	30.6	29.9	29.2	28.5	24.9	21.0	16.4	11.0	3.9
37.5	36.9	36.3	35.7	35.1	34.4	33.8	33.2	32.5	31.9	31.2	27.7	24.0	19.8	14.9	8.9
40.0	39.4	38.8	38.2	37.6	37.0	36.4	35.8	35.1	34.5	33.9	30.5	26.9	23.0	18.5	13.3

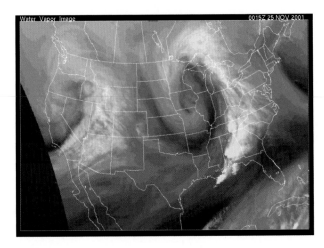

FIGURE 6.8
Water vapor imagery portrays moisture at altitudes above 3000 m (10,000 ft). Milky white indicates relatively high concentrations of water vapor; black indicates little or no water vapor; and bright white blotches are clouds.

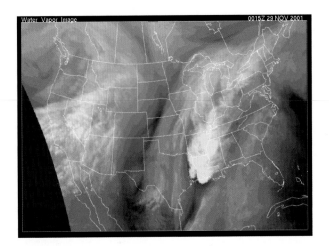

FIGURE 6.9
A water vapor image showing a water vapor plume extending from Texas northeastward to the Great Lakes region.

How Air Becomes Saturated

As the relative humidity nears 100%, condensation or deposition of water vapor becomes more and more likely. Condensation or deposition within the atmosphere produces clouds so the probability of cloud development increases as the relative humidity nears saturation. A **cloud** is a visible aggregate of tiny water droplets and/or ice crystals suspended in the atmosphere. (Under special circumstances, the relative humidity can rise slightly higher than 100% without water vapor changing phase; in that case air is *supersaturated*.)

What causes the relative humidity to increase? This is an important question because clouds are required for precipitation and without precipitation, the global water cycle would not exist (and neither would we). The relative humidity of unsaturated air increases: (1) When air is cooled; the saturation vapor pressure decreases while the actual vapor pressure remains constant. (2) When water vapor is added to the air at constant temperature; the vapor pressure increases while the saturation vapor pressure remains constant. In this chapter, we focus on the first process and cloud formation by expansional cooling.

EXPANSIONAL COOLING AND COMPRESSIONAL WARMING

Expansional cooling is the principal means whereby clouds form in the atmosphere. Whenever a gas (or mixture of gases) expands, the temperature of the gas drops; this process is known as **expansional cooling**. Expansional cooling explains why air released through the open valve of a bicycle tire is cool to the touch. Air pressure within the tire (perhaps 60 lbs per square in.) is much greater than the atmospheric pressure outside the tire (around 14 lbs per square in.) Air streaming through the tire valve and into the atmosphere does work as it pushes aside air that formerly occupied the volume into which it expands. Work requires energy, and the energy used in the work of expansion is drawn from the internal energy of the air so the temperature of the expanding air drops.

Air that ascends within the atmosphere also undergoes expansional cooling in response to falling air pressure. As noted in Chapter 5, air pressure is determined by the weight per unit area of the column of air above and so always decreases with increasing altitude. As an air parcel ascends in the atmosphere, it expands in the same way as a helium-filled balloon does as it drifts skyward. A rising air parcel expands and its temperature drops; if unsaturated, the air parcel's relative humidity increases.

Conversely, as the pressure on an air parcel increases, the air parcel is compressed and warms, a process known as **compressional warming**. A familiar example is the warming of the cylinder wall of a tire pump as air is pumped (compressed) into a tire. The work of compressing air is converted into heat. As an air parcel descends within the atmosphere, it is compressed, its temperature rises, and its relative humidity drops.

In a more general sense, imagine upward or downward currents of air as consisting of continuous streams of individual air parcels. The temperature behavior of an air current is the same as that of a component air parcel. In summary, ascending unsaturated currents of air cool and the relative humidity increases, and descending air currents warm and the relative humidity decreases.

ADIABATIC PROCESS AND LAPSE RATES

During a so-called **adiabatic process**, no heat is exchanged between a parcel and its environment. Within the atmosphere, expansional cooling and compressional warming of unsaturated air are essentially adiabatic processes. That is, while ascending or descending within the atmosphere, an unsaturated air parcel is neither heated nor cooled by radiation, conduction, phase changes of water, or mixing with the environment. The temperature of an ascending or descending unsaturated air parcel changes in response to expansion or compression only.

Adiabatic cooling of ascending unsaturated air amounts to 9.8 Celsius degrees per 1000 m (5.5 Fahrenheit degrees per 1000 ft) of ascent. This is the **dry adiabatic lapse rate** (Figure 6.10). The *dry* designation can be misleading; ascending and descending air parcels contain some water vapor. The dry adiabatic lapse rate applies to any parcel that is not saturated with water vapor (because no condensation or evaporation can take place). On the other hand, the atmosphere compresses descending air, and the air temperature rises 9.8 Celsius degrees for every 1000 m (5.5 Fahrenheit degrees per 1000 ft) of descent. Thus, the dry adiabatic lapse rate enables us to accurately predict the temperature change of unsaturated air as it moves vertically (up or down) within the atmosphere.

Should rising air parcels cool to the point that the relative humidity nears 100% and condensation or deposition takes place, the ascending air no longer cools

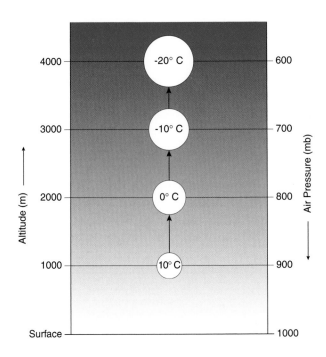

FIGURE 6.10
The dry adiabatic lapse rate describes the expansional cooling of ascending unsaturated air parcels. Here the dry adiabatic lapse rate is rounded off to 10 Celsius degrees per 1000 m.

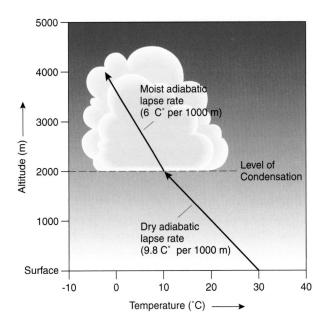

FIGURE 6.11
Ascending unsaturated (clear) air cools at the dry adiabatic lapse rate whereas ascending saturated (cloudy) air cools at the smaller moist adiabatic lapse rate.

at the dry adiabatic rate (Figure 6.11). Latent heat that is released to the environment during condensation or deposition *partially* counters expansional cooling. Consequently, an ascending saturated (cloudy) air parcel cools more slowly than an ascending unsaturated (clear) air parcel. Rising saturated air parcels cool at the **moist adiabatic lapse rate**.

Although the magnitude of the dry adiabatic lapse rate is constant, the moist adiabatic lapse rate varies with temperature. As noted earlier, the saturation vapor pressure (or saturation mixing ratio) increases systematically with rising temperature. That is, warm saturated air has a higher vapor pressure (or mixing ratio) than cool saturated air. For the same temperature change, condensation or deposition in warm saturated air releases more latent heat than condensation or deposition in cool saturated air. The greater the quantity of latent heat released, the more the expansional cooling is offset, and the smaller is the magnitude of the moist adiabatic lapse rate. The moist adiabatic lapse rate ranges from about 4 Celsius degrees per 1000 m (2.2 Fahrenheit degrees per 1000 ft) for very warm saturated air to very near 9.8 Celsius degrees per 1000 m (5.5 Fahrenheit degrees per 1000 ft) for very cold saturated air. For

convenience, we use a value of 6 Celsius degrees per 1000 m (3.3 Fahrenheit degrees per 1000 ft) for the moist adiabatic lapse rate.

Regardless of temperature, the relative humidity of ascending saturated (cloudy) air remains constant at about 100%. As saturated air expands and cools, however, its saturation mixing ratio declines and so does its actual mixing ratio. That is, some water vapor is converted to water droplets (condensation) or ice crystals (deposition). Consider an illustration. Saturated air at 10 °C (50 °F) has a mixing ratio of about 7.8 g/kg (Table 6.4). If that air is lifted 1000 m, it cools moist adiabatically to 4 °C (39 °F). At that temperature, the saturation mixing ratio is about 5.1 g/kg. Because the relative humidity remains at 100%, the mixing ratio must drop from 7.8 g/kg to about 5.1 g/kg. This means that 2.7 g/kg of water vapor condensed to cloud droplets. Some of these cloud droplets may grow into raindrops and fall out of the cloud (Chapter 7).

As long as an air parcel continues ascending and expanding, its temperature drops. Forces arising from density differences between the rising and surrounding air may either enhance or suppress vertical motion of air. The net effect depends on the stability of the atmosphere.

Atmospheric Stability

An air parcel is subject to buoyant forces that arise from density differences between the parcel and the surrounding (*ambient*) air. At constant pressure, the higher the temperature of an air parcel, the lower is its density. Parcels that are warmer (lighter) than the ambient air tend to rise, whereas parcels that are cooler (denser) than the ambient air tend to sink. An air parcel continues to rise or sink until it reaches ambient air of the same temperature (and density).

Meteorologists determine **atmospheric stability** by comparing the temperature change of an ascending or descending air parcel with the temperature profile, or *sounding*, of the ambient air layer in which the parcel ascends or descends. As we have seen, the cooling rate of a rising air parcel depends on whether the parcel is saturated (moist adiabatic lapse rate) or unsaturated (dry adiabatic lapse rate), whereas the warming rate of any descending air parcel is essentially 9.8 Celsius degrees per 1000 m. Recall from Chapter 2 that balloon-borne radiosondes routinely monitor the vertical temperature profile (sounding).

Within a **stable air layer**, an ascending air parcel becomes cooler (denser) than the ambient air, and a descending air parcel becomes warmer (less dense) than the ambient air (Figure 6.12). Any upward or downward displacement of an air parcel in stable air gives rise to forces that will tend to return the parcel to its original altitude. Within an **unstable air layer**, on the other hand, an ascending air parcel remains warmer (less dense) than the ambient air and continues to ascend, and a descending air parcel remains cooler (denser) than the ambient air and continues to descend (Figure 6.13).

SOUNDINGS

Although lapse rates for both unsaturated and saturated air parcels are essentially fixed, the temperature profile (sounding) of the ambient air changes significantly from season to season, from day to day, and sometimes even from one hour to the next. Changes in the sounding translate into changes in atmospheric stability. Soundings (and stability) can change as a consequence of (1) local radiational heating or cooling, (2) air mass advection, or (3) large-scale ascent or descent of air.

On a clear and calm night, radiational cooling chills the ground, which in turn, cools and stabilizes the overlying air. On the other hand, during the day, intense

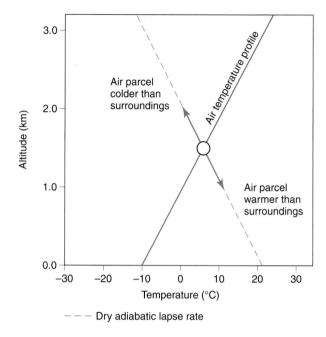

— — — Dry adiabatic lapse rate

FIGURE 6.12
Upward and downward displacements of an unsaturated air parcel within a stable air layer. The displaced parcel is subject to a force that restores it to its original altitude; that is, stable air inhibits vertical motion.

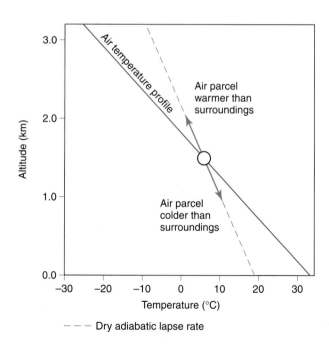

— — — Dry adiabatic lapse rate

FIGURE 6.13
Upward and downward displacements of an unsaturated air parcel within an unstable air layer. The displaced parcel accelerates away from its original altitude; that is, unstable air enhances vertical motion.

solar radiation heats the ground, which in turn, heats and destabilizes the overlying air. An air mass is stabilized as it travels over a colder surface (snow-covered ground, for example), and it is destabilized as it flows over a warmer surface. Whether by radiation or by advection, atmospheric stability changes because the temperature profile (i.e., the sounding) changes. Air is stabilized when cooling from below reduces the temperature lapse rate (the rate at which air temperature decreases with altitude), and air is destabilized when heating from below increases the temperature lapse rate.

Generally, an air layer becomes more stable when it descends (subsides) and less stable when it ascends. When an air layer subsides, the upper portion undergoes more compressional warming than does the lower portion so that the lapse rate within the air layer decreases and stability increases. On the other hand, the lower portion of an ascending air layer is usually more humid than the upper portion and so achieves saturation sooner. Consequent release of latent heat in the lower portion of the air layer increases the lapse rate and stability decreases. Further complicating matters, stability can change with altitude. For example, a stable air layer may be sandwiched between unstable air layers. Different types of air mass advection occurring at different altitudes within the troposphere are often responsible for such variations in stability.

Figure 6.14 summarizes atmospheric stability conditions for a variety of soundings, along with the dry adiabatic and average moist adiabatic lapse rates. If the sounding indicates that the temperature of the ambient air is dropping more rapidly with altitude than the dry adiabatic lapse rate (that is, more than 9.8 Celsius degrees per 1000 m), then the ambient air is unstable for both saturated and unsaturated air parcels. This situation is called **absolute instability**. If the sounding lies between the dry adiabatic and moist adiabatic lapse rates, **conditional stability** prevails. That is, the air layer is stable for unsaturated air parcels and unstable for saturated air parcels. This situation is relatively common. Conditional stability means that unsaturated air must be forced upward in order to reach saturation. But once saturation is achieved, the now cloudy air is buoyed upward; that is, the cloud builds vertically.

An air layer is stable for both saturated and unsaturated air parcels when the sounding indicates any of the following conditions: (1) The temperature of the ambient air drops more slowly with altitude than the moist adiabatic lapse rate. (2) The temperature does not change with altitude (*isothermal*). (3) The temperature

increases with altitude (*temperature inversion*). Any one of these three types of temperature profiles indicates **absolute stability**.

What happens when a sounding coincides with either the dry or moist adiabatic lapse rate? A sounding that equals the dry adiabatic lapse rate is neutral for unsaturated air parcels and unstable for saturated air parcels. A sounding that is the same as the moist adiabatic lapse rate is neutral for saturated air parcels and stable for unsaturated air parcels. Within a **neutral air** layer, a rising or descending air parcel always has the

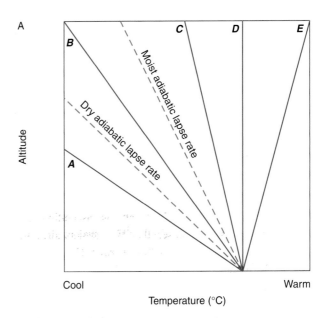

Sounding	Clear air	Cloudy air	Category	
A	Unstable	Unstable	Absolute instability	
B	Stable	Unstable	Conditional stability	
C	Stable	Stable	Lapse	Absolute stability
D	Stable	Stable	Isothermal	
E	Stable	Stable	Inversion	

FIGURE 6.14

Air stability is determined by comparing the temperature (density) of an ascending or descending air parcel with the temperature (density) of the surrounding air (sounding). (A) An air parcel at the Earth's surface initially has the same temperature as its surroundings. If the parcel is lifted, its temperature drops either dry adiabatically if it is unsaturated (clear air), or moist adiabatically if it is saturated (cloudy air). The temperature of the surrounding air is determined by radiosonde measurements. Depending on the specific sounding (A through E), the atmosphere is either stable or unstable. Stability conditions for each of the five sample soundings are summarized in B.

same temperature (and density) as its surroundings. Hence, a neutral air layer neither impedes nor spurs upward or downward motion of air parcels.

Atmospheric stability influences weather by affecting the vertical motion of air. Stable air suppresses vertical motion whereas unstable air enhances vertical motion, convection, expansional cooling, and cloud development. Because stability also affects the rate at which polluted air mixes with clean air, stability is an important factor in assessing air pollution potential. We examine this issue in the Essay, Atmospheric Stability and Air Quality.

STÜVE DIAGRAM

Soundings and adiabatic processes in the atmosphere can be displayed graphically on a thermodynamic diagram. Although several types are available, in this course we use the **Stüve thermodynamic diagram**.

The Stüve diagram is designed to permit evaluation of various atmospheric processes using graphical techniques (Figure 6.15). Coordinates on this diagram consist of temperature (horizontal axis), increasing from left to right, and pressure (vertical axis),

decreasing upward. Also marked on the diagram are altitudes that correspond to pressure levels, based on the standard atmosphere (Chapter 5). Dry adiabatic lapse rates are represented by a family of straight red lines slanting from lower right toward upper left. Often these lines are simply called *dry adiabats*. An ascending unsaturated air parcel follows a dry adiabat (either along one of the drawn adiabats or one that parallels the printed lines) from its initial temperature and pressure to its final temperature and pressure. Consider an example.

Suppose that the temperature of an unsaturated air parcel is 20 °C (68 °F) and the air pressure is 1000 mb. These initial conditions plot as point A on the Stüve diagram in Figure 6.16. Now simulate expansional cooling by lifting the air parcel to the 850-mb level by moving along a dry adiabat. The temperature of the parcel at 850 mb is about 8 °C (46 °F) (point B in Figure 6.16). Now continue along the dry adiabat to a pressure of 700 mb where the air parcel temperature is about –8 °C (18 °F) (point C in Figure 6.16). Finally, move back along the same dry adiabat to 1000 mb to simulate descent of the air parcel and compressional warming. Because it warms at the same rate it cooled, the temperature of the parcel is now at its original value of 20 °C (68 °F). In this case, the parcel remained unsaturated; water vapor did not change phase so that latent heat was not a factor in altering the parcel's temperature.

Moist adiabatic lapse rates appear on a Stüve diagram as a set of gently curving dashed lines, also known as *moist adiabats*. Moist adiabats become nearly indistinguishable from dry adiabats at very low pressure and low temperature, conditions where there is very little water vapor and latent heat release. Suppose that a parcel of saturated air at the base of a cloud has a temperature of –8 °C (18 °F) at a pressure of 700 mb (point A in Figure 6.17).

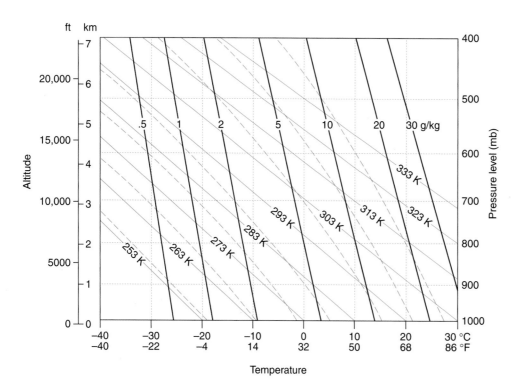

FIGURE 6.15
A Stüve thermodynamic diagram.

FIGURE 6.16
On a Stüve thermodynamic diagram, an unsaturated air parcel at point A is subject to a dry adiabatic expansion to point B (850 mb) and then to point C (700 mb).

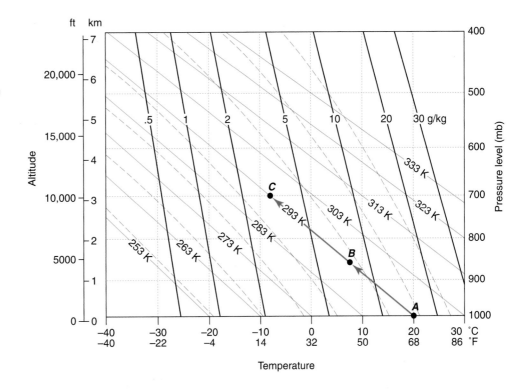

FIGURE 6.17
On a Stüve thermodynamic diagram, a saturated air parcel at point A (700 mb) is subject to a moist adiabatic expansion to point B (500 mb). The parcel is then subject to a dry adiabatic compression to point C (700 mb).

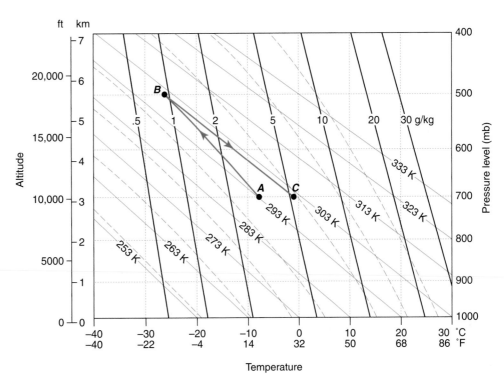

As the parcel ascends along a moist adiabat to 500 mb, its temperature drops to –27 °C (-17 °F) (point B in Figure 6.17). Assume that the water that condensed during the ascent immediately fell from the parcel (precipitated). If the parcel then descends back to the 700-mb level, it follows a dry adiabat and is compressionally warmed to -1 °C (30 °F) (point C in Figure 6.17). The parcel is now some 7 Celsius degrees (12 Fahrenheit degrees) warmer than it was when it started its ascent because of the release of latent heat during its ascent.

Plotting a sounding on a Stüve diagram enables meteorologists to determine the stability of various layers within the atmosphere and assess the potential for deep convection and cloud development. Meteorologists also use plotted values of saturation mixing ratio (slanted black lines labeled in gm per kg) to determine relative humidity and saturation levels of rising air parcels.

Lifting Processes

Ascending unsaturated air cools and its relative humidity increases. With sufficient ascent and expansional cooling, the relative humidity nears 100% and condensation or deposition begins; that is, clouds form. What causes air to rise? Air rises (1) as the ascending branch of a convection current, (2) along the surface of a front, (3) up the slopes of a hill or mountain, or (4) as the consequence of converging surface winds.

As we saw in Chapter 4, sensible heating (conduction and convection) is an important means of heat transfer between Earth's surface and atmosphere. The sun heats Earth's surface, which then heats the overlying air by conduction. The heated air expands and its density decreases. Cool, dense air sinks toward Earth's surface and displaces the less dense heated air. The heated air rises, expands, and cools. Cumulus clouds may form where convection currents ascend and the sky is generally cloud-free where convection currents descend (Figure 6.18). The higher the altitude reached by ascending convection currents, the greater the amount of expansional cooling, and the more likely it is that clouds (and precipitation) will form. Convection currents that soar to great altitudes within the troposphere typically spawn thunderstorms (*cumulonimbus clouds*).

Clouds and precipitation are often triggered by frontal uplift, which occurs where contrasting air masses meet. Recall that a **front** is a narrow zone of transition

between two air masses that differ in temperature and/or humidity. A warm and humid air mass is less dense than a cold and dry air mass and hence, as a cold air mass retreats, the warm air advances by riding up and over the cold air (Figure 1.4A). The leading edge of the advancing warm air at Earth's surface is known as a **warm front**. In contrast, cold and dry air displaces warm and humid air by sliding under it and forcing the

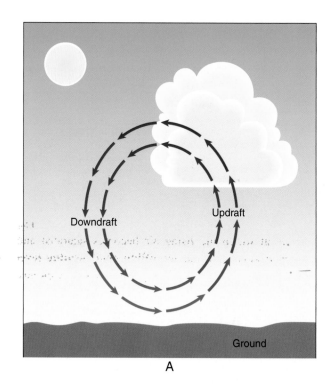

A

B

FIGURE 6.18
A convection current consists of (A) an updraft and a downdraft; (B) cumuliform clouds form where air ascends and the sky is cloud-free where the air descends.

warm air upward (Figure 1.4B). The leading edge of advancing cold air at Earth's surface is known as a **cold front**. The net effect of the replacement of one air mass by another air mass is the lifting and expansional cooling of air, cloud development, and perhaps rain or snow. Hence, clouds and precipitation are often (but not always) associated with fronts. We have much more to say about fronts and associated weather in Chapter 10.

Orographic lifting occurs where air is forced upward by topography, the physical relief of the land. Horizontal winds sweeping across the landscape alternately ascend hills and descend into valleys. If relief is sufficiently great, the resulting expansional cooling and compressional warming of air affects cloud and precipitation development.

A mountain range that intercepts the prevailing winds forms a natural barrier that is responsible for a cloudier, wetter climate on one side of the range than on the other side (Figure 6.19). Air that is forced to ascend the *windward* slopes (facing the oncoming wind) expands and cools, which increases its relative humidity. With sufficient cooling, clouds and precipitation develop. The altitude at which the rising air becomes saturated and clouds form is known as the **lifting condensation level (LCL)** and is marked usually by the base of the clouds. Meanwhile, air that descends the *leeward* slopes (downwind side) is compressed and warms, raising its saturation vapor pressure. Existing clouds vaporize and its relative humidity decreases so that precipitation is less likely. In this way, mountain ranges induce contrasting climates: moist climates on the windward slopes and dry

climates on the leeward slopes. Dry conditions often extend many hundreds of kilometers downwind of a prominent mountain range; this region is known as a **rain shadow**. The Rocky Mountain rain shadow, for example, extends eastward to about the 100th meridian (100 degrees W longitude), traditionally considered the boundary between irrigated agriculture to the west and rain-fed agriculture to the east.

An orographically-linked contrast in precipitation is particularly apparent in the Pacific Northwest, where the north-south trending Coastal and Cascade Mountain Ranges intercept the prevailing west-to-east flow of humid air from off the Pacific Ocean. Exceptionally rainy conditions prevail in western Washington and Oregon, whereas semiarid conditions characterize much of the eastern portion of those states. For example, the average annual precipitation (rain plus melted snow) at Astoria, OR on the Pacific coast is 172 cm (67.7 in.) but only 20.8 cm (8.2 in.) at Yakima, WA in the rain shadow of the Cascade Mountains. This climate contrast affects the indigenous plant and animal communities, domestic water supply, demand for irrigation water, types of crops that can be grown, and requirements for human shelter.

The influence of topography on precipitation patterns is also dramatic on the mountainous islands of Hawaii, where a few volcanic peaks top 4000 m (13,000 ft). Over the ocean waters surrounding the islands, estimated mean annual rainfall is uniformly between about 560 mm (22 in.) and 700 mm (28 in.). On the islands, however, mean annual precipitation ranges from only 190 mm (7.5 in.) leeward of the Kohala Mountains on the Big Island of Hawaii to 11,990 mm (39.3 ft) on the windward slopes of Mount Waialeale on the island of Kauai.

Another mechanism that leads to uplift is convergence of surface winds. Associated upward motion means expansional cooling, increasing relative humidity, and eventually cloud and perhaps precipitation formation. In later chapters, we describe weather systems in which convergence of surface winds plays an important role in triggering stormy weather. For example, converging surface winds are largely responsible for cloudiness and precipitation in a low-pressure system (cyclone). Also, converging of sea breezes is a major factor in the relatively high frequency of thunderstorms in central Florida.

Fronts, topography, or converging winds can force stable unsaturated air upward to the point that clouds develop. If the air is conditionally stable, clouds

FIGURE 6.19
Winds ascend along the windward slopes of a mountain range and descend along the leeward slopes. The climate is wetter along the windward slopes than the leeward slopes.

will surge upward and precipitation may be heavy. On the other hand, if air is absolutely stable, clouds will show little vertical development and precipitation is likely to be light if it occurs at all.

Conclusions

The atmosphere is one of many reservoirs of water in the global water cycle. Water vapor enters the atmosphere from Earth's surface via evaporation of liquid water, transpiration by plants, and sublimation of ice and snow. Water moves from the atmosphere to Earth's surface via condensation (dew), deposition (frost), and precipitation. The amount of water vapor in air (humidity) is expressed as vapor pressure, mixing ratio, specific humidity, absolute humidity, relative humidity, dewpoint, or precipitable water. The saturation vapor pressure and saturation mixing ratio increase with rising air temperature so that there is more water vapor in saturated warm air than in saturated cold air. On the other hand, the relative humidity of unsaturated air varies inversely with temperature.

Expansional cooling is the principal means whereby air becomes saturated with respect to water and clouds form. Ascending air expands and cools, its relative humidity nears 100%, and water vapor condenses (or deposits) as aggregates of tiny water droplets (or ice crystals) that are visible as clouds. Conversely, descending air is compressed and warms, its relative humidity drops, and clouds vaporize or fail to develop in the first place.

Ascending air parcels cool at the dry adiabatic lapse rate if unsaturated and at the moist adiabatic lapse rate if saturated. Atmospheric stability influences vertical motion in the atmosphere such that a stable air layer suppresses ascent of air and cloud development, whereas an unstable air layer enhances ascent of air and favors cloud development. Atmospheric stability is determined by comparing the cooling rate of ascending air parcels with the temperature profile (sounding) of the ambient air through which air parcels ascend. Air ascends in convection currents, along fronts, up the slopes of mountain ranges, and in response to converging surface winds.

We continue our study of moisture in the atmosphere in the next chapter with a focus on cloud and precipitation forming processes.

Basic Understandings

- The global water cycle is the ceaseless circulation of a fixed quantity of water among Earth's oceanic, atmospheric, and terrestrial reservoirs. As part of this cycle, water and heat energy move between Earth's surface and the atmosphere via evaporation, condensation, transpiration, sublimation, deposition, and precipitation.
- The global water budget indicates an annual surplus of water on the continents and a deficit of water at sea, implying a net flow of water from land to sea. Precipitation falling on land can vaporize, infiltrate the ground, be taken up by plant roots and transpired to the atmosphere, and/or run off into rivers or streams.
- Vapor pressure, mixing ratio, specific humidity, and absolute humidity are direct measures of the water vapor component of air. Relative humidity compares the actual amount of water vapor in air to the amount of water vapor at saturation.
- Saturation vapor pressure and saturation mixing ratio signal equilibrium in the flux of water molecules entering and leaving the vapor phase. Both parameters increase with rising temperature because temperature largely regulates the rate of vaporization of water.
- When the water vapor concentration of air equals the water vapor concentration at saturation, the relative humidity is 100%. For unsaturated air, the relative humidity varies inversely with air temperature. On a calm day, the relative humidity is highest when the air temperature is lowest (near sunrise), and the relative humidity is lowest when the air temperature is highest (early to mid-afternoon).
- Dewpoint is the temperature to which air is cooled at constant pressure to achieve saturation. Dew consists of tiny droplets of water that condense onto cold surfaces from air that has been chilled to its dewpoint. When cooling at constant pressure produces saturation at air temperatures below freezing, water vapor deposits on cold surfaces as frost. The air temperature at which frost forms is the frost point.
- Precipitable water is the depth of water that would be produced if all the water vapor in a vertical column of air, extending from Earth's surface to the tropopause, were condensed into liquid water. Generally, the amount of precipitable water decreases with increasing latitude in response to

lower average temperatures and reduced evaporation.

- Hygrometers are instruments that directly measure the water vapor concentration of air. They include dewpoint hygrometers, hair hygrometers, and electronic hygrometers. A somewhat less direct measure of humidity is provided by a psychrometer consisting of dry-bulb and wet-bulb thermometers mounted side-by-side. The greater the wet-bulb temperature depression, the lower is the relative humidity. When the wet-bulb and dry-bulb temperatures are equal, the relative humidity is 100%.

- Water vapor imagery portrays water vapor (and clouds) at altitudes above about 3000 m (10,000 ft) and is derived from infrared radiation emitted by atmospheric water at a wavelength of 6.7 micrometers. Water vapor plumes are particularly valuable for tracking the long-distance flow of moisture within the atmosphere.

- The relative humidity of unsaturated air increases when the air is cooled (lowering the saturation vapor pressure) or when water vapor is added to the air (increasing the vapor pressure) at constant temperature. As the relative humidity nears 100%, clouds are increasingly likely to form.

- Expansional cooling and compressional warming of unsaturated air parcels are essentially adiabatic processes. That is, no net heat exchange occurs between an air parcel and its surroundings as that parcel ascends or descends within the atmosphere.

- Meteorologists determine atmospheric stability by comparing the temperature (or density) of an air parcel moving vertically (up or down) within the atmosphere with the temperature (or density) of the surrounding (ambient) air. A radiosonde determines the variation of ambient air temperature with altitude (a sounding).

- Stable air inhibits vertical motion, convection, and cloud formation. Unstable air enhances vertical motion, convection, and cloud formation. Local radiational heating or cooling, air mass advection, or large-scale ascent or subsidence can change the stability of an air mass.

- Expansional cooling and cloud development occur through uplift of air in convection currents, along fronts or mountain slopes, or where surface winds converge.

- Clouds form in the updraft of convection currents and dissipate or fail to develop in the downdraft.

- The ascent of warmer, less dense air along a front may give rise to clouds and precipitation.

- Winds ascend along the windward slopes of a mountain range and descend along the leeward slopes. For this reason, the windward slopes tend to be cloudier and wetter than the leeward slopes. A rain shadow may extend hundreds of kilometers downwind of a prominent mountain range.

ESSAY: Humidity and Human Comfort

The relationship between the vapor pressure of ambient air and evaporative cooling helps explain why people usually are more uncomfortable when the weather is both hot and humid. At air temperatures above about 25 °C (77 °F), perspiring and consequent evaporative cooling enhance heat loss from the body. But if the air is also humid, the rate of evaporation is reduced, hampering the body's ability to maintain a constant core temperature of 37 °C (98.6 °F). The body's *core* refers to vital organs such as the heart and lungs. In contrast, drier air facilitates evaporative cooling. At air temperatures above about 25 °C (77 °F), most people feel more comfortable when the air is dry rather than humid.

A combination of high temperature and high humidity adversely affects everyone to some extent. People generally become more irritable and are less able to perform physical and mental tasks. During hot and humid weather, the efficiency of factory workers declines, and students do not concentrate as well on their studies. Extreme heat and humidity can be a lethal combination when these weather conditions overtax the body's capacity to maintain a constant core temperature.

In mid-July 1995, high air temperature combined with unusually high relative humidity pushed the so-called *heat index* to record high levels over the Midwest. Chicago was particularly hard hit. The Chicago Board of Health set the number of excess deaths (presumably because of the heat) over the average mortality for the month of July at 733. Most victims of heat and humidity were elderly people who lived alone or were bedridden. Although the 1995 Chicago heat wave was an extreme weather event with an unusually high mortality, National Weather Service scientists estimate that episodes of extreme heat (usually accompanied by high humidity) on average take the lives of almost 150 people annually in the U.S. (based on records from 1985 to 2000).

Scientists have experimented with a variety of indexes that attempt to gauge the combined influence of temperature and humidity on human wellbeing and advise people of the potential hazards of heat stress. Since the summer of 1984, the National Weather Service has regularly reported the *heat index,* or *apparent temperature index,* which was developed by R.G. Steadman in 1979.

The heat index accounts for the increasing inability of the human body to dissipate heat to the environment as the relative humidity rises. As illustrated in Table 1, at an air temperature of 90 °F (32 °C) and a relative humidity of 60%, the body loses heat to the environment at the same rate as if the air temperature were 105 °F (41 °C) with a relative humidity of 10%. In both cases, the apparent temperature is 100 °F (38 °C). Examination of Table 1 reveals that various combinations of air temperature and relative humidity produce the same apparent temperature, implying that all these temperature/relative-humidity combinations have the same impact on human comfort and wellbeing.

TABLE 1
Apparent Temperature Index

Air Temp. (°F)	0	5	10	15	20	25	30	35	40	45	50	55	60	65	70	75	80	85	90	95	100
									Relative Humidity (%)												
120	107	111	116	123	130	139	148														
115	103	107	111	115	120	127	135	143	151												
110	99	102	105	108	112	117	123	130	137	143	150										
105	95	97	100	102	105	109	113	118	123	129	135	142	149								
100	91	93	95	97	99	101	104	107	110	115	120	126	132	138	144						
95	87	88	90	91	93	94	96	98	101	104	107	110	114	119	124	130	136				
90	83	84	85	86	87	88	90	91	93	95	96	98	100	102	106	109	113	117	122		
85	76	79	80	81	82	83	84	85	86	87	88	89	90	91	93	95	97	99	102	105	108
80	73	74	75	76	77	77	78	79	79	80	81	81	82	83	85	86	86	87	88	89	91
75	69	69	70	71	72	72	73	73	74	74	75	75	76	76	77	77	78	78	79	79	80
70	64	64	65	65	66	66	67	67	68	68	69	69	70	70	70	70	71	71	71	71	72

The heat index is divided into four categories based on the severity of potential health impacts; these categories and associated symptoms of heat stress are listed in Table 2. Apparent temperatures in category IV pose the least hazard. When the apparent temperature is between 80 °F and 90 °F (27 °C and 32 °C), physical activity is more fatiguing than usual and caution is advised. At the opposite end of the scale, apparent temperatures in category I pose the greatest health risk. When the apparent temperature is 130 °F (54 °C) or higher, the danger is extreme and heatstroke (*hyperthermia*) may be imminent. During the heat wave of 10-16 July 1995, the highest heat index values were in the range of 115 °F to 120 °F (46 °C to 49 °C) in parts of Kansas, Iowa, Wisconsin, Illinois, and lower Michigan.

TABLE 2
Hazards Posed by Heat Stress by Range of Apparent Temperature[a]

Category	Apparent temperature	Heat syndrome
I	130 °F (54 °C) or higher	Extreme danger; heat stroke imminent.
II	105 to 130 °F (41 to 54 °C)	Danger; heat cramps or heat exhaustion likely. Heat stroke possible with prolonged exposure and physical activity.
III	90 to 105 °F (32 to 41 °C)	Heat cramps or heat exhaustion possible with prolonged exposure and physical activity.
IV	80 to 90 °F (27 to 32 °C)	Fatigue possible with prolonged exposure and physical activity.

Source: NOAA, National Weather Service
[a]Apparent temperature combines the effects of heat and humidity on human comfort.

The significance of a specific apparent temperature for personal wellbeing depends upon (1) the assumptions that form the basis of Steadman's index, and (2) age, general health, and body characteristics. Steadman's heat index assumes a person at rest in the shade with light winds. Personal experience tells us that exercise and exposure to the direct rays of the sun add considerably to the heat load that the body must dissipate to maintain a normal core temperature. For example, just standing in the sun causes the apparent temperature to climb as much as 8 Celsius degrees (14 Fahrenheit degrees). Furthermore, if the ambient air temperature is higher than skin temperature, the wind will convey more heat to the body than away from the body. Any combination of physical activity, exposure to direct sunlight, or hot winds elevates the danger of high temperature and high humidity more than suggested by the apparent temperature.

Where do people experience the most uncomfortable (and potentially life-threatening) summer weather? In view of the adverse impact of high humidity on evaporative cooling, many of us would probably point to locales bordering the Gulf of Mexico where high temperature and high humidity are the norm. For example, during a summer heat wave, a typical temperature/relative humidity combination in New Orleans is 90 °F (32 °C)/60%. According to Tables 1 and 2, these conditions translate into an apparent temperature of 100 °F (38 °C), a category III hazard. During such weather conditions, most people would experience some discomfort and everyone is well advised to closely monitor his or her level of physical activity.

Perhaps surprisingly, North America's most stressful summer weather typically does not occur near the Gulf of Mexico, but instead in the deserts of Arizona and California. Although the relative humidity is considerably lower, much higher ambient air temperatures translate into greater heat stress than along the Gulf Coast. For example, residents of Phoenix, AZ experience summer afternoon temperatures that often approach 110 °F (43 °C) with a relative humidity of 20%. This combination of heat and humidity gives an apparent temperature of 112 °F (44 °C), a category II hazard. In general, residents of the eastern half of North America associate greater discomfort with the muggy days of summer whereas residents of the American Southwest ascribe their days of greatest discomfort to the very hot, albeit dry, conditions of the desert in summer.

Low humidity also influences human comfort. As the relative humidity declines, the vapor pressure gradient between the skin and the surrounding air rises, and the evaporative cooling rate increases. The lower the relative humidity, the greater the vapor pressure gradient, and consequently, the higher the rate of heat loss from the body via evaporative cooling. As shown in Table 1, for a specified ambient air temperature and successively lower values of relative humidity, the difference between apparent temperature and ambient air temperature decreases. Below some threshold value of relative humidity, the apparent temperature is actually lower than the ambient air temperature. At low relative humidity, people experience the apparent temperature as lower than the ambient air temperature because the higher rate of evaporation causes an excessive loss of heat. That is, enhanced evaporative cooling of the skin at low relative humidity gives the sensation that the air is cooler than it actually is.

ESSAY: Measuring Evapotranspiration

The simplest and least expensive instrument for measuring evapotranspiration is an *evaporation pan*, a large cylindrical metal pan filled with fresh water and equipped with a water height gauge and floating thermometer (that monitors temperature at the air/water interface). The standard National Weather Service evaporation pan measures 121.9 cm (48 in.) in diameter and is 25.4 cm (10 in.) deep (see Figure). Changes in water level are recorded for as long as temperatures remain high enough that the water surface stays ice-free. The change in water level from one day to the next (adjusted for any rainfall) is recorded as evaporation.

A more sophisticated instrument for estimating evapotranspiration is the lysimeter. A *lysimeter* consists of an enclosed block of soil with a natural vegetation cover (usually grass) that is seated on a scale. Continuous measurements are made of rainfall and water percolating through the soil. Day-to-day changes in the weight of the soil block that are not accounted for by rainfall, runoff, or percolation are attributed to evapotranspiration.

A lysimeter can also be used to measure potential evapotranspiration, a useful parameter in arid and semiarid localities. *Potential evapotranspiration* refers to the maximum possible water loss (to vapor) given the available heat energy (required for latent heating). Potential evapo-transpiration is measured in the same way as

FIGURE
Standard National Weather Service evaporation pan.

soil/vegetation surface of the lysimeter is continually irrigated. Under these conditions, evapotranspiration proceeds at the maximum possible rate. Lysimeters are even sunk into a snow pack to monitor sublimation and evaporation (of meltwater).

Numerical models are used to compute evapotranspiration based on direct measurements or estimates of selected environmental factors that influence the rate of vaporization of water. Evapotranspiration increases with rising temperature, increasing wind speed, and as the difference in vapor pressure between a water surface and the overlying air increases. Transpiration also varies with dewpoint, intensity of incident solar radiation, length of daylight, type of vegetation, stage in a plant's life cycle, and soil moisture.

ESSAY: Atmospheric Stability and Air Quality

On the morning of 26 October 1948, a fog blanket reeking of pungent sulfur dioxide (SO_2) fumes spread over the town of Donora in Pennsylvania's Monongahela Valley. Before the fog lifted five days later, almost half of the area's 14,000 inhabitants had fallen ill and 20 had died. That killer fog resulted from a combination of mountainous topography and weather conditions that trapped and concentrated deadly effluents from the community's steel mill, zinc smelter, and sulfuric acid plant.

Air pollutants are especially dangerous when atmospheric conditions inhibit their dilution. Once pollutants enter the atmosphere through smokestacks or exhaust pipes, their concentrations usually begin to decline as they mix with cleaner air. The more thorough the mixing, the more rapid is the rate of dilution. When conditions in the atmosphere favor rapid dilution, the impact of air pollution is usually minor. On other occasions, called *air pollution episodes*, conditions in the atmosphere minimize dilution, and the impact can be severe, especially on human health. The two weather conditions that most influence the rate of dilution are wind speed and atmospheric stability.

Intuitively, we know that air is likely to mix more thoroughly on a windy day than on a calm one. When it is windy, turbulent eddies accelerate dilution by mixing polluted air and cleaner air. But when the air is calm, dilution is by the much slower process of *molecular diffusion*, that is, dispersal at the molecular level. As a general rule, doubling the wind speed cuts the concentration of air pollutants in half (Figure 1).

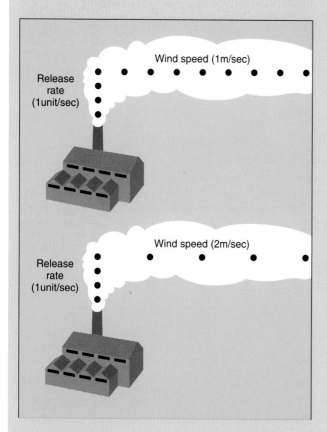

FIGURE 1
Doubling the wind speed from 1 m per sec to 2 m per sec increases the spacing between puffs of smoke by a factor of two, thereby reducing pollutant concentrations by one half.

Some weather patterns favor light winds and thus inhibit dispersal of contaminants. Over a broad region about the center of an anticyclone (a *high*), for example, winds are light or the air is calm, and pollutants do not disperse readily. On the other hand, within a cyclone (a *low*), stronger winds mean more rapid dilution of air pollution. In addition, rain or snow associated with a cyclone cleans the air by washing pollutants to the ground.

The stability of air is a second atmospheric factor that influences the rate at which polluted air mixes with clean air. Stability affects vertical motion within the atmosphere so that convection and turbulence are enhanced when the air is unstable and inhibited when the air is stable. A parcel of polluted air emitted into unstable air undergoes more mixing than does the same parcel of polluted air emitted into stable air. Stable air inhibits the vertical transport of air pollutants, and a layer of stable air aloft may act as a lid over the lower troposphere, trapping air pollutants near Earth's surface. Steady emission of contaminants into stable air results in the accumulation of pollutants.

Mixing depth is the vertical distance between Earth's surface and the altitude to which convection currents (that is, mixing) reach. When mixing depths are great (many kilometers, for example), the relative abundance of clean air allows pollutants to readily mix and dilution is enhanced. When mixing depths are shallow (less than 1000 m, for example), air pollutants are restricted to a smaller volume of air, and concentrations may build to unhealthy levels. When air is stable, convection and turbulence are suppressed and mixing depths are shallow. When air is unstable, convection and turbulence are enhanced and mixing depths increase. Because solar heating triggers convection, mixing depths tend to be greater in the afternoon than in the morning, during the day than at night, and in summer than in winter.

We can sometimes estimate the stability of air layers by observing the behavior of a plume of smoke belching from a smokestack. If the smoke enters an unstable air layer, the plume undulates. In general, this plume behavior indicates that polluted air is mixing readily with the surrounding cleaner air and being diluted. The net effect is improved air quality (except where the plume temporarily loops to the ground). On the other hand, a plume of smoke that flattens and spreads slowly downwind indicates very stable conditions and minimal dilution.

An air pollution episode is most likely when a persistent *temperature inversion* develops. Within an air layer characterized by a temperature inversion, air temperature increases with altitude; that is, warmer, lighter air overlies cooler, denser air. (A temperature inversion is the *inverse* of the usual temperature profile of the troposphere.) This extremely stable stratification strongly inhibits mixing and dilution of pollutants. A temperature inversion can form by (1) subsidence (sinking) of air, (2) extreme radiational cooling, or (3) some cases of air mass advection. The inversion may occur aloft or near the surface.

A *subsidence temperature inversion* forms a lid over a broad area, often encompassing several states at once (Figure 2). It develops during a period of fair weather when a warm anticyclone stalls. An anticyclone produces subsiding air that is warmed by compression. Subsiding warm air is prevented from reaching Earth's surface by the *mixing layer*, the portion of the troposphere in which convection thoroughly mixes the air. The air temperature within the mixing layer drops with altitude, but air just above the mixing layer, having been warmed by sinking and compression, is warmer than air at the top of the mixing layer. A temperature inversion marks a transition between the mixing layer below and the compressionally warmed air above. During these atmospheric conditions, air pollutants are distributed throughout the mixing layer, but no higher than the temperature inversion. Pollutants are confined to a relatively small volume of air, and continual emission will elevate concentrations.

A *radiational temperature inversion* is usually more localized than a subsidence temperature inversion (Figure 3). With clear night skies, radiational cooling chills both the ground and the air above the ground. However, the ground is a better emitter of infrared radiation than is the overlying air, so the ground surface cools faster than the air above it. Heat is conducted from the warmer overlying air to the cooler ground.

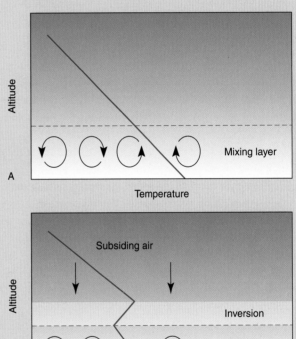

FIGURE 2

An elevated temperature inversion can develop through subsidence of air. Compare a sounding prior to subsidence (A) with a sounding during subsidence (B). The temperature inversion acts as lid over the lower atmosphere, trapping air pollutants.

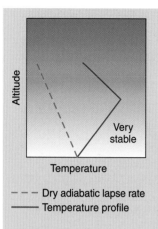

FIGURE 3
A low-level temperature
inversion caused by
radiational cooling.

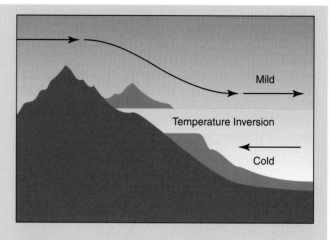

FIGURE 4
A temperature inversion develops aloft adjacent to the leeward
slopes of the Rocky Mountains. A west-to-east air flow
descends the leeward slopes, is warmed by compression, and
overlies a layer of colder air advected southward by north winds.

Chilling of the air from below means that the coldest air is next to the ground and the air temperature rises with increasing altitude; that is, a low-level temperature inversion forms. This process requires light winds or calm conditions because stronger winds would mix the air and prevent the stratification of warm air over cold air.

A radiational temperature inversion usually disappears within a few hours after sunrise as the sun heats the ground and the wind strengthens. The ground heats the overlying air and eventually reestablishes a normal sounding in which air temperature decreases with altitude. In winter, however, where the sun's rays are relatively weak and a highly reflective layer of snow covers the ground, a radiational temperature inversion may persist for days or even weeks.

Advection of air masses can also give rise to temperature inversions and affect air quality. This sometimes happens at the base of the Rocky Mountains (Figure 4). A west wind is compressionally warmed as it descends the leeward slopes of the mountain range but along the foot of the mountains surface winds from the north advect a shallow layer of cold air southward. Hence, a temperature inversion aloft separates the warm air above from the surface layer of cold air. Although temperature inversions also characterize warm and cold fronts, these inversions have little adverse effect on air quality because fronts are moving and the accompanying precipitation washes pollutants from the air.

CHAPTER 7

CLOUDS, PRECIPITATION, AND WEATHER RADAR

I am the daughter of earth and water,
And the nursling of the sky:
I pass through the pores of the ocean and shores;
I change, but I cannot die.

PERCY BYSSHE SHELLEY, The Cloud

Case-in-Point

A familiar sight in the daytime sky almost anywhere is *contrails*, bright white lines of ice crystals that form in the exhaust of jet aircraft (Figure 7.1). Contrails are modifying the cloud cover along heavily traveled air corridors between major urban areas with possible implications for weather and climate. According to some studies, contrails already cover about 5% of the sky in some portions of the eastern United States. A contrail, short for "condensation trail," develops when the hot humid air in jet engine exhaust mixes with the cold, drier air at high altitudes. Turbulence in the exhaust causes the mixing. A similar process on a cold winter day enables us to "see our breath." For details on this process, see the Essay, "Clouds by Mixing."

Depending on atmospheric conditions, contrails may dissipate (sublimate) within minutes or hours or they may spread laterally forming wispy cirrus clouds that persist for a day or longer. Increased cloud cover caused by contrails can affect the local radiation balance in two ways. Contrails and their cirrus products reflect incoming solar radiation thereby cooling Earth's surface. They also absorb and emit terrestrial infrared radiation thereby enhancing the greenhouse effect and elevating Earth's surface temperature. Except during times of low sun angle (just after sunrise and before sunset), the warming effect dominates. Furthermore, contrails may stimulate precipitation locally by being a source of ice crystal nuclei for lower clouds.

Model studies indicate that so far warming due to contrails amounts to only a tiny fraction (perhaps 1%)

of the magnitude of warming projected to accompany the build-up of greenhouse gases. However, with the anticipated increase in global aircraft traffic, some scientists expect contrail-related warming to be more significant in the future (perhaps up to 0.1 Celsius degree) with the average contrail sky coverage expected to more than double by 2050.

Driving Question:

How do clouds and precipitation form?

Clouds whisk across the sky in ever-changing patterns of white and gray; fog lends an eerie silence to a dreary day. Clouds and fog are products of condensation or deposition of water vapor within the atmosphere. Most clouds are the consequence of saturation brought about by uplift and expansional cooling of air. Fog is a cloud that forms near or in contact with Earth's surface and most fogs develop when the lowest layers of air are chilled to saturation via radiational or advective cooling.

Cloud formation is an essential part of the global water cycle because without clouds there would be no rain or snow. And yet most clouds—even most of those associated with large storm systems—do not produce precipitation. The focus of this chapter is the development and classification of clouds and fog, and the formation and types of precipitation. We also examine weather radar, a valuable tool for locating and tracking areas of precipitation and for monitoring air motion within weather systems. This chapter closes with a summary of the various direct and remote methods for measuring precipitation.

Cloud Formation

Water vapor is an invisible component of air, but its condensation and deposition products (water droplets and ice crystals) are visible. A **cloud** is the visible product of condensation or deposition of water vapor within the atmosphere; it consists of an aggregate of minute water droplets and/or ice crystals suspended in the atmosphere. This section covers the process of cloud development.

Clouds are increasingly likely to form as air

FIGURE 7.1
Contrails form when the hot, humid exhaust from jet engines mixes with the cold, drier ambient air.

nears saturation. But when scientists try to simulate this process in the laboratory using a clean-air chamber, a cloud does not form even if the relative humidity is elevated well above 100%, that is, to *supersaturation* values. A clean-air chamber is an enclosed container of air from which all solid and liquid particles (aerosols) have been filtered out. Introducing water vapor into the chamber raises the relative humidity but even if the relative humidity is elevated to as high as 200%, no cloud forms. How is that possible?

We expect that once air becomes saturated (relative to a flat-water surface), condensation would initially produce extremely small droplets and those droplets would eventually grow into cloud droplets through additional condensation. However, droplets will not form in the first place unless the relative humidity rises to extraordinarily high levels of supersaturation.

THE CURVATURE EFFECT

Tiny droplets cannot form or exist without a supersaturated environment because of the so-called **curvature effect**: The curvature of a water surface affects the ability of water molecules to escape (vaporize) from a water surface. The smaller a droplet, the greater the concentration of surrounding water vapor that is necessary for the droplet to grow. The curvature of the surface of a spherical water droplet increases as the radius of the droplet decreases; that is, the surface of a small droplet has a greater curvature than the surface of a large droplet. (At the limit, a flat-water surface has no curvature.) With increasing curvature, molecules that form the surface of a droplet have fewer neighboring molecules and are more weakly bonded. Molecules composing a flat-water surface have the most neighboring molecules and are most strongly bonded. Bond strength affects the flux of water molecules from the liquid to vapor phase. For this reason, water molecules more readily escape a water droplet than a flat-water surface and water molecules more readily escape a small water droplet than a large water droplet. Consequently, at the same temperature, the saturation vapor pressure surrounding a small water droplet is greater than that surrounding a large droplet.

The values of saturation vapor pressure listed in Table 6.3 are defined for air over a flat surface of fresh water and are the basis for computing relative humidity (Chapter 6). The saturation vapor pressure increases with increasing curvature (and decreasing droplet size). For droplets having a radius of 0.001 micrometer, the saturation vapor pressure at a given temperature is about 3.4 times that specified in Table 6.3. This computes to a relative humidity of 340%. On the other hand, for droplets having a radius greater than 1.0 micrometer, the difference between the saturation vapor pressure surrounding the droplet and the values listed in Table 6.3 is minimal and computes to a relative humidity only slightly higher than 100%.

The bottom line: In clean air, water vapor will not condense into tiny water droplets unless the air is supersaturated to very high levels. In the real atmosphere, the relative humidity never even approaches such great supersaturation values, so tiny droplets cannot form or grow. What then makes it possible for clouds to form?

ROLE OF NUCLEI

The clean-air chamber experiment described above failed to produce a cloud because the air was not sufficiently supersaturated. In the real atmosphere, supersaturated air is not needed for clouds to develop; in fact, cloud droplets readily form as the relative humidity nears 100%. And clouds are abundant; at any given time, about 60% of the planet is shrouded in clouds. How then do cloud droplets form? The simple answer is that they have a head start. Suspended in air are solid and liquid particles that provide relatively large surface areas on which condensation or deposition initially takes place. These naturally occurring particles that promote condensation or deposition in the atmosphere are known as **nuclei**. Nuclei have radii greater than 1.0 micrometer and favor condensation (or deposition) of water vapor. Once condensation (or deposition) starts, they become comparably sized cloud droplets (or ice crystals). Additional growth of these droplets (or ice crystals) by condensation (or deposition) is more likely because they can grow in an environment with a relative humidity that exceeds 100% by only 1% or so. Most cloud particles have radii of 2 to 10 micrometers.

Nuclei, essential for cloud formation, are abundant in the atmosphere and are continually cycled into the atmosphere from Earth's surface. Sources of nuclei include volcanic eruptions, wind erosion of soil, forest fires, and ocean spray. When sea waves break, drops of salt water enter the atmosphere and evaporate leaving behind tiny sea-salt crystals that function as nuclei. In addition, emissions from domestic and industrial chimneys contribute nuclei to the atmosphere.

Depending on the product (liquid water droplets or ice crystals), a distinction is made between cloud condensation nuclei and ice-forming nuclei. **Cloud** condensation nuclei (CCN) promote condensation of water vapor at temperatures both above and below the freezing point of water. Within the atmosphere, water vapor can condense into cloud droplets that remain liquid even at temperatures well below 0 °C (32 °F). Droplets at such temperatures are described as *supercooled* (discussed below). **Ice-forming nuclei (IN)** are much less common than CCNs and promote formation of ice crystals only at temperatures well below freezing. The most efficient ice-forming nuclei are substances having a crystal structure similar to that of ice and most are almost insoluble in water. Clayey soil particles are excellent ice-forming nuclei. The two types of ice-forming nuclei are freezing nuclei and deposition nuclei. *Freezing nuclei* are particles on which water vapor condenses and subsequently freezes; they are active at temperatures below about –9 °C (16 °F). *Deposition nuclei* are particles on which water vapor deposits directly as ice (without first becoming a liquid) and are not fully active until the temperature drops below about –20 °C (–4 °F).

Hygroscopic nuclei are a special category of cloud condensation nuclei because they possess a chemical attraction for water molecules. Condensation begins on hygroscopic nuclei at relative humidities under 100%. Magnesium chloride ($MgCl_2$), a salt in sea-spray, can promote condensation at a relative humidity as low as 70%! Clouds form more efficiently where hygroscopic nuclei are abundant. Urban-industrial areas have many sources of hygroscopic nuclei and this helps to explain why localities downwind of large cities typically are somewhat cloudier and rainier than upwind localities. Findings from the *Metropolitan Meteorological Experiment (METROMEX)* conducted in the 1970s indicated that average summer rainfall was about 25% greater downwind of St. Louis, MO than upwind of the city. Besides being a source of hygroscopic nuclei, cities also spur cloud and precipitation development by contributing water vapor (raising the relative humidity) and heat (adding to the buoyancy of air).

SUPERCOOLED WATER

About 1783, the Swiss naturalist Horace de Saussure discovered that fresh water could be cooled below 0 °C (32 °F) and remain liquid. Sixty-seven years later while on a balloon ascent the French chemist Jean Barrel confirmed the existence of supercooled cloud droplets. Cloud condensation nuclei promote condensation of water vapor at temperatures both above and below 0 °C. How can cloud droplets stay liquid even at temperatures well below freezing?

Freezing of fresh water in the absence of already existing ice begins with the linking of water molecules to form a tiny ice structure known as an *ice embryo*. Once an ice embryo grows larger than a certain critical size, its growth continues and freezing begins. Within the very small droplets that compose clouds, kinetic molecular activity is sufficient to inhibit growth of ice embryos to a critical size and water droplets remain liquid even at temperatures well below freezing. As the droplet temperature falls, however, the average kinetic molecular activity diminishes, the number of ice embryos increases, and the probability of an ice embryo growing to critical size increases. The formation of ice embryos of critical size due to the chance aggregation of water molecules is known as *homogeneous* (or *spontaneous*) *nucleation*. If no foreign particles are present that would act as ice-forming nuclei, a cloud droplet can cool to as low as –39 °C (–38.2 °F) without freezing. Any additional cooling, however, always leads to homogeneous nucleation. This threshold temperature is called the *Schaefer point*, named for Vincent J. Schaefer, a pioneer in cloud physics research at the General Electric Research Laboratory in Schenectady, NY.

If a supercooled water droplet contains (or comes in contact with) foreign particles that are ice-forming nuclei, the cloud droplet will freeze at a temperature below freezing but well above the Schaefer point. Water molecules collect on the surface of a foreign particle and form an ice embryo. The ice embryo is the same size as the particle and may be close to critical size so that any additional growth of the embryo causes the droplet to freeze. This process is referred to as *heterogeneous nucleation*.

Classification of Clouds

Clouds occur in a variety of sizes and shapes, something that is obvious from even a cursory glance at the sky. Characteristics that are common to different clouds form the basis for their classification. Cloud classification is desirable because different types of clouds tell us something about atmospheric processes responsible for their formation and may provide clues as to future weather. British naturalist Luke Howard was among the first to devise a classification of cloud types. Formulated in 1802-1803, the essentials of Howard's scheme are still used today.

TABLE 7.1
Summary of Cloud Classification Schemes

		Altitude	
Temperature	*Composition*	*Stratiform*	*Cumuliform*
Cold	ice crystals	high	clouds with
	supercooled droplets	middle	vertical
Warm	water droplets	low	development

Clouds are classified on the basis of (1) general appearance, (2) altitude of cloud base, (3) temperature, or (4) composition (Table 7.1). Based on a cloud's general appearance, the simplest distinction is among cirriform, stratiform, and cumuliform clouds. A **cirriform cloud** is wispy or fibrous, a **stratiform cloud** is layered, and a **cumuliform cloud** is heaped or puffy. On the basis of altitude of cloud base, a distinction is made among high, middle, and low clouds, and clouds having significant vertical development. High, middle, and low clouds generally occur in layers and are produced by gentle

TABLE 7.2
International Cloud Classification

Cloud group	*Altitude of base (m)*[a]
High clouds	
Cirrus (Ci)	5000-13,000
Cirrostratus (Cs)	5000-13,000
Cirrocumulus (Cc)	5000-13,000
Middle clouds	
Altostratus (As)	2000-7000
Altocumulus (Ac)	2000-7000
Low clouds	
Stratus (St)	surface-2000
Stratocumulus (Sc)	surface-2000
Nimbostratus (Ns)	surface-2000
Clouds with vertical development	
Cumulus (Cu)	to 3000
Cumulonimbus (Cb)	to 3000

[a]Average for middle latitudes.

uplift of air (typically less than 5 cm per sec or 1.0 mi per hr) over a broad geographical area. Clouds having significant vertical development usually cover smaller areas but are associated with much more vigorous uplift (sometimes in excess of 30 m per sec or 70 mi per hr).

Clouds can be described as warm or cold. The temperature in a **warm cloud** is above 0 °C (32 °F) whereas the temperature in a **cold cloud** is at or below 0 °C (32 °F). Average air temperature decreases from sea level to the tropopause and ascending cloudy air expands and cools so that high clouds are colder than low clouds. Clouds that surge to great altitudes, such as thunderstorm clouds (cumulonimbus) may be warm near their base but cold aloft. To a large extent, temperature dictates the composition and appearance of clouds. Cold clouds are made up of ice crystals and/or supercooled water droplets; ice crystal clouds appear wispy especially at their edges. Warm clouds are made up of water droplets and their boundaries are more sharply defined.

Here we present brief descriptions of the most common cloud types, organized by altitude of cloud base in middle latitudes (Table 7.2). The actual altitude of cloud base varies seasonally and with latitude. For example, the base of high clouds may be as low as 3000 m (10,000 ft) in polar latitudes and as high as 18,000 m (60,000 ft) in the tropics. Furthermore, certain qualifying terms are applied to clouds having special characteristics; some of these are listed in Table 7.3.

HIGH CLOUDS

High clouds have bases at altitudes above 5000 m (16,000 ft) where average temperatures are typically below –25 °C (-13 °F). At such low temperatures, clouds are almost exclusively composed of ice crystals, a composition that helps explain their fibrous appearance in the presence of sunlight. Their names include the prefix *cirro* from the Latin meaning "a curl of hair."

Cirrus (Ci) clouds are nearly transparent and

TABLE 7.3
Special Characteristics of Cloud Types

Characteristic	Meaning	Applied to
Castellanus	Tower-like vertical development	Cirrocumulus, altocumulus
Congestus	Crowded in heaps	Cumulus
Fractus	Broken, ragged, torn	Stratus, cumulus
Humilis	Little vertical development, flat base	Cumulus
Incus	Anvil-shape	Cumulonimbus
Lenticularis	Lens-shaped	Cirrostratus, altocumulus, Stratocumulus
Mammatus	Hanging protuberances	Cumulonimbus
Pileus	Cap-shaped cloud above	Cumuliform
Uncinus	Hook-shaped	Cirrus

occur as delicate silky strands, sometimes called mares' tails (Figure 7.2A). Strands are actually streaks of falling ice crystals blown laterally by strong winds. Like cirrus clouds, **cirrostratus (Cs)** clouds are also nearly transparent, so the sun or moon readily shines through them. They form a thin, white veil or sheet that partially or totally covers the sky (Figure 7.2B). **Cirrocumulus (Cc)** clouds consist of small, white, rounded patches arranged in a wavelike or mackerel pattern (Figure 7.2C). Rarely do these clouds cover the entire sky. As a rule, no high cloud is thick enough to prevent objects on the ground from casting shadows during daylight hours.

MIDDLE CLOUDS

Middle clouds have bases at altitudes typically between 2000 and 7000 m (6500 and 23,000 ft). These clouds, which feature temperatures generally between 0 °C and –25 °C (32 °F and –13 °F), are composed of supercooled water droplets or a mixture of supercooled water droplets and ice crystals. Their names include the prefix *alto*. **Altostratus (As)** clouds occur as uniformly gray or bluish white layers that totally or partially cover the sky (Figure 7.3A). They are usually so thick that the sun is only dimly visible, as if viewed through frosted glass. (As a rule, clouds must have a thickness of at least 50 m or about 150 ft to completely block out the sun.) **Altocumulus (Ac)** clouds consist of roll-like patches or puffs that often form waves or parallel bands (Figure 7.3B). They are distinguished from cirrocumulus by the larger size of the cloud patches and by sharper edges indicating the presence of water droplets rather than ice crystals. Altocumulus may occur in several distinct

A   B   C

FIGURE 7.2
High stratiform clouds are composed of mostly ice crystals: (A) cirrus, (B) cirrostratus, and (C) cirrocumulus.

A

B

FIGURE 7.3
Middle stratiform clouds are composed of ice crystals, water droplets, or a combination of the two: (A) altostratus, (B) altocumulus.

layers simultaneously and rarely produce precipitation that reaches the ground. If they appear early on a warm, humid summer day, there is a reasonably good chance of an afternoon thunderstorm.

LOW CLOUDS

Low clouds have bases that range from Earth's surface (fog) up to an altitude of perhaps 2000 m (6600 ft). Low clouds that form at temperatures above -5 °C (23 °F) are composed of mostly water droplets. **Stratocumulus (Sc)** clouds consist of large, irregular puffs or rolls separated by areas of clear sky (Figure 7.4A). Seldom does significant precipitation fall from stratocumulus clouds. **Stratus (St)** appears as a uniform gray layer stretching from horizon to horizon (Figure 7.4B). Fog occurs where stratus meets the ground, and

when fog *lifts*, the stratus cloud deck may break up into stratocumulus.

Usually only drizzle falls from stratus clouds, but significant amounts of rain or snow may fall from the much thicker **nimbostratus (Ns)** clouds. Nimbostratus resemble stratus except that they are darker gray and have a less uniform, more ragged base. Precipitation from nimbostratus tends to be light to moderate and continuous for 12 or more hours. By contrast, relatively brief but heavy precipitation is often associated with a cumulonimbus (thunderstorm) cloud. Gentle rainfall from nimbostratus more readily infiltrates the soil, whereas the intense downpour of a thunderstorm quickly saturates the upper soil and then mostly runs off, perhaps causing flash flooding (Chapter 11).

CLOUDS HAVING VERTICAL DEVELOPMENT

Clouds having significant vertical development form in the updrafts of convection currents that can surge to great altitudes in the troposphere. The altitude at which condensation begins to occur through convection

A

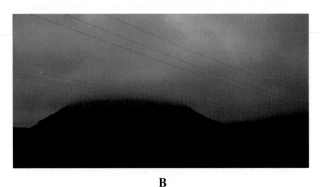

B

FIGURE 7.4
Low stratiform clouds: (A) stratocumulus, (B) stratus.

is known as the **convective condensation level (CCL)** and coincides with the altitude of cumuliform cloud base, typically between 1000 and 2000 m (3600 and 6600 ft). **Cumulus (Cu)** clouds are relatively small puffy clouds that resemble balls of cotton dotting the sky on a fair weather day (Figure 7.5A). Solar heating drives convection so that cumulus cloud development often follows the daily variation of incoming solar radiation. On a fair day, cumulus clouds begin forming by middle to late morning, after the sun has warmed the ground and initiated convection. Cumulus sky cover is most extensive by mid afternoon, usually the warmest time of day. If cumulus clouds show some vertical growth, these normally *fair-weather* clouds may produce a brief, light shower of rain or snow. As sunset approaches, convection weakens, and cumulus clouds begin dissipating (that is, they vaporize).

Where convection is suppressed, so too is the development of cumulus clouds. Relatively cold surfaces chill and stabilize the overlying air and inhibit convection so that cumulus clouds do not readily form over snow-covered surfaces. Cold water also suppresses convection. This effect is sometimes observed along the shores of the Great Lakes during late spring and early summer afternoons when, on average, lake surface temperatures are lower than temperatures of the adjacent land surface. Fair-weather cumulus clouds develop over the warm land but not over the relatively cold lake surface (Figure 7.6). In fact, rows of cumulus clouds may form a distinct boundary along the shoreline.

Once cumulus clouds form, the stability profile of the troposphere determines the extent of vertical cloud development and whether cumulus clouds build into more ominous thunderstorm clouds. If ambient air aloft is stable, vertical motion is inhibited and cumulus clouds exhibit little vertical growth. Under these conditions, the weather is likely to remain fair. On the other hand, if ambient air aloft is unstable for saturated (cloudy) air, then vertical motion is enhanced, and the tops of cumulus clouds surge upward. If the ambient air is unstable to great altitudes, the entire cloud mass takes on a cauliflower appearance as it builds into a **cumulus congestus** cloud (Figure 7.5B) and perhaps a **cumulonimbus (Cb)** cloud (Figure 7.5C). Cumulonimbus clouds have tops that sometimes reach altitudes of 20,000 m (60,000 ft) or higher. The upper portion of a cumulonimbus cloud, with its familiar anvil shape, is typically composed of mostly ice crystals, the middle portion may be supercooled water droplets or a mixture of supercooled water droplets and ice crystals,

A

B

C

FIGURE 7.5
Cumuliform clouds: (A) cumulus, (B) cumulus congestus, (C) cumulonimbus.

FIGURE 7.6
In this visible satellite image, clusters of cumulus clouds develop over the relatively warm land surface but not over the adjacent cold surfaces of the Great Lakes.

while the lower portion of the cloud consists of ordinary water droplets.

Although this cloud classification category is usually reserved for cumuliform clouds, altostratus and nimbostratus may also be considered to be clouds having vertical development. Normally classified as middle and low clouds, respectively, because of cloud-base altitude, altostratus and nimbostratus often are so thick that they extend through more than one level.

SKY WATCHING

Cumuliform and stratiform clouds are often observed in the sky at the same time. For example, a layer of high cirrus may overlie low-level cumulus clouds. The cirrus may have developed in advance of an approaching surface warm front and is too thin to significantly weaken the incoming solar radiation. The sun heats the ground, triggering convection and development of cumulus clouds.

When more than one cloud layer appears in the sky at the same time, it can be instructive to visually track the direction of movement of each cloud layer. Often clouds at different altitudes move in different directions at different speeds. Because most clouds are displaced with the wind, this observation indicates that, at least within the troposphere, the horizontal wind changes direction and magnitude with altitude. Any change in wind direction or wind speed with distance is known as **wind shear**. Strong vertical shear in horizontal wind speed or direction may cause clouds (or cloud elements) to line up into rows that often extend for hundreds of kilometers. These so-called **cloud streets**

are observed in cumulus, stratocumulus, and cirrocumulus clouds. They develop where the air is humid and the axes of convective cells align with the average wind direction.

Another feature to watch for are well-defined holes or canals in an otherwise uniform cloud layer. Holes in clouds are not unusual; numerous published reports and photographs of them have appeared over the years. P.V. Hobbs of the University of Washington attributes at least some holes in clouds to aircraft flying through thin cloud layers composed of supercooled water droplets (probably altocumulus). Turbulence produced in the wake of an aircraft induces expansional cooling of air; water droplets freeze to ice crystals that fall out of the cloud, thereby creating a hole in the cloud. Aircraft ascending or descending through a cloud may produce nearly circular holes, whereas aircraft flying horizontally through a cloud layer may create cloud-free canals.

UNUSUAL CLOUDS

Chances are that all of us have seen most of the clouds described above because the circumstances leading to their formation are quite common. Other, more unusual clouds are formed by special atmospheric conditions that in some cases arise from orographic effects.

Many people have reported certain disk-shaped clouds as UFOs (*Unidentified Flying Objects*). Observing such clouds over a period of time reveals another intriguing characteristic besides their shape: They are almost motionless. These are **altocumulus lenticularis** clouds (lens-shaped altocumulus) that are generated when a mountain range disturbs large-scale horizontal winds. As strong horizontal winds encounter a mountain range, the wind is deflected up the windward slopes and down the leeward slopes. This occurs as the prevailing westerlies, blowing from west to east, cross the Sierra Nevada Mountains of California or the Front Range of the Rocky Mountains. If the air ascending the windward slopes is stable, then the wind is deflected into a wave-like pattern that can extend hundreds of kilometers downwind of the mountain crest (Figure 7.7).

If air is sufficiently humid, lenticular clouds form at the crests of the waves and are absent in the troughs. Where air ascends within a wave, expansional cooling leads to condensation (or deposition) and cloud formation. Where air descends within a wave, compressional warming causes clouds to vaporize. Clouds formed in this way directly over the mountain are known as **mountain-wave clouds**. (If these clouds ex-

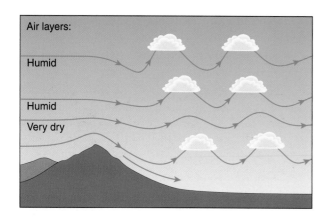

FIGURE 7.7
A mountain range deflects the horizontal wind into a wave-like pattern that extends downwind of the summit. If the air is sufficiently humid, clouds develop in the wave crests where ascending air expands and cools but clouds are absent in wave troughs where descending air is compressed and warms.

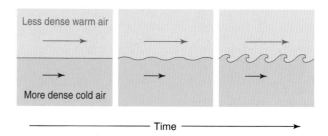

FIGURE 7.8
Formation of Kelvin-Helmholtz waves along the interface between two layers of air that contrast in density. Billow clouds may form in the wave crests.

tend just downwind of the mountain peak, they are referred to as *banner clouds*.) Lenticular clouds formed in the wave crests downwind of the mountain are known as **lee-wave clouds**.

Mountain-wave and lee-wave clouds are stationary, even though winds are strong because the wave itself is stationary. A stationary wave within the atmosphere is known as a *standing wave*. A waterfall on a river presents a good analogy. In the plunge pool at the base of the waterfall, the water is very turbulent, but if we follow the river downstream, we soon leave the turbulent region behind. The river water flows ceaselessly, but the turbulent segment remains stationary at the foot of the waterfall, just as a standing wave in the atmosphere remains stationary to the lee of a mountain range.

In some cases, a rotor-like circulation develops under the crest of a standing wave. Air rotates about a horizontal axis that roughly parallels the mountain range and sometimes a cloud (*rotor cloud*) forms in the ascending branch. Turbulence associated with the rotor can be a major hazard to aviation.

In a stable atmosphere, waves and wave-type clouds also can develop along the interface between two layers of air that are moving horizontally at different speeds. In Figure 7.8, a layer of relatively warm air overlies a layer of colder, denser air. The warm air is moving horizontally more rapidly than the cold air. This vertical shear in the horizontal wind gives rise to waves, know as *Kelvin-Helmholz waves*, along the boundary between the two air layers. (Waves formed in this way are analogous to wind-driven water waves that form

along the boundary between the ocean and atmosphere). Clouds that develop at the crests of Kelvin-Helmholz waves consist of a train of regularly spaced curls and are know as *billow clouds*.

Waves traveling horizontally within the atmosphere are also responsible for patterns exhibited by stratocumulus and cirrocumulus (as well as by varieties of altocumulus in addition to lenticularis). These waves are not linked to mountain ranges and propagate at different altitudes within the troposphere. Such waves produce bands of clouds that are aligned either parallel or perpendicular to the wind direction.

Almost all water vapor is confined to the troposphere, so it is not surprising that most clouds occur below the tropopause. Mountain-wave clouds sometimes develop in the stratosphere above high mountain crests and occasionally the top of an intense cumulonimbus cloud surges through the tropopause and into the lower stratosphere. Other examples of clouds in the upper atmosphere are nacreous and noctilucent clouds.

Nacreous clouds (or *polar stratospheric clouds*) occur in the stratosphere at altitudes between about 20 and 30 km (12 and 19 mi) where temperatures favor water in either the solid or supercooled state. These are cirrus or altocumulus lenticularis clouds that form on sulfuric acid nuclei possibly of volcanic origin. Because of their soft, pearly luster, these rarely seen clouds are also known as *mother-of-pearl clouds*. They are best viewed at high latitudes in winter when illuminated by the setting sun.

Somewhat mysterious are cirrostratus-like **noctilucent clouds** that form in the upper mesosphere above an altitude of about 80 km (50 mi). They are seldom observed, and then only at high latitudes (usually north of 47 degrees N) in summer (mid-May to mid-August) around midnight. From well below the horizon, the sun illuminates noctilucent clouds, imparting to them an ethereal silvery-white appearance that stands out against the black night sky. They are probably composed of mostly tiny water-ice crystals formed on meteoric dust particles.

How is it possible for ice clouds to form in the upper atmosphere where there is very little water vapor? The key is the exceptionally low air temperature in the upper mesosphere. As noted in Chapter 2, temperatures are lowest at the mesopause. Whereas January temperatures are generally in the range of –53 °C (–64 °F) to –43 °C (–46 °F), July temperatures may drop as low as –173 °C (–279 °F). At such low temperatures, ice clouds can form at extremely low vapor pressures.

Nonetheless, this begs the question of where the water vapor comes from. Some is ejected into the upper atmosphere during violent volcanic eruptions, but the most important source may be chemical reactions involving methane (CH_4). Atmospheric circulation transports methane from the troposphere into the stratosphere where it is broken down by solar radiation, releasing hydrogen that is oxidized to water vapor through a complex series of chemical reactions. Some of that water vapor is transported to the upper mesosphere and deposited as the ice crystals in noctilucent clouds.

It is a curious fact that noctilucent clouds were only first reported in 1885 (in Germany and Russia) and their frequency of sighting appears to be on the increase. During the same period, the concentration of atmospheric methane has also been increasing, especially since the mid- to late-1800s. G.E. Thomas of the University of Colorado's Laboratory for Atmospheric and Space Physics, suspects that there may be a connection between upward trends in atmospheric methane and noctilucent cloud sightings.

Fog

Fog is a visibility-restricting suspension of tiny water droplets or ice crystals in an air layer next to Earth's surface. Simply put, fog is a cloud in contact with the ground. By international convention, fog is defined as restricting visibility to 1000 m (3250 ft) or less; otherwise the suspension is called **mist**. (The popular definition of mist is *light drizzle*.) Fog may develop when air becomes saturated through radiational cooling, advective cooling, the addition of water vapor, or

FIGURE 7.9
Radiation fog develops as a consequence of radiational cooling on a night when there is some air movement and the atmosphere is humid at low levels.

expansional cooling.

With a clear night sky, light winds, and an air mass that is humid near the ground and relatively dry aloft, radiational cooling may cause the air near the ground to approach saturation. When this condition occurs, a ground-level cloud forms and is called **radiation fog** (Figure 7.9). High humidity at low levels within the air mass is usually due to evaporation of water from a moist surface. Hence, radiation fog is most common over marshy areas or where the soil has been saturated by a recent period of heavy rainfall or snowmelt.

Light winds, rather than calm air, favor development of radiation fog. Light winds produce a slight mixing that more effectively transfers heat from throughout the layer of humid air to the relatively cold ground. Consequently, the entire air layer is chilled below the dewpoint (Chapter 6). If air is calm, however, there is no mixing and heat transfer is by conduction alone. Because motionless air is a poor conductor of heat, only a very thin layer of air immediately in contact with the ground is cooled to saturation. Hence, calm conditions favor dew or frost rather than radiation fog. On the other hand, if winds become too strong, the humid air at low levels mixes with the drier air aloft, the relative humidity drops, and radiation fog disperses or fails to develop. Air that is chilled by radiational cooling is relatively dense and, in hilly terrain, drains downslope and settles in low-lying areas such as river valleys and wetlands. This air movement is known as *cold-air drainage*. Hilltops may thus be clear of fog, while fog is thick and persistent in deep valleys.

Once a fog bank forms at night, radiational cooling becomes a little more complicated. The top surface of the fog bank very efficiently emits infrared (IR) radiation skyward. The fog droplets absorb some of the IR emitted by the ground and the droplets, in turn, emit IR to the ground so that the ground cools more slowly. The maximum cooling rate shifts to the top of the fog bank, the fog layer thickens, and visibility continues to drop.

Often radiation fog persists for only a few hours after sunrise. Fog gradually thins and disperses as saturated air at low altitudes mixes with drier air above the fog bank; the relative humidity drops and fog droplets vaporize. Mixing may be caused by convection triggered by solar heating of the ground or by a strengthening of regional winds. In winter, however, the upper surface of the fog layer readily reflects the weak rays of the sun and radiation fog may be persistent. For example, in the valleys of the Great Basin and in California's San Joaquin Valley, winter radiation fogs may persist for many days to weeks at a time.

Fog sometimes accompanies air mass advection. The temperature and the water vapor concentration of an air mass depend on the nature of the surface over which the air mass forms and travels. As an air mass moves from one place to another, termed *air mass advection*, those characteristics change, partly because of the modifying influence of the surfaces over which the air mass travels. When the advecting air passes over a relatively cold surface, the air mass may be chilled to saturation in its lowest layers. This type of cooling is known as *advective cooling* and occurs, for example, in early spring in the northern states when mild, humid air flows over relatively cold, snow-covered ground. Snow on the ground may chill the overlying air to the dewpoint and fog develops. Fog formed by advective cooling is known as **advection fog**.

Advection fog also develops when warm and humid air streams over the relatively cold surface of a lake or the ocean. Persistent, dense fogs develop in this way over the Great Lakes in summer (Figure 7.10). Thick *sea fog* forms over the frigid waters of the Grand Banks of Newfoundland when mild maritime air from over the warm Gulf Stream flows northward.

Nocturnal radiational cooling sometimes combines with air mass advection to produce fog. An initially dry air mass becomes more humid in its lower levels following a long trajectory over the open water of a lake or the ocean. Over land areas, downwind from the water body, if the night sky is clear, the modified air

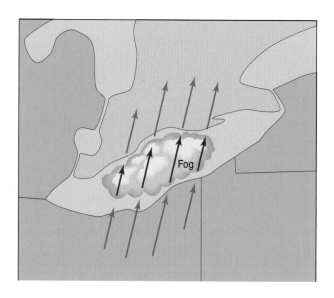

FIGURE 7.10
In the summer, fog forms over the Great Lakes when relatively warm and humid air passes over the cold water surface and is chilled to its dewpoint.

mass is subjected to extreme radiational cooling and fog develops. San Francisco's famous fog forms in this way. Onshore winds transport cool, humid air into the city and nocturnal radiational cooling chills the air to saturation.

Steam fog (also known as *Arctic sea smoke*) is fog that develops in late fall or winter when extremely cold and dry air flows over a large unfrozen body of water. Evaporation and sensible heating cause the lower portion of the air mass to become more humid and warmer than the air above. Heating from below destabilizes the air and the consequent mixing of mild, humid air with cold, dry air brings the air to saturation and fog forms. Because the air is destabilized, fog appears as rising filaments or streamers that resemble smoke or steam. Steam fog also develops on a cold day over a heated outdoor swimming pool or hot tub and sometimes over a wet highway or field when the sun comes out after a rain shower.

Another way of explaining the origin of steam fog is to consider the distribution of vapor pressure over the water surface. Within about 1 to 2 m (3 to 7 ft) of the water surface, the vapor pressure is relatively high because of evaporation of water into the heated air. Above this shallow humid air layer, the air is colder and the vapor pressure is relatively low. A vertical vapor pressure gradient develops over the water body and, in response, water molecules stream (diffuse) upward. (As we will see in Chapter 8, gases move from high toward low pressure in response to a pressure gradient.) As the humid air streams upward, it expands and cools and the

FIGURE 7.11
Upslope fog formed when humid air is forced to ascend the windward slopes of a hill or mountain. Expansional cooling brings the air to saturation.

water vapor condenses into tiny droplets that we see as steam. About 5 to 10 m (15 to 35 ft) above the water surface, the rising streamers encounter drier air and evaporate.

Fog also develops on hillsides or mountain slopes as a consequence of the upslope movement of humid air. Ascending humid air undergoes expansional cooling and eventually reaches saturation. Any further ascent of the saturated air produces fog. Fog formed in this way is called **upslope fog** (Figure 7.11). In the coastal mountain ranges of California, upslope fog sometimes overtops the range crest and spreads as stratus clouds over the leeward valleys. A cloud layer formed in this way is known as *high fog*.

Precipitation Processes

Appearance of clouds is no guarantee that it will rain or snow. Nimbostratus and cumulonimbus clouds produce the bulk of precipitation, but most clouds do not yield any significant rain or snow. This is because a special combination of circumstances is required for precipitation to develop. Key to understanding the mechanisms of precipitation formation is the concept of terminal velocity, the speed of a falling particle. We begin this section by examining terminal velocity and then cover the two mechanisms of precipitation formation: the collision-coalescence process and the Bergeron-Findeisen process.

TERMINAL VELOCITY

The speed of a falling cloud droplet or ice crystal (or any other particle) in calm air is regulated by (1) *gravity*, the force that accelerates the particle directly downward towards Earth's surface, and (2) the resistance offered by the air through which the particle falls. A downward accelerating particle meets increasing air resistance while gravity remains essentially constant. The magnitude of the resisting force eventually equals gravity; that is, the forces come into balance. If the forces become balanced, then according to Newton's first law of motion, the downward moving particle attains a constant speed. (*Newton's first law of motion* states that an object in constant straight-line motion or at rest remains that way unless acted upon by an unbalanced force.) In this case, air resistance (directed upward) balances gravity (acting downward) and the particle continues downward at constant speed. That speed is the particle's **terminal velocity**.

Terminal velocity helps explain why the tiny particles that compose clouds (water droplets and ice crystals) will not fall to Earth's surface as precipitation unless they undergo considerable growth. Generally, terminal velocity increases with the size of the particle (Figure 7.12). For a particle to descend from a cloud to Earth's surface as precipitation, the particle's terminal velocity must be greater than the rising motion of air within the cloud (updrafts). Cloud water droplets and ice crystals are so small (typically 10 to 20 micrometers in diameter) and their terminal velocities are so low (usually only 0.3 to 1.2 cm per sec), that even weak updrafts will keep them suspended in clouds. Even if droplets or ice crystals fall through the base of a cloud, their terminal velocity is so slow that it would take 24 hours or longer for the cloud particles to reach Earth's surface. Long before that happens, the particles would vaporize in the unsaturated air under the cloud. (The relative humidity below a cloud is typically under 100%.)

For clouds to precipitate, cloud particles must grow large enough that their terminal velocities overwhelm the updrafts. This is no minor task! It takes the water content of about 1 million cloud droplets (having diameters of 10 to 20 micrometers) to form a single raindrop (about 2 mm in diameter). Condensation or deposition alone cannot account for the formation of raindrop-sized particles. Cloud droplets grow into raindrops through collision-coalescence of cloud droplets in warm clouds whereas a combination of the Bergeron-Findeisen process and collision-coalescence causes ice crystals in cold clouds to grow into snowflakes.

FIGURE 7.12
The terminal velocity of a
particle falling through air
increases with the size
of the particle.

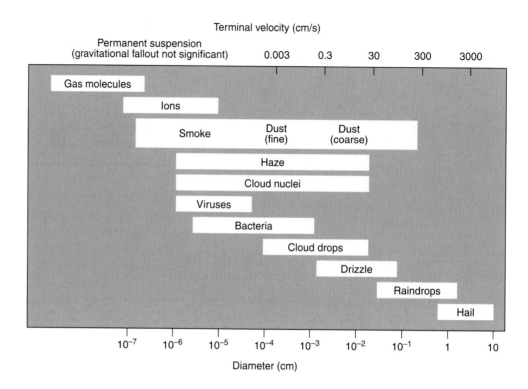

WARM-CLOUD PRECIPITATION

In a *warm cloud* (at temperatures above 0 °C),
droplets may grow by colliding and coalescing (merging)
with one another, the so-called **collision-coalescence
process**. The collision-coalescence process takes place
in a cloud made up of droplets of different sizes. Cloud
droplets of unequal diameters have different terminal
velocities, greatly increasing the likelihood that faster
falling (larger) droplets will overtake, and then collide
and coalesce with slower falling (smaller) droplets in
their paths. Through repeated collisions and coalescence,
droplets grow in size (Figure 7.13). (On the other hand,
cloud droplets of uniform diameter have essentially the
same terminal velocities so that collisions are infrequent.)
Through collision-coalescence, a warm cloud droplet
may grow so large that it becomes unstable and splits into
smaller droplets that go on to coalesce with other
droplets in a kind of chain reaction. Eventually some
droplets become sufficiently large that they survive a fall
from clouds to Earth's surface as raindrops.

A key factor in precipitation formation through
collision-coalescence is collision efficiency. *Collision
efficiency* refers to the fraction of all droplets in the path
of a falling larger droplet that comes in contact with the
larger droplet. Collision efficiency varies with the size of
the larger droplet and is minimal (less than 10%) for
droplets less than 40 micrometers in diameter but

approaches 100% for droplets greater than 80
micrometers in diameter. Larger droplets form on larger
nuclei (e.g., sea salt particles), through random collisions
between smaller droplets, or via mixing between cloud
droplets and the surrounding drier air.

COLD-CLOUD PRECIPITATION

Although precipitation development through
collision-coalescence in warm clouds is important,
especially in the tropics, most precipitation in middle and
high latitudes originates in cold clouds (clouds or
portions of clouds at temperatures below 0 °C).
Precipitation is most likely to fall from a cloud that
initially is composed of a mixture of supercooled water
droplets and ice crystals regardless of season. In
response to differences in vapor pressure relative to water
and ice surfaces present in such a cloud, ice crystals grow
at the expense of the surrounding supercooled water
droplets. This is the basis of the so-called **Bergeron-
Findeisen process**, also known as the *ice-crystal
process*, named for the Swedish meteorologist Tor
Bergeron who first proposed the idea around 1930, and
the German physicist Walter Findeisen who subsequently
elaborated on the concept.

As noted above, cloud condensation nuclei are
much more abundant than ice-forming nuclei so that a
mixed cloud (at least initially) consists of far more super-

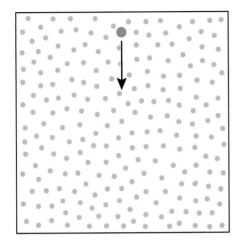

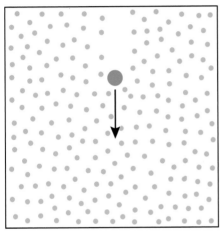

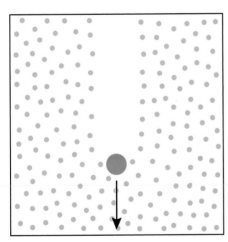

- Cloud droplets
- Rain drop

FIGURE 7.13
Within a warm cloud, a relatively large droplet falls through a cloud of much smaller droplets. The larger droplet falls faster and collides with the smaller droplets in its path and grows by coalescence. This is the collision-coalescence process.

cooled water droplets than ice crystals. Supercooled water droplets quickly vaporize as ice crystals grow. At subfreezing temperatures, water molecules more readily vaporize from a liquid (supercooled) water surface than from an ice surface because water molecules are more strongly bonded in the solid phase than the liquid phase. At the same subfreezing temperature, the saturation vapor pressure is greater over supercooled water than over ice. For temperatures below 0 °C (32 °F), two values of the saturation vapor pressure are given in Table 6.3: one over supercooled water, the other over ice.

In a cloud composed of a mixture of ice crystals and supercooled water droplets, a vapor pressure that is *saturated* for water droplets is actually *supersaturated* for ice crystals at the same temperature. Suppose, for example, that a cloud of supercooled water droplets at –15 °C (5 °F) is saturated relative to water (relative humidity = 100%). According to Table 6.3, the saturation vapor pressure is 1.91 mb. Imagine the sudden appearance of an ice crystal in the same region of the cloud. Table 6.3 indicates that the saturation vapor pressure over ice at –15 °C (5 °F) is 1.65 mb so that air surrounding the ice crystals is *supersaturated* (relative humidity greater than 100%). Water vapor deposits on the ice crystals and the crystals grow. Deposition lowers the vapor pressure surrounding the water droplets to unsaturated conditions and water droplets evaporate. Also, a lower vapor pressure in the air surrounding the ice crystals produces a pressure gradient such that water vapor molecules diffuse toward the ice. Through the Bergeron-Findeisen process, then, ice crystals grow at the expense of supercooled water droplets.

Growth of ice crystals is accompanied by an increase in their terminal velocities. Larger, faster falling ice crystals overtake, collide and coalesce with smaller, slower falling ice crystals and supercooled water droplets (Figure 7.14). Ice crystals grow still larger and may fall out of the cloud base. If air temperature is below freezing at least most of the way to the ground, ice crystals reach Earth's surface as snowflakes. If air temperatures in the lower reaches of the cloud or between the cloud and Earth's surface are above freezing, snowflakes melt into raindrops.

Once a falling raindrop or a snowflake leaves the base of a cloud, it enters unsaturated air and begins to vaporize. The longer the distance to the Earth's surface, (i.e., the higher the cloud base) and the lower the relative humidity of air beneath the clouds, the greater is the quantity of rain or snow that vaporizes. This is one of the reasons why highlands, being closer to the base of clouds, usually receive more precipitation than lowlands.

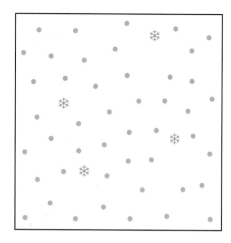

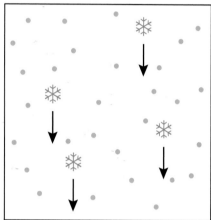

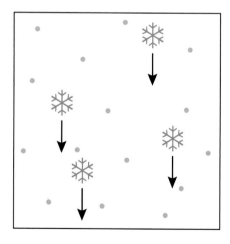

● Supercooled cloud droplets

❄ Ice crystals

FIGURE 7.14
Within a cold cloud, ice crystals grow at the expense of supercooled water droplets. As they grow larger, ice crystals fall faster and collide with droplets and other ice crystals in their paths. Eventually, they may grow large enough to fall out of the cloud as snowflakes. This is the Bergeron-Findeisen process.

FIGURE 7.15
Much of the rain falling from the base of a distant thunderstorm evaporates in the relatively dry air beneath the cloud. The shaft of falling precipitation is known as virga.

Viewed from a distance, a shaft of water or ice particles falling from a cloud may appear as dark curtains or streaks against a bright background (Figure 7.15). **Virga** is the name given to such streaks when the water or ice particles vaporize before reaching Earth's surface. This is a common sight in the Great Basin of the American West where much potential precipitation vaporizes while falling through the very dry air beneath cloud base.

For a discussion of efforts to increase precipitation by stimulating natural precipitation processes, refer to the Essay, Rainmaking.

Forms of Precipitation

Within the atmosphere, some water vapor condenses or deposits forming clouds, and through the collision-coalescence and Bergeron-Findeisen processes, some clouds precipitate. **Precipitation** is water in solid or liquid form that falls from clouds to Earth's surface. Included are liquid precipitation (rain, drizzle), freezing precipitation (freezing rain, freezing drizzle), and frozen precipitation (snow, snow pellets, snow grains, ice pellets, hail).

Rain consists of liquid water drops that fall mostly from nimbostratus and cumulonimbus clouds. Air resistance causes raindrops to fall as slightly flattened spheres (not teardrops as popularly portrayed) with diameters generally in the range of 0.5 to 6 mm (0.02 to 0.2 in.). At greater diameters, drops are unstable and break up into smaller drops. (Imagine what life would be like if raindrops were the size of basketballs!) Water's surface tension keeps small drops together, but once a critical size is reached, surface tension is incapable of maintaining the drop. Research meteorologists sampling warm-cloud raindrops over Hilo, HI discovered an exception to this rule. They reported raindrops as large as 8 mm (0.3 in.) in diameter (about the size of a pea). In this case, strong updrafts in convective clouds prolonged the period of collision-coalescence, thereby enabling warm drops to grow to extraordinary sizes. At mid and high latitudes, raindrops usually begin as snowflakes (or sometimes would-be hailstones) that melt and coalesce as they descend through air at temperatures above 0 °C (32 °F).

Drizzle consists of liquid water drops having diameters between 0.2 and 0.5 mm (0.01 and 0.02 in.) that drift very slowly toward Earth's surface. Drizzle originates mostly in stratus clouds that are so low and thin that droplets undergo only limited growth by collision-coalescence. Drizzle often occurs with fog and contributes to poor visibility.

Snow is an agglomeration of ice crystals in the form of flakes. Although constituent ice crystals are hexagonal (six-sided), snowflakes vary in shape and size depending on water vapor concentration and temperature in the portion of the cloud where they form and grow. Crystals may consist of needles, dendrites, plates, or columns (Figure 7.16). Snowflake size depends in part on the availability of water vapor during crystal growth. At very low air temperatures, water vapor concentration is relatively low, and snowflakes are small. Snowflake size also depends on collision efficiency as flakes fall through the atmosphere. At air temperatures near or slightly above freezing, snowflakes more readily stick together after colliding and their aggregate diameters have been known to reach 5 to 10 cm (2 to 4 in.). Appearance of such large flakes may indicate that snow is about to turn to rain.

Snow pellets and snow grains are related to snowflakes. **Snow pellets** (formerly called *soft hail* or *graupel*) are soft conical or spherical white particles of ice having diameters of 2 to 5 mm (0.08 to 0.2 in.). They form when supercooled cloud droplets collide and freeze on an ice crystal. **Snow grains** (also known as *granular snow*) are flat or elongated opaque white particles of ice, usually less than 1.0 mm (0.04 in.) in diameter. They originate in much the same way as drizzle except that they freeze prior to reaching the ground. Is it ever too cold to snow? When is it too warm to snow? For

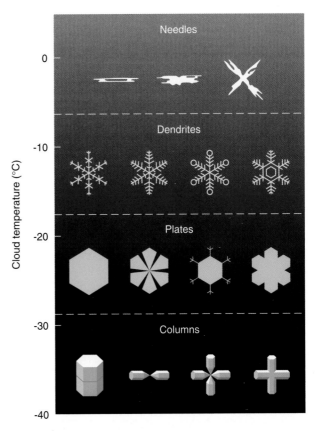

FIGURE 7.16
Snowflakes take on a variety of forms depending to a certain extent on cloud temperature.

answers to these questions, see the Essay.

Ice pellets, commonly called *sleet,* are spherical or irregularly shaped transparent or translucent particles of ice that are 5 mm (0.2 in.) or less in diameter. Ice pellets form when snowflakes partially or completely melt as they fall through above-freezing air beneath cloud base (Figure 7.17). These raindrops or partially melted snowflakes then fall into a relatively thick layer of subfreezing air and refreeze into ice particles prior to striking the ground.

Freezing rain (or *freezing drizzle*) consists of rain (or drizzle) drops that become supercooled and at least partially freeze on contact with cold surfaces (at subfreezing temperatures), forming a coating of ice (*glaze*) on roads, tree branches, and other exposed surfaces. Even a thin coating of ice can create hazardous driving and walking conditions. Sometimes glaze grows so thick that its weight brings down tree limbs and snaps power lines. Freezing rain develops in much the same way as sleet except that the layer of subfreezing air at Earth's surface is shallower (Figure 7.17). If the surface

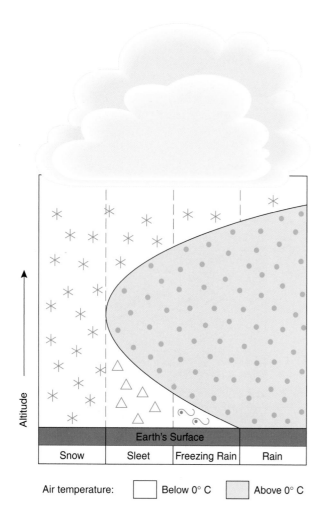

FIGURE 7.17
Schematic cross-section of the lower atmosphere showing temperature conditions required for formation of frozen, freezing, and liquid forms of precipitation.

temperature is close to freezing, the supercooled drops partially freeze, producing a wet and very slippery surface. We can readily distinguish ice pellets from freezing rain, because ice pellets bounce when striking the ground whereas freezing raindrops do not bounce.

Freezing rain can be highly localized and persistent, especially where the terrain is hilly or mountainous such as the Appalachians of Pennsylvania and West Virginia. Cold, dense air drains into valley bottoms. Rain that falls from warm air aloft into this shallow layer of subfreezing air becomes supercooled and forms a glaze on cold surfaces. This situation often persists until the wind shifts and flushes the subfreezing air out of the valley bottom.

Glaze can develop on a road surface even without supercooling of raindrops. A prolonged period

of extremely cold winter weather will chill a road surface to temperatures well below freezing. A sudden shift in weather pattern causes the air temperature to rise above the freezing point while the road surface temperature remains below freezing. Rain (not supercooled) freezes on the road surface forming a glaze that is just as hazardous as if it were produced by freezing rain or drizzle.

Hail consists of balls or jagged lumps of ice, often characterized by concentric internal layering resembling the structure of an onion. Hail develops within intense thunderstorms (cumulonimbus clouds) characterized by vigorous updrafts, an abundant supply of supercooled water droplets, and great vertical cloud development. Updrafts transport ice pellets into the middle and upper portions of the cumulonimbus cloud. Along the way, ice pellets grow by collecting super-cooled water droplets, and eventually become too heavy to be supported by updrafts. Ice pellets then descend through the cloud, exit the cloud base, and enter air that is typically above the freezing point. Pellets begin melting, but if large enough initially, some ice will survive the journey to Earth's surface as *hailstones*.

Most hailstones are harmless granules of ice less than 1 cm in diameter, but violent thunderstorms may spawn destructive hailstones the size of golf balls or larger. In fact, large hail having a diameter of 0.75 in. (1.9 cm) or greater is one of the criteria whereby the National Weather Service designates a thunderstorm as severe. Hail is usually a spring or summer phenomenon that is particularly devastating to crops. We have more to say about hail in Chapter 11.

Weather Radar: Locating Precipitation

Weather radar (acronym for *r*adio *d*etection *a*nd *r*anging) is a valuable remote sensing tool for determining the location, movement, and intensity of areas of precipitation. Weather radar emits microwave signals and receives reflected signals from targets as it continually scans a large volume of the lower atmosphere. Today's weather radar can detect a tornado's circulation as it develops within its parent thunderstorm cloud, the spiral bands of a hurricane before they sweep onshore, and microbursts that can be hazardous to aircraft. Weather radar also monitors rainfall rates and cumulative rainfall totals that may forewarn of flash flooding.

The first reported use of radar for meteorological purposes was on 20 February 1941, when a radar unit on the south coast of England tracked a thunderstorm a distance of 11 km (7 mi). More widespread application of radar for weather analysis began shortly after World War II. Early weather radars were short-range surplus military units. Not until the mid-1950s, following major tornado and hurricane disasters, did the U.S. Congress allocate funds for purchase and installation of new long-range radar units designed specifically for meteorological purposes. In 1959, these units, called *WSR-57* radars, went into service for hurricane detection along the Atlantic and Gulf coasts and for tornado and severe thunderstorm surveillance over the central United States. By 1964, *WSR-57* radars were operating at 32 sites and by the late 1960s were a routine feature on televised weathercasts in many parts of the nation. Weather radar was upgraded in 1974 with installation of *WSR-74* radars. During the early and mid-1990s, as a key component of the $4.5 billion modernization of the National Weather Service, the obsolete *WSR-57* and *WSR-74* radars were replaced by the more technologically sophisticated *WSR-88D* radars. (*WSR* stands for Weather Surveillance Radar; the number refers to the year of development; and, *D* is for Doppler.) As of this writing a network of 113 *WSR-88D* weather radar units monitors much of the nation.

The *WSR-88D* weather radar operates in either a reflectivity or velocity mode. In the *reflectivity mode*, the radar signal detects the location, movement, and intensity of areas of precipitation, and in the *velocity mode*, the radar determines air motions directly toward or away from the radar associated with the circulation within a weather system.

REFLECTIVITY MODE

The *WSR-88D* weather radar emits short pulses of microwave energy having wavelengths of 10.0 to 11.1 cm (Figure 7.18). At these wavelengths, radar signals are reflected (or more precisely, scattered) by rain, snow, or hail but not significantly by the tiny water droplets or ice crystals that compose clouds. That is, weather radar is designed to detect (*see*) precipitation-size particles but not the parent clouds. Falling precipitation particles reflect some of the radar signal back to a receiving unit where it is electronically processed and displayed on a computer screen as a **radar echo** (electrical pulse).

Even when no precipitation is falling, the *WSR-88D* weather radar can detect boundaries within the atmosphere. Operating in the very sensitive *clear-air mode*, radar signals detect dust particles or swarms of insects that tend to collect along boundaries within air masses (e.g., an outflow boundary from a distant thunderstorm). Birds feeding on the insects also produce radar echoes and thereby enhance boundaries. These boundaries may be important because they are potential sites for future development of thunderstorms (Chapter 11).

The weather radar sending and receiving unit is a dish-type antenna housed in a radome (Figure 7.19). A *radome* is a spherical structure composed of fiberglass that protects the radar antenna from wind and weather but is transparent to the radar signal. Pulsed radar signals are sent out and received hundreds of times each second as the antenna sweeps out a spiraling volume scan of the sky every 5 minutes. The radar combines several single

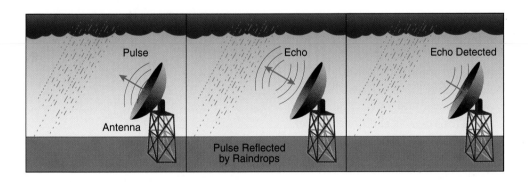

FIGURE 7.18
Operating in the reflectivity mode, a weather radar continually sends out pulses of microwave energy that are scattered by raindrops, hailstones, and snowflakes back to a receiving unit that processes and displays radar echoes on a television-type screen.

FIGURE 7.19
This radome for a *WSR-88D* radar unit houses a rotating radar dish antenna.

elevation scans (ranging from 0.5° to 19.5° above the horizon) with a 360-degree sweep in the horizontal direction to complete one volume scan. The product is a map of radar echoes representing the precipitation distribution (if any) in the area surrounding the radar unit. The speed of the radar pulse is known so that the elapsed time between emission and reception of a radar signal can be calibrated to give the distance to the precipitation.

The strength (intensity) of a radar echo depends on the reflectivity of the targeted precipitation, with the greatest reflectivity for hailstones and large raindrops. The concentration of raindrops along the path of the radar beam also affects echo intensity but to a lesser extent. Echo intensity is an index of rainfall rate and can be used to identify severe thunderstorm cells, which often contain large hail. Radar systems have electronic devices that display echo intensity on a color scale. By one convention (often seen on televised weathercasts), red indicates very heavy rain (or hail) and, at the other end of the scale, light green signifies very light rain (Figure 7.20). Other color schemes are sometimes used to portray snow or a mixture of rain and snow. As described in more detail later in this chapter, weather radar can also produce maps of rainfall totals over a specified time period.

Not all radar echoes are caused by meteorological phenomena. Nearby objects fixed on the ground, such as buildings or smokestacks, also reflect radar signals; such radar echoes constitute **ground clutter**. A ground clutter pattern produced in this way is unique to a particular radar site, appears all the time, and hence is readily distinguished from precipitation echoes. On the other hand, special atmospheric conditions sometimes develop that give rise to ground clutter that may be mistaken for precipitation. For example, a strong temperature inversion may cause the outgoing radar signal to bend downward so that it strikes the ground. Such ground clutter differs from precipitation echoes in that it is stationary, stronger, and grainier in appearance.

Sometimes a radar screen shows a broad area of echoes, but weather stations located in the same area report no precipitation. In this case, the radar is detecting echoes from rain and/or snow that is falling from clouds but vaporizing completely in the relatively dry air that is beneath cloud base. (Recall our earlier discussion of *virga*.)

VELOCITY (DOPPLER) MODE

WSR-88D is the first weather radar to operate in both the velocity and reflectivity modes. In the velocity mode, the radar system applies the Doppler principle to determine air motions within a weather system. The **Doppler effect** is named for Johann Christian Doppler, the Austrian physicist who was the first to explain the phenomenon in 1842. For this reason, radar with velocity-detection capability is often referred to as a

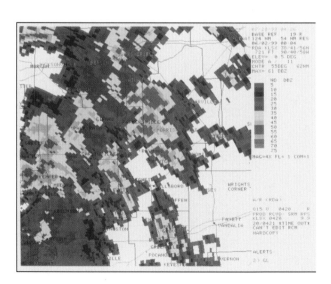

FIGURE 7.20
A sample radar reflectivity product in which echo intensity is graduated by color.

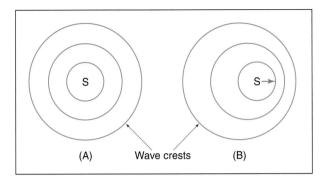

FIGURE 7.21
The Doppler effect is the shift in the frequency of sound or electromagnetic waves that accompanies the motion of the wave source(s) or wave receiver. (A) A sound wave source (e.g., train whistle) is stationary and wave frequency is uniform everywhere. (B) The wave source is in motion so that wave frequency is greater ahead of the source than behind the source.

Doppler radar. The Doppler effect is also employed in devices that measure the speed of a pitched ball or motor vehicle traffic.

The **Doppler effect** refers to a shift in the frequency of sound waves or electromagnetic waves emanating from a moving source. For example, the pitch (frequency) of a train whistle sounds higher as a train approaches and drops off as the train pulls away. As shown schematically in Figure 7.21A, the crests of sound waves radiate outward from a stationary source as evenly spaced concentric circles; that is, the wave frequency is uniform. If the source is moving, however, as in Figure 7.21B, wave crests become more closely spaced in the direction in which the source is moving so that the frequency is higher ahead of the source and lower behind the source.

Doppler radar monitors the motion of precipitation particles moving *directly toward* or *away from* the radar antenna. The frequency of the radar signal shifts slightly between emission and the return signal (echo) and this frequency shift is calibrated in terms of the motion of the particle (Figure 7.22). No frequency shift occurs for components of particle motion that are perpendicular to the radar beam; that is, Doppler radar cannot detect air motion at right angles to the radar beam. When air is moving in directions other than directly toward or away from the path of the radar signal, only the component of motion toward or away from the radar is detectable. Doppler radar displays are color-coded so that greens and blues (cold colors) indicate motion directly toward the radar and reds and yellows (warm colors) indicate motion directly away from the radar (Figure 7.23).

The principal advantage of Doppler radar is that meteorologists can use it to provide the public with advance warning of hazardous weather including severe thunderstorms, tornadoes, and potentially flooding rains. The geographical range of radar is limited by the curvature of the Earth; that is, radar signals travel along downward-bending paths somewhat less curved than Earth's surface. Radar sees well beyond the horizon but its beam gains altitude as distance from the radar increases so that when operating in the reflectivity mode, the radar's maximum range is about 460 km (285 mi). For other reasons when operating in the velocity (Doppler) mode, the radar's maximum range decreases to about 230 km (143 mi).

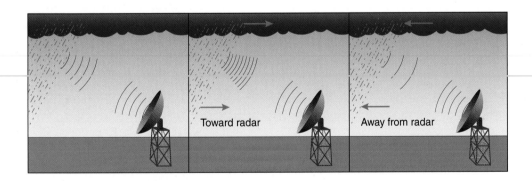

FIGURE 7.22
Doppler radar operating in three different weather situations. (A) With a stationary rain shower, the frequency of the emitted signal is the same as that of the return echo. (B) If the rain shower approaches the radar, the frequency of the return echo is greater than that of the emitted signal. (C) If the rain shower moves away from the radar, the frequency of the return echo is less than that of the emitted signal. [Adapted from SAM, NOAA/Environmental Research Laboratories/Forecast Systems Laboratory]

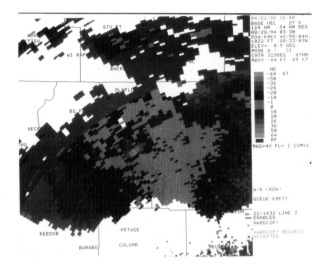

FIGURE 7.23
A sample Doppler weather radar image. Greens and blues (cold colors) indicate motion directly toward the radar whereas reds and yellows (warm colors) indicate motion directly away from the radar.

Measuring Precipitation

Rainfall and snowfall are routinely measured in terms of depth of accumulation over a specified interval of time, usually hourly and every 6 hrs and 24 hrs. Measurements are made directly by collecting a volume of rain or snow in a rain or snow gauge or estimated remotely by weather radar or satellite sensors.

RAIN AND SNOW GAUGES

Today precipitation is collected and measured using essentially the same instrument that was developed centuries ago: a container open to the sky. The first mention of a rain gauge in European literature appeared in 1639, although gauges were used in Asia much earlier. About 400 B.C., the Indian author Kautilya described the first known rain gauge in his manuscript *Arthastra*. The gauge consisted of a 45-cm (20-in.) diameter bowl, and measurements were taken regularly as a way of estimating the annual crop to be sown.

A standard **rain gauge** design consists of a cylinder equipped with a cone-shaped funnel at the top (Figure 7.24). The funnel directs rainwater into a narrower cylinder seated inside the larger outer cylinder. The funnel and narrow cylinder magnify the scale of measurement so that the instrument can resolve rainfall into increments as small as 0.01 in. (0.25 mm). A simple graduated stick is used to measure the depth of water that

accumulates in the inner cylinder. (The stick is graduated in inches in the U.S. but in millimeters in Canada and most other nations.) Rainfall of less than 0.005 in. (0.1 mm) is recorded as a *trace*. Rainfall is usually measured at some fixed time once every 24 hrs, and the gauge is then emptied. This type of rain gauge must be read manually.

Continuous monitoring of rainfall is often useful, especially in flood-prone areas. Accordingly, some rain gauges are designed to provide a cumulative record of rainfall through time. Two common designs are the weighing-bucket and tipping-bucket rain gauges. These recording rain gauges can be used in remote locations. A **weighing-bucket rain gauge** calibrates the weight of accumulating rainwater in terms of water depth. This instrument has a device that marks a chart on a clock-driven drum or sends an electronic signal to a computer for processing. At subfreezing temperatures, antifreeze in the collection bucket or a heater melts snow or ice pellets producing a cumulative record of melt water.

Compared to a weighing-bucket rain gauge, a **tipping-bucket rain gauge** does not perform as well at subfreezing air temperatures or at very high rainfall rates.

FIGURE 7.24
A standard National Weather Service non-recording rain gauge.

This instrument consists of a free-swinging container partitioned into two compartments, each of which can collect the equivalent of 0.01 in. of rainfall. Each compartment alternately fills with water, tips and spills its contents, and trips an electric switch that either marks a chart on a clock-driven drum or sends an electrical pulse to a computer for recording. A heated tipping-bucket gauge is a standard component of the NWS Automated Surface Observing System (ASOS).

For snow, meteorologists are interested in (1) the depth of snow that falls during the period between observations, (2) the melt water equivalent of that snowfall, and (3) the depth of snow on the ground at observation time. New snowfall accumulates on a simple wooden board placed on top of the old snow cover. Using a ruler graduated in tenths of an inch (or a meter stick graduated in centimeters), snow depth is measured to the board. The board is then swept clean and moved to a new location. Snowfall measurements, reported to the nearest 0.1 in., are made at regularly scheduled times, typically once daily.

The melt water equivalent of new snowfall can be determined by measurements taken by a weighing-bucket gauge or by melting the snow collected in a non-recording gauge (with the funnel and inner cylinder removed). The average density of fresh snow is about 0.1 gram per cubic centimeter so that as a very general rule, 10 cm of fresh snow melts to 1 cm of water. However, the actual snow/melt-water ratio varies considerably depending on the temperature at which the snow falls and the crystalline form of the snowflakes. Wet snow falling at surface air temperatures at or above 0 °C (32 °F) has much greater water content than dry snow falling at very low surface air temperatures. The ratio of snowfall depth to melt-water depth may vary from 3 to 1 for very wet snow to 30 to 1 or higher for dry fluffy snow. Snow composed of columnar crystals can be relatively dense whereas snow made up of dendritic (star-like) crystals (typical of lake-effect snow) has very low density with ratios as great as 50 to 1. As a snow cover ages, snow converts to tiny granules of ice, air-filled space is reduced, and snow density increases.

Snow depth on the ground is usually determined using a graduated rule inserted vertically into the snow at several representative locations. These measurements are averaged, rounded to the nearest inch, and reported daily. Care must be taken to avoid areas where the wind has produced snowdrifts.

Rainfall and snowfall can be highly variable from one place to another, especially when produced by convective showers and thunderstorms. Precipitation gauges are sited in order that measurements are accurate and as representative as possible. Instruments must be sheltered from strong winds, which cut the collection efficiency of rain gauges. Winds tend to accelerate over the open top of a gauge, reducing the amount of rain that enters the instrument. One study found that winds blowing at 16 km (10 mi) per hr reduced the amount of rainwater collected by 10% compared to the amount that would enter the gauge if the air were calm. Gauges should not be placed in windswept locations, but must be sited well away from buildings and tall vegetation that might shield the instrument from precipitation. As a general rule, obstacles should be no closer to the gauge than four times their height.

Traditionally, meteorologists derive spatial patterns of atmospheric variables from measurements made by instruments at discrete points, that is, at weather stations. This approach works reasonably well for variables such as air temperature and pressure that vary continuously from one place to another. Hence, the temperature is likely to be near 65 °F at a location half way between two weather stations, one reporting 60 °F and the other reporting 70 °F. Basing spatial patterns on measurements at discrete points does not work as well for discontinuous variables such as precipitation. Precipitation is typically highly variable spatially. Hence, it does not always follow that moderate amounts of rain fell at a location halfway between two weather stations, one reporting heavy rainfall and the other reporting light rainfall. Furthermore, regular precipitation measurements are missing from vast stretches of the ocean where most precipitation falls. To help remedy this situation, scientists developed radar technologies and satellite-borne instruments that provide more continuous measurements of precipitation over broad areas.

REMOTE SENSING OF PRECIPITATION

The *WSR-88D* weather radar determines not only the location and intensity of rainfall but also provides estimates of the total amount of rainfall. Whereas networks of precipitation gauges measure rainfall at discrete locations often more than 100 km (60 mi) apart, radar measures rainfall within continuous volume scans of the atmosphere. A special computer algorithm uses radar monitoring of rainfall rate to generate color-coded maps of rainfall totals over a specific period of time (Figure 7.25).

To estimate rainfall, the *WSR-88D* radar

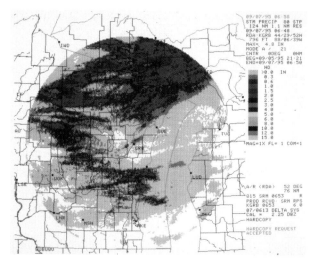

FIGURE 7.25
Cumulative rainfall as determined by computer analysis of radar echoes.

measures the reflectivity obtained from a volume scan that combines the lowest four tilt angles (0.5°, 1.5°, 2.5°, and 3.5° above the horizon) as frequently as every six minutes. Reflectivity is proportional to the *surface area* of raindrops whereas rainfall rate is proportional to the *volume* of the raindrops. Reflectivity data are converted to rainfall rate by estimating the raindrop size distribution and results are then checked against selected precipitation gauge measurements. If properly calibrated, radar can better represent hourly precipitation amounts over broad geographical areas than networks of precipitation gauges. This guidance information is particularly valuable in flash flood forecasting (Chapter 11).

Satellite sensors are also valuable remote sensing tools for precipitation. The Tropical Rainfall Measuring Mission (TRMM) utilizes satellite technology primarily to measure rainfall in the tropics. Knowledge of tropical rainfall and the associated flow of heat energy promises to improve our understanding of the global climate system and make possible more realistic global climate models. About two-thirds of all precipitation falls in the tropics and much of the planetary-scale heat transport originates in the tropics. TRMM is a joint venture of the National Aeronautics and Space Administration (NASA) and Japan's National Space Development Agency.

The TRMM satellite, in orbit about 350 km (215 mi) above Earth's surface, utilizes three sensors to measure rainfall remotely: radar, a microwave imager, and a visible/IR scanner. Satellite-based precipitation radar (PR) is similar to ground-based radar except that

the signal has a much shorter wavelength. PR obtains vertical profiles of rain and snow from Earth's surface to an altitude of about 20,000 m (65,500 ft) and has a horizontal resolution on Earth's surface of 4 km (2.5 mi). The TRMM Microwave Imager (TMI) measures the minute flux of microwave energy emitted by the Earth-atmosphere system. Variations in the amount of energy received at different wavelengths are interpreted to measure water vapor, cloud water, and rainfall intensity. Similar sensors have been flown aboard polar-orbiting Defense Meteorological Satellites since 1987. The TRMM Visible and Infrared Scanner (VIRS) monitors five spectral channels from visible to infrared (between 0.63 and 12 micrometers). The height of convective clouds (determined from IR-derived cloud-top temperature) is the basis for rainfall estimates; that is, the higher the cloud, the greater is the rainfall. The same type of precipitation-measuring scanner is used routinely on Polar Orbiting Environmental Satellites (POES) and Geostationary Operational Satellites (GOES) for weather analysis and forecasting.

Meteorologists also measure snow cover remotely, information that is valuable for water resource managers and flood forecasters. In mountainous areas where snowfall is considerable, the melt water equivalent of snow can be determined remotely using snow pillows. A *snow pillow* is a device that is filled with antifreeze solution and fitted with a manometer (an instrument that measures pressure changes) that is calibrated to give the water equivalent from the weight of the overlying snow cover. Data are radioed to a satellite and from there to a data center for downloading and analysis. The *Airborne Gamma Radiation Snow Survey Program* also provides remote sensing of the water content of snow packs. At designated times, low-flying aircraft measure terrestrial gamma ray emission along selected flight lines. Gamma radiation emitted by Earth materials attenuates as the snow pack thickens. Comparison of gamma emission over the same flight line with and without snow cover is calibrated in terms of the melt water equivalent of snow (or soil moisture).

Conclusions

As air nears saturation, water vapor begins condensing or depositing on nuclei. With continued cooling, usually due to expansion of rising air, clouds form. Clouds are distinguished on the basis of appearance (cirriform, stratiform, cumuliform), altitude of their base (high, middle, low, having significant vertical development), temperature (cold, warm), and composition (water droplets, ice crystals). Fog is a cloud in contact with Earth's surface and forms when air becomes saturated at low levels through radiational cooling, warm humid air moving over a relatively cold surface, cold air advection over a relatively warm, wet surface, or expansional cooling that accompanies upslope motion of air in mountainous areas.

Cloud droplets and ice crystals are much too small to fall to Earth's surface as precipitation. Through the collision-coalescence and Bergeron-Findeisen processes, cloud droplets and ice crystals grow large enough to survive the fall to Earth's surface as precipitation. Precipitation occurs as rain, drizzle, snow, ice pellets (sleet), freezing rain (or freezing drizzle), and hail. Weather radar sends out pulses of microwave energy that locate areas of precipitation and determine the intensity of rainfall. Operating in the Doppler mode, weather radar can monitor air motions within weather systems, making it possible to provide the public with advance warning of the potential for severe weather. A variety of direct and remote sensing techniques are employed by scientists to measure precipitation.

Atmospheric circulation plays a key role in bringing air to saturation and triggering cloud development. The next five chapters describe the many atmospheric circulation systems and their associated weather. We begin in Chapter 8 with a discussion of the forces that drive and shape atmospheric circulation.

Basic Understandings

- Clouds, the visible product of condensation or deposition within the atmosphere, are composed of tiny water droplets, ice crystals, or a combination of the two.
- As the relative humidity of air nears 100%, condensation and deposition occur on nuclei, tiny solid and liquid particles suspended in the atmosphere. Cloud condensation nuclei are much more abundant than ice-forming nuclei (i.e., freezing nuclei and deposition nuclei). Most ice-forming nuclei are active at temperatures well below the freezing point.
- Many condensation nuclei are hygroscopic; that is, they have a chemical affinity for water molecules

and induce condensation at relative humidities less than 100%.

- Clouds are classified by general appearance (cirriform, stratiform, cumuliform), altitude of base (high, middle, low, significant vertical development), temperature (cold, warm), and composition (water droplets, ice crystals).
- High, middle, and low clouds are caused by relatively gentle uplift of air over a broad geographical area. These clouds are layered, that is, stratiform. Clouds having significant vertical development are the consequence of more vigorous uplift and are heaped or puffy in appearance, that is, cumuliform.
- High clouds are composed of mostly ice crystals, whereas middle and low clouds are mostly water droplets or a mixture of ice crystals and supercooled water droplets.
- Nimbostratus and cumulonimbus are the principal precipitation-producing clouds. Precipitation from nimbostratus is typically lighter and of longer duration than precipitation from cumulonimbus.
- Atmospheric stability determines whether cumulus clouds build vertically into cumulus congestus or cumulonimbus clouds. The convective condensation level corresponds to the base of cumuliform clouds.
- Lee-wave clouds develop downwind of a prominent mountain range and are nearly stationary. Nacreous and noctilucent clouds are among the very few clouds that occur above the troposphere. Nacreous clouds form in the upper stratosphere where temperatures favor water in either the solid or supercooled state. They are cirrus or altocumulus lenticularis that form on sulfuric acid nuclei possibly of volcanic origin. Noctilucent clouds develop in the upper mesosphere and may be composed of ice deposited on meteoric dust particles.
- Fog is a visibility-restricting suspension of tiny water droplets or ice crystals in an air layer in contact with Earth's surface. Essentially, fog is a cloud at ground level. Based on mode of origin, fog is classified as radiation fog, advection fog, steam fog, or upslope fog.
- The relatively low terminal velocities of cloud droplets and ice crystals mean that they remain suspended in the atmosphere indefinitely unless they vaporize or undergo significant growth.
- The collision-coalescence and Bergeron-Findeisen processes are mechanisms whereby cloud particles grow large enough to counter updrafts and fall to

Earth's surface as precipitation.

- The collision-coalescence process occurs in warm clouds and requires the presence of relatively large cloud droplets that grow through collision and coalescence with smaller cloud droplets.

- The Bergeron-Findeisen process requires the coexistence of ice crystals, supercooled water droplets, and water vapor. At the same temperature, the saturation vapor pressure surrounding a supercooled water droplet is higher than the saturation vapor pressure surrounding an ice crystal. Hence, air that is saturated for supercooled droplets is supersaturated for ice crystals. Ice crystals grow while water droplets vaporize.

- The bulk of precipitation that falls at middle and high latitudes originates in cold clouds composed of a mixture of ice crystals and supercooled water droplets. Once the precipitation process begins, collision and coalescence promote growth of precipitation particles.

- Principal forms of precipitation are rain, drizzle, snow, ice pellets (sleet), freezing rain, and hail. The form of precipitation depends on the source cloud and the temperature profile (sounding) of the air beneath the cloud.

- Rain and snow fall from relatively thick clouds (nimbostratus or cumulonimbus) whereas drizzle falls from relatively thin clouds (stratus). Ice pellets are raindrops that freeze prior to reaching the ground, whereas freezing rain consists of supercooled drops that freeze on contact with surfaces at subfreezing temperatures. Hail is produced by intense thunderstorms having vigorous updrafts, an abundant supply of supercooled water droplets, and great vertical development.

- Weather radar is used to determine the location and movement of areas of precipitation. Operating in the reflectivity mode, radar echo strength increases with precipitation intensity. Operating in the Doppler mode, weather radar can be used to determine the detailed motion of air within a weather system.

- Precipitation is measured directly by collecting a volume of rain or snow in a rain or snow gauge or estimated remotely using weather radar or satellite sensors.

ESSAY: Clouds by Mixing

Outdoors on a cold winter day, we can sometimes "see our breath." We are actually seeing a small cloud formed by the mixing of our warm, humid breath with the colder, drier ambient air. Mixing of air masses (or air layers within the same air mass) that differ in temperature and vapor pressure is another mechanism whereby clouds form. Surprisingly, a cloud may form even though the two air masses are initially unsaturated.

As emphasized earlier, the saturation vapor pressure increases rapidly with rising temperature. This relationship is depicted schematically in the Figure below where temperature is plotted on the horizontal axis and vapor pressure is plotted on the vertical axis. The average temperature and average vapor pressure of a specific air mass plot as a single point on the diagram. An air mass that plots below the saturation vapor pressure curve is unsaturated, on the curve is saturated, and above the curve is supersaturated.

Suppose that two different unsaturated air masses plot as points *A* and *B* on the diagram. It is reasonable to assume that mixing the two air masses would produce a new air mass having properties (average temperature and vapor pressure) that would plot somewhere along a straight line connecting the points *A* and *B*. The precise location of that point (the mixture) along the line depends on the relative volumes of the two air masses. In any case, the mixture of the two air masses is unsaturated so that no cloud forms. Consider, however, two other unsaturated air masses, *C* and *D*, one cold and dry, the other warm and humid. In this case the straight line linking *C* and *D* intersects the saturation vapor pressure curve. The new air mass resulting from the mixing of equal volumes of air masses *C* and *D* is saturated and a cloud forms.

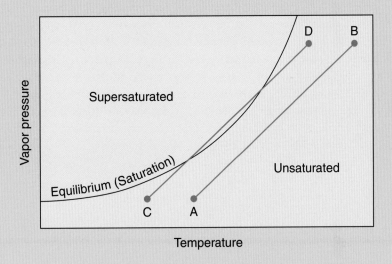

Figure
Variation of saturation vapor pressure with temperature. An air mass that plots as a point above the curve is supersaturated, on the curve is saturated, and below the curve is unsaturated.

This description of cloud formation by mixing provides insight as to the cause of jet aircraft contrails (Figure 7.1). The exhaust from jet engines mixes with the ambient air and may form a *contrail* (i.e., condensation trail) behind the aircraft. Heat and water are among the combustion products of jet engines so that the exhaust is hot and humid. Turbulence in the wake of a jet engine promotes the mixing of the exhaust with ambient air. If the ambient air has the appropriate combination of vapor pressure and temperature, the mixture will be saturated and a contrail will form. Such conditions are most likely in the upper troposphere where commercial jetliners travel.

ESSAY: When Is It Too Cold or Too Warm to Snow?

During an episode of particularly frigid winter weather, some people argue that it is too cold to snow. In fact, the coldest weather is often accompanied by fair skies, and the climate records of the northern United States and Canada indicate that snowfall totals decline with falling temperature. For some northern U.S. cities (e.g., Billings, MT, Huron, SD, and Minneapolis-St. Paul, MN), March on average is both the snowiest and the mildest of the winter months (December through March). In Canada, outside of the mountains, average annual snowfall declines from more than 400 cm (160 in.) in parts of Newfoundland to less than 100 cm (40 in.) along the frigid shores of the Arctic Ocean. But even though total snowfall decreases with falling temperature, snow is possible even at extremely low air temperatures. In bitter cold air, however, snowflakes are small and accumulations are usually meager.

The relatively small amount of water vapor in very cold air means that comparatively little water is available for precipitation (i.e., low amounts of *precipitable water*). Recall from Chapter 6 that the saturation vapor pressure drops rapidly as air temperature falls. For example, the water vapor concentration in saturated air at -30 °C (-22 °F) is only about 12% of the water vapor concentration in saturated air at -5 °C (23 °F). Hence, the amount of water potentially available for precipitation decreases with dropping air temperature.

Heaviest snowfalls typically occur when the temperature of the lower atmosphere is within a few degrees of the freezing point, because at that temperature, the potential amount of water that can precipitate in the solid form (as snow) is at a maximum. On the other hand, moderate or heavy snowfall is very unlikely when the temperature of the lower atmosphere falls below -20 °C (-4 °F). But even in the coldest regions of the globe where precipitable water amounts are lowest, some snow falls. For example, an estimated average annual 5 cm (2 in.) of snow falls on the high interior plateau of Greenland where average annual temperatures are below -30 °C (-22 °F).

When is it too warm to snow? Surprisingly, snow can fall even when the near-surface air temperature is as high as 10 °C (50 °F). The essential requirement is that the wet-bulb temperature remains below 0 °C (32 °F); that is, the relative humidity must be very low. For example, if the air temperature is 5 °C (41 °F), the relative humidity must be under 32% for the wet-bulb temperature to be subfreezing (see Table 6.5). The first snowflakes vaporize or melt as they fall through relatively dry, above-freezing air and reach the ground as raindrops. Vaporization and melting of snowflakes tap heat from the ambient air; that is, sensible heat is converted to latent heat, and the air temperature drops. With sufficient vaporization and melting, the air eventually cools to the wet-bulb temperature, that is, below the freezing point. Hence, at the Earth's surface, what started out as rain (or a mixture of rain and snow) turns to snow. This is most likely to happen if the precipitation is moderate to heavy—more vaporizing and melting snowflakes means more cooling.

ESSAY: Rainmaking

Since the mid-1940s, considerable research has gone into methods of enhancing precipitation by cloud seeding. *Cloud seeding* is an attempt to stimulate natural precipitation processes by injecting nucleating agents into clouds. (Cloud seeding also has been used to modify hail, fog, and lightning.) Most cloud seeding is directed at cold clouds although in recent years warm cloud seeding has received significant interest.

The objective of seeding cold clouds is to stimulate the Bergeron-Findeisen process (described elsewhere in this chapter) in clouds that are deficient in ice crystals. The seeding (nucleating) agent typically is either silver iodide (AgI), a substance with crystal properties similar to ice, or dry ice, solid carbon dioxide (CO_2), which sublimates at a temperature of about -80 °C (-112 °F). (Extremely cold liquid propane is also sometimes used.) Silver iodide crystals are freezing nuclei that are active at -5 °C (23 °F) and below. Dry ice pellets are so cold that within a cloud they cause surrounding supercooled water droplets to freeze. Frozen droplets then grow into snowflakes through the Bergeron-Findeisen process followed by collision and coalescence. Furthermore, latent heat released when supercooled water droplets freeze makes the cloud slightly warmer and more buoyant, stimulating additional cloud growth that can lead to more precipitation.

Clouds can be seeded via aircraft or from the ground. In either case, silver iodide is released into updrafts at cloud base from liquid fuel generators or pyrotechnic flares. Only one gram of silver iodide can produce as many as 10^{15} particles. Dry ice pellets are dropped through an opening in the aircraft floor directly into a cloud's updraft. Dry ice pellets are less efficient at producing ice crystals than silver iodide; as much as 2000 grams of dry ice are needed to match the output of just one gram of silver iodide. Of the two methods of seeding, cloud seeding from ground-based generators, while relatively inexpensive, can be less satisfactory than airborne seeding because the seeding agent may not diffuse adequately and reach sufficient altitudes to be effective. Nonetheless, both methods are used.

The traditional method of seeding warm clouds is to inject hygroscopic substances (e.g., sea salt crystals) into clouds to stimulate formation of relatively large cloud droplets that grow into raindrops by collision and coalescence. In the early 1990s, this technique was given a boost by the development of a new cloud-seeding flare (based on a U.S. Navy fog-producing flare) that when ignited releases large quantities of microscopic salt particles (mostly potassium chloride). A series of flares mounted on an aircraft wing releases salt particles. The plane flies just below cloud base so that updrafts carry salt particles into the (cumuliform) cloud. Initial results from experimental use of this technique in South Africa and Mexico suggest that flares are more effective than the old methods of dispensing hygroscopic nucleating agents in liquid sprays. In August 2000, Roelof Bruintjes and his colleagues at the National Center for Atmospheric Research (NCAR) reported satistically significant rainfall enhancement using this technique during a 3-year cloud seeding experiment carried out over drought-prone northern Mexico.

The Sierra Cooperative Pilot Project (SCPP) is an example of a long-term, ongoing study of cloud seeding. Scientists seed *orographic clouds* over the windward western slopes of California's Sierra Nevada Mountains with the goal of increasing snowfall and thereby thickening the mountain snow pack. The consequent increase in spring runoff is intended to help California meet its growing domestic and agricultural water demands. The American River Basin, just west of Lake Tahoe, is the principal SCPP study site, and January through March is the primary seeding season. Clouds targeted for seeding are rich in supercooled water droplets but deficient in ice crystals. To identify clouds best suited for seeding, SCPP scientists employ an aircraft that is outfitted with sophisticated instruments that measure the size and concentration of cloud and precipitation particles. Another aircraft seeds selected (cold) clouds with silver iodide or dry ice pellets. Meanwhile, on the ground, an array of precipitation gauges, weather radar, and other weather instruments monitor the effectiveness of seeding.

Does cloud seeding work? In the 1970s and early 1980s, NOAA scientists conducted a statistically rigorous experiment designed to test the effectiveness of cloud seeding. The experiment was carried out over southern Florida and involved the seeding of cumulus clouds. The testing period was about evenly divided between days when clouds were seeded with silver iodide crystals and days when clouds were seeded with inert (chemically inactive) sand grains. Only after the experiment was completed and the results analyzed were participating scientists informed of the specific days when silver iodide was the seeding agent. This procedure was intended to ensure both an unbiased selection of clouds to be seeded and an unbiased interpretation of results. The Florida seeding experiment consisted of two phases. Results from an initial *exploratory* phase were to be either verified or rejected by a later *confirmatory* phase when seeding was repeated. Results of the first phase were very encouraging, showing a 25% increase in rainfall on days when silver iodide was the seeding agent, compared to days when sand was the bogus seeding agent. This finding was statistically significant at the 90% level, meaning that there is only a 10% probability that the increased rainfall was a chance occurrence. The success of the initial phase was not repeated during the second phase when seeding brought no statistically significant increase in rainfall.

Although cloud seeding is successful in some instances, the actual amount of additional precipitation produced by cloud seeding and the advisability of large-scale seeding efforts are controversial. Some cloud seeders claim to increase precipitation by 15% to 20% or more, but the question remains as to whether the rain or snow that follows cloud seeding would have fallen anyway. Even if apparently successful, some people argue that cloud seeding may merely redistribute a fixed supply of precipitation, so that an increase in precipitation in one area might mean a compensating reduction in another. For example, rainmaking might benefit agricultural interests on the High Plains of eastern Colorado but also deprive wheat farmers of rain in the adjacent downwind states of Kansas and Nebraska. Such conflicts can erupt in legal battles between residents of adjacent counties, states, and provinces.

The potential for cloud seeding success is greatest in areas already receiving considerable precipitation, as in the SCPP effort described above. Unfortunately, the potential for successful cloud seeding in arid or drought-prone areas is less than promising because the necessary atmospheric conditions are usually not present either because of insufficient atmospheric moisture or lack of suitable lifting mechanism to produce clouds.

CHAPTER 8

WIND AND WEATHER

No one can tell me,
Nobody knows,
Where the wind comes from,
Where the wind goes.

A.A. MILNE
"Wind on the Hill" Now We Are Six

Case-in-Point

A popular ballad by Gordon Lightfoot memorialized the November 1975 sinking of the ore carrier *Edmund Fitzgerald* with a loss of its crew of 29 in the storm-tossed waters of Lake Superior. At the time, the 222-m (729-ft) long ship was the largest ore carrier on the Great Lakes. Early on the afternoon of 9 November, the ship was fully loaded with 26,116 tons of iron ore (taconite) pellets when it left the Duluth-Superior harbor at the far western end of Lake Superior on a northeast course at about 20 knots (23 mi per hr). The ship's destination was a steel plant on Zug Island in the Detroit River.

At 6 a.m. (CST) on 9 November, a cyclone (low-pressure system) began organizing over central Kansas. From there, the intensifying storm tracked toward the northeast and at 6 a.m. on 10 November passed near La Crosse, WI, and at noon was centered just west of Marquette, MI. By this time the storm's central pressure had dropped to 982 mb and gale-force northeast winds were sweeping the eastern end of Lake Superior. Winds gusted to 115 km (71 mi) per hr at Sault Ste Marie, MI.

At 1 a.m. (CST) on 10 November, the *Edmund*

Fitzgerald reported northeast winds at 97 km (60 mi) per hr with waves to 3 m (10 ft). At 7 a.m., the ship was about 73 km (45 mi) north of Copper Harbor, MI and reporting northeast winds at 65 km (40 mi) per hr. With the storm bearing down, the *Edmund Fitzgerald* changed course to the east and then southeast hugging the north shore of Lake Superior. The ship's captain, Ernest McSorley, chose a course that would take the ship through waters that were sheltered from the strong northeast winds. That afternoon, the storm passed over the *Edmund Fitzgerald*. In the evening, as the storm center neared Moosonee, ON along the shore of James Bay, winds on Lake Superior shifted from northeast to north and then northwest and west. The longer fetch of the northwest and west wind over the lake caused the waves to grow. Another ship within several kilometers of the *Edmund Fitzgerald* estimated winds at 95 km (58 mi) per hr gusting to 137 km (85 mi) per hr and waves of 3.5 m to 5 m (12 to 16 ft).

Sometime between 6:15 and 6:25 p.m. (CST), the *Edmund Fitzgerald* foundered and sank in 163 m (535 ft) of water about 27 km (17 mi) northwest of Whitefish Point, MI. More than 1000 ships have sunk in the Great Lakes and most of these wrecks were weather-related. The *Edmund Fitzgerald* was the largest ever to go down.

Driving Question:

What forces control the wind?

Some weather systems are responsible for clear skies, light winds, and frosty mornings, whereas others bring ominous clouds, precipitation, and biting winds. Some weather systems trigger brief showers, and others are accompanied by persistent fog and drizzle. Certain weather systems dominate the weather over thousands of square kilometers for weeks at a time. Different weather systems bring different types of weather depending on the air circulation (wind) that characterizes each system.

Wind, the principal focus of this chapter, is the local motion of air measured relative to the rotating Earth. The atmosphere is coupled to the planet and rotates with it. Once every 24 hrs, all points on Earth's surface and in the atmosphere (except right at the poles) complete a circular path in space. The circumference of that path decreases with increasing latitude; hence, the speed at which fixed points on Earth progress eastward

decreases with increasing latitude. The speed drops from 1670 km (1035 mi) per hr at the equator to 835 km (517 mi) per hr at 60 degrees N and S, and to zero at the poles. We are not aware of this rapid motion because we also move with the rotating planet and its atmosphere at the same speed. In meteorology, we are interested in air motion measured *relative to* Earth's continually moving surface (i.e., the wind).

In Chapter 4, we saw how unequal rates of radiational heating and cooling produced temperature gradients within the Earth-atmosphere system. In response to temperature gradients between Earth's surface and the troposphere as well as between low latitudes and polar latitudes, the atmosphere circulates and heat is redistributed. This chapter covers the various forces that either initiate or modify atmospheric circulation, that is, the wind. First, each force is examined separately as if it acted independently of all the other forces. Forces are then combined to demonstrate how together they drive atmospheric circulation. We begin by learning how wind speed and direction are measured.

Monitoring Wind Speed and Direction

The velocity of the wind is a *vector* quantity, that is, it has both magnitude (speed) and direction. A distinction is usually made between *horizontal* (east-west and north-south) and *vertical* (up-down) components of the wind. Except in small, intense weather systems such as thunderstorms, the magnitude of vertical air motion is typically only 1 % to 10 % of the horizontal wind speed. Nonetheless, as demonstrated in Chapter 6, the vertical component of the wind plays the key role in cloud formation by promoting expansional cooling. Furthermore, as shown later in this chapter, vertical and horizontal wind components are linked so that a change in one may be accompanied by a change in the other.

The most common wind-monitoring instruments are designed to measure only the horizontal component of the wind. For some specialized research purposes, very sensitive instruments are available that measure vertical wind speeds or a combination of vertical and horizontal wind components. An ordinary **wind vane** consists of a free-swinging horizontal shaft with a vertical plate at one end and a counterweight (arrowhead) at the

FIGURE 8.1
An airport windsock gives wind direction and a general indication of wind speed. The sock points downwind.

or in degrees. Measured clockwise from true north, an east wind is specified as 90 degrees, a south wind as 180 degrees, a west wind as 270 degrees, and a north wind as 360 degrees. Wind direction is reported as 0 degrees only during calm conditions (when the wind speed is zero).

Wind speed can be estimated by observing the wind's effect on lake or ocean surfaces or on land-based flexible objects such as trees. Such observations are the basis of the **Beaufort scale**, which is a graduated sequence of wind strength ranging from 0 for calm conditions to 12 for hurricane-strength winds (Table 8.1). The scale bears the name of Sir Francis Beaufort, who developed it in the early 1800s while a ship's commander in the British Navy. Beaufort's goal was to standardize terms used by sailors in describing the state of the sea under various wind conditions. In 1838, after some revi-

other end (Figure 1.5). The counterweight always points directly into the wind. Another design is the airport **windsock**, which consists of a cone-shaped cloth bag that is open at both ends (Figure 8.1). The larger end of the sock is held open by a metal ring that is attached to a pole and is free to rotate. Air enters the larger opening and stretches the sock downwind.

Wind direction is always designated as the direction *from which* the wind blows. A wind blowing from the east toward the west is described as an *east* wind and a wind blowing from the northwest to the southeast is a *northwest* wind. A wind vane may be linked electronically or mechanically to a dial that is calibrated to read in points of the compass

TABLE 8.1
Beaufort Scale of Wind Force

Beaufort Number	General description	Land and sea observations for estimating wind speed	Wind speed 10 m above ground (km/hr)
0	Calm	Smoke rises vertically; sea like mirror	<1
1	Light air	Smoke but not wind vane shows direction of wind; slight ripples at sea	1-5
2	Light breeze	Wind felt on face, leaves rustle, wind vane moves; small, short wavelets.	6-11
3	Gentle breeze	Leaves and small twigs move constantly, small flags extended; large wavelets, scattered whitecaps.	12-19
4	Moderate breeze	Dust and loose paper raised, small branches moved; small waves, frequent whitecaps.	20-28
5	Fresh breeze	Small leafy trees swayed; moderate waves.	29-38
6	Strong breeze	Large branches in motion, whistling heard in utility wires; large waves, some spray.	39-49
7	Near gale	Whole trees in motion; white foam from breaking waves.	50-61
8	Gale	Twigs break off trees; moderately high waves of great length.	62-74
9	Strong gale	Slight structural damage; crests of waves begin to roll over, spray may impede visibility.	75-88
10	Storm	Trees uprooted, considerable structural damage; sea white with foam, heavy tumbling of sea.	89-102
11	Violent storm	Very rare, widespread damage; unusually high waves.	103-118
12	Hurricane	Very rare, disastrous; much foam and spray, greatly reduced visibility.	≥119

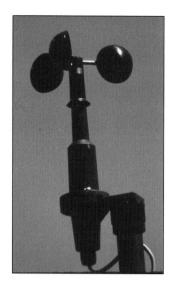

FIGURE 8.2
A cup anemometer measures wind speed; the greater the wind speed, the faster the cups spin.

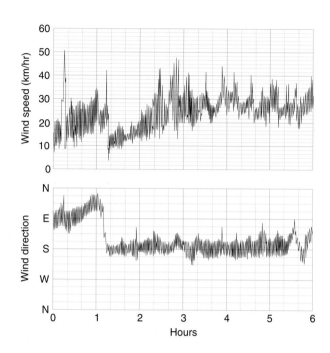

FIGURE 8.3
Continuous trace of time variations in wind speed and direction over a six-hour period.

sion, the British Navy adopted the Beaufort scale, and in 1853, it was sanctioned for international use by seafarers. Later, when the scale was extended from sea to land, it was necessary to develop wind speed equivalents for each Beaufort number; this was done in 1926. The scale is still used today. Although crude measures, some modern-day mariners prefer Beaufort numbers to onboard instrument measurements of wind speed.

A **cup anemometer** consists of 3 or 4 open hemispheric cups mounted to spin horizontally on a vertical shaft (Figure 8.2). At least one open cup faces the wind at any time. The rotation rate of the cups is calibrated to read in m per sec, km per hr, or knots. (One knot = 1 nautical mi per hr = 0.515 m per sec = 1.15 mi per hr.) Several other types of anemometers are available, including the very sensitive **hot-wire anemometer**. In this instrument, the wind blows past a heated wire or wires and the heat lost to moving air is calibrated in terms of wind speed.

Recording a continuous trace of wind speed and direction can be informative. Wind vanes and anemometers are linked to pens that record on a paper chart that is attached to a clock-driven drum. For example, the trace in Figure 8.3 indicates considerable variation in both wind direction and wind speed with time. The spectrum of wind gusts and lulls indicates turbulent air. More commonly today, a computer records the output from wind vanes and anemometers.

Ideally, a wind vane or anemometer system should be mounted on a tower so that the instruments

monitor horizontal winds 10 m (33 ft) above the ground. (This is the standard height for National Weather Service anemometers.) Rooftop locations should be avoided because winds tend to accelerate over buildings. In addition, the system should be sited well away from (1) structures that might shelter the instruments and (2) any obstacles that might channel (and thus accelerate) the wind.

Wind vanes and anemometers monitor winds near Earth's surface. In Chapter 2, we saw how meteorologists measure winds aloft by tracking the movements of a balloon-borne radiosonde (a rawinsonde observation). Over the past decade, the Doppler effect has been used in a new technology designed to monitor winds up to an altitude of about 16,000 m (52,500 ft). For more on these *wind profilers*, refer to this chapter's first Essay.

This chapter's second Essay, Wind Power, describes how modern technology is tapping the energy of air in motion.

Forces Governing the Wind

Several forces influence the wind. *A force* is defined as some push or pull that can cause an object at rest to move or that alters the movement of an object already in motion. Like the wind, a force has both direction and magnitude (a *vector* quantity).

In examining the various forces that govern the wind, it is useful to apply those forces to an *air parcel* containing a unit mass (e.g., single kilogram) of air. Imagine the wind to be a continuous stream of air composed of discrete air parcels. Assume that any force acting on an air parcel has the same influence on a stream of air parcels, in other words, on the wind. Hence, in describing each force that influences air motion, we examine the force per unit mass of air.

In meteorology, *force* and *acceleration* are terms that are sometimes used interchangeably. This practice follows from **Newton's second law of motion,** where

force = (mass) × (acceleration).

If the air parcel is a unit mass, a force per unit mass is numerically equivalent to acceleration. Although the terms are numerically equivalent, acceleration (a change in velocity) is actually a response to a force. A force acts on an air parcel to change the acceleration of that parcel. Furthermore, acceleration involves a change in speed or direction or both. We have more on this later.

Forces acting on air parcels, which either initiate or modify motion, are the consequence of (1) air pressure gradients, (2) the centripetal force, (3) the Coriolis effect, (4) friction, and (5) gravity. Actually, the centripetal force is not an independent force but occurs as a consequence of other forces.

PRESSURE GRADIENT FORCE

A gradient is simply a change in some property over distance. An **air pressure gradient** exists whenever air pressure varies from one place to another. As noted in Chapter 5, spatial variations in air pressure can arise from contrasts in air temperature (principally), from differences in water vapor concentration, or from both. In addition, diverging and converging winds can bring about air pressure changes and thereby induce an air pressure gradient.

Air pressure gradients occur both horizontally and vertically within the atmosphere. A horizontal pressure gradient refers to air pressure changes along a surface of constant altitude (mean sea level, for example). A vertical pressure gradient refers to the air pressure change directly above a point on Earth's surface and is a permanent feature of the atmosphere because air pressure is greatest at the Earth's surface and always

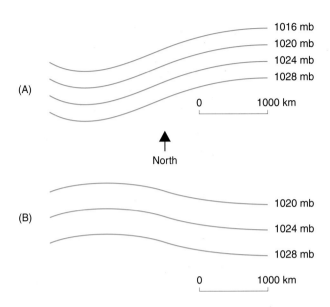

FIGURE 8.4
In this map view, the horizontal air pressure gradient is relatively steep (A) where isobars are closely spaced and weaker (B) where isobars are farther apart. Isobars are lines joining places of equal air pressure. Here the contour interval (difference between successive isobars) is 4 mb.

decreases with altitude. Horizontal air pressure gradients can be determined on weather maps from patterns of **isobars**, lines joining points having the same air pressure. But first, to eliminate the influence of station elevation on air pressure, the barometer reading at each weather station is adjusted to sea level (Chapter 5). By convention, isobars are drawn at 4-mb (4-hPa) intervals and interpolation between weather stations is always necessary.

An isobaric analysis is used to locate centers of high and low pressure and to determine the magnitude of the horizontal air pressure gradient between weather systems. Where isobars are closely spaced (Figure 8.4A), air pressure changes rapidly with distance, and the pressure gradient is described as steep or strong. More widely spaced isobars (Figure 8.4B) indicate that air pressure changes more gradually with distance, and the pressure gradient is weaker. Note that air pressure gradients are always measured in the direction of greatest change, that is, *perpendicular* to isobars.

How do air pressure gradients influence the motion of air? Consider an analogous situation. Suppose a bathtub is partially filled with water, as shown in Figure 8.5. Sloshing the water back and forth from one end of the tub to the other creates rapidly changing water pressure gradients along the tub bottom. At any instant, the water surface is not horizontal, and the bottom water

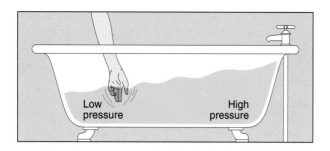

FIGURE 8.5
Sloshing water back and forth from one end of a tub to the other end creates water pressure gradients along the bottom of the tub. The water pressure gradient along the tub bottom is analogous to a horizontal air pressure gradient in the atmosphere. In response to a pressure gradient, water (or air) flows from an area of higher pressure toward an area of lower pressure.

pressure is high where the water level is high, and low where the water level is low. If we stop agitating the water, the sloshing back and forth dampens, and the undulating water surface returns to horizontal. The water pressure along the bottom is then the same from one end of the tub to the other. Hence, in response to a water pressure gradient, water flows from one end of the tub (where the water pressure is greater) to the other end (where the water pressure is less), thereby eliminating the pressure gradient.

Similarly, when an air pressure gradient develops, air flows to eliminate the pressure gradient. The force that causes air parcels to move as the consequence of an air pressure gradient is known as the **pressure gradient force** and it always acts directly across isobars and towards low pressure. The magnitude of the pressure gradient force is inversely related to isobar spacing. The wind is strong where the pressure gradient is steep (closely spaced isobars), and light or calm where the pressure gradient is weak (widely spaced isobars).

CENTRIPETAL FORCE

Isobars plotted on surface weather maps are almost always curved, indicating that the direction of the pressure gradient force changes from one place to another. Consequently, the wind blows in curved paths. Curved motion indicates the influence of the centripetal force.

The centripetal force can be illustrated by a simple demonstration. A rock is tied to a string and then whirled about so that the tethered rock describes a circular orbit of constant radius (Figure 8.6). The string

exerts a net force on the rock by confining it to a curved (circular) path. At any instant, the net force is directed inward, perpendicular to the direction of motion, and toward the center of the circular orbit. For this reason, the net force is known as the **centripetal** (*center-seeking*) **force**. If we cut the string, the centripetal force no longer operates; that is, there is no longer a net force confining the rock to a curved path. The rock flies off in a straight line as described by **Newton's first law of motion**; that is, an object in straight-line, unaccelerated motion remains that way unless acted upon by an unbalanced force.

A net force (i.e., an unbalanced force) causes acceleration. We usually think of acceleration as a change in speed, as when an automobile speeds up. But acceleration is a vector quantity; that is, it has both magnitude and direction, and acceleration may consist of a change in either speed or direction, or both. In our rock-on-a-string example, centripetal force is responsible only for a continual change in the direction of the rock (curved rather than straight-line path); the rock neither speeds up nor slows down. The acceleration imparted to a unit mass by the centripetal force is directly proportional to the square of the velocity and inversely proportional to the radius of curvature.

The centripetal force is not an independent force; rather, it arises from the action of other forces and may be the consequence of imbalances in other forces. In our rock-on-a-string example, the tension of the string is responsible for the centripetal force. Consider another example. Suppose you are a passenger in an auto that

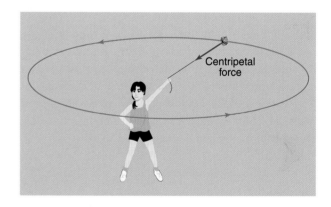

FIGURE 8.6
A rock tied to a string follows a circular path as it is whirled about. Centripetal (*center-seeking*) is the name of the force that confines an object to a curved path. If the string is cut, the centripetal force is eliminated, and the rock moves in a straight line (tangent to the circular path).

rounds a curve at a high speed. You feel a force that pushes you outward from the turning auto. Actually, you experience the tendency for your body to continue moving in a straight path while the auto follows a curved path. In this case, the frictional resistance of the tires against the pavement provides the centripetal force.

A centripetal force operates whenever air parcels follow a curved path. As we will see later in this chapter, the centripetal force arises from an imbalance in other forces operating in the atmosphere.

CORIOLIS EFFECT

Imagine that you are located far away at some fixed point in space, looking back at planet Earth. Over many hours, you follow the track of a storm, which is clearly identifiable by a slowly swirling mass of clouds. From your perspective, the storm center appears to move in a straight line at constant speed. At the same time, an observer on Earth is tracking the storm using radar and surface weather observations. From that observer's perspective, the storm center appears to follow a curved path. Surprisingly, both descriptions of the storm's track are correct!

The two descriptions are correct because the two observers used different frames of reference in following the storm's movements. The Earthbound observer's frame of reference is the solid Earth to which the familiar north-south, east-west, and up-down

coordinate system is attached. To the Earthbound observer, it is not obvious that this coordinate system is rotating because it and the observer rotate along with the turning Earth. Viewed from space, it is evident that the Earth-bound coordinate system actually shifts as Earth rotates (Figure 8.7). In fact, Earth and the coordinate system rotate under the storm (or any other object moving freely over Earth's surface). Meanwhile, from your location in space, you followed the storm's movement with respect to a non-rotating coordinate system, fixed in space with respect to the background stars. In summary, the difference in observed storm tracks (curved versus straight) arise from the differences in coordinate systems (rotating versus non-rotating).

Recall our earlier discussion of *Newton's first law of motion* and the centripetal force. We saw that curved motion implies that a net (or unbalanced) force is operating, whereas unaccelerated, straight-line motion implies a balance of forces. Applying this law to our storm track example, we conclude that a net force operates when we use the Earthbound rotating coordinate system, whereas forces are balanced when we use the non-rotating coordinate system fixed in space. Hence, changing our frame of reference (coordinate system) from non-rotating to rotating gives rise to a net force responsible for curved motion. This deflective force is known as the **Coriolis effect**, named for Gaspard Gustav de Coriolis, who first described the phenomenon

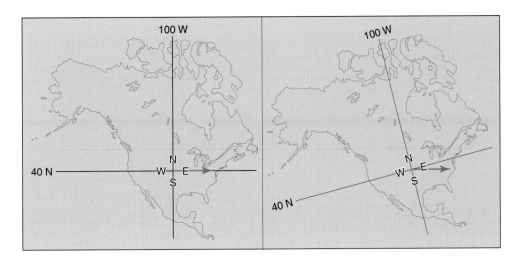

FIGURE 8.7
Viewed from a fixed point in space, the familiar north-south, east-west frame of reference rotates eastward in space as Earth rotates on its axis. Rotation of this coordinate system gives rise to the Coriolis deflection. In this case, a west wind shifts to the northwest.

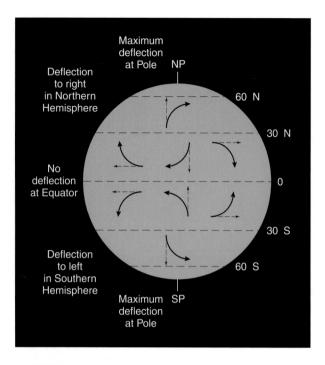

FIGURE 8.8
Large-scale winds are deflected to the right of their initial direction in the Northern Hemisphere and to the left of their initial direction in the Southern Hemisphere. This deflection, known as the Coriolis effect, is maximum at the poles and zero at the equator.

mathematically in 1835. Wind direction and speed are measured with respect to the north-south, east-west, and up-down frame of reference fixed to the rotating planet. Therefore, we must take the Coriolis effect into account in any explanation of air circulation. The Coriolis force deflects the horizontal wind to the right of its initial direction in the Northern Hemisphere and to the left of its initial direction in the Southern Hemisphere (Figure 8.8).

Why does the Coriolis deflection reverse direction between the hemispheres so that large-scale winds in the Southern Hemisphere swerve to the left rather than to the right? This reversal is related to the difference in an observer's sense of Earth's rotation in the two hemispheres. To an observer looking down from high above the North Pole, the planet rotates counterclockwise, whereas to an observer high above the South Pole, the planet rotates clockwise. For the observer measuring motion relative to a coordinate system fixed to the rotating with Earth, this reversal in the direction of rotation between the two hemispheres translates into a reversal in Coriolis deflection.

Although the Coriolis force influences the wind regardless of its direction, the amount of deflection varies significantly with latitude (as the *sine* of the latitude).

Earth's rotation on its axis imparts a rotation to our Earth-bound frame of reference that is maximum at the poles and decreases in magnitude with decreasing latitude to zero at the equator. This variation with latitude can be understood by visualizing the daily rotation of towers situated at different latitudes. In a 24-hr day, Earth makes one complete rotation, as would a tower if located at the North or South Pole. In the same period, a tower at the equator would not rotate at all because of its orientation perpendicular to Earth's axis of rotation. At any latitude in between, some rotation of a tower occurs but not as much as at the poles. The Coriolis force is thus latitude-dependent: the Coriolis deflection is zero at the equator and increases with latitude to a maximum at the poles.

The magnitude of the Coriolis force also varies with wind speed and spatial scale of atmospheric circulation. The Coriolis deflection increases as the wind strengthens because, in the same period of time, faster moving air parcels cover greater distances than slower air parcels. The longer the trajectory, the greater is the shift of the rotating coordinate system with respect to a moving air parcel. For practical purposes, the Coriolis force significantly influences the wind only in large-scale weather systems, that is, systems larger than ordinary thunderstorms. Large-scale weather systems also have longer life expectancies than small-scale systems so that air parcels cover greater distances over longer periods of time.

A rotational motion usually accompanies the draining of water from a sink or a bathtub. It is a popular misconception that the direction of this rotation (clockwise or counterclockwise) is consistently in one direction in the Northern Hemisphere and in the opposite direction in the Southern Hemisphere, presumably because of the Coriolis effect. At the very small scale represented by the water in a sink or bathtub, the magnitude of the Coriolis force is simply too small to significantly influence the direction of rotation. Drainage direction is more likely a consequence of some already present motion of the water or the shape of the sink or bathtub and may be either clockwise or counterclockwise.

FRICTION

Friction is the resistance that an object or medium encounters as it moves in contact with another object or medium. We are all familiar with the friction associated with solid objects, as when we attempt to slide a heavy appliance across the floor. But friction also

FIGURE 8.9
Rocks in a riverbed are obstacles that break the current into turbulent eddies downstream from the rocks, slowing the stream. A similar frictional interaction takes place as wind encounters obstacles on Earth's surface.

FIGURE 8.10
A snow fence slows the wind, reducing its ability to transport snow. Hence, snow accumulates immediately downwind from the snow fence.

affects fluids, both liquids and gases. The friction of fluid flow is known as **viscosity** and is of two types, *molecular viscosity* and *eddy viscosity*. One source of fluid friction is the random motion of molecules composing a liquid or gas; this type of fluid friction is called **molecular viscosity**. Considerably more important, however, is fluid friction that arises from much larger irregular motions, called *eddies*, which develop within fluids; this type of fluid friction is known as **eddy viscosity**.

A swiftly flowing stream illustrates the effects of eddy viscosity. Rocks in the streambed obstruct the flow of water and cause the current to break into eddies immediately downstream of the rocks (Figure 8.9). Eddies, visible as swirls of water, tap some of the stream's kinetic energy so that the stream slows. In an analogous manner, obstacles on Earth's surface such as trees, houses, and telephone poles break the wind into eddies of various sizes to the lee of each obstacle. Consequently, the wind slows.

A snow fence provides a practical illustration of the frictional slowing of the wind (Figure 8.10). Snow fences are designed to trap wind-blown snow, in some instances to prevent snow from drifting onto a nearby highway and in others to keep the soil snow covered. (Snow that accumulates downwind of a snow fence ensures a supply of soil moisture and prevents the soil from freezing to great depths.) A snow fence taps some of the wind's kinetic energy by breaking the wind into small eddies. The wind speed diminishes, losing some of its snow-transporting ability, and snow accumulates on the downwind side of the fence.

The rougher the surface of the Earth, the greater is the eddy viscosity of the wind. A forest thus offers more frictional resistance to the wind than does the smoother surface of a freshly mowed lawn. Eddy viscosity diminishes rapidly with altitude above Earth's surface, away from obstacles mainly responsible for frictional resistance. Hence, horizontal wind speed increases with altitude. This explains the advantage of siting a windmill at as high an elevation above surrounding land as possible. Above an average altitude of about 1000 m (3300 ft) above Earth's surface, friction is a minor force that has little impact on the smooth flow of air. The atmospheric zone to which frictional resistance (eddy viscosity) is essentially confined is called the **atmospheric boundary layer**.

Turbulence is fluid flow characterized by eddy motion. Various obstacles on Earth's surface exert a drag on the wind and are sources of eddies. Irregular fluid flow that originates in this way is known as *mechanical turbulence*. In addition, eddies develop in air as a consequence of solar heating of the ground; irregular fluid flow that originates in this way is known as *thermal turbulence*. Convection currents are examples of thermal turbulence within the atmosphere. In actual practice, it is virtually impossible to distinguish between the two sources of turbulent eddies (mechanical or thermal).

Regardless of source, we experience turbulent eddies as gusts of wind. The gustiness of the wind often varies with the time of day; that is, gusts tend to be strongest during the warmest hours of the day.

GRAVITY

Air parcels are subject to the same force that holds all objects on Earth's surface, that is, **gravity**. Gravity is the net result of two forces working together: gravitation and centripetal force. *Gravitation* is the force of attraction between Earth and all other objects; its magnitude is directly proportional to the product of the masses of the objects and inversely proportional to the square of the distance between their centers of mass. The much weaker centripetal force is imparted to all objects because of their rotation with Earth on its axis. Combined as gravity, the two forces accelerate a unit mass of any object directly downward toward Earth's surface at the rate of 9.8 m per sec each second.

Gravity always acts directly downward. For this reason, gravity, unlike the Coriolis effect and friction, does not modify the horizontal wind. Gravity influences air that is ascending or descending, such as in convective currents, and gravity is responsible for the downhill drainage of cold, dense air.

SUMMARY

We have now examined individually the various forces that influence horizontal and vertical air motion, and we can draw the following conclusions:

1. The *pressure gradient force* accelerates air parcels perpendicular to isobars away from regions of high air pressure and toward regions of low air pressure. Acceleration is directly proportional to the pressure gradient; that is, the closer the spacing of isobars, the steeper the pressure gradient, and the greater is the acceleration.

2. A *centripetal force* is an imbalance of actual forces and exists whenever the wind describes a curved path. It is responsible for a change in wind direction, but not wind speed.

3. The *Coriolis effect* arises from the rotation of Earth on its axis and deflects large-scale winds to the right of their initial direction in the Northern Hemisphere and to the left of their initial direction in the Southern Hemisphere. The Coriolis deflection increases with latitude from zero at the equator to a maximum at the poles and is directly proportional to wind speed.

4. *Friction* acts opposite to the wind direction and increases with increasing surface roughness. Friction slows horizontal winds blowing within about 1000 m (3300 ft) of Earth's surface.

5. *Gravity* accelerates air downward toward the Earth's surface but it does not modify horizontal winds.

Joining Forces

To this point, we have examined forces operating in the atmosphere as if each force acted independently of all the others. In reality, these forces interact with one another in governing both the direction and speed of the wind. In some cases, two or more forces achieve a balance or equilibrium. From *Newton's first law of motion*, when the forces acting on an air parcel are in balance, there is no net force, and the parcel either remains stationary or continues to move along a straight path at constant speed. When forces are balanced, net acceleration is zero.

Let us now examine how forces interact in the atmosphere to control the vertical and horizontal flow of air, that is, the wind. These interactions result in (1) hydrostatic equilibrium, (2) the geostrophic wind, (3) the gradient wind, and (4) surface winds, horizontal winds within the atmospheric boundary layer.

HYDROSTATIC EQUILIBRIUM

Air pressure always declines with increasing altitude so that a vertical air pressure gradient is a permanent feature of the atmosphere. As shown schematically in Figure 8.11, the force due to this pressure gradient is directed upward from higher air pressures at Earth's surface toward lower air pressures aloft. If this force acted alone, the vertical pressure gradient force would accelerate air away from Earth, and

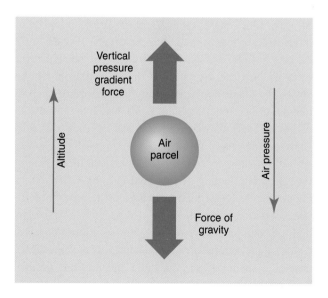

FIGURE 8.11
In the atmosphere, hydrostatic equilibrium is the balance between the upward-directed pressure gradient force and the downward-directed force of gravity.

we would be left gasping for breath. However, except for relatively brief periods in some small-scale violent weather systems (e.g., severe thunderstorms), the atmosphere's vertical pressure gradient force is almost balanced by the equal but oppositely directed force of gravity. An actual balance of these two forces is known as **hydrostatic equilibrium**.

Whenever forces are in balance, net acceleration is zero; that is, there is no change in speed or direction. Hydrostatic equilibrium, however, does not preclude vertical (up or down) motion of air. With balanced forces, ascending air parcels continue moving upward at *constant* velocity, and descending air parcels continue moving downward at *constant* velocity. Slight deviations from hydrostatic equilibrium cause air parcels to change speed (accelerate) vertically.

GEOSTROPHIC WIND

The **geostrophic wind** is an unaccelerated, horizontal wind that follows a straight path at altitudes above the atmospheric boundary layer. It results from a balance between the horizontal pressure gradient force and the Coriolis force. The Coriolis deflection is significant only in broad-scale circulations so that the geostrophic wind develops only in large-scale weather systems.

Consider the traditional description of the evolution of geostrophic equilibrium: An air parcel is placed in a preexisting horizonal pressure field where isobars are straight and parallel (Figure 8.12). In response to the horizontal pressure gradient force (P_H), the air parcel initially accelerates directly across isobars, away from high pressure and toward low pressure. As the air parcel accelerates, however, the Coriolis force (C) strengthens and causes the air parcel to turn gradually to the right of its initial flow direction (in the Northern Hemisphere). The Coriolis force changes direction as the parcel turns, remaining at right angles to the parcel's direction of motion. The two forces eventually act in opposite directions and attain a balance, known as *geostrophic equilibrium*, so that the geostrophic wind blows at a constant speed in a straight path parallel to isobars with the lowest air pressure to the left of the direction of air motion.

Numerical simulations of the interactions of the pressure gradient force and the Coriolis force conducted by J.A. Knox and colleagues at NASA's Goddard Institute for Space Studies show that air parcels actually undergo an oscillatory motion along the curved path shown in Figure 8.12. The oscillatory motion gradually dampens as air parcels approach geostrophic equilibrium.

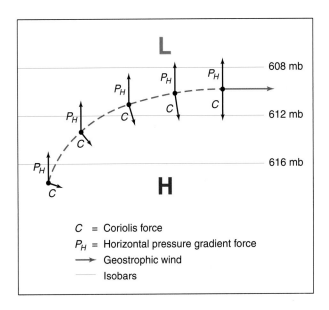

FIGURE 8.12
This diagram approximates the evolution of geostrophic equilibrium, a balance between the horizontal pressure gradient force and the Coriolis force. The geostrophic wind blows parallel to straight isobars at altitudes above the atmospheric boundary layer.

This *inertial oscillation* is an interaction of the pressure field with the motion of the air parcels and may be an important feature of some weather systems.

GRADIENT WIND

The **gradient wind** has many characteristics in common with the geostrophic wind. It is also large-scale, horizontal, and frictionless, and blows parallel to isobars. The important distinction between the two is that the geostrophic wind blows in a straight path, whereas the path of the gradient wind is curved. Forces are not balanced in the gradient wind because a net centripetal force constrains air parcels to a curved trajectory. Recall from earlier in this chapter that the centripetal force changes only the direction and not the speed of an air parcel. The horizontal pressure gradient force, the Coriolis force, and the centripetal force interact in the gradient wind.

A gradient wind develops at altitudes above the atmospheric boundary layer around a dome of high air pressure, called an **anticyclone** (or *High*), or around a center of low air pressure, called a **cyclone** (or *Low*). In an idealized anticyclone, isobars form a series of concentric circles about the location of highest air pressure, as shown in Figure 8.13. The horizontal pressure gradient force (P_H) is directed radially outward,

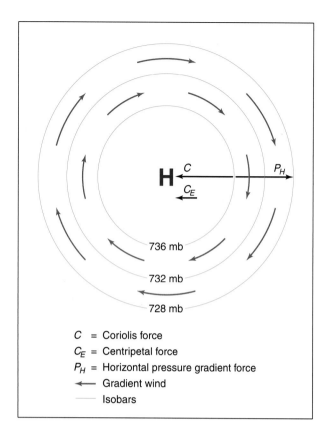

FIGURE 8.13
Viewed from above in the Northern Hemisphere, the gradient wind blows clockwise and parallel to isobars in an anticyclone. In this idealized case, isobars form a pattern of concentric circles.

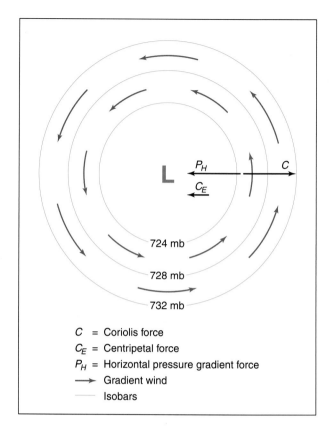

FIGURE 8.14
Viewed from above in the Northern Hemisphere, the gradient wind blows counterclockwise and parallel to isobars in a cyclone. In this idealized case, isobars form a pattern of concentric circles.

away from the center of the high. The Coriolis force (C) is directed inward. The Coriolis force is slightly greater than the pressure gradient force, with the difference between the two forces giving rise to the inward-directed centripetal force (C_E). (This is what was meant earlier when we indicated that a centripetal force results from an imbalance of other forces.) Viewed from above, in a Northern Hemisphere anticyclone above the atmospheric boundary layer, the gradient wind blows clockwise and parallel to isobars.

In an idealized cyclone, isobars form a series of concentric circles about the location of lowest air pressure. As indicated in Figure 8.14, the horizontal pressure gradient force (P_H) is directed inward toward the cyclone center, and the Coriolis force (C) is directed radially outward from the center of the low. The pressure gradient force is slightly greater than the Coriolis force, with the difference being equal to the net inward-directed centripetal force (C_E). Viewed from above, in a Northern

Hemisphere cyclone above the atmospheric boundary layer, the gradient wind blows counterclockwise and parallel to isobars.

The geostrophic and gradient winds are models (Chapter 2) that only approximate the actual behavior of horizontal winds above the atmospheric boundary layer. These approximations are nonetheless quite useful, and meteorologists routinely rely on such approximations in their analysis of weather maps.

SURFACE WINDS

Geostrophic winds and gradient winds are frictionless; that is, they occur at altitudes where frictional resistance is insignificant. How does friction affect horizontal winds within the atmospheric boundary layer? Intuitively, we know that friction should slow the wind, but in addition, friction interacts with the other forces and alters the wind direction.

For large-scale air motion along a straight path,

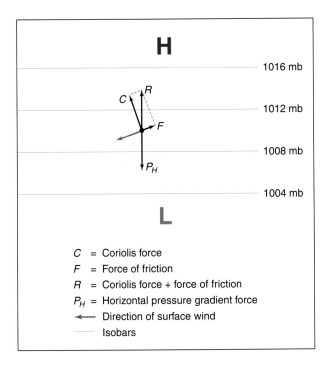

FIGURE 8.15
Within the atmospheric boundary layer, friction slows the wind and shifts the wind across isobars toward low pressure.

the frictional force (*F*) combines with the Coriolis force (*C*) to balance the horizontal pressure gradient force (P_H) as shown in Figure 8.15. Friction acts directly opposite (180 degrees) to the wind direction whereas the Coriolis force is always at a right angle (90 degrees) to the wind direction. Friction slows the wind and thereby weakens the Coriolis deflection so that the Coriolis no longer balances the horizontal pressure gradient force. The horizontal pressure gradient force (P_H) is balanced by the resultant (*R*) of the Coriolis (*C*) plus friction (*F*). Friction (due to the roughness of Earth's surface) slows the horizontal wind and shifts the wind direction across isobars and toward low pressure.

The angle between wind direction and isobars depends on friction, which in turn depends on the roughness of Earth's surface. That angle varies from 10 degrees or less over relatively smooth surfaces, where friction is minimal, to almost 45 degrees over rough terrain, where friction is greater. As we noted earlier, friction's influence on the horizontal wind diminishes with altitude and where it becomes negligibly small marks the top of the atmospheric boundary layer. Horizontal winds strengthen with altitude through the atmospheric boundary layer. Furthermore, the angle between wind direction and isobars is maximum near

Earth's surface, decreases with altitude, and is essentially zero at the top of the atmospheric boundary layer (Figure 8.16). Above the atmospheric boundary layer, the horizontal wind is either geostrophic (where isobars are straight) or gradient (where isobars are curved).

How does surface roughness affect horizontal surface winds blowing in an anticyclone and cyclone? As with straight-line surface winds, friction slows cyclonic and anticyclonic winds and combines with the Coriolis force to shift winds so that they blow across isobars and toward low pressure. Viewed from above, surface winds in a Northern Hemisphere anticyclone blow clockwise and spiral outward, as shown in Figure 8.17, and surface winds in a Northern Hemisphere cyclone blow counterclockwise and spiral inward, as shown in Figure 8.18.

Characteristic surface winds in a cyclone are the basis for a simple rule of thumb for locating cyclone centers. If you, in your Northern Hemisphere location, stand with your back to the wind and then turn approximately 45 degrees to your right, the cyclone

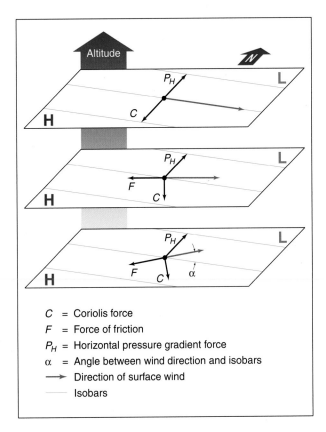

FIGURE 8.16
For the same horizontal air pressure gradient, the angle between the wind direction and isobars decreases with altitude within the atmospheric boundary layer.

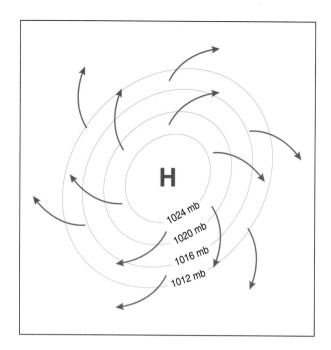

FIGURE 8.17
Viewed from above in the Northern Hemisphere, surface winds blow clockwise and outward in an anticyclone.

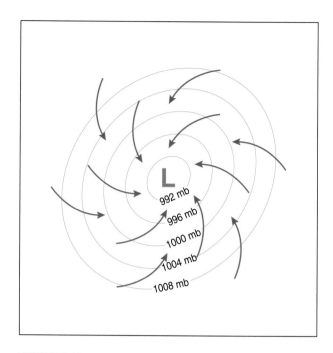

FIGURE 8.18
Viewed from above in the Northern Hemisphere, surface winds blow counterclockwise and inward in a cyclone.

center will be located to your left. This rule is a modification of an observation first stated in 1857 by Dutch meteorologist Christopher H. D. Buys-Ballot. It must be applied with caution, however, because isobars are not always circular in cyclones and large-scale surface winds may be modified by local air circulation such as a sea breeze.

In the Southern Hemisphere, anticyclonic and cyclonic winds are opposite their Northern Hemisphere counterparts. The difference is due to the Coriolis deflection acting to the left of the direction of motion in the Southern Hemisphere, opposite to its effect in the Northern Hemisphere. Viewed from above in the Southern Hemisphere, surface winds in a cyclone blow clockwise and spiral inward, whereas surface winds in an anticyclone blow counterclockwise and spiral outward. Above the atmospheric boundary layer, Southern Hemisphere anticyclonic winds blow counterclockwise and parallel to isobars, and Southern Hemisphere cyclonic winds blow clockwise and parallel to isobars.

A glance at almost any national weather map reveals that isobars seldom describe lengthy straight segments or circular patterns (Figure 8.19). Isobars often form more complicated patterns of *ridges* (anticyclonic curves) and *troughs* (cyclonic curves). In ridges and troughs, winds tend to parallel isobars above the

atmospheric boundary layer and cross isobars toward low pressure near Earth's surface. An additional consideration in analyzing isobaric patterns for wind is the spacing of isobars. As noted earlier, the steeper the air pressure gradient, the faster is the wind. Where isobars are closely spaced, the geostrophic and gradient winds are relatively strong. Where isobars are widely spaced, these winds are weak. The same rule applies to surface winds.

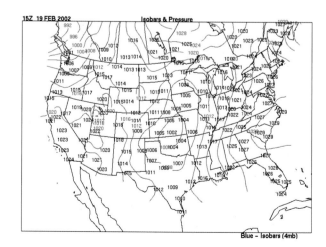

FIGURE 8.19
On a typical surface weather map, isobars exhibit clockwise curvature (*ridges*) and counterclockwise curvature (*troughs*).

Continuity of Wind

Air is a continuous fluid, and this *continuity* implies a link between the horizontal and vertical components of the wind. For example, surface winds are forced to follow Earth's undulating topography, ascending hills and descending into valleys. In addition, uplift occurs along frontal surfaces as one air mass advances and either overrides or pushes under another retreating air mass (Chapter 6). Having examined the horizontal circulation in anticyclones and cyclones, we can identify other important connections between the horizontal and vertical components of the wind.

As noted earlier in this chapter, surface winds in a Northern Hemisphere anticyclone spiral clockwise and outward from its high pressure center. Consequently, horizontal surface winds diverge from the center of the high. A vacuum does not develop at the center, however, because air slowly descends toward Earth's surface and replaces the air that is diverging. Aloft, horizontal winds converge above the center of the surface high (Figure 8.20). Recall that adiabatic compression raises the temperature and saturation vapor pressure, causes existing clouds to vaporize, and lowers the relative humidity of descending clear air. Skies therefore tend to be clear within anticyclones, and anticyclones are appropriately described as *fair weather* systems. Furthermore, within an anticyclone, the horizontal air pressure gradient is typically very weak over a broad area around the center of the system. Light winds or calm air coupled with clear skies and low water vapor content favor intense nocturnal radiational cooling. Air adjacent

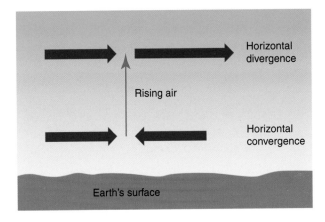

FIGURE 8.21
In this idealized vertical cross-section of a cyclone, surface winds converge, air ascends, and winds aloft diverge.

to the ground may be chilled to saturation so that dew, frost, or radiation fog may develop (Chapter 7).

Surface winds in a Northern Hemisphere cyclone spiral counterclockwise and inward. Surface winds therefore converge toward the center of a low. Air does not simply pile up at the center; rather, air ascends in response to converging surface winds and diverging winds aloft (Figure 8.21). Recall that adiabatic expansion lowers the temperature and saturation vapor pressure, thereby increasing the relative humidity of ascending air. Clouds and precipitation may eventually develop, so that cyclones are typically *stormy weather* systems.

Continuity of the wind also means that vertical motion can be induced by downwind changes in surface roughness. The rougher the Earth's surface, the greater the resistance it offers to horizontal winds. When the horizontal wind blows from a rough surface to a relatively smooth surface, as when it blows from land to sea, the wind accelerates. As shown in Figure 8.22, this acceleration causes the wind to diverge (stretch), thereby inducing downward motion of air (an example of *speed divergence*). In contrast, when the horizontal wind blows from a smooth to a rough surface, the wind slows and converges (piles up), thereby inducing upward air motion (an example of *speed convergence*). This is one reason why, along a coastline, cumuliform clouds (cumulus) tend to develop with an onshore wind (directed from sea to land) and tend to dissipate with an offshore wind (directed from land to sea). Frictionally-induced convergence of surface winds also plays an important role in the development of *lake-effect snow*, as discussed in this chapter's third Essay.

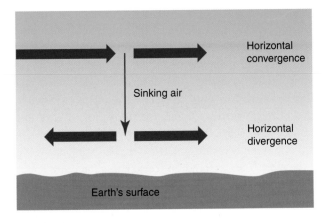

FIGURE 8.20
In this idealized vertical cross-section of an anticyclone, horizontal winds converge aloft, air descends, and surface winds diverge.

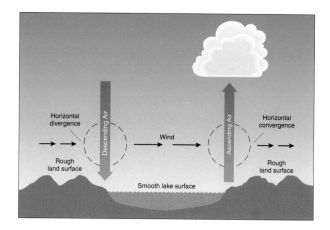

FIGURE 8.22
Surface winds accelerate and undergo horizontal divergence when blowing from a rough to a smooth surface (e.g., from land to water). Surface winds slow and undergo horizontal convergence when blowing form a smooth to a rough surface (e.g., from water to land). Divergence of surface winds causes air to descend whereas convergence of surface winds causes air to ascend.

Scales of Weather Systems

Although the atmosphere is a continuous fluid, for convenience of study we subdivide atmospheric circulation into discrete weather systems that operate at various spatial and temporal scales (Table 8.2). The large-scale wind belts encircling the planet (polar easterlies, midlatitude westerlies, and trade winds) are global or **planetary-scale systems**. **Synoptic-scale systems** are continental or oceanic in scale; migrating cyclones, hurricanes, and air masses are examples. **Mesoscale systems** include thunderstorms and sea and lake breezes, circulation systems that are so small that they may influence the weather in only a portion of a large city or county. A weather system covering a very small area (e.g., a weak tornado) represents the smallest spatial

TABLE 8.2
Scales of Atmospheric Circulation

Circulation	Space scale	Time scale
Planetary scale	10,000 to 40,000 km	weeks to months
Synoptic scale	100 to 10,000 km	days to week
Mesoscale	1 to 100 km	hours to day
Microscale	1 m to 1 km	seconds to hour

subdivision of atmospheric motion, **microscale systems**.

Circulation systems not only differ in spatial scale, they also contrast in life expectancy. Patterns in the planetary-scale circulation may persist for weeks or even months. Synoptic-scale systems typically last for several days to a week or so. Mesoscale systems usually complete their life cycles in a matter of hours to perhaps a day, whereas microscale systems might persist for minutes or less. Other differences exist among the various scales of atmospheric circulation systems. At the micro- and mesoscale, vertical wind speeds may be comparable in magnitude to horizontal wind speeds. At the synoptic and planetary scales, however, horizontal winds are considerably stronger than vertical flow. Furthermore, at the micro- and mesoscale, the Coriolis effect is usually negligibly small. In contrast, the Coriolis effect is very important in synoptic- and planetary-scale circulation systems.

Each smaller scale weather system is part of, and dependent on, larger scale atmospheric circulation; that is, the various scales of atmospheric motion form a kind of hierarchy. For example, extreme nocturnal radiational cooling requires a synoptic weather pattern that favors clear skies and light or calm winds. At the microscale, such weather conditions may be accompanied by formation of frost or radiation fog in a river valley. In Chapters 9 through 12, we apply basic understandings of the atmosphere in examining weather systems of all scales, starting with those operating at the planetary scale.

Conclusions

Unequal rates of radiational heating and radiational cooling within the Earth-atmosphere system give rise to gradients in temperature. In response to those temperature gradients, the atmosphere circulates and thereby, heat energy is converted to kinetic energy. In this chapter, we have examined the various forces that initiate and shape atmospheric circulation (the wind). Note that the pressure gradient force and gravity would exist even if air were not in motion, and the other forces (centripetal, Coriolis, and friction) come into play only after air is in motion.

From everyday experience, we are aware of the end product of the forces described in this chapter, namely, the wind, but we are not readily aware of the individual forces. In learning about atmospheric forces and how they interact, we see that each force is bound by certain constraints. For example, friction is important only in the lower troposphere, and the Coriolis effect always shifts the wind to the right in the Northern Hemisphere. In the following chapters, our awareness of these and other constraints will aid in understanding the characteristics of the various weather systems; for example, why hurricanes do not form at the equator, and why winds in a tornado may blow in either a clockwise or counterclockwise direction. We are now ready to begin our discussion of the characteristics of the various circulation systems, beginning with those operating at the planetary scale.

Basic Understandings

- Wind is the movement of air measured relative to Earth's surface. Wind has both direction and magnitude and is usually divided into horizontal and vertical components. A wind vane and anemometer are the usual instruments for monitoring surface winds.
- The horizontal wind is governed by interactions of the pressure gradient, Coriolis, friction, and centripetal forces.
- The pressure gradient force initiates air motion and arises in part from spatial variations in air temperature and, to a lesser extent, water vapor concentration. In response to gradients in pressure, air accelerates away from areas of relatively high air pressure, directly across isobars, and toward areas of relatively low air pressure.
- The centripetal force arises from an imbalance in actual forces and operates whenever the wind follows a curved path. The centripetal force is responsible for a change in wind direction and not a change in wind speed.
- The Coriolis force arises from Earth's rotation on its axis. Wind is deflected to the right of its initial direction in the Northern Hemisphere and to the left in the Southern Hemisphere. The force employed to describe the deflective effect is zero at the equator and increases with latitude to a maximum at the poles. The Coriolis force increases as wind speed increases and is most important in large-scale (planetary and synoptic-scale) circulation systems.
- Friction affects horizontal winds blowing within about 1000 m (3300 ft) of Earth's surface (the

atmospheric boundary layer). Obstacles on Earth's surface slow the wind by breaking it into turbulent eddies.
- Gravity always accelerates objects directly downward and is important in vertical motion of air (e.g., cold air drainage).
- Hydrostatic equilibrium is the balance between the upward-directed pressure gradient force resulting from the decrease of air pressure with altitude and the downward-directed force of gravity. Slight deviations from hydrostatic equilibrium cause air to accelerate upward or downward.
- The geostrophic wind is an unaccelerated, horizontal wind that blows in a straight path parallel to isobars at altitudes above the atmospheric boundary layer. The geostrophic wind results from a balance between the horizontal pressure gradient force and the Coriolis force.
- The gradient wind is a horizontal wind that parallels curved isobars at altitudes above the atmospheric boundary layer. The centripetal force operates in the gradient wind as the result of an imbalance between the horizontal pressure gradient force and the Coriolis force. Viewed from above in the Northern Hemisphere, the gradient wind blows clockwise in anticyclones and counterclockwise in cyclones.
- In large-scale (synoptic- and planetary-scale) circulation systems, friction slows the near-surface wind and interacts with the Coriolis force to shift the wind direction across isobars toward lower pressure.
- Within the atmospheric boundary layer, viewed from above, horizontal winds blow clockwise and outward in Northern Hemisphere anticyclones but counterclockwise and inward in Northern Hemisphere cyclones.
- In an anticyclone, horizontal divergence of surface winds causes air above to descend and warm by compression. Hence, an anticyclone is generally a fair-weather system. In a cyclone, horizontal convergence of surface winds causes ascending air, which expands and cools. Hence, a cyclone is frequently a stormy weather system.
- Along a coastline, surface winds blowing offshore speed up and undergo horizontal divergence (inducing descending air), whereas onshore winds are slowed by friction and undergo horizontal convergence (inducing ascending air).
- Atmospheric circulation is typically divided into four spatial/temporal scales: planetary, synoptic, mesoscale, and microscale.

ESSAY: Wind Profilers

A recent application of Doppler radar is providing meteorologists with much more detailed surveillance of winds aloft than is possible by tracking radiosondes (a *rawinsonde* observation). A *wind profiler* consists of a wire-mesh antenna about half the size of a tennis court that sends radar signals (having wavelengths of 33 cm to 6 m) upward into the atmosphere (see Figure). The signal detects changes in atmospheric density caused by turbulent mixing of volumes of air that differ slightly in temperature and humidity. Fluctuations in the index of refraction are used as a tracer of both the horizontal and vertical components of the wind.

In May 1992, the National Oceanic and Atmospheric Administration (NOAA) announced that the nation's first wind profiler network was up and operating. The network consists of 29 wind profilers in 15 Midwestern States. Each wind profiler monitors wind speed and direction at 72 different levels up to an altitude of 16,000 m (52,500 ft). Data are automatically collected every 6 minutes and transmitted to a quality-control hub in Boulder, CO. Wind data are then averaged over 1-hr periods and made available to National Weather Service Forecast Offices across the nation.

The principal advantage of a wind profiler over a radiosonde is the much greater frequency of wind observations. As noted in Chapter 2, radiosondes are launched once every 12 hrs. Winds aloft can change significantly in much shorter periods. Such changes could, for example, alter the track of a cyclone so that a city near the storm's path receives snow instead of rain. More detailed monitoring of winds aloft also permits airlines to better plan flight routes that avoid strong headwinds and take advantage of tailwinds. And, in recent years, temperature profiling and moisture-sensing capabilities have been added to some wind profilers.

FIGURE
A wind profiler located in northwestern Missouri.

ESSAY: Wind Power

Harnessing the kinetic energy of the wind is a technology that was well established as early as the 12th century in portions of the Middle East where water power was not available. In North America, the energy crisis of the 1970s spurred renewed interest in this ancient technology. More recent efforts to reduce our nation's dependency on fossil fuels in view of concerns over possible global climate change has rekindled interest in wind power. Today, scientists and engineers employ modern aerodynamic principles and space-age materials in designing and constructing modern wind-driven turbines that convert some of the wind's kinetic energy into electricity.

In Chapter 4, we saw how the sun drives the atmosphere. Only about 2% of the solar energy that reaches Earth is ultimately converted to the kinetic energy of wind, but that is still a tremendous amount of energy. Theoretically, windmill blades can convert a maximum of about 60% of the wind's energy into mechanical energy. In practice, however, wind generators extract at best only about 25% of the wind's energy. Furthermore, minimum wind speeds must be at least 15 km (9 mi) per hour for most wind-powered electricity generating systems to operate economically. The amount of power that a wind turbine can extract from air in motion varies directly with (1) the cube of the instantaneous wind speed at hub height (V^3), (2) the area swept out by the windmill blade, and (3) air density.

Wind speed is by far the most important consideration in evaluating a location's wind energy potential. Even small changes in wind speed translate into great variations in energy harvested. For example, doubling the wind speed (a common occurrence) multiplies the available wind power by a factor of eight: $2 \times 2 \times 2$. Wind speed usually increases with altitude above Earth's surface, especially in the lowest meters of the atmosphere, so that a wind turbine is mounted on the top of a tower to take advantage of the stronger winds at that elevation. As a general rule, a minimum of several years of detailed wind monitoring is needed for a preliminary evaluation of wind power potential at any site. Also, the long-term climate record should also be consulted to check for frequency of potentially destructive winds. Wind data from nearby weather stations can be very useful, as long as care is taken in extrapolating data between localities and from one elevation to another. (Recall that standard National Weather Service anemometers are located 10 m or 33 ft above the ground.)

At ordinary speeds, wind is a relatively diffuse energy source, comparable in magnitude to incoming solar radiation. Hence, a wind turbine's power generation potential also depends on the area swept out by the windmill blades; larger windmill blades can harvest more energy. Until recently, design and strength-of-materials problems limited the size of wind turbines. Today, availability of stronger and lighter materials for blades are making possible larger wind turbines that can generate as much as 0.5 megawatt of electricity. One wind turbine design uses three 23-m (75-ft) blades (see Figure).

FIGURE
This windmill farm in northeastern Wisconsin is designed to meet the electrical energy needs of about 3600 households.

Tapping the wind's energy for electric power generation faces several challenges. Both wind speed and direction vary with exposure of the site, roughness of the terrain, and season of the year. The most formidable obstacle to the development of wind power potential stems from the inherent variability of the wind. The electrical output of wind turbines varies as a consequence, and a wind power system must include a means of storing the energy generated during gusty periods for use when the wind is light or the air is calm. Banks of batteries may serve this purpose. Wind power has its greatest immediate potential in regions where average winds are relatively strong and consistent in direction. Favorable regions include the western High Plains, the Pacific Northwest coast, portions of coastal California, the Great Lakes region, the south coast of Texas, and exposed summits and passes in the Rockies and Appalachians.

Economy of scale favors centralized arrays of many wind turbines, called *windmill farms*, over individual household wind turbines. Windmill farms consist of a dozen or more super wind turbines, each capable of producing as much as 0.5 megawatt of electricity. Windmill farms currently operating in California supplement conventional sources by feeding electricity directly to power grids. Since 1981, more than 16,000 turbines have been installed in windy mountain passes in California and by 2005, windmill farms are expected to supply about 10% of the state's total electrical needs.

ESSAY: Lake-Effect Snow

A highly localized fall of snow immediately downwind of an open lake in mid-latitudes of the Northern Hemisphere is known as *lake-effect snow*. Typically, such snows extend inland only a few tens of kilometers. In fact, the snowfall is often confined to such a small area that weather stations do not detect the event except by radar or direct observer reports. Residents of the affected area, however, may be swamped by snow. On 21 January 1994, Adams, NY at the eastern end of Lake Ontario, received 152 cm (60 in.) of road-clogging snow in only 18 hrs. Such extreme snowfall is sometimes called a *snowburst*.

Lake-effect snow is most common in autumn and early winter when lake surface temperatures are still relatively mild. As early season outbreaks of cold (arctic) air stream over the lake, water readily evaporates and raises the vapor pressure of the lowest portion of the advecting air mass. (This same process may also produce *steam fog*.) In addition, the warmer lake water heats the advecting cold air from below, reducing its stability and enhancing convection and cloud development. Often this is all that is needed to trigger snowfall over the lake.

As the modified (milder, more humid, and less stable) air flows toward the lake's lee (downwind) shore, the contrast in surface roughness between the lake and land becomes important. The rougher land surface slows onshore winds, and the consequent horizontal convergence induces ascent of air, further development of clouds, and lake-effect snow. The topography of the downwind shore also affects the amount of precipitation: Hilly terrain forces greater uplift and enhances snowfall.

Ultimately, the frequency and intensity of lake-effect snows hinge on the degree of air mass modification, which in turn, depends on (1) the temperature contrast between the relatively warm lake surface and overlying cold air, and (2) the distance the wind travels over open water (called *fetch*). Field studies conducted in the Great Lakes region indicate that lake-effect snow is most likely when the temperature difference between the lake-surface and an altitude of about 1500 m (5000 ft) is greater than 13 Celsius degrees (23 Fahrenheit degrees) and the temperature contrast between the lake-surface and the adjacent land surface exceeds 10 Celsius degrees (18 Fahrenheit degrees). Ideally, the fetch of the wind must be at least 160 km (100 mi) and the horizontal wind direction must exhibit very little shear with altitude [varying by less than 30 degrees between the lake-surface and an altitude of about 3000 m (10,000 ft)].

To the lee of the Great Lakes, the bulk of lake-effect snows falls between mid-November and mid-January, the period when the temperature contrast between lake-surface and overlying air is usually greatest. Cold air usually sweeps into the Great Lakes region on west and northwest winds. Considering the maximum possible fetch, the greatest potential for substantial lake-effect snows is along the downwind southern and eastern shores of the Great Lakes. In these so-called *snowbelts*, lake-effect snow accounts for a substantial portion of total seasonal snowfall (see Figure 1 map).

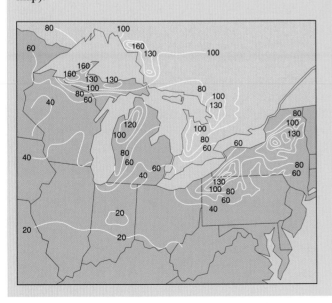

Figure 1
Average annual snowfall in the Great Lakes region (in inches). Note that the greatest snowfall totals occur downwind of the lakes and are largely due to lake-effect snows. [NOAA data]

Occasionally, however, lake-effect snow develops on the western shores of the lakes which are normally upwind. For example, an early winter cyclone tracking through the lower Great Lakes region may produce strong northeast winds which blow onshore on the western shores. In this case, sometimes distinguished as *lake-enhanced snow*, the lake's snow-generating mechanism adds to the snowfall produced by the cyclone. This can mean paralyzing accumulations of snow for the Milwaukee-Chicago metropolitan area.

Strong cold winds blowing over the Great Lakes from the west and northwest produce lake-effect snow that falls from cloud bands that are oriented parallel to the wind direction. These *wind-parallel snow bands* are shown in the visible satellite image in Figure 2. Lake-effect snow can also develop with relatively weak regional winds. With a sufficient temperature contrast between water and land surfaces (more than 10 Celsius degrees or 18 Fahrenheit degrees), a horizontal air pressure gradient develops between the land and lake with higher pressure over the colder land surface and lower pressure over the warmer lake surface. In response to this pressure gradient, a cold wind develops that is directed offshore (called a *land breeze*). The leading edge of the land breeze is like a miniature cold front that forces the warmer lake air upward; clouds form and snow develops along bands oriented parallel to the shoreline. *Shore-parallel snow bands* are shown in the visible satellite image in Figure 3. Where shore-parallel snow bands drift ashore, they usually produce greater snowfalls than wind-parallel snow bands.

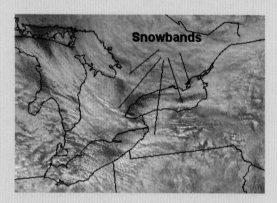

Figure 2
A visible satellite image showing wind-parallel snow bands.

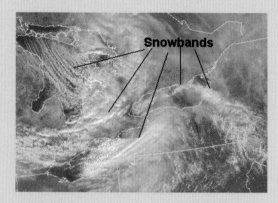

Figure 3
A visible satellite image showing shore-parallel snow bands.

In addition to the Great Lakes region, lake-effect snows also affect the valleys southeast of Utah's Great Salt Lake. Each year perhaps a half dozen significant lake-effect snowfalls occur in the Tooele and Salt Lake valleys, bowl-shaped depressions that slope up and away from Great Salt Lake. Both valleys parallel the northwest-southeast-trending long axis of the lake. Which of the two valleys receives the heavier snowfall depends on wind direction. Cold air on northwest winds is moderated by contact with the relatively warm lake surface, and downwind, the mountain valley topography forces horizontal convergence and uplift. Clouds billow upward and locally release bursts of heavy snow.

Great Salt Lake snowbursts are generally less intense than their Great Lakes counterparts. Major reasons are the smaller area of the lake and the shorter fetch of winds blowing over the lake, so that air masses do not modify as much. The waters of Great Salt Lake cover an area that is only about 13% of that of Lake Ontario, the smallest of the five Great Lakes. Maximum fetch on Great Salt Lake is only about 120 km (75 mi), whereas maximum fetches on the Great Lakes are many hundreds of kilometers.

CHAPTER 9

ATMOSPHERE'S PLANETARY CIRCULATION

The fair breezes blew, the white foam flew,
The furrow follow'd free;
We were the first that ever burst
Into that silent sea.
Down dropt the breeze, the sails dropt down,
'Twas sad as sad could be
And we did speak only to break
The silence of the sea!
All in a hot and copper sky,
The bloody Sun, at noon,
Right up above the mast did stand,
No bigger than the Moon.
Day after day, day after day,
We stuck, nor breath nor motion;
As idle as a painted ship
Upon a painted ocean.

SAMUEL TAYLOR COLERIDGE

Case-In-Point

In 1982-83, the weather seemed to go wild in many parts of the world and after it was over the total worldwide impact included 2000 deaths and an estimated $13 billion in property damage. Excessive rains between mid-November 1982 and late January 1983 caused the worst flooding of the century in usually arid Ecuador. Strong winds and torrential rains associated with six tropical storms lashed French Polynesia in a span of only three months. At the other extreme, drought parched eastern Australia, Indonesia, and southern Africa. Huge drought-related wildfires broke out in Australia and Borneo. Australia's drought was the region's worst in 200 years, causing a $2 billion crop loss and the deaths of millions of sheep and cattle. Meanwhile, drought in sub-Saharan Africa grew worse. Over North America, the winter storm track shifted hundreds of kilometers southeast of its usual location, bringing episodes of destructive high winds and heavy rains to portions of California. Flooding rains also caused havoc across the southeastern United States. Meanwhile ski resorts in the northern United States experienced a snow drought and considerable economic loss.

Just prior to these worldwide weather extremes, the ocean circulation off the northwest coast of South America changed drastically with dire implications for marine productivity. Along the coast of Ecuador and Peru, the quantity of plankton plunged 20-fold. The decline in plankton reduced the numbers of anchovy, which feed on plankton, to a record low. Other fish dependent on anchovy, such as jack mackerel, suffered a similar fate. Commercial fisheries off the coast of Ecuador and Peru collapsed. With the decline in fish populations, marine birds (e.g., frigate birds and terns) and marine mammals (e.g., fur seals and sea lions) also experienced major population declines as insufficient food supplies caused breeding failures as well as starvation of adults.

The impact of these 1982-83 environmental changes was particularly severe on the ecology of the remote island of Kiritimati (Christmas Island) at 2 degrees N, 157 degrees W, in the central tropical Pacific. An estimated 17 million seabirds, which feed on fish and squid, normally nest on the island. But scientists arriving on Kiritimati in November 1982 discovered that almost all adult birds had abandoned the island, leaving their young to starve. Apparently, sharp declines in food sources forced fish and squid to search for better feeding grounds, and the adult birds followed their prey. By July 1983, with the return of more typical ocean circulation and a resurgence in plankton populations, adult birds returned to Kiritimati, nesting began again, and sea bird populations started to recover.

What caused these drastic changes in atmospheric and oceanic conditions? Early on, some scientists attributed the weather extremes to the violent eruption of the Mexican volcano El Chichón on 4 April 1982. But it quickly became apparent that the worldwide weather extremes were linked to large-scale atmosphere/ocean circulation changes in the tropical Pacific and soon a new scientific term was added to the

public's vocabulary, El Niño. The 1982-83 El Niño spurred research on atmosphere-ocean interactions and the deployment of an array of direct and remote sensing instruments in the tropical Pacific to provide early warning of a developing El Niño. In the late 1990s, when El Niño returned in full fury, the global community was better able to prepare, lessening the impact.

Driving Question:

What are the principal features of the planetary-scale atmospheric circulation?

Ultimately, the planetary-scale atmospheric circulation is responsible for the development and displacement of smaller scale weather systems. Appropriately, our description of weather systems begins with those that are global in scale. This chapter covers the pressure systems and wind belts that operate at the planetary scale, with special emphasis on circulation patterns of the midlatitude westerlies. We also cover El Niño and La Niña, atmosphere/ocean interactions in the tropical Pacific that have ramifications for weather and climate in many regions of the world.

Idealized Circulation Pattern

Planetary-scale atmospheric circulation is a complex pattern of winds and pressure systems. To understand how this circulation is shaped, we begin with an idealized model of planet Earth. Imagine Earth as a nonrotating sphere with a uniform solid surface. Also assume that the sun heats the equatorial regions more intensely than polar regions as it does now. A temperature gradient develops between the equator and poles. In response, two huge convection cells form, one in each hemisphere (Figure 9.1A). Cold, dense air sinks at the poles and flows at the surface toward the equator, where it forces warm, less dense air to rise. Aloft, the equatorial air flows toward the poles, completing the convective circulation.

If our idealized planet begins to rotate, the Coriolis effect comes into play (Figure 9.1B). In the Northern Hemisphere, surface winds observed relative to Earth's surface shift to the right and blow toward the southwest, and in the Southern Hemisphere, surface winds shift to the left and blow toward the northwest.

FIGURE 9.1

Planetary-scale air circulation on an idealized spherical model of Earth featuring a uniform solid surface. (A) If the sphere is initially non-rotating, huge convection currents develop on the sunlit portion of the planet's two hemispheres so that air circulates between the hot equator and cold poles. (B) With a rotating Earth, surface winds blow from the northeast in the Northern Hemisphere and from the southeast in the Southern Hemisphere owing to the Coriolis effect. (C) In reality, surface winds divide into three belts in each hemisphere of the rotating planet. (D) Zones of converging and diverging surface winds give rise to east-west belts of low pressure and high pressure.

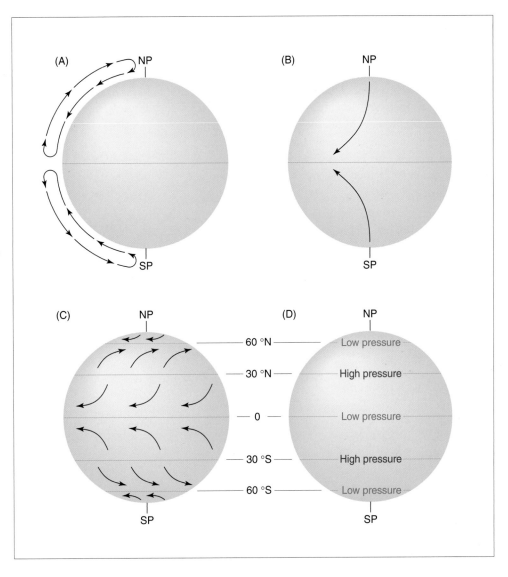

Hence, on our hypothetical planet, surface winds blow counter to the planet's direction of rotation, which is from west to east. As we know from the Coriolis effect, the curved flow results from Earth's surface turning because of rotation.

Circulation is maintained in the atmosphere of our idealized Earth because the planetary-scale winds split into three belts in each hemisphere, so that some winds blow with and some winds blow against the planet's rotational direction (Figure 9.1C). In the Northern Hemisphere, average surface winds are from the northeast between the equator and 30 degrees latitude, from the southwest between 30 and 60 degrees, and from the northeast between 60 degrees and the North Pole. In the Southern Hemisphere, surface winds blow from the southeast between the equator and 30 degrees, from the northwest between 30 and 60 degrees, and from the southeast between 60 degrees and the South Pole.

Surface winds converge along the equator and along the 60-degree latitude circles. Convergence leads to rising air and expansional cooling, cloud development, and precipitation. These convergence zones are belts of low surface air pressure (Figure 9.1D). On the other hand, surface winds diverge at the poles and along the 30-degree latitude circles. In these regions, air descends, is compressed and warms, and the weather is generally fair. These divergence zones are belts of high surface air pressure.

If the continents and ocean are added to our idealized Earth, the temperature characteristics of Earth's surface become more complicated, and so do the planetary-scale air pressure pattern and winds. Some of the pressure belts break into separate cells, and important contrasts in air pressure develop over land versus sea. Now our idealized model of planetary-scale atmospheric circulation more closely approximates the actual circulation pattern. In the next section, we describe the principal features of the planetary-scale atmospheric circulation.

Features of the Planetary-Scale Circulation

Maps of average air pressure at sea level for January and July reveal several areas of relatively high and low air pressure (Figure 9.2). These are the **semipermanent pressure systems**. Although these systems are persistent features of the planetary-scale circulation, they undergo important seasonal changes in both location and strength, hence the modifier *semipermanent*. Pressure systems

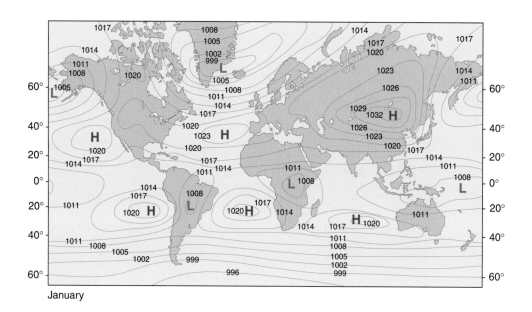

January

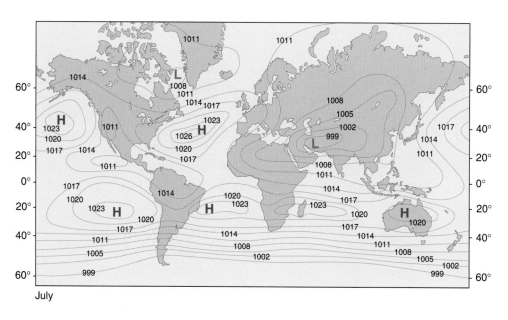

July

FIGURE 9.2
Mean sea-level air pressure for January and July. Contour lines are isobars in millibars (mb).

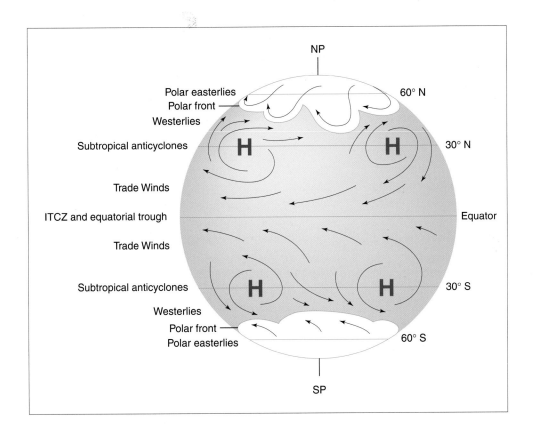

FIGURE 9.3
Schematic representation
of the planetary-scale
surface circulation of the
atmosphere.

include subtropical anticyclones, the intertropical convergence zone (ITCZ), subpolar lows, and polar highs. These pressure systems are in turn, linked by planetary-scale wind belts (Figure 9.3).

PRESSURE SYSTEMS AND WIND BELTS

Subtropical anticyclones are imposing features of the planetary-scale circulation that are centered over subtropical latitudes (on average, near 30 degrees N and S) of the North and South Atlantic, the North and South Pacific, and the Indian Ocean. These highs extend from the ocean surface up to the tropopause and have a major influence on weather and climate of vast areas of the ocean and land.

Stretching from the center of each subtropical high, outward over its eastern flank, are extensive areas of subsiding stable air. Subsiding air undergoes compressional warming, which produces low relative humidities and fair skies. The world's major subtropical deserts, including the Sahara of North Africa and the Sonora of Mexico and southwest United States, are located under the eastern flanks of subtropical anticyclones. On the far western portions of the subtropical highs, however, subsidence is less, the air is not as stable, and episodes of cloudy, stormy weather are more frequent. The contrast in climate between the eastern and western flanks of a subtropical high is apparent across southern North America. The weather of the American Southwest (on the eastern side of the North Pacific high, also called the *Hawaiian high)* is considerably drier than the weather of the American Southeast (on the western side of the North Atlantic high, also called the *Bermuda-Azores high*).

As is typical of anticyclones, a subtropical high features a weak horizontal air pressure gradient over a broad area surrounding the system's center. Hence, surface winds are very light or the air is calm over extensive areas of the subtropical ocean. This situation played havoc with ancient sailing ships, which were becalmed for days or even weeks at a time. Ships setting sail from Spain to the New World were often caught in this predicament, and crews were forced to jettison their cargo of horses when supplies of water and food ran low. For this reason, early mariners referred to this region of calm air as the **horse latitudes**, a name now applied to all latitudes between about 30 and 35 degrees N and S under subtropical highs.

In the Northern Hemisphere, viewed from above, surface winds blow clockwise and outward, away from the centers of the subtropical highs, forming the westerlies and trade winds. Surface winds north of the horse latitudes constitute the highly variable **midlatitude westerlies** (which on average actually blow from the southwest). Surface winds blowing from the northeast out of the southern flanks of the anticyclones are known as the **trade winds**. The trades are the most persistent winds on the planet, in some regions blowing from the same direction more than 80% of the time. Analogous winds develop in the Southern Hemisphere, but recall that the Coriolis deflection is to the left in the Southern Hemisphere. A counterclockwise and outward surface airflow thus causes southeast trade winds on the northern flanks of the Southern Hemisphere subtropical highs and a belt of northwesterlies on the southern flanks.

Mariners were aware of the westerlies and trades of the North Atlantic at least as early as the 15th century. On his venture to find a route to India, Christopher Columbus took advantage of what we now know is the circulation about the Bermuda-Azores sub-tropical high. In his westward voyage, Columbus first sailed southward into the northeast trade winds that eventually took his ships to the Caribbean. On his return trip to Spain, he sailed north into the westerlies.

Trade winds of the two hemispheres converge into a broad east-west equatorial belt of light and variable winds, called the **doldrums**. In that belt, ascending air induces cloudiness and rainfall. The most active weather develops along the **intertropical convergence zone (ITCZ)**, a discontinuous low-pressure belt with thunderstorms paralleling the equator. The mean position of the ITCZ approximately corresponds to the latitude of Earth's highest mean annual surface temperature, the so-called **heat equator**. Primarily because there is more land in the Northern Hemisphere than in the Southern Hemisphere, the world-wide average location of the heat equator is near 10 degrees N latitude.

On the poleward side of the subtropical highs, surface westerlies flow into regions of low pressure. In the Northern Hemisphere, there are two separate **subpolar lows**: the *Aleutian low* over the North Pacific Ocean and the *Icelandic low* over the North Atlantic. These pressure cells mark the convergence of the mid-latitude southwesterlies with the polar northeasterlies. In contrast, in the Southern Hemisphere, the midlatitude northwesterlies and the polar southeasterlies converge along a nearly continuous belt of low pressure surrounding the Antarctic continent (the *circumpolar vortex*).

Surface westerlies meet and override the polar easterlies along the **polar front**. Recall that a *front* is a narrow zone of transition between air masses that differ in temperature, humidity, or both. In this case, dense, cold air masses flowing toward the equator meet milder, less dense midlatitude air masses moving poleward. The polar front is not continuous around the globe; rather, it is well-defined in some areas and not in others, depending on the temperature contrast across the front. Where that temperature gradient is steep, the front is well defined and is a potential site for development of synoptic-scale cyclones (Chapter 10). On the other hand, where the air temperature contrast is minimal, the polar front is poorly defined and inactive or non-existent. The polar front is usually apparent on a surface weather map of North America, dividing colder air to the north from warmer air to the south. At any time, segments of the polar front may be stationary or moving northward as a *warm front* or southward as a *cold front*.

In polar regions, air subsides and flows away at the surface from the centers of shallow, cold anticyclones. In the Northern Hemisphere, **polar highs** are well developed only in winter over the continental interiors. In the Southern Hemisphere, cold highs persist over the glacier-bound Antarctic continent year-round.

WINDS ALOFT

What is the pattern of the planetary-scale winds aloft, that is, in the middle and upper troposphere? As noted earlier, air subsides in subtropical anticyclones, sweeps toward the equator as the surface trade winds, and then ascends in the ITCZ. Aloft, in the middle and upper troposphere, winds blow poleward, away from the doldrums and into the subtropical highs. The Coriolis effect shifts these upper-level winds toward the right in the Northern Hemisphere (southwest winds), and toward the left in the Southern Hemisphere (northwest winds). In the tropics, therefore, winds aloft blow in a direction opposite that of the surface trade winds.

The north-south vertical profile of this low-latitude circulation resembles a huge convection current (Figure 9.4). This circulation is known as the **Hadley cell**, named for George Hadley, who first proposed its existence in 1735. Hadley cells are situated on either side of the ITCZ and extend poleward to the subtropical highs. It was once proposed that separate cells, similar to Hadley cells, occurred in both midlatitudes and polar latitudes. But upper-air monitoring does not provide definitive evidence of such well-defined cells.

Aloft in midlatitudes, winds blow from west to east in a wavelike pattern of ridges and troughs, as shown in Figure 9.5. These winds are responsible for the

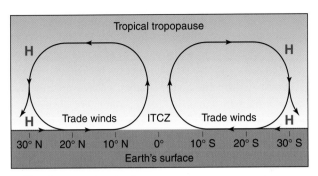

FIGURE 9.4
Idealized north-south cross-section of the Hadley cell circulation in tropical latitudes of the Northern and Southern Hemispheres. Air rises in the Intertropical Convergence Zone (ITCZ) and sinks in the subtropical anticyclones.

development and movement of the synoptic-scale weather systems (highs, lows, air masses, and fronts) discussed in Chapter 10. Also, their north/south components contribute to poleward heat transport described in Chapter 4. Because these winds are so important for midlatitude weather, we examine the upper-air westerlies more extensively in a separate section of this chapter.

Figure 9.6 is a vertical cross-sectional profile of prevailing winds in the troposphere from the North Pole to the equator. In this perspective, we are viewing only the north-south and vertical (up-down) components of the wind and are neglecting the west-east component. Note that the altitude of the tropopause does not steadily increase from pole to equator but occurs in discrete steps. Thus, the polar tropopause is at a lower altitude than the

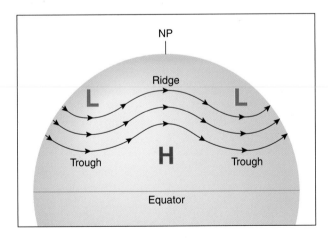

FIGURE 9.5
In the middle and upper troposphere, the Northern Hemisphere westerlies blow from west to east in a wave-like pattern of ridges (clockwise turns) and troughs (counterclockwise turns).

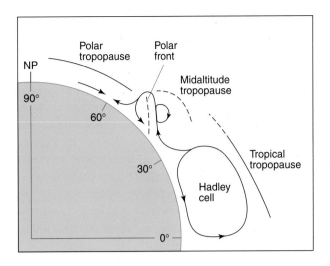

FIGURE 9.6
Vertical cross-section of the north-south (meridional) component of the atmospheric circulation in the Northern Hemisphere. Note that the tropopause occurs in three segments and the vertical scale is greatly exaggerated.

midlatitude tropopause, which, in turn, is at a lower altitude than the tropical tropopause. The altitude of the tropopause is directly related to the mean air temperature of the troposphere; that is, the lower the mean temperature, the lower is the altitude of the tropopause.

TRADE WIND INVERSION

The circulation on the eastern flank of subtropical anticyclones gives rise to the **trade wind inversion**, a persistent and climatically significant feature of the planetary-scale circulation over the eastern portions of tropical ocean basins. Key to formation of the trade wind inversion is the descending branch of the Hadley cell. As air subsides in the subtropical highs as part of the Hadley circulation, it is compressionally warmed and its relative humidity decreases. Descending air encounters the *marine air layer*, a shallow layer of air over the ocean surface. Where sea-surface temperatures are relatively low, the marine air layer is cool, humid, and stable. Where sea-surface temperatures are relatively high, the marine air layer is warm, more humid, less stable, and well mixed by convection.

Air subsiding from above is warmer (and much drier) than the upper portion of the marine air layer. A *temperature inversion*, known as the trade wind inversion, forms at the altitude where air subsiding from above meets the top of the marine air layer. As noted in Chapter 6, in a temperature inversion, the air temperature increases with increasing altitude. Air characterized by a temperature inversion is extremely stable and strongly inhibits vertical motion of air. The trade wind inversion is an elevated temperature inversion that essentially acts as a lid over the lower atmosphere.

As noted earlier in this chapter, most subsidence occurs on the eastern flank of a subtropical high. A trade wind inversion develops to the east and southeast of the center of a subtropical high, in the region dominated by the trade winds. The inversion is highest near the center of the high and gradually slopes downward toward the western coasts of the continents. For example, on average the trade wind inversion is at an altitude of about 2000 m (6560 ft) over Hawaii and about 800 m (2400 ft) over coastal southern California.

Because of its extreme stability, the trade wind inversion acts as a cap on the vertical development of convective clouds and rainfall. As we will see in Chapter 12, this inhibits development of tropical storms and hurricanes. The inversion also limits orographic precipitation (Chapter 6). For example, some high volcanic peaks on the Hawaiian Islands poke through the trade wind inversion into the dry air above. Because of orographic lifting, rainfall generally increases with elevation along windward slopes of the volcanic mountains but only up to the trade wind inversion. Above the inversion, conditions are dry.

SEASONAL SHIFTS

Between winter and summer, important changes take place in the planetary-scale circulation. Pressure systems, the polar front, the planetary wind belts, and the ITCZ follow the sun, shifting toward the poles in spring and toward the equator in autumn. Because the seasons

are reversed in the two hemispheres, the planetary-scale systems of both hemispheres move north and south in tandem. In addition, the strength of pressure cells varies between seasons. Subtropical anticyclones, such as the Bermuda-Azores high, exert higher surface pressures in summer than in winter. The Icelandic low deepens in winter and greatly weakens in summer, and, though well developed in winter, the Aleutian low disappears in summer.

Seasonal reversals in surface air pressure also occur over the continents. These pressure shifts stem from the contrast in the effects of solar heating between land and sea. For reasons presented in Chapter 4, the ocean surface exhibits smaller temperature variations over the course of a year than does Earth's land surface. Hence, continents are dominated by relatively high pressure in winter and relatively low pressure in summer. In winter, in response to extreme radiational cooling, cold anticyclones develop over northwestern North America and over the interior of Asia, the most prominent of which is the massive *Siberian high*. In summer, in response to intense solar heating, a belt of low pressure forms across North Africa and from the Arabian peninsula eastward into Southeast Asia. Warm low-pressure cells also develop in summer over arid and semi-arid regions of Mexico and the Southwestern United States.

Seasonal shifts in the planetary-scale wind belts, pressure systems, and the ITCZ leave their mark on the world's climates. Northward migration of the ITCZ triggers summer monsoon rains in Central America, North Africa, India, and Southeast Asia. As shown in Figure 9.7, north-south movements of the ITCZ are

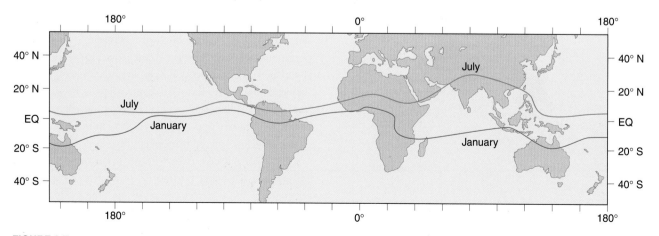

FIGURE 9.7
The Intertropical Convergence Zone (ITCZ) follows the sun, reaching its most northerly location in July and its most southerly location in January.

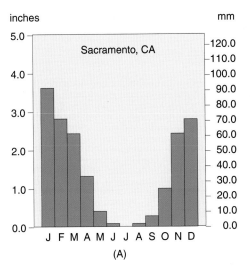

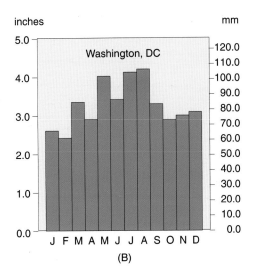

FIGURE 9.8
Mean monthly precipitation (rain plus melted snow) in inches and millimeters for (A) Sacramento, CA, and (B) Washington, DC for the period 1961–90.

greater over continents than over the ocean. Over the ocean, the mean latitude of the ITCZ varies by only about 4 degrees through the year (between 4 degrees N in April and 8 degrees N in September). Anchoring of the ITCZ over the ocean is a consequence of the ocean's greater thermal inertia.

The influence of the subtropical anticyclones on precipitation regimes is illustrated by the contrast in the march of mean monthly precipitation between Washington, DC and Sacramento, CA (Figure 9.8). Although the two cities have about the same latitude, Sacramento has a distinct rainy season (winter) and dry season (summer), whereas Washington, D.C., has abundant precipitation throughout the year. In summer, Sacramento is under the dry eastern flank of the Hawaiian high, while Washington is on the receiving end of the humid flow on the western flank of the Bermuda-Azores high. Consequently, Sacramento is dry but Washington has frequent episodes of convective showers. After the subtropical highs shift equatorward in autumn, the weather of both cities is influenced by widespread precipitation associated with west-to-east moving cyclones so that winters are wet in both places.

Monsoon Circulation

Monsoon is derived from the Arabic word (*mausim*) for season. A **monsoon circulation** characterizes regions where seasonal reversals in prevailing winds cause wet summers and relatively dry winters. Although a weak monsoon develops over central and eastern North America, we are concerned here with the much more vigorous monsoon circulations over Africa and Asia, where more than 2 billion people depend on monsoon rains for their drinking water and agriculture. Over much of India, monsoon rains (falling mostly between June and September) account for 80% or more of total annual precipitation.

What causes monsoons? Monsoons are linked to seasonal shifts in the planetary-scale circulation, specifically north-south shifts of the intertropical convergence zone (ITCZ). As first proposed in 1686 by Edmund Halley, monsoons also depend on seasonal contrasts in the heating of land and ocean. As we have seen, the ocean has a greater thermal inertia than does the land (Chapter 4). Beginning in spring, relatively cool air over the ocean and relatively warm air over the land give rise to a horizontal air pressure gradient directed from sea to land, that produces an onshore flow of humid air. Over the land, intense solar heating triggers convection. Hot, humid air rises, and consequent expansional cooling leads to condensation, clouds, and rain. Release of latent heat intensifies the buoyant uplift, triggering even more rainfall. Aloft, the air spreads seaward and subsides over the relatively cool ocean surface, thus completing the monsoon circulation. Monsoon winds have sufficiently long trajectories and persist long enough to be influenced by the Coriolis effect. Surface monsoon winds are deflected to the right in the Northern Hemisphere and to the left in the Southern Hemisphere.

By early autumn, radiational cooling chills the land more than the adjacent ocean surface, resulting in higher surface pressures over land and setting up a horizontal air pressure gradient directed from land to sea. Air subsides over the land, and dry surface winds sweep seaward. Air rises over the relatively warm ocean surface and aloft drifts landward, completing the winter monsoon circulation. Over land, therefore, the summer monsoon is wet, and the winter monsoon is dry.

Topography complicates the monsoon circulation and the geographical distribution of rainfall. For example, the massive Tibetan plateau strongly influences the Asian monsoon because its elevation tops 4000 m (13,000 ft) over a large area. In winter, the westerly jet stream (a corridor of very strong winds in the westerlies) splits into two branches, one to the south and the other to the north of the plateau. The southerly branch steers cyclones that originate over the Mediterranean across northern India and brings significant precipitation to that region. Meanwhile, the rest of India experiences the dry monsoon flow. In spring, the southern branch weakens and by late May shifts northward over the plateau. It is not until this happens that the moist monsoon flow begins over India.

Monsoon rainfall is neither uniform nor continual. On the contrary, the rainy season typically consists of a sequence of active and dormant phases. During a *monsoon active phase*, the weather is mostly cloudy with frequent deluges of rain, but during a *monsoon dormant phase*, the weather is sunny and hot. The monsoon shifts from active to dormant phases as bands of heavy rainfall surge inland. Heavy rains first strike coastal areas and soak the ground. As the soil becomes saturated with water, more of the available solar radiation is used for evaporation and less for sensible heating. Coastal areas cool, uplift weakens, and skies partially clear (monsoon dormant phase). Meanwhile, the area of maximum heating, most vigorous uplift, and heavy rains shifts inland. Back in the coastal areas, however, the hot sun eventually dries the soil, sensible heating intensifies, uplift strengthens, and rains resume (monsoon active phase). This sequence of active and dormant phases is repeated about every 15 to 20 days during the wet monsoon.

Solar radiation, land and water distribution, and topography impose some regularity on the monsoon circulation; that is, summers are wet and winters are dry. The planetary-scale circulation (especially shifts of the ITCZ) and the strength and distribution of convective activity, however, vary from one year to the next. These variations mean that the intensity and duration of monsoon rains change from year to year. As a consequence, drought is always possible in monsoon climates. The Essay, "Drought in Sub-Saharan Africa," considers a particularly tragic example.

Waves in the Westerlies

The midlatitude westerlies of the Northern Hemisphere merit special attention here because they govern the weather over North America. As noted earlier, in the middle and upper troposphere, the westerlies flow about the hemisphere in wavelike patterns of ridges and troughs. Winds exhibit a clockwise (anticyclonic) curvature in ridges, and counterclockwise (cyclonic) curvature in troughs. Between two and five waves typically encircle the hemisphere at any one time. These long waves are called **Rossby waves**, after Carl G. Rossby, the Swedish-American meteorologist who discovered them in the late 1930s. Rossby waves characterize the westerlies above the 500-mb level, that is, above the altitude where the pressure drops to 500 mb. Below this level, waves are distorted somewhat by friction and topographic irregularities of the Earth's surface.

Meteorologists describe the upper-air westerlies in terms of (1) wavelength (distance between successive troughs or, equivalently, successive ridges), (2) amplitude (north-south extent), and (3) the number of waves encircling the hemisphere. The westerlies exhibit changes in all three characteristics, and, as a direct consequence, the weather changes. The westerlies are more vigorous in winter than in summer. In winter, they strengthen and exhibit fewer waves of longer length and greater amplitude. These seasonal changes stem from variations in the north-south air pressure gradient, which is steeper in winter because of the greater temperature contrast between north and south at that time of year. In summer, north-south temperature differences are less, air pressure gradients are weaker, and so are the westerlies.

ZONAL AND MERIDIONAL FLOW

The *weaving westerlies* have two components of motion: a north-south wind superimposed on a west-to-east wind. The north-south airflow is the westerlies' *meridional component* and the west-to-east airflow is its *zonal component*. The meridional component of Rossby waves brings about a north-south exchange of air masses and poleward transport of heat. In the Northern Hemisphere, winds blowing from the southwest carry warm air masses northeastward, and winds from the northwest transport cold air masses southeastward. Cold air is exchanged for warm air and heat is transported poleward. However, as Rossby waves change length, amplitude, and number, concurrent changes take place in

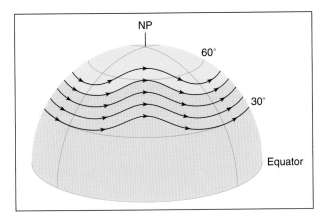

FIGURE 9.9
Midlatitude westerlies exhibit a zonal flow pattern when winds blow almost directly from west to east, with only a small meridional (north-south) component.

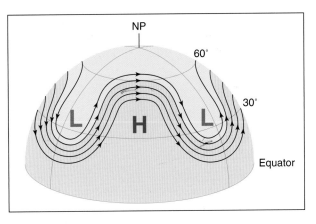

FIGURE 9.10
Midlatitude westerlies exhibit a meridional flow pattern when west to east winds have a strong meridional (north-south) component.

the advection of air masses. Consider some examples.

Occasionally, the westerlies flow almost directly from west to east, nearly parallel to latitude circles, with a weak meridional component (Figure 9.9). This is a **zonal flow pattern** in which the north-south exchange of air masses is minimal. Cold air masses stay to the north, and warm air masses remain in the south. At the same time, air that originated over the Pacific Ocean floods the coterminous United States and southern Canada. *Pacific air* dries out to some extent as it passes over the western mountain ranges, and is compressed and warmed as it descends onto the Great Plains, spreading relatively mild and generally fair weather east of the Rocky Mountains.

At other times, the westerlies exhibit considerable amplitude and flow in a pattern of deep troughs and sharp ridges (Figure 9.10). In this **meridional flow pattern**, masses of cold air surge southward and warm air streams northward. Greater temperature contrasts develop over the United States and southern Canada. Where contrasting air masses collide, warm air overrides cold air, and the stage is set for the development of synoptic-scale cyclones that are then swept along by the westerlies.

These two illustrations of Rossby wave configurations are the opposite extremes of a wide range of possible westerly wind patterns, each featuring different components of meridional and zonal flow. The westerly wave pattern is more complicated in a **split flow pattern**, in which westerlies to the north have a wave configuration that differs from that of westerlies to the south. For example, winds may be zonal over central Canada and meridional over much of the coterminous United States.

The westerly wind pattern typically shifts back and forth between dominantly zonal and dominantly meridional flow. For example, zonal flow might persist for a week and then give way to a more meridional flow that lasts for a few weeks, and then return to zonal flow again. The transition from one wave pattern to another is usually abrupt and sometimes takes place within a single day. This abruptness poses a challenge to weather forecasters because a sudden shift in the upper-air winds may divert a cyclone toward or away from a locality, or cause an unanticipated influx of colder air that could, for example, change rain to snow.

Unfortunately for the long-range weather forecaster, the shifts between westerly wave patterns have no regularity; that is, there is no predictable zonal/meridional cycle. The only observation useful to forecasters is that meridional patterns tend to persist for longer periods than zonal patterns. During the winter of 1976-77, for example, a strong meridional wave pattern persisted over North America from late October through mid-February. Northwesterly flow aloft brought surge after surge of bitterly cold arctic air into the midsection of the United States and resulted in one of the coldest winters of the 20th century for that region. Meanwhile, southwesterly winds aloft brought unseasonably mild air to the far-western North America, including Alaska.

BLOCKING SYSTEMS AND WEATHER EXTREMES

For the continent as a whole, North American weather is more dramatic when the westerly wave pattern is strongly meridional. Sometimes undulations of the westerlies become so great that huge, whirling masses of

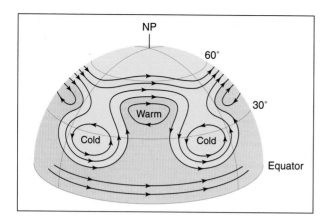

FIGURE 9.11
Blocking pattern in the midlatitude westerlies in which huge pools of rotating air are cut off from the main west to east flow. The pool of cool air rotating in a counterclockwise direction is a cutoff low, and the pool of warm air rotating in a clockwise direction is a cutoff high. The latter system is sometimes referred to as an *omega block* because of its resemblance to the Greek letter.

air actually separate from the main westerly air flow. This situation, shown schematically in Figure 9.11, is analogous to the whirlpools that form in rapidly flowing rivers. In the atmosphere, cutoff masses of air whirl in either a cyclonic or an anticyclonic direction. A *cutoff low* or a *cutoff high* that prevents the usual west-to-east movement of weather systems is referred to as a **blocking system**. Because a blocking circulation pattern tends to persist for extended periods (usually several weeks or longer), extremes of weather such as drought or flooding rains or excessive heat or cold often result. Consider some examples.

In the spring and early summer of 1986, an unusually severe and prolonged drought affected a ten-state region in the southeastern United States. (For a discussion of what constitutes a drought, refer to this chapter's second Essay.) In the hardest hit areas, precipitation totals for the eight months prior to August 1986 were less than 40% of the long-term average. The primary reason for the May-to-early-July episode of drought was a persistent westerly long-wave pattern that featured a trough anchored off the East Coast near longitude 65 degrees W and, upstream (to the west), a ridge over the Southeast. In effect, the center of the Bermuda-Azores high extended its influence well west of its usual domain. This persistent circulation pattern brought day after day of subsiding stable air that suppressed the convective activity that normally brings the Southeast abundant spring and summer rainfall.

The drought that affected the Midwest and Great Plains during the spring and summer of 1988 is another example of the linkage between a weather extreme and a blocking weather pattern. In Figure 9.12, the major upper-air circulation features that dominated the summer of 1988 are contrasted with long-term average conditions. From early May through mid-August, the prevailing westerlies were more meridional than usual and featured a huge stationary, warm high-pressure system over the nation's midsection and troughs over both the west and east coasts. The belt of strongest westerlies was displaced north of its usual position so that moisture-bearing weather systems were diverted into central Canada, well north of their usual paths. In the Corn Belt, the May through June period was the driest since 1895. By late July, drought was categorized as either severe or extreme over 43% of the land area of the contiguous United States. By the end of the growing season, the impact on the nation's grain harvest was severe:

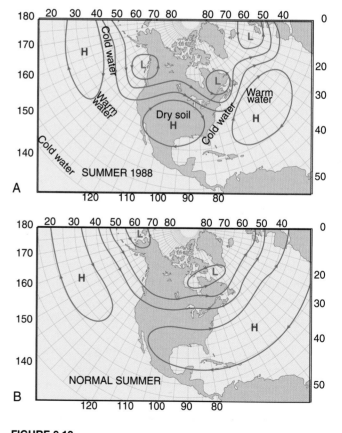

FIGURE 9.12
Prevailing circulation pattern in the mid to upper troposphere (A) during the summer of 1988 as compared to (B) the long-term average circulation pattern. The blocking warm anticyclone over the central United States contributed to severe drought.

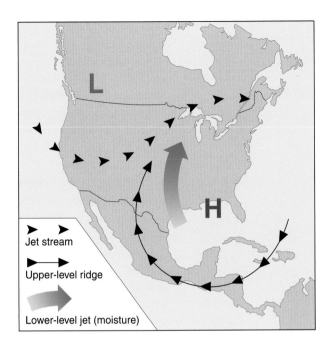

FIGURE 9.13
Principal features of the prevailing atmospheric circulation pattern during the summer of 1993.

Corn production was down by 33%, soybeans by 20%, and spring wheat by more than 50%.

A blocking circulation pattern during the summer of 1993 was responsible for record flooding in the Midwest and drought over the Southeast (Figure 9.13). A cold upper-air trough stalled over the Pacific Northwest and northern Rocky Mountains, bringing unseasonably cool weather to those regions. Meanwhile, the Bermuda-Azores high shifted west of its usual location over the subtropical Atlantic, causing the worst drought since 1986 over the Carolinas and Virginia. Between the northwestern trough and the subtropical high to the southeast, the principal storm track and an unseasonably strong jet stream weaved over the Midwest. As this circulation pattern persisted through June, July, and part of August, a nearly continual procession of disturbances brought heavy thunderstorms and record rainfall to the Missouri and Upper Mississippi River valleys. Across Iowa, Illinois, and Wisconsin, the June-July period was the wettest on record. Meanwhile over the Southeast, subsiding, stable air inhibited thunderstorm development so that hot and dry conditions prevailed through most of the summer.

Heavy rains falling on the drainage basins of the Missouri and Upper Mississippi River valleys saturated soils and triggered excessive runoff, all-time record river crests, and flooding that impacted all or part of nine states. Setting the stage for flooding in the upper Mississippi River valley was the wet autumn of 1992, heavy winter snowfall, and abundant spring snowmelt. The worst flooding occurred between Minneapolis, MN and Cairo, IL, on the Mississippi River, and between Omaha, NE and St. Louis, MO, on the Missouri River. Flooding was not only unprecedented in magnitude but also in persistence with some places recording more than one record river crest (Figure 9.14). At St. Louis, the Mississippi River remained above the previous record crest for three weeks. And the societal and economic impacts of the 1993 flood were devastating. In many cities (such as Des Moines, IA and St. Joseph, MO), the fresh water supply was cut off. Barge traffic was halted on the Upper Mississippi River and a portion of the Missouri River from late June to early August. At least 50,000 homes were damaged or destroyed; and more than 4 million hectares (10 million acres) of cropland were inundated. All told, property damage totaled $15-20 billion, about one-third in crop losses. The death toll from flooding was 48.

While some areas in the Midwest experienced more than twice the average summer rainfall in 1993, parts of the Southeast received less than half of the long-term average (Figure 9.15). As a whole, the region experienced its second driest July on record, and from Alabama and Georgia north to Tennessee and Virginia, July 1993 was the hottest on record. The combination of drought and heat stress caused severe crop damage,

FIGURE 9.14
Sign marks the crest of floodwaters at Missouri City, MO during the summer of 1993.

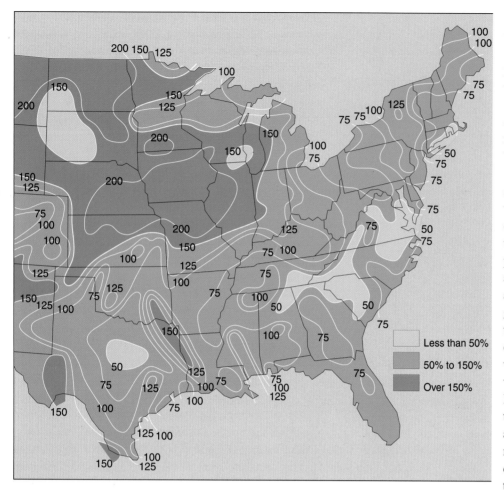

FIGURE 9.15
Total rainfall for the period June-August 1993, expressed as a percentage of the long-term average. [From NOAA, Climate Analysis Center]

waves. Although Rossby waves usually drift very slowly eastward, short waves propagate rapidly through the Rossby waves. Whereas five or fewer long waves encircle the hemisphere, the number of short waves may be a dozen or more.

Both short waves and long waves in the westerlies can contribute to cyclone development. For the same isobar spacing (air pressure gradient), gradient and geostrophic wind speeds are not equal. Anticyclonic gradient winds are stronger than geostrophic winds, and cyclonic gradient winds are weaker than geostrophic winds. Hence, as shown in Figure 9.16, westerly winds tend to strengthen in a ridge and weaken in a trough. The result is horizontal divergence of mid- to upper-tropospheric winds to the east of a trough (and west of a ridge). Short and long waves thus favor cyclone development by inducing horizontal divergence aloft. Figure 9.17 portrays the relationship between an upper-air wave and surface cyclone and anticyclone.

especially in South Carolina where over 95% of the corn crop and 70% of the soybean crop were lost. Total crop losses were estimated at $1 billion. Deaths due to heat stress along the Eastern Seaboard topped 100.

As illustrated by weather events of the summers of 1986, 1988, and 1993, a persistent meridional flow pattern in the westerlies can block the usual movement of weather systems and cause weather extremes such as drought or flooding rains. In fact, any westerly wave pattern, meridional or zonal, can cause extremes in weather if the pattern persists for a sufficiently long period of time. A persistent westerly wave pattern means the same type of air mass advection, the same storm tracks, and basically the same weather type.

SHORT WAVES

Another feature of the weaving westerlies are **short waves**, ripples superimposed on Rossby long

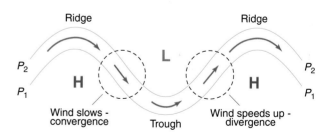

FIGURE 9.16
Westerlies speed up in ridges and slow in troughs, inducing horizontal convergence aloft ahead of ridges and horizontal divergence aloft ahead of troughs. Solid lines are isobars; P_1 is greater than P_2.

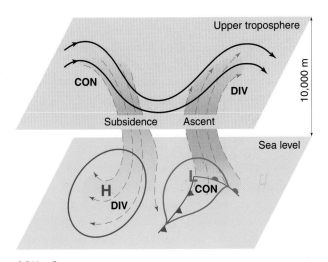

CON = Convergence
DIV = Divergence

FIGURE 9.17
Schematic representation of the relationship between waves in the westerlies and a surface high and low.

Jet Streams

Within the atmosphere are relatively narrow corridors of very strong winds, known as *jet streams*. In middle latitudes, the most prominent jet stream is located directly above the polar front in the upper troposphere between the midlatitude tropopause and the polar tropopause. Because of its close association with the polar front, it is known as the **polar front jet stream**. This jet follows the meandering path of the planetary westerly waves and attains wind speeds that frequently top 160 km (100 mi) per hr. Westbound aircraft understandably avoid this jet stream because it is a head wind, whereas eastbound flights seek it because it is a tail wind.

Why is a jet stream associated with the polar front? The link between a jet stream near the tropopause and a horizontal temperature gradient at the Earth's surface hinges on the influence of air temperature on air density. The polar front is a narrow zone of transition between relatively cold and warm air masses. Cold air is denser than warm air, so air pressure drops more rapidly with altitude in a column of cold air than it does in a column of warm air (Figure 9.18). Hence, even if the air pressure at the Earth's surface is nearly the same everywhere, with increasing altitude a horizontal pressure gradient develops between cold and warm air masses. At any specified altitude within the troposphere, the pressure

is higher in the warm air mass than in the cold air mass. In response to this horizontal air pressure gradient, air accelerates away from high pressure (the warm air column) and towards low pressure (the cold air column). Simultaneously, the Coriolis force comes into play and eventually balances the horizontal pressure gradient force. The net effect is for the wind to blow parallel to isotherms (that is, parallel to the front) with the cold air to the left of the direction of motion.

Strengthening of the horizontal pressure gradient with increasing altitude means that the horizontal wind speed increases with altitude. Wind speed is highest near the tropopause because at altitudes above the tropopause (in the stratosphere) the horizontal temperature gradient reverses direction. In the troposphere, the coldest air is at higher latitudes, but in the stratosphere, the coldest air is at lower latitudes. In the stratosphere, the horizontal pressure gradient weakens with increasing altitude so that the horizontal wind speed also weakens with altitude above the tropopause.

Like the front with which it is associated, the polar front jet stream is not uniformly well defined around the globe. Where the polar front is well defined, that is, where surface horizontal temperature gradients are particularly steep, jet stream winds are stronger. Such a segment, in which the wind may strengthen by as much as an additional 100 km (62 mi) per hr, is known as a **jet streak**. The strongest jet streaks develop in winter along the east coasts of North America and Asia where the contrast in temperature between snow-covered land

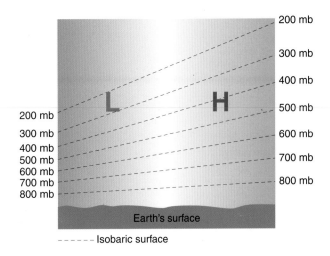

FIGURE 9.18
Air pressure drops more rapidly with altitude in cold air than in warm air giving rise to a horizontal pressure gradient directed from warm air toward cold air.

and ice-free sea surface is particularly great. Over those areas, jet streak wind speeds on rare occasions have exceeded 350 km (217 mi) per hr. A typical jet streak is 160 km (100 mi) wide, 2 to 3 km (1 to 2 mi) thick, and several hundred kilometers in length.

A very important role is played by jet streaks in the generation and maintenance of synoptic-scale cyclones. Air flowing through a jet streak changes speed and sometimes direction, and these changes induce a pattern of horizontal divergence and horizontal convergence aloft. As described in Chapter 8, patterns of divergence and convergence aloft are linked to vertical air motion and divergence and convergence of surface winds.

Consider for example Figure 9.19, a straight jet streak viewed from above. Dashed lines are *isotachs*, lines of equal wind speed (in km per hr). Air accelerates as it enters the jet streak and decelerates as it exits the jet streak. Viewed from above, a jet streak can be considered to be divided into four quadrants: left-rear, right-rear, left-front, and right-front. Horizontal divergence occurs in both the left-front and right-rear quadrants, and horizontal convergence takes place in both the right-front and the left-rear quadrants. In a synoptic-scale cyclone, horizontal winds diverge aloft and converge at the surface. A jet streak provides *upper-air support* for a cyclone by contributing horizontal divergence aloft. For an essentially straight jet streak, the strongest horizontal divergence is in the left-front quadrant, so it is under this sector of the jet streak that a cyclone typically has the best chance of developing. This rule also applies to a cyclonically curved jet streak; for an anticyclonically curved jet streak, the strongest divergence is in the right-rear quadrant.

Although the polar front jet stream weaves with the westerlies, a jet streak typically progresses from west to east at a faster pace than does the west to east displacement of the troughs and ridges in the westerlies. The strongest divergence aloft occurs when a jet streak is on the east side of a trough so that cyclone development is much more likely when a jet streak is in that position.

Like the polar front, the jet stream undergoes important seasonal shifts. The jet stream strengthens in winter, when the north-south air temperature contrast is greatest, and weakens in summer, when temperature contrasts are less. As shown in Figure 9.20, the average summer location of the polar front jet stream is across

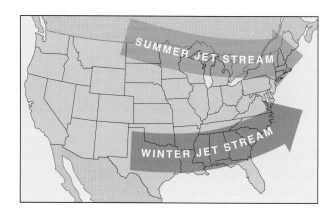

FIGURE 9.20
Approximate average location of the polar front jet stream in winter (December through March) and summer (June through October).

southern Canada, and the average winter position is across the southern United States. These locations represent long-term averages; the jet stream actually weaves over a considerable range of latitude from week to week, and even from one day to the next. As a general rule, when the polar front jet stream is south of your location, the weather tends to be relatively cold, and when the polar front jet stream is north of your location, the weather tends to be relatively warm.

The polar front jet stream is not the only jet stream. The **subtropical jet stream** occurs near the break in the tropopause between tropical and middle latitudes, on the poleward side of the Hadley cell. It is strongest in winter and less variable with latitude than its northerly counterpart. Other jet streams include the

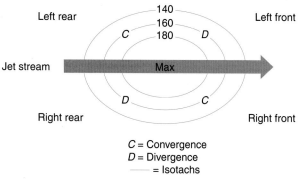

C = Convergence
D = Divergence
——— = Isotachs

FIGURE 9.19
Map view of a jet streak, a segment of accelerated winds within the jet stream, with associated regions of horizontal divergence (*D*) and horizontal convergence (*C*). In this view from above, a jet streak is outlined by *isotachs*, lines of equal wind speed (in km per hr). In this case of a straight jet streak, the strongest horizontal divergence is in the left-front quadrant.

tropical easterly jet that is a feature of the summer circulation at about 15 degrees N over North Africa, and south of India and Southeast Asia. In addition, in summer a *low-level jet*, perhaps several hundred meters above Earth's surface, surges up the Mississippi River Valley and contributes to the development of nocturnal thunderstorms (Chapter 11).

El Niño and La Niña

Some mid latitude weather extremes such as drought and episodes of unusually heavy rains are linked to changes in atmospheric and oceanic circulation in the tropics. One of the most extensively studied of these circulation anomalies occurs in the tropical Pacific and is known as **El Niño**. (Other atmosphere/ocean oscillations are described in this chapter's third Essay.) During El Niño, trade winds weaken, sea-surface temperatures (SSTs) rise well above long-term averages over the central and eastern tropical Pacific, and areas of heavy rainfall shift from the western into the central tropical Pacific. **La Niña** sometimes (but not always) follows El Niño and is a period of exceptionally strong trade winds in the tropical Pacific with lower than usual SSTs in the central and eastern tropical Pacific. Based on changes in SSTs in the central and eastern tropical Pacific, some scientists refer to El Niño as the *warm phase* and La Niña as the *cold phase* of this tropical atmosphere/ocean interaction.

Broad-scale changes in sea-surface temperature patterns over the tropical Pacific that accompany El Niño and La Niña influence the prevailing circulation of the atmosphere in mid latitudes, especially in winter. Weather extremes that most often accompany El Niño are essentially opposite those that usually occur during La Niña.

HISTORICAL PERSPECTIVE

Originally, El Niño was the name Peruvian fishermen gave to a period of unusually warm waters and poor fishing off the northwest coast of South America that often coincided with the Christmas season. (*El Niño* is a Spanish reference to the Christ child). Typically, warm water episodes along the Peruvian coast are relatively brief, lasting perhaps a month or two before sea-surface temperatures and the fisheries return to normal. Occasionally, however, El Niño persists for 12 to 18 months or even longer and is accompanied by significant changes in SSTs over vast expanses of the

tropical Pacific, major shifts in planetary-scale atmospheric and oceanic circulation patterns, and the collapse of important fisheries. Today, atmospheric and oceanic scientists reserve the term El Niño for these long-lasting major atmosphere/ocean anomalies.

An important step in understanding El Niño came with the discovery of the so-called southern oscillation by Sir Gilbert Walker in 1924. Monsoon failure in 1877 and 1899 caused terrible drought and famines in India. In 1904, Walker was charged with developing a method to predict the monsoon and he set out on an extensive search for any possible relationship between monsoon rains and weather conditions in various parts of the world. One of his discoveries was the **southern oscillation**, a seesaw variation in surface air pressure between the western and central tropical Pacific. When air pressure is low at Darwin (on the north coast of Australia), air pressure is high at Tahiti (an island in the south Pacific at about 18 degrees S and 149 degrees W). Conversely, when air pressure is high at Darwin, it is low at Tahiti. The air pressure gradient across the tropical Pacific thus changes as air pressure rises to the west and falls to the east (and vice versa). Walker found that drought in India was associated with high air pressure at Darwin (and low air pressure at Tahiti).

The full significance of Walker's discovery of the southern oscillation was not recognized for more than four decades. In 1966, the Norwegian meteorologist Jacob Bjerknes, while at the University of California at Los Angeles, demonstrated a relationship between El Niño and the southern oscillation. Using oceanic/atmospheric observations obtained from the tropical Pacific during 1957-58, Bjerknes found that an El Niño episode begins when the air pressure gradient across the tropical Pacific begins to weaken, heralding the slackening of the trade winds. Scientists now refer to this link between El Niño and the southern oscillation by the acronym **ENSO**.

Not until the El Niño of 1982-83, one of the two most intense of the 20th century, did the scientific community and general public realize the potential worldwide impact of ENSO. Some of the effects of the 1982-83 El Niño are described in this chapter's Case-in-Point. That event spurred development of numerical models to simulate ENSO and deployment of a network of instrumented buoys and satellites to provide early warning of a developing El Niño. Also, the last two decades have seen increasing interest in La Niña, the name given to an atmosphere/ocean interaction that is essentially the opposite of El Niño. Some scientists refer to the warm phase/El Niño and cold phase/La Niña as opposite extremes of the so-called *ENSO-cycle*.

So important are El Niño and La Niña in year-to-year climatic variability that signs of the development of one or the other are routinely incorporated into long-range seasonal weather outlooks (Chapter A). Such outlooks identify areas of expected positive or negative anomalies in temperature and precipitation and inform regional agricultural, energy, and water management strategies. These strategies are designed to help lessen the impact of attendant weather extremes on water supply and food production in various parts of the world.

LONG-TERM AVERAGE CONDITIONS IN THE TROPICAL PACIFIC

El Niño and La Niña represent significant departures from the long-term average atmosphere/ocean conditions in the tropical Pacific. Understanding these phenomena first requires a look at those long-term average (or *neutral*) conditions, beginning with a summary of Ekman transport and upwelling.

Through a coupling of surface winds with ocean surface waters, known as **Ekman transport**, the southeast trade winds of the Southern Hemisphere drive warm surface waters westward, away from the northwest coast of South America (Figure 9.21). If Earth did not rotate, frictional coupling between moving air and the ocean surface would push a thin layer of water in the same direction as the wind. This surface layer in turn would drag the layer beneath it, putting it into motion. This interaction would continue downward through successive ocean layers, like cards in a deck, each moving forward at a slower speed than the layer above. Because Earth rotates, the shallow layer of surface ocean water set in motion by the wind is deflected to the right of the wind direction in the Northern Hemisphere and to the left of the wind direction in the Southern Hemisphere (the *Coriolis effect*). Except at the equator, where the Coriolis effect is zero, each layer of water put into motion by the layer above shifts direction because of Earth's rotation.

Viewed from above, the decreasing horizontal motion of successively lower layers of water produces a spiral, the so-called **Ekman spiral**, with the current gradually turning and weakening with increasing depth (Figure 9.22). The surface ocean layer moves at an angle of up to 45 degrees to the right or left (depending on the hemisphere) of the surface wind direction. But through the top 100 m (330 ft) or so of the ocean, the Ekman

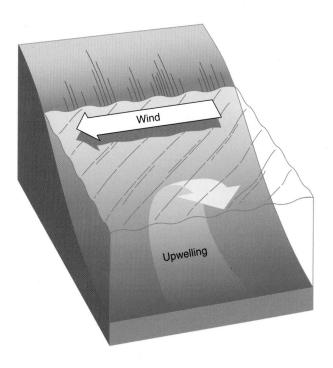

FIGURE 9.21
Upwelling of cold, nutrient rich bottom waters off the northwest coast of South America.

spiral causes a *net* transport of water at an angle of about 90 degrees to the wind direction, to the right in the Northern Hemisphere and to the left in the Southern Hemisphere. The Ekman spiral and associated Ekman transport are named for the Scandinavian physicist, Vagn Walfred Ekman who developed an idealized model to describe these phenomena in 1905. He based his theory on data that showed how the research vessel *Fram* drifted in relation to the prevailing winds while frozen in the pack ice near Spitzbergen.

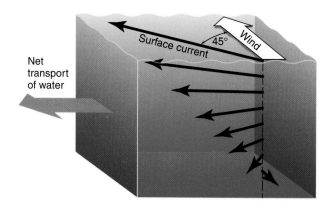

FIGURE 9.22
The Ekman spiral in the Southern Hemisphere.

Warm surface waters driven away from the northwest coast of South America by Ekman transport are replaced by cold, nutrient-rich waters that well up from depths of 200 to 1000 m (about 650 to 3300 ft), a process known as **upwelling**. Although the zone of upwelling is relatively narrow (typically less than 200 km or 125 mi wide), the abundance of nutrients conveyed into the sunlit surface waters spurs an explosive growth of phytoplankton populations which, in turn, supports a diverse marine ecosystem and highly productive fishery. Peru's fishing industry boomed in the 1950s and 1960s and by 1970 was one of the world's largest commercial fisheries. But a combination of over-fishing plus the 1972 El Niño sent the Peruvian fishery into a tailspin from which it has not recovered.

Surface waters are also relatively cool along the equator in the eastern tropical Pacific because of upwelling. Near the equator, the northeast trade winds of the Northern Hemisphere converge with the southeast trade winds of the Southern Hemisphere. The associated Ekman transport (although weak because of a minimal Coriolis effect) causes surface waters to diverge away from the equator and cooler water wells up from below replacing the departing surface waters. *Equatorial upwelling* is responsible for a strip of relatively low sea surface temperatures extending along the equator from the South American coast westward to about the international dateline (180 degrees W).

Meanwhile, the trade winds drive a pool of relatively warm surface waters westward toward Indonesia and northern Australia. The wedge of warm water raises sea level (about 0.5 m or 18 in.) and lowers the thermocline in the western tropical Pacific. (The *thermocline* is the transition zone between warm surface water and cold deep water.) The contrast in sea surface temperature between the western and eastern tropical Pacific (averaging about 8 Celsius degrees or 14.5 Fahrenheit degrees) has implications for rainfall across the tropical Pacific. Relatively cool surface waters in the central and eastern tropical Pacific chill the overlying air and suppress convection so that rainfall is light in that region and along the adjacent western coastal plain of South America. Over the western tropical Pacific, warm surface waters heat the overlying air, strengthening convection and giving rise to heavy rainfall.

THE WARM PHASE

As long as high pressure and low SSTs persist over the eastern Pacific basin and low pressure and high SSTs persist over the western Pacific basin, the trade

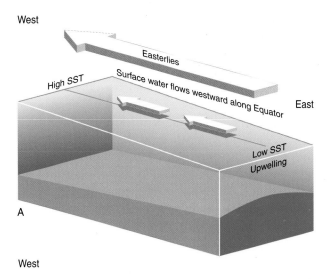

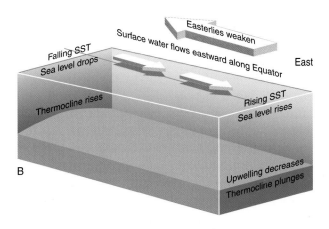

FIGURE 9.23
Long-term average conditions (A) and (B) development of El Niño in the tropical Pacific Ocean.

winds are strong and upwelling remains vigorous along the coast of Ecuador and Peru and the equator east of the international dateline (Figure 9.23). With the onset of El Niño, air pressure falls over the eastern tropical Pacific and rises over the western tropical Pacific as part of the southern oscillation. The air pressure gradient across the tropical Pacific weakens and the trade winds slacken. During a particularly intense El Niño, trade winds west of the international dateline reverse direction and blow toward the east.

In response to these shifts in atmospheric circulation over the tropical Pacific, changes occur in surface ocean currents, SSTs, sea surface elevation, depth to the thermocline, and upwelling. Warm surface water in the west drifts slowly eastward across the tropical Pacific until deflected northward and southward by the continental landmasses, sometimes reaching as far north as the coast of western Canada and as far south as central Chile. Eastward drift of warm water is so slow that it may take several months to reach the west coasts of North and South America. In the western tropical Pacific, SSTs drop, sea-level falls, and the depth to the thermocline decreases whereas in the eastern tropical Pacific, SSTs increase, sea-level rises, and the depth to the thermocline increases. The impacts of these environmental changes can be severe on marine ecosystems. Arrival of warm surface waters in the eastern tropical Pacific effectively blocks upwelling along the coast of Ecuador and Peru. Deprived of nutrients, phytoplankton production declines and the commercial fish harvest plummets.

Lower than usual SSTs in the western tropical Pacific and higher than usual SSTs in the central and eastern tropical Pacific coupled with the change in trade wind circulation give rise to anomalous weather patterns in the tropics and subtropics. Long-term average winds over Indonesia are onshore so that rainfall is normally abundant, but during El Niño, prevailing winds are offshore and the weather of Indonesia is dry. El Niño-related droughts may also grip India and eastern Australia. Meanwhile, warmer than usual surface waters off the west coast of South America spur convection and heavy rainfall along the normally arid coastal plain, causing flash flooding. Also, stronger than usual winds aloft tend to inhibit tropical storm and hurricane development over the Atlantic Basin (Chapter 12).

El Niño usually brings dry weather to the Hawaiian Islands. Almost all of Hawaii's major droughts during the 20th century coincided with an El Niño event. As part of the usual atmospheric circulation changes that occur during El Niño, the North Pacific subtropical anticyclone shifts so that its associated region of subsiding air moves closer to the Islands. The greater than usual frequency of subsiding air over the Islands is responsible for a persistent dry weather pattern.

El Niño also has a ripple effect on mid latitude weather and climate, especially in winter. A linkage between changes in atmospheric circulation occurring in widely separated regions of the globe, often thousands of kilometers apart, is known as a **teleconnection**. Higher than usual SSTs over the central and eastern tropical Pacific heat and destabilize the troposphere. Deep convection generates towering thunderstorms which help drive atmospheric circulation, altering the course of jet streams, storm tracks, and moisture transport at higher latitudes. Events in the ocean and atmosphere are

somewhat analogous to the effect of large boulders on a swiftly flowing stream. Boulders in the streambed induce a train of waves that extend downstream. Moving the boulders displaces the train of waves. Thunderstorm clouds building high into the troposphere deflect upper air winds in much the same manner as boulders in a stream so that a shift in location of the principal area of convection eastward over the tropical Pacific redirects the atmospheric circulation.

During typical El Niño winters, prevailing storm tracks bring abundant rainfall and cooler than usual conditions to the Gulf Coast states, from Texas to Florida. Over the northern United States and Canada, prevailing winds tend to blow from west to east with little meridional transport so that extremely cold air masses tend to move eastward across the Arctic and northern Canada (Figure 9.24). Persistence of this circulation pattern prevents exceptionally cold air masses from invading the northern United States. This dominantly zonal (west to east) flow in the westerlies favors mild weather over western Canada, Alaska, and parts of the northern United States. Zonal flow in the westerlies also diminishes the usual spring clashing of warm, humid air masses moving northeastward from the Gulf of Mexico with cold, dry air masses sweeping southeastward from Canada, reducing the number of severe thunderstorms and tornadoes in the Ohio and Tennessee River Valleys.

Although some weather extremes almost always accompany El Niño, no two events are exactly the same because El Niño is only one of many factors that influence short-term interannual variations in climate. In southern California, for example, heavy winter rains (snows at higher elevations) have occurred during some but not all El Niño events. Record heavy rainfall in Southern California during January 1995 was linked to a shift of the jet stream (and storm track) south of its usual position over the eastern Pacific. In that case, a change in atmospheric circulation associated with El Niño was the culprit. Whereas the 1982-83 El Niño brought severe drought to eastern Australia, the 1997-98 El Niño drought was far less severe.

The 1997-98 El Niño rivaled its 1982-83 predecessor as the most intense of the 20th century. Associated weather extremes worldwide caused an estimated 2100 deaths (many due to flooding) and property damage amounting to at least $33 billion. On the other hand, some locations benefited from the anomalous weather. Winter weather hazards were lessened in the northern states of the United States and in southern Canada where there were also considerable

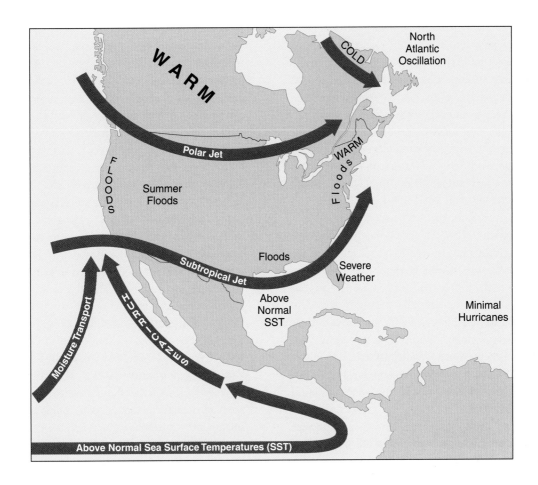

FIGURE 9.24
Principal features of the atmospheric circulation over North America during a typical El Niño.

energy savings for space heating. This event developed rapidly with the trade winds weakening and eventually reversing direction in the western tropical Pacific in early 1997. A pool of exceptionally warm surface waters (SSTs greater than 29 °C or 84 °F) in the western tropical Pacific migrated eastward. Equatorial upwelling ceased during the Northern Hemisphere summer of 1997. Warming was rapid in the eastern tropical Pacific with SSTs setting new record highs each successive month from June through December 1997. By late 1997, the thermocline essentially flattened across the tropical Pacific basin, rising some 20 to 40 m (65 to 130 ft) in the west and falling more than 90 m (300 ft) in the east. Somewhat lower surface-water temperatures in the west and much higher than usual surface water temperatures in the east weakened the usual east-west SST gradient. By further reducing the east-west pressure gradient, the weaker SST gradient contributed to additional slackening of the trade winds.

The 1997-98 El Niño ended abruptly in mid-May 1998. Trade winds strengthened rapidly, upwelling resumed along the equator and off the northwest coast of South America, and SSTs over the eastern tropical Pacific plummeted in response to upwelling of very cold water. At one location along the equator (125 degrees W), the SST fell 8 Celsius degrees (14 Fahrenheit degrees) in only four weeks.

THE COLD PHASE

Sometimes, but not always, La Niña follows El Niño. **La Niña** is a period of unusually strong trade winds and exceptionally vigorous upwelling in the eastern tropical Pacific. During La Niña, SST anomalies (departures from the long-term average) are essentially opposite those observed during El Niño; that is, surface waters are colder than usual over the central and eastern tropical Pacific but warmer than usual over the western tropical Pacific. Although scientists disagree on the precise threshold sea surface temperatures for El Niño and La Niña, SST anomalies over the eastern tropical

Pacific typically have a greater magnitude during El Niño than during La Niña. SSTs usually rise about 5 to 6 degrees Celsius (9 to 11 Fahrenheit degrees) above the long-term average during an intense El Niño and drop 2 to 3 Celsius degrees (3.6 to 5.4 Fahrenheit degrees) below the long-term average during a strong La Niña.

Accompanying La Niña are worldwide weather extremes that are often opposite those observed during El Niño. And as with El Niño, the most consistent teleconnections in mid latitudes show up in winter. In the tropical Pacific, lower than usual SSTs inhibit rainfall in the east and higher than usual SSTs enhance rainfall in the west, including Indonesia, Malaysia, and northern Australia during the Northern Hemisphere winter and the Philippines during the Northern Hemisphere summer. Elsewhere around the globe, the Indian monsoon rainfall (in summer) tends to be heavier than average (especially in northwest India) and wet conditions prevail over southeastern Africa and northern Brazil (during the Northern Hemisphere winter). The winter is dry from Southern Brazil to central Argentina. Meanwhile, weak winds aloft favor tropical storm and hurricane formation over the Atlantic Basin.

Across mid latitudes of the Northern Hemisphere, westerlies tend to be more meridional during La Niña. These winds steer cold air masses toward the southeast and warm air masses toward the northeast. Occasionally, a meridional flow pattern becomes so extreme that a blocking pattern (discussed earlier in this chapter) develops. For example, a blocking pattern was responsible for the severe drought that afflicted the central United States during the La Niña year of 1988.

In spring, a more meridional flow pattern in the westerlies increases the likelihood of severe thunderstorms and tornadoes across the central United States. Such a flow pattern brings together air masses that contrast in temperature and humidity, a key ingredient for severe weather development. Also in the United States, La Niña tends to be accompanied by below average winter precipitation and mild temperatures in a band from the Southwest, through the central and southern Rockies, and eastward to the Gulf Coast. Winter in the Pacific Northwest tends to be cooler and wetter than usual. Lower than usual winter temperatures also occur in the northern Intermountain West and north-central states.

PREDICTION AND MONITORING

Scientists have developed numerical models that simulate El Niño and La Niña. These models approximate oceanic processes that alter sea surface temperatures, and the atmospheric response, including convection, clouds, and winds. Forecasters rely on two basic types of numerical models to predict the onset, evolution, and decay of El Niño: empirical (or statistical) models and dynamical models. An *empirical model* compares current atmospheric and oceanic conditions with comparable observational data for periods preceding El Niño (or La Niña) episodes over the prior 40 years of record. A match or at least some resemblance between past and present conditions is the basis for a prediction. A *dynamical model* consists of a series of mathematical equations that quantify physical interactions among atmosphere, ocean, and land. These so-called *coupled* models are more sophisticated than empirical models and run on supercomputers.

Two of the earliest El Niño models were developed at Scripps Institution of Oceanography in La Jolla, CA and at Columbia University's Lamont-Doherty Earth Observatory in Palisades, NY. The Scripps model is empirical whereas the Lamont-Doherty model is a coupled dynamical model. The Lamont-Doherty model successfully forecasted the 1986-87 El Niño and both models gave advance warning of the 1991-92 El Niño. But neither model successfully predicted the rapid development or intensity of the 1997-98 El Niño.

An assessment of the performance of 12 models (6 empirical and 6 dynamical) in predicting the onset, evolution and demise of the 1997-98 El Niño was disappointing. The coupled dynamical model operated by the Climate Prediction Center (CPC) of the National Centers for Environmental Prediction (NCEP) in Camp Springs, MD did well in predicting the onset of the 1997-98 El Niño. But the NCEP and all other models greatly underestimated (by about 50%) the magnitude of sea-surface warming in the eastern tropical Pacific and demonstrated little skill in forecasting the decay of El Niño. The NCEP model called for gradual cooling in the eastern tropical Pacific through 1998 when, in fact, sea-surface temperatures plunged in May.

Official forecasts of a pending El Niño appeared in the spring of 1997 (after the warming of the eastern tropical Pacific had already begun). Model output strongly influenced the CPC's winter outlook issued in November 1997 for December 1997 through February 1998. That outlook correctly predicted heavier than usual precipitation across the southern United States and anomalous warmth over the northern one-third of the nation. Predictions of unusually dry conditions in

Indonesia, northern South America, and southern Africa also verified. Above average rainfall predicted for Peru, east Africa, and northern Argentina was also correct. But extreme drought expected for northeast Australia did not materialize.

Reliable observational data from the tropical Pacific Ocean and atmosphere are essential for detecting a developing El Niño or La Niña and for initializing numerical models. The accuracy of dynamical models depends not only on how well their constituent equations simulate the coupled ocean/atmosphere/land system, but also the reliability of observational data. Numerical models use those initial conditions as a starting point for predicting future states of the ocean and atmosphere.

At least part of the reason why dynamical models performed well in simulating the evolution of the 1997-98 El Niño was the increasing amount of atmosphere/ocean observational data from the tropical Pacific. Those data were obtained by monitoring systems put in place as part of the 10-year (1985-94) Tropical Ocean Global Atmosphere (TOGA) program. The *ENSO Observing System,* consisting of an array of instrumented buoys, current meters, and satellites, was fully operational by December 1994. Almost 70 Tropical Atmosphere Ocean (TAO) moored buoys measure air temperature, relative humidity, surface winds, sea surface temperatures, and subsurface temperatures at 10 different levels down to a depth of about 500 m (1600 ft). Buoys are scattered over the area bounded by latitude 8 degrees N and 8 degrees S, and longitude 95 degrees W and 137 degrees E. Observational data are telemetered to NOAA's Pacific Marine Environmental Laboratory in Seattle, WA via a NOAA polar-orbiting satellite. Those data are available on the Internet in near real-time.

Remote sensing by satellite plays an important role in providing early warning of the development of El Niño or La Niña. NASA and NOAA satellites monitor cloud cover and map sea surface temperatures. The TOPEX/Poseidon satellite, a joint mission of the United States and France, was launched in 1992 and is monitoring global changes in ocean surface currents related to El Niño and La Niña. Radar (microwave) altimeters onboard the satellite (orbiting at an altitude of 1340 km or 830 mi) accurately measure changes in ocean surface elevation and these changes are used to monitor surface ocean currents. In December 2001, NASA and its French counterpart launched Jason 1, which is designed to replace TOPEX/Poseidon and promises more accurate ocean height measurements.

Another satellite mission that is helping scientists detect the onset and follow the evolution of El Niño and La Niña is the joint U.S.-Japanese Tropical Rainfall Measuring Mission (TRMM) launched in 1997. TRMM uses active radar plus passive microwave sensors to monitor clouds, precipitation, and radiation processes over the area between 40 degrees N and 40 degrees S. One major advantage of the satellite is its TRMM Microwave Imager (TMI), an instrument that "sees through" clouds and measures sea-surface temperatures. The TMI uses the microwaves emitted by the sea surface to characterize its IR emission spectrum and determine its radiation temperature. Microwaves pass through clouds with little attenuation but are strongly absorbed and reflected by rainfall so that TMI can measure SSTs only during fair weather.

More accurate predictions of El Niño and La Niña are enabling humankind to better cope with the impacts of short-term climate variability. Such forecasts inform strategic planning in agriculture, fisheries, and water resource management. Consider an example. Peru's economy, like that of all developing nations, is very sensitive to climate. In Peru, El Niño is bad for fishing and often accompanied by destructive flooding. La Niña benefits fishing but may also mean drought and crop failure. Warning of a pending El Niño or La Niña prior to the start of the growing season allows agricultural interests and government officials to consult on what crops to plant to optimize food production. For example, if the forecast calls for El Niño, rice is favored over cotton. Rice thrives during a wet growing season whereas cotton is more drought-tolerant and is more suitable during La Niña. Also, in anticipation of heavy rains that are likely to accompany a full-blown El Niño, water resource managers can gradually draw down reservoirs to reduce the amount of flooding.

El Niño can be expected about once every 2 to 7 years (Figure 9.25). Although El Niño does not always give way to La Niña, La Niña appears more likely to follow an intense El Niño rather than a weak one. During an intense El Niño, the global water cycle conveys a great amount of heat out of the eastern tropical Pacific and into higher latitudes. Unusually cold water located just below the warm surface water is poised to upwell to the surface as soon as trade winds strengthen. A significant La Niña followed the intense 1997-98 El Niño, for example.

During the second half of the 20th century, El Niño conditions prevailed 31% of the time and La Niña occurred 23% of the time. Ten El Niño episodes occurred over the past 40 years; the most recent one was

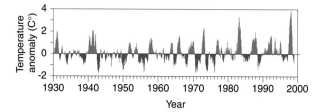

FIGURE 9.25
Sea-surface temperature anomalies (departures from long-term averages) from the region of the tropical Pacific between 5 degrees N and 5 degrees S and from 90 degrees W to 150 degrees W. Warm anomalies (greater than about 0.5 Celsius degree) generally indicate El Niño conditions whereas cold anomalies (less than –0.5 Celsius degree) generally indicate La Niña. [From NOAA, Pacific Marine Environmental Laboratory, Seattle, WA]

in 1997-98. Prior ones took place in 1994-95, 1991-93, 1986-87, and 1982-83. During the 1980s and 1990s, La Niña events were less frequent than El Niño events. There was La Niña in 1998-2000 and prior ones in 1995-96, 1988, 1975, and 1973.

Documentary records of the impacts of El Niño date from as early as 1525 A.D. in Peru. Geological and archaeological evidence indicates that El Niño has been occurring at least as far back as late in the Pleistocene Ice Age and provides some perspective on recent El Niño events. An El Niño signal shows up in a 4000-year record of lake sediments extracted from Glacial Lake Hitchcock, which occupied the Connecticut River valley during the waning phase of the last ice age. The lake sediment record spans the period from about 17,500 to 13,500 years ago, during the final retreat of glacial ice from New England. Analysis of lake sediments from the Ecuadorian Andes and archaeological deposits from northern Peru points to a lull in El Niño activity from about 12,000 to 5000 years ago. The New England record appears to include both large and small El Niños (with periods of 2.5 to 5 years) whereas the Ecuadorian and Peruvian evidence likely reflects only the most intense El Niño events. These ancient records suggest that El Niño has occurred regularly since the late Pleistocene but El Niño intensity has varied over thousands of years (i.e., the tempo is roughly the same but the beat has alternately intensified and weakened). We are now in a period of particularly intense El Niño events.

Conclusions

We have examined the time-averaged characteristics of the planetary-scale circulation. The main features of that circulation are the ITCZ, trade winds, subtropical highs, the westerlies, polar front, subpolar lows, polar easterlies, and polar highs. We saw that aloft (in the middle and upper troposphere), winds blow counter to the surface trades in tropical latitudes and in a west-to-east wave pattern at mid and high latitudes. Through the course of a year, these components of the planetary-scale circulation shift in tandem north and south with the sun.

In this chapter, we also focused on the linkages between the westerly winds aloft and the weather of mid-latitudes. Specifically, the westerlies (1) are the source of horizontal divergence aloft for the development of cyclones, (2) steer storms, and (3) control air mass advection and poleward heat transport. Our examination of atmospheric circulation continues in the next chapter with a focus on components of midlatitude synoptic-scale weather systems, i.e., air masses, fronts, cyclones, and anticyclones.

Basic Understandings

- The principal features of the planetary-scale circulation are the intertropical convergence zone (ITCZ), trade winds, subtropical anticyclones, westerlies, subpolar lows, polar front, polar easterlies, and polar highs. The ITCZ, wind belts, subtropical highs, and polar front follow the sun, shifting poleward in spring and equatorward in autumn.
- Trade winds blow out of the equatorward flanks of the subtropical anticyclones, and the westerlies blow out of their poleward flanks. The east side of a subtropical anticyclone features subsiding air and dry conditions, whereas the west side is more humid and receives more rainfall.
- Aloft, in the middle and high troposphere, winds reverse direction, completing the Hadley cell circulation. At higher latitudes, upper-air winds weave from west to east as Rossby long waves, consisting of ridges (clockwise turns) and troughs (counterclockwise turns).
- Contrasts in Earth's surface temperatures in winter favor relatively high air pressure over continents and

low air pressure over the ocean. In summer, this pattern reverses, with relatively low pressure prevailing over continents and high pressure over the ocean.

- In subtropical latitudes, a monsoon circulation is responsible for wet summers and dry winters. Differences in solar heating of land versus sea, topography, and seasonal shifts in the planetary-scale circulation (especially the ITCZ) play key roles in the monsoon circulation.

- The wave pattern exhibited by the upper-air westerlies varies in length, amplitude, and number. At one extreme, westerlies can be strongly zonal, that is, they blow west-to-east with little amplitude. At the other extreme, westerlies can be strongly meridional, that is, they blow west-to-east with considerable amplitude. Shifts between zonal and meridional flow patterns alter north-south air mass exchange, poleward heat transport, and cyclone development and movement.

- Cutoff cyclones and cutoff anticyclones block the usual west to east progression of weather systems and may lead to extended periods of extreme weather conditions such as drought, excessive rainfall, or periods of unusually high or low temperature.

- The midlatitude jet stream is a narrow corridor of high-speed winds within the westerlies; it is located near the tropopause and above the polar front. A jet streak is a region of particularly strong jet stream winds situated over a well-defined segment of the polar front.

- Cyclone development is most likely under the left front quadrant of a jet steak and to the east of an upper-level trough. Upper-air support is strongest in that quadrant because of the strongest horizontal divergence.

- El Niño refers to a lengthy period of unusually high sea surface temperatures over a vast area of the central and eastern tropical Pacific. El Niño is accompanied by weather extremes in tropical latitudes and in other areas of the globe. La Niña is the term coined for atmosphere/ocean conditions essentially opposite El Niño.

- Ekman transport plays an important role in governing sea-surface temperatures (SSTs) by affecting upwelling of ocean water. Viewed from above, the wind-driven horizontal motion of successively lower layers of ocean water forms a spiral with the current gradually turning and

weakening with increasing depth. The Ekman spiral transports surface waters (the upper 100 m or so) at an angle of about 90 degrees to the right of the wind flow in the Northern Hemisphere and to the left in the Southern Hemisphere.

- During neutral (long-term average) conditions, relatively cool surface waters in the central and eastern tropical Pacific chill the overlying air and suppress convection so that rainfall is light in that region as well as along the adjacent western coastal plain of South America. Meanwhile, over the western tropical Pacific, relatively warm surface waters heat the overlying air, strengthening convection and giving rise to heavy rainfall.

- As El Niño evolves, sea-surface temperatures rise, sea level climbs, and depth to the thermocline increases in the eastern tropical Pacific. Meanwhile, sea-surface temperatures fall, sea level falls, and depth to the thermocline decreases in the western tropical Pacific. Conditions are relatively dry in the western tropical Pacific and wet in the central tropical Pacific.

- La Niña is a period of unusually strong trade winds, exceptionally vigorous upwelling, and lower than usual sea surface temperatures in the eastern tropical Pacific.

ESSAY: Drought in Sub-Saharan Africa

Perhaps nowhere in the world has the variability of rainfall caused more human misery than in sub-Saharan Africa, much of which is known as the Sahel. The Sahel is the transition zone between the Sahara Desert to the north and Africa's humid savanna and rainforest to the south and includes all or part of the poorest nations in the world: Mauritania, Senegal, Mali, Burkina Faso, Niger, and Chad (see map). Relatively low mean annual rainfall, considerable year-to-year variability in rainfall, and prolonged drought have brought considerable hardship to residents of the Sahel. The people of the Sahel are particularly vulnerable to the effects of drought because most of them depend on agriculture for their livelihood. Droughts forced them off their lands that, even in the best of times, are marginal for survival of crops and livestock. They migrated to cities in search of food and work, and many ended up in refugee camps. In the worst of times, horrible scenes of starving children, emaciated livestock, and withered crops were televised to a worldwide audience.

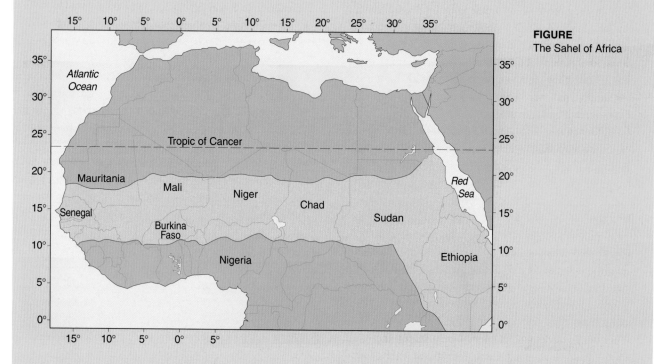

FIGURE
The Sahel of Africa

The Earth system perspective is useful in understanding drought in the Sahel because drought is the product of interactions involving the atmosphere, ocean, land, and people. The Sahel has a dryland climate, featuring a distinct rainy season (Northern Hemisphere summer) and dry season (Northern Hemisphere winter). The average length of the rainy season and mean annual rainfall increases from north to south across the Sahel. At the northern edge of the Sahel (about 18 degrees N latitude), the rainy season usually does not begin until June and may last only a month or two; mean annual rainfall is less than 100 mm (about 4 in.). In the extreme southern Sahel (about 10 degrees N), the rainy season is underway as early as April and persists for up to 5 or 6 months; mean annual rainfall exceeds 500 mm (about 20 in.). Shifts between the dry season and rainy season are linked to seasonal changes in prevailing winds associated with shifts in the location of the ITCZ. As the ITCZ follows the sun, its northward surge in spring triggers convective rains, and its southward shift in fall brings the rainy season to an end. During the rainy season, surface winds blowing from the south and southwest transport humid air from over both the Atlantic and Indian Oceans. During the dry season, the Sahel is under the eastern flank of the North Atlantic subtropical high and surface winds blow from the dry north and northeast.

In a summer when the ITCZ does not penetrate as far north as usual, or arrives late, or shifts southward early, rainfall is below average. A succession of such summers means drought. In the 20th century, the people of the Sahel endured three major long-term droughts: 1910-1914, 1940-1944, and 1970-1985. Rainfall during the 1961-1990 period was 20% to 40% lower than it was during the prior three decades, constituting the greatest 30-year anomaly in precipitation recorded anywhere in the world. Drought and famine claimed over 600,000 lives in 1972-1975 and again in 1984-1985. The reconstructed long-term climate record (based on lake-level fluctuations and historical accounts of landscape changes) indicates that droughts lasting 1 or 2 decades are the norm in the Sahel.

Some scientists argue that the principal reason for drought in the Sahel is changes in atmospheric circulation patterns that are linked to anomalies in sea-surface temperatures (SSTs). When SSTs are higher than normal in the eastern tropical Atlantic south of the equator and southwest of West Africa (e.g., the Gulf of Guinea), rainfall is below the long-term average in the Sahel. Apparently, SST patterns affect the movements of the ITCZ. The unusual persistence of drought in sub-Saharan Africa may be the consequence of human activities that alter the land-cover and thereby reinforce dry conditions. Overgrazing and deforestation may contribute to aridity by altering the regional radiation balance. These activities denude the soil's vegetative cover and degrade the quality and moisture-holding capacity of soils. Soil albedo increases, convection weakens, and rainfall is less likely. What has happened in parts of the Sahel may be *desertification*, the name applied to the conversion of arable land to desert due to some combination of climate change and human mismanagement of the land.

ESSAY: Defining Drought

Drought is an extended period of moisture deficit that is of particular concern to agricultural interests. Drought also affects the supply of water for domestic use and hydroelectric power generation. Soils dry out, crops wither and die, lakes and other reservoirs shrink, and the flow of rivers and streams slackens, perhaps impeding navigation. Unusually high temperatures often accompany summer drought adding to the stress on crops.

A drought usually begins gradually and it is difficult if not impossible to tell whether a spell of dry weather actually constitutes the onset of drought. Similarly, the end of a drought is always uncertain because one rain event, even if substantial, does not necessarily break a drought. Furthermore, whether a dry spell is a drought depends on its impact so that a distinction is made among hydrologic drought, agricultural drought, and meteorological drought.

For water resource interests, a *hydrologic drought* is a period of moisture deficit that reduces stream or river discharge and groundwater supply to levels that seriously impede water-based activities such as irrigation, barge traffic, or hydroelectric power generation. Hydrologic drought develops when the water supply is inadequate during one or more successive water years. A *water year* is defined to extend from 1 October of one year through 30 September of the next year. *Agricultural drought* depends on the shorter-term supply of rainfall and soil moisture for crops during the growing season. Complicating the criterion for agricultural drought is the fact that different crops have different water requirements, and the water needs of a specific crop species change as the crop progresses through its life cycle. Inadequate moisture at a critical stage of crop growth and maturation, especially over successive growing seasons, may constitute agricultural drought. Hydrologic and agricultural droughts do not always occur at the same time.

Varying criteria have been used to define *meteorological drought*. Some meteorologists define drought as a period when the seasonal or annual precipitation falls below a certain threshold percentage (85%, for example) of the long-term (e.g., 30-year) average. But basing the criterion for drought on precipitation alone ignores the influence of temperature and wind on the evaporation rate. Higher temperatures and/or stronger winds increase the evaporation rate.

One of the most popular drought criteria is incorporated in the *Palmer Drought Severity Index*. The Palmer Index uses temperature and rainfall data in a formula that gauges unusual dryness or wetness over extended intervals from months to years. The National Weather Service and the U.S. Department of Agriculture jointly compute the Palmer Index every week for each of the 344 climatic divisions across the United States. A Palmer Index map portrays those divisions experiencing drought with negative values, while those regions receiving excess precipitation have positive index values. Index values range from greater than +4.00 for extremely wet conditions to under –4.00 for extreme drought; zero indicates long-term average moisture levels.

ESSAY: Atmosphere/Ocean Oscillations

Research on El Niño spurred interest in other regular oscillations involving the ocean and atmosphere, including the North Atlantic Oscillation (NAO) and Pacific Decadal Oscillation (PDO). In general, NAO and PDO impact more restricted geographical areas and operate over longer time periods than either El Niño or La Niña.

Recall from elsewhere in this chapter, the time-averaged planetary-scale circulation over the North Atlantic features low pressure near Iceland (the *Icelandic low*) and a massive subtropical anticyclone centered near 30 degrees N and stretching from Bermuda eastward to near the Azores. The *North Atlantic Oscillation (NAO)* refers to a seesaw variation in air pressure between Iceland and the Azores. When air pressure is higher than the long-term average over the Azores, it is lower than the long-term average over Iceland and vice versa. The air pressure gradient between the Bermuda-Azores subtropical anticyclone and the Icelandic low drives winds and steers storms from west to east over the North Atlantic.

The North Atlantic Oscillation influences precipitation patterns and winter temperatures over eastern North America and much of Europe and North Africa. The so-called *NAO Index* is directly proportional to the strength of the North Atlantic air pressure gradient. When the NAO Index is relatively high, stronger than usual winter winds blow across the North Atlantic steering cold air masses over eastern Canada and the United States so that winters tend to be colder than usual in that region. But cold air masses modify considerably as they move over the relatively mild ocean surface, warming and becoming more humid so that winters are milder and wetter than usual downstream over Europe. Meanwhile, dry winters prevail in the Mediterranean region. On the other hand, when the NAO Index is low, steering winds over the North Atlantic shift southward so that winters are colder than usual over northern Europe and wet and mild conditions prevail from the Mediterranean eastward into the Middle East. The eastern United States and Canada tend to experience relatively mild winters while winters in the southeast United States are colder than usual.

The NAO Index varies from one year to the next and from decade to decade but is much less regular than ENSO. The NAO Index was generally low in the 1950s and 1960s and high from about the mid-1970s through the 1990s (with the exception of 1997-98). Changes in winter moisture supply associated with NAO have had varied impacts in Europe and North Africa. In recent decades of relatively high NAO-Index, wetter winters have increased hydroelectric power potential in the Scandinavian nations, lengthened the growing season over northern Eurasia, but also reduced the snow cover for winter recreation. Meanwhile, moisture deficit has been the problem in the Iberian Peninsula, the watershed of the Tigris and Euphrates Rivers, and the Sahel of North Africa.

The *Pacific Decadal Oscillation (PDO)* is a long-lived El Niño-like variation in climate over the North Pacific and North America. Sea surface temperatures fluctuate between the north central Pacific and the western coast of North America. During a PDO *warm phase*, SSTs are lower than usual over the broad central interior of the North Pacific and above average in a narrow strip along the coasts of Alaska, western Canada, and the Pacific Northwest. In an interesting parallel to what happens off the coast of Ecuador and Chile during El Niño, the shallow layer of relatively warm surface waters off the Pacific Northwest Coast suppresses upwelling of nutrient-rich bottom water. Populations of phytoplankton and zooplankton plummet and juvenile salmon migrating to coastal areas from streams and rivers starve. On the other hand, during a PDO *cold phase*, SSTs are higher in the North Pacific interior and lower along the coast.

Key to the climatic impact of PDO is the strength of the *Aleutian low*, which prevails through the winter off the Alaskan coast. During a PDO *warm phase*, the Aleutian low is well developed and its strong counterclockwise winds steer mild and relatively dry air masses into the Pacific Northwest. Winters tend to be mild and dry and water supplies suffer from reduced mountain snow pack. But during a PDO *cold phase*, a weaker Aleutian cyclone allows more cold, moist air masses to invade the Pacific Northwest so that winters are colder and wetter, and the mountain snow pack is thicker. PDO phases tend to last for 20 to 30 years. Cold phases persisted from 1890 to 1924 and again from 1947 to 1976 whereas warm phases prevailed from 1925 to 1946 and 1977 through the 1990s.

CHAPTER 10

WEATHER SYSTEMS OF MIDDLE LATITUDES

The mere formulation of a problem is far more often essential than its solution, which may be merely a matter of mathematical or experimental skill. To raise new questions, new possibilities, to regard old problems from a new angle requires creative imagination and marks real advances in science.

UNKNOWN

Case-in-Point

Daniel Defoe, the English journalist and novelist, was the first to propose that midlatitude storms generally track from west to east. Defoe drew this conclusion from his study of a great storm that lashed the British Isles on 7-8 December 1703. He received reports that, several days earlier, a similar storm had ravaged the East Coast of North America. Benjamin Franklin is generally credited as the first American to discover that storms usually move in an easterly or northeasterly direction. On 21 October 1743, storm clouds over Philadelphia prevented Franklin from viewing a lunar eclipse. Later, through correspondence with his brother, Franklin learned that the eclipse was visible in Boston (about 500 km or 310 mi northeast of Philadelphia) and the next day was stormy. Franklin concluded that a storm and its associated cloud cover had tracked northeastward along the coast.

Driving Question:

What weather systems shape the weather of middle latitudes?

$\mathbf{T}$he ceaseless succession of synoptic-scale weather systems is largely responsible for the considerable day-to-day variability of weather in middle latitudes. Middle latitudes of the Northern Hemisphere extend from the Tropic of Cancer north to the Arctic Circle. In this chapter, we examine the weather conditions that accompany the major weather systems of midlatitudes: air masses, fronts, cyclones (lows), and anticyclones (highs). In addition, this chapter covers selected local and regional circulation systems. All these weather systems are interrelated and their development and movements are ultimately governed by the pattern of planetary-scale circulation described in the previous chapter.

Air Masses

An **air mass**, a huge expanse of air covering thousands of square kilometers, is relatively uniform horizontally in temperature and water vapor concentration (humidity), with both quantities typically decreasing with altitude.

The properties of an air mass depend upon the type of surface over which it develops, that is, its source region. An air mass *source region* features nearly homogeneous surface characteristics over a broad area with little topographic relief, such as a great expanse of snow-covered ground or a vast stretch of ocean. To become uniform in temperature and humidity, an air mass must reside in its source region for several days to weeks.

Simply put, air masses are either cold (polar, abbreviated as *P*) or warm (tropical or *T*), and either dry (continental or *c*) or humid (maritime or *m*). Air masses that form over cold snow-covered surfaces of high latitudes are relatively cold, whereas those that develop over the warm surfaces of low latitudes are relatively warm. Air masses that form over land tend to be relatively dry, and those that develop over the ocean are relatively humid. Based on combinations of temperature and humidity, the four basic types of air masses are: cold and dry, *continental polar* (*cP*); cold and humid, *maritime polar* (*mP*); warm and dry, *continental tropical* (*cT*); and warm and humid, *maritime tropical (mT)*. A fifth air mass type, *arctic* (*A*) air, is dry like continental polar air but colder.

NORTH AMERICAN TYPES AND SOURCE REGIONS

All the air mass types listed above occur over North America, forming over either the continent or surrounding ocean waters in certain characteristic regions. Source regions for these air masses are plotted in Figure 10.1.

Continental tropical air (cT) develops over the subtropical deserts of Mexico and southwestern United States primarily in summer and is hot and dry. **Maritime tropical air (mT)** is very warm and humid because its source regions are tropical and subtropical seas (e.g., Gulf of Mexico). It retains these properties year-round and is responsible for oppressive summer heat and humidity east of the Rocky Mountains. The source regions for **maritime polar air (mP)** are the cold ocean waters of the North Pacific and North Atlantic, especially north of 40 degrees N. Along the West Coast, *mP* air brings heavy winter rains (snows in the mountains) and persistent coastal fogs in summer. Dry **continental polar air (cP)** develops over the northern interior of North America. In winter, *cP* air is typically very cold because the ground in its source region is often snow covered, daylight is short, solar radiation is weak, and

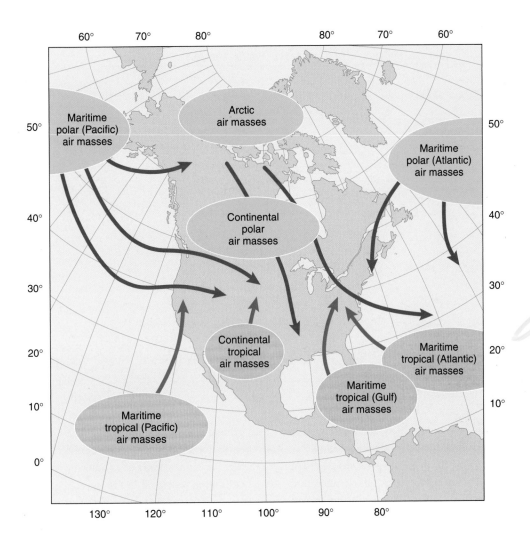

FIGURE 10.1
Source regions of North American air masses.

nocturnal radiational cooling is extreme. In summer, when the snow-free source regions warm in response to extended hours of bright sunshine, *cP* air is quite mild and pleasant.

Arctic air (*A*) forms over the snow- or ice-covered regions of Siberia, the Arctic Basin, Greenland, and North America north of about 60 degrees N in much the same way as continental polar air but in a region that receives very little solar radiation in winter, although it still radiates strongly to space in the infrared. These exceptionally cold and dry air masses are responsible for the bone-numbing winter cold waves that sweep across the Great Plains, at times penetrating as far south as the Gulf of Mexico and central Florida. For example, the so-called *Siberian Express* is the name given to an arctic air mass that forms over Siberia, crosses over the North Pole into Canada, and plunges into the Great Plains and then

south and eastward.

Air masses differ not only in temperature and humidity but also in stability. As noted in Chapter 6, stability is an important property of air because it influences vertical motion and the consequent development of clouds and precipitation. Table 10.1 is a list of the usual stability, temperature, and humidity characteristics of North American air masses within their source regions.

MODIFICATION OF AIR MASSES

Air masses do not stay in their source regions indefinitely. As they move from place to place, their properties modify, those of some air masses changing more than others. Changes may occur in temperature, humidity, and/or stability. **Air mass modification** occurs primarily by (1) exchange of heat or moisture, or both, with the surface over which the air mass travels; (2)

TABLE 10.1
Stability, Temperature, and Humidity Characteristics of North American Air Masses

Air mass type	Source region stability		Characteristics	
	Winter	Summer	Winter	Summer
A	Stable		Bitter cold, dry	
cP	Stable	Stable	Very cold, dry	Cool, dry
cT	Unstable	Unstable	Warm, dry	Hot, dry
mP (Pacific)	Unstable	Unstable	Mild, humid	Mild, humid
mP (Atlantic)	Unstable	Stable	Cold, humid	Cool, humid
mT (Pacific)	Stable	Stable	Warm, humid	Warm, humid
mT (Atlantic)	Unstable	Unstable	Warm, humid	Warm, humid

radiational heating or cooling; and (3) adiabatic heating or cooling associated with large-scale vertical motion.

In winter, as a cP air mass travels southeastward from Canada into the lower 48-states, its temperature usually modifies rapidly. Although daily minimum temperatures in the Northern Plains might dip well below -18 °C (0 °F), minimum temperatures may not drop much below the freezing point after the polar air arrives in the southern states. This rapid air mass modification occurs because, outside of its source region, polar air is usually colder than the ground over which it travels. The sun warms the snow-free ground, and the warmer ground heats the bottom of the air mass, destabilizing it and triggering convective currents that distribute heat vertically into the air mass.

A similar process of heating from below and destabilization occurs when a cP air mass crosses the East Coast and travels over the relatively warm waters of the western Atlantic. In addition, evaporation from the sea surface increases the water vapor concentration of the air mass. Saturation is readily achieved in the cold, humid, and relatively unstable lower levels of the air mass. Extensive low-level cloudiness and fog develop. These clouds usually are organized into narrow rows, known as *cloud streets*, oriented parallel to the wind direction.

When cP air travels over snow-covered ground,

however, modification is less rapid because much of the incoming solar radiation is reflected rather than being converted to heat by absorption. The relatively cold snow surface chills the air, increasing stability and weakening convection. Extreme nocturnal radiational cooling plays an important role in inhibiting modification of the cP air mass, especially when nights are long.

A tropical air mass does not modify as readily as a polar air mass because, outside of its source region, tropical air is often warmer than the ground over which it travels. The bottom of the air mass cools by contact with the ground; this cooling stabilizes the air and suppresses convective currents. Hence, cooling is restricted to the lowest portion of the air mass. By contrast, if a tropical air mass moves over a warmer surface, the air mass can become even warmer. Thus, a cold wave loses much of its punch as it pushes southward, but a summer heat wave can retain its warmth as it journeys from the Gulf of Mexico northward into southern Canada.

Air masses undergo significant changes in temperature and humidity through orographic lifting. When cool and humid mP air sweeps inland from off the Pacific Ocean, the air is forced up the windward slopes of coastal mountain ranges and expands and cools adiabatically. Cooling triggers condensation or deposition (cloud development) and precipitation. Latent heat released during condensation (and deposition) partially offsets adiabatic cooling. Then, as the air mass descends the leeward slopes into the Great Basin, it warms adiabatically and clouds dissipate. Some evaporative cooling is associated with the dissipation of clouds, but net heating (and drying) occurs because the water vapor that was condensed and precipitated out on the windward side of the mountains is no longer available for evaporation. The same processes are repeated as the air mass is forced to flow up and over the Rockies.

Eventually, the air mass emerges on the Great Plains considerably milder and drier than the original *mP* air mass. East of the Rockies, such an air mass is described as modified **Pacific air**.

When the westerly wave pattern aloft is dominantly zonal (Chapter 9), Pacific air floods the eastern two-thirds of the United States and southern Canada. Polar air masses stay far to the north, and tropical air masses keep to the south. Consequently, much of that region experiences an episode of mild and generally dry weather.

Frontal Weather

A **front** is a narrow zone of transition between air masses that differ in density. Density differences are usually due to temperature contrasts; for this reason we use the nomenclature *cold* and *warm* fronts. However, density differences may also arise from contrasts in humidity. Although the transition zone associated with an actual front may have a width of a hundred or more kilometers, traditionally a line representing a front is drawn on a map along the warm (less dense) side of the transition zone. Air temperatures are nearly constant on the warm side of the front and fall with distance from the front to a region of nearly uniform temperatures in the cold air mass.

A front is also associated with a trough in the sea-level pressure pattern, a corresponding wind shift, and converging winds. Where contrasting air masses meet, the colder (denser) air forces the warmer (less dense) air to ascend, often cooling the air sufficiently for clouds and precipitation to develop. Depending on the slope of the front in the vertical plane and the motion of air relative to the front, frontal weather may be confined to a very narrow band, or it may extend over a broad region. In addition, the slope of the front influences the types of clouds and precipitation forming along the front. In this section, we discuss the four basic synoptic-scale fronts: stationary, warm, cold, and occluded.

STATIONARY FRONT

A **stationary front** is just that, a front that exhibits essentially no lateral movement. For example, a stationary front develops along the Front Range of the Rocky Mountains when a shallow pool of polar air (typically about 1000 m thick) surges south and southwestward out of the Prairie Provinces of Canada. The cold dry air mass abuts the mountain range and can

push no further westward; therefore, its leading edge is marked by a stationary front paralleling the crest of the mountain range. Milder air remains in the Great Basin to the west of the mountains. (On occasion, the cold air mass may be sufficiently thick that cold air pours westward through mountain passes into the Great Basin.) Similarly, a stationary front can form when any type of preexisting front becomes parallel to the upper-level flow pattern. Under the proper conditions, a stationary front can also form along a boundary in the surface temperature pattern, such as a coastline or the demarcation between a regional snow cover and bare ground.

Many features of a stationary front are common to all fronts. As shown in vertical cross section in Figure 10.2, a front slopes back from Earth's surface toward colder air, or more precisely, toward denser air. A front lies in a trough in the pressure pattern on any horizontal surface intersecting the front; this is especially evident in the sea-level isobars. Recall that the wind blows approximately parallel to the isobars, with friction turning the wind somewhat toward lower pressure. The wind typically changes direction rather abruptly across the front. The change in wind direction and speed across a front is usually associated with convergence, which leads to upward motion, clouds, and perhaps precipitation.

A stationary front does not always have a broad region of associated clouds and precipitation as depicted in Figure 10.2. Frontal weather can vary considerably from case to case, depending on the supply of water vapor and the specifics of air motion relative to the front. In cases that do produce precipitation, the rain or snow falls mostly on the cold side of the stationary front. Warm humid air flows up and over the cooler air mass, more or less along the frontal surface. Ascending air cools by expansion, which triggers condensation and perhaps precipitation. This situation is often referred to as **overrunning** and can result in an extended period of relatively widespread cloudiness and drizzle, light rain, or light snow.

WARM FRONT

If a stationary front begins to move in such a way that the cold (more dense) air retreats so that the warm (less dense) air can advance, the front changes in character and becomes a **warm front**. The overall characteristics of a warm front are very similar to those of the stationary front, as shown in Figure 10.3.

Differences between warm fronts and stationary

fronts are evident in a comparison of Figures 10.3 and 10.2. The slope of the warm front (ratio of vertical rise to horizontal distance) is shallower near Earth's surface because surface roughness (friction) has slowed the front. Winds on the warm side of the front are quite similar in both instances, but air on the cold side of the front is retreating in the case of the warm front. Thus, the warm air advances relative to Earth's surface, rather than just gliding up and over the cold air.

As a warm front approaches some locality, clouds develop and gradually lower and thicken in the following general sequence: cirrus, cirrostratus, altostratus, nimbostratus, and stratus. This sequence of stratiform clouds reflects gentle uplift associated with overrunning in a relatively stable environment. The initial wispy cirrus clouds may appear more than 1000 km (620 mi) in advance of the surface warm front (Figure 10.4). Slowly, clouds spread laterally and form thin sheets of cirrostratus, turning the sky a bright milky white. The tiny ice crystals comprising these high clouds (bases above 5000 m or 16,500 ft) may reflect and refract sunlight to produce halos or sundogs (Chapter B). Appearance of these optical phenomena may herald the

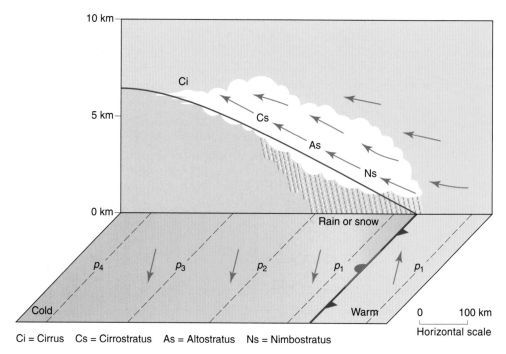

Ci = Cirrus   Cs = Cirrostratus   As = Altostratus   Ns = Nimbostratus

FIGURE 10.2
A stationary front. The vertical scale is greatly exaggerated.

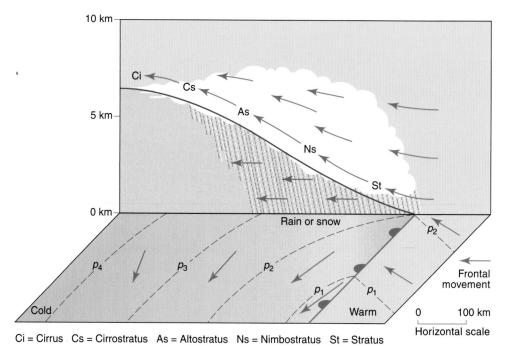

Ci = Cirrus   Cs = Cirrostratus   As = Altostratus   Ns = Nimbostratus   St = Stratus

FIGURE 10.3
A warm front. The vertical scale is greatly exaggerated.

FIGURE 10.4
Thin wispy cirrus clouds may appear more than 1000 km (620 mi) in advance of a surface warm front.

approach of stormy weather a few days in advance.

In time, cirrostratus clouds give way to altostratus, thin clouds with bases at altitudes of 2000 to 7000 m (6500 to 23,000 ft). Soon after altostratus clouds thicken enough to block out the sun, additional lowering of the cloud deck often occurs, accompanied by light rain or snow. Steady precipitation falls from low, gray nimbostratus clouds and persists until the warm front finally passes, a period that may exceed 24 hrs. Copious amounts of rain may fall ahead of the surface warm front, and because precipitation intensity is usually only light to moderate, much of the water infiltrates the soil (as long as the ground is unfrozen and not already saturated with water). Farmers prefer this type of rain. If the air is cold enough for the precipitation to fall in the form of snow, then accumulations may be substantial.

Just ahead of the surface warm front, steady precipitation usually gives way to drizzle falling from low stratus clouds (bases below 2000 m or 6500 ft) and sometimes to fog. So-called **frontal fog** develops when rain falling into the shallow layer of cool air at the ground evaporates and increases the water vapor concentration to saturation. After the warm front finally passes, frontal fog dissipates and skies at least partially clear because the zone of overrunning has also passed. The weather usually turns warmer and more humid (higher dewpoints).

The cloud and precipitation sequence just described for a warm front applies when the advancing warm air is stable. The sequence changes somewhat when the advecting warm air is unstable. In that situation, uplift is more vigorous and often gives rise to cumulonimbus clouds (thunderstorms) embedded within

the zone of overrunning ahead of the surface warm front. Lightning, thunder, and brief periods of heavy rainfall, or perhaps snowfall, may punctuate the otherwise steady fall of light-to-moderate precipitation.

COLD FRONT

A stationary front becomes a **cold front** if it begins to move in such a way that colder (more dense) air displaces warmer (less dense) air. Over North America in winter, the temperature contrast is typically greater across a cold front than across stationary or warm fronts. In summer, however, maximum air temperatures on either side of a cold front are sometimes essentially the same. When this occurs, the density contrast between the two air masses arises from differences in humidity rather than differences in temperature. At the same temperature and pressure, lower humidity air is denser than higher humidity air (Chapter 6). Following passage of a cold front, both temperature and humidity typically drop. Dramatic temperature drops are less likely in summer when humidity changes can be large.

Although surface roughness causes the slope of a warm front to become shallower, it steepens a cold front close to the Earth's surface into a characteristic nose shape, shown in Figure 10.5. The slope of a cold front is steeper (1:50 to 1:100) than the slope of a warm front (1:150 on average). Because of the steep frontal slope and the typical flow aloft across the front from the cold to the warm side, uplift is restricted to a narrow zone at or near the cold front's leading edge. Low-level air motion is also quite different from that in warm and stationary fronts; the low-level air motion in the cold air is at least in part toward the front and forces the warm air aloft.

If the cold front advances slowly but steadily— say 30 km (19 mi) per hr—the type of frontal weather depends on the stability of the warmer air. Any precipitation is likely to be showery and brief, occurring in a narrow band at or just ahead of the front. If the warm air is relatively stable, then nimbostratus and altostratus clouds may form. If the warm air is unstable, the uplift is more vigorous, giving rise to towering cumulonimbus clouds with cirrus clouds blown downstream from the cumulonimbus by winds at high altitudes. These thunderstorms may be accompanied by strong and gusty surface winds, hail, or other violent weather. If a well-defined cold front moves along at a rapid pace, say 45 km (28 mi) per hr, then a **squall line**, a band of intense thunderstorms, may develop either at the front or up to 300 km (180 mi) ahead of the front. Squall lines are discussed in more detail in Chapter 11, where we

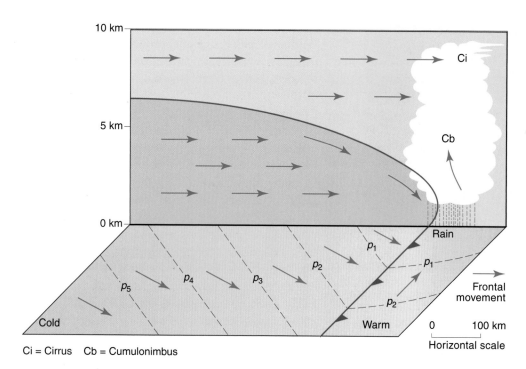

Ci = Cirrus   Cb = Cumulonimbus

FIGURE 10.5
A cold front. The vertical scale is greatly exaggerated.

consider thunderstorms and severe weather systems.

A typical cold front trails south or southwestward from the center of an extratropical cyclone, as discussed later in this chapter, and sweeps along from west to east. In winter, cold fronts often drop

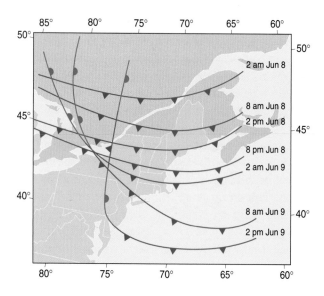

FIGURE 10.6
A back-door cold front advances across New England toward the south and southwest.

southward out of Canada into the Great Plains, even traveling as far south as the Gulf of Mexico. Southward or southwestward moving cold fronts occur east of the Appalachians in New England, but in that region, they are seen most frequently in the summer and fall and are referred to as **back-door cold fronts**. These fronts often usher in welcome relief from hot weather, in the form of *cP* air from Canada or *mP* air from off the Atlantic. The Appalachians tend to impede the westward progress of the colder air, which is then free to move farther south on the eastern side of the Appalachians along the Piedmont and coastal plains (Figure 10.6).

OCCLUDED FRONTS

Late in its life cycle, as a wave cyclone moves into colder air, a front forms, known as an **occluded front**, or simply an *occlusion*. There are three types of occlusions, distinguished by the temperature contrast between the air behind the cold front and the air ahead of the warm front. They are cold occlusion, warm occlusion, and neutral occlusion. If air behind the advancing cold front (*cP*) is colder than the cool air (*mP*) ahead of the warm front, the cold air slides under and lifts the air in the warm sector, the cool air, and the warm front (Figure 10.7). The resulting front, known as a *cold occlusion*, has the characteristics of a cold front at the surface, but the temperature contrast between the cold and cool air masses is less than across a typical winter cold front. Weather ahead of the occlusion is similar to that in advance of a warm front, but the actual frontal passage may be marked by more showery conditions, similar to a cold front. This type of occlusion is most common over the eastern half of the North American continent.

A *warm occlusion* develops when air behind the advancing cold front is not as cold as the air ahead of the

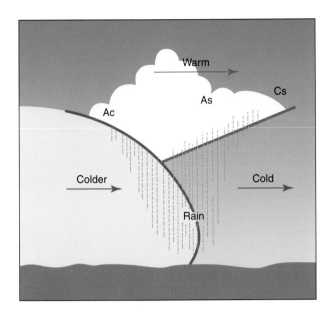

FIGURE 10.7
Schematic vertical cross-section of a cold-type occlusion with the vertical scale greatly exaggerated.

warm front (Figure 10.8). This type of occlusion often occurs in the northerly portions of western coasts, such as in Europe or in the Pacific Northwest. In this case, the air behind the cold front is relatively mild (*mP*), having traversed ocean waters, whereas the air ahead of the warm front is relatively cold (*cP*), having traveled over land. With this type of occlusion, the air behind the cold

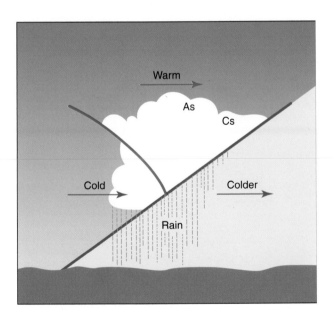

FIGURE 10.8
Schematic vertical cross-section of a warm-type occlusion with the vertical scale greatly exaggerated.

front slides under the warm air but rides over the colder air. Weather ahead of a warm occlusion is similar to that ahead of a warm front, with the surface front behaving as a warm front.

A *neutral occlusion* is quite common and features little temperature difference between air ahead of the warm front and air behind the cold front. Although not obvious in the surface temperature field, this type of occlusion is marked by a trough, wind-shift line, and band of cloudiness and precipitation.

Regardless of type, an occlusion can be difficult to locate from surface weather observations, because the temperature contrast across the front is often small, precipitation occurs over a broad region masking the front, and the associated trough may not be as pronounced as it is for a cold or warm front. However, satellite imagery shows that an occluded front can be as sharply defined as a cold front in oceanic weather systems.

SUMMARY

At middle and high latitudes, the lower troposphere is comprised of air masses that differ in temperature and humidity. Narrow transition zones, called fronts, separate air masses. The motion of a front is related to the lower tropospheric motion on the cold side of the front and fronts are classified on the basis of the movement of the cold air mass. Cold air retreats ahead of a warm front, advances behind a cold front, and moves parallel to a stationary front. Clouds and precipitation develop along fronts only when a significant density contrast exists between air masses, and the supply of water vapor is adequate. With little difference in temperature and humidity across the front, the front may pass virtually unnoticed except for a shift in wind direction.

Properties that define a front (differences in temperature and humidity, wind shift, convergence, and pressure trough) change with time similar to the way air masses modify. If processes in the atmosphere increase the density contrast between air masses (e.g., through convergence), then a front forms or grows stronger; this is called *frontogenesis*. On the other hand, if the density contrast between air masses lessens, the front weakens; this is called *frontolysis*. Precipitation associated with the front also tends to increase or diminish in intensity as the front strengthens or weakens. Frontal weather occurs in combination with cyclones. We explore this relationship next.

Midlatitude Cyclones

The **extratropical cyclone**, or low-pressure system (*low*), is a major weather maker of middle latitudes. Viewed from above in the Northern Hemisphere, surface winds blow counterclockwise and inward toward the center of a cyclone. Surface winds converge, air rises, expands, and cools, resulting in clouds and precipitation. This section explores the life cycle and characteristics of these synoptic-scale weather systems.

LIFE CYCLE

During World War I, researchers at Bergen, Norway first described the basic stages in the life cycle of an extratropical cyclone and for this reason this conceptual model is referred to as the **Norwegian cyclone model**. The model was derived primarily from surface weather observations. Subsequent advances in atmospheric monitoring, especially remote sensing by satellite, have verified the Norwegian model with only minor alterations. Amazingly, the model remains a close approximation of our current understanding of midlatitude cyclones, even though individual cyclones may not follow the model precisely.

With adequate *upper-air support*, an extratropical cyclone can form and intensify. **Cyclogenesis**, the birth of a cyclone, usually takes place along the polar front directly under an area of strong horizontal divergence in the upper troposphere. As noted in Chapter 9, strong divergence aloft occurs to the east of an upper-level trough and under the left-front quadrant of an upper-level jet streak. If divergence aloft removes more mass from a column of air than is brought in by converging surface winds, the air pressure at the bottom of the column falls. Consequently, a horizontal pressure gradient develops, and a cyclonic circulation begins. A storm is born. Westerlies aloft then steer and support the cyclone as it progresses through its life cycle (Figure 10.9).

Just prior to the formation of a prototypical cyclone, the polar front is often stationary, and surface winds blow parallel to the front. As the surface air pressure drops, surface winds converge and the front begins to move (Figure 10.9A). West of the low center (the point of lowest sea-level air pressure), the polar front pushes toward the southeast as a *cold front*. East of the low center, the polar front advances northward as a *warm front*. The minimum pressure for the incipient low might be 1000 mb, with a single closed isobar on a standard

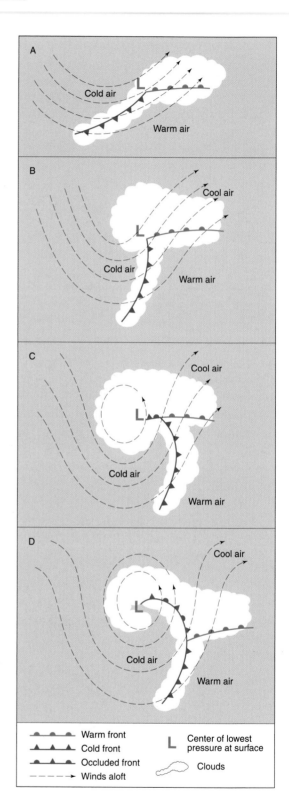

FIGURE 10.9
The life cycle of a midlatitude cyclone (low-pressure system): (A) incipient cyclone, (B) wave cyclone, (C) beginning of occlusion, and (D) bent-back occlusion. As a wave cyclone, the system's center is located east of the upper-level trough; at occlusion, the system's center is under the upper-level trough. Air temperatures refer to conditions at Earth's surface.

surface weather map, which uses a 4-mb interval between isobars. Satellite imagery shows that the narrow cloud band associated with the stationary front develops a bulge at the low center and extends along the developing warm front. The upper-level circulation pattern depicted in Figure 10.9A shows a trough to the west of the surface low, a position that is favorable for further development (*deepening*) of the system. With more horizontal divergence aloft than convergence near the surface, the surface pressure at the cyclone center continues to fall.

If the upper-tropospheric circulation pattern does not favor further development of the cyclone, the low center typically ripples along the stationary front without deepening, producing light precipitation along the way. Such cyclones affect only a small area and typically travel along the front at 50 to 70 km (31 to 43 mi) per hr. On the other hand, if atmospheric conditions favor development of the incipient cyclone, its central pressure continues to drop, the associated horizontal pressure gradient becomes steeper, and counterclockwise winds strengthen. The upper-level trough also frequently deepens while remaining to the west of the low center, as cold air is brought into the system from the northwest.

The so-called *warm sector* of the cyclone (the area between the warm and cold fronts, occupied by warm air at the surface) becomes better defined during the early stage of storm development. At this stage in the cyclone's life cycle, fronts form a pronounced wave pattern (Figure 10.9B), hence the descriptive name *wave cyclone*. The surface cyclone may now have a central pressure of 992 mb, with perhaps three closed isobars, and typically is moving eastward or northeastward at 40 to 55 km (25 to 35 mi) per hr. A large-scale comma-shaped cloud pattern, known as a **comma cloud**, becomes apparent in satellite images at this stage, reflecting the strengthening of the system's circulation (Figure 10.10). The head of the comma extends from the low center to the northwest, with its tail trailing southward or southwestward along the cold front. Extensive stratiform cloudiness caused by overrunning appears north of the warm front.

The cold front generally moves faster than the warm front so that the angle between the two fronts gradually closes; that is, the area of the surface warm sector decreases. As the cold front closes in on the warm front, an *occluded front* begins forming near the low center, forcing the warm air aloft and causing the warm sector at the surface to occupy a smaller area away from the surface low center. Figure 10.9C represents the beginning of the occlusion stage of the cyclone. The

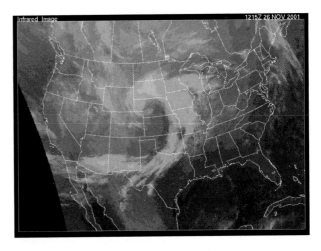

FIGURE 10.10
The comma-shaped pattern of clouds on this infrared satellite image is characteristic of a well-developed midlatitude cyclone. Note the *dry slot* to the south of the system's center over Nebraska.

upper-level pattern now features a closed circulation almost above the surface cyclone. When the upper-level low center is directly over the surface low center, the system is described as *vertically stacked*. Dry air descending behind the cold front is drawn nearly into the center of the cyclone as a *dry slot* that separates the cloud band along the cold front from the comma head, now more west than northwest of the low center. The central pressure of the cyclone has dropped significantly to perhaps 985 mb in the time that elapsed between Figures 10.9B and 10.9C.

Some cyclones continue deepening even after formation of an occluded front. Typically, the upper-air low center is not directly over the surface low center while deepening is occurring, but it may be fairly close. These systems develop a closed, vertically stacked (or nearly so) circulation that is troposphere-deep and typically move much more slowly than previously, say 30 km (20 mi) per hr, and may even stall completely. The circulation tends to draw the occluded front around the low center into a configuration sometimes called a *bent-back occlusion*. The central pressure of the low may now be 980 mb or lower, with a half dozen or more closed isobars quite close together, and surface winds that can exceed 75 km (45 mi) per hr. Such intense cyclones occur over the North Atlantic or North Pacific Oceans. Although still present, the surface warm sector is detached from the cyclone center and an occluded front forms as the warm sector is pushed aloft.

The cold, warm, and occluded fronts all come

together at the point of occlusion, or **triple point**, where conditions are favorable for formation of a new cyclone, sometimes called a *secondary cyclone*. (Note the similarity in appearance between the triple point with its fronts in Figure 10.9D and the incipient low with its fronts in Figure 10.9A). At this stage, the cloud pattern typically becomes a spiraling swirl with enhanced bands associated with the fronts. A very intense system with a central pressure of 960 mb or lower can cause the spiraling cloud band to circle the low center several times.

Eventually, the cyclone weakens, that is, its central pressure rises, the horizontal pressure gradient weakens, and winds diminish; this process is called **cyclolysis** or *filling*. The cyclone can weaken during any of the stages described above if its upper-air support decreases to the point that horizontal divergence aloft becomes less than horizontal convergence near the surface. Then the surface pressure at the low center rises, and the surrounding horizontal pressure gradient weakens. As the central pressure rises, the cyclone loses its identity in the sea-level pressure field and is marked only by an area of cloudy skies and drizzle. Such weakening is inevitable once the system is vertically stacked, but the filling process may not begin for many hours and may proceed slowly, allowing an intense circulation to persist for days.

The cyclone life cycle described above may occur over several days, or it may be completed in a much shorter period. If upper-air support is less favorable, the storm may spend a longer time in any one of the early stages, even weaken temporarily, and still become fully occluded. Sometimes cyclones develop with meager upper-air support (weak divergence aloft) and are poorly defined. At other times, widespread cloudiness and precipitation are linked to an upper-air or surface *trough*, and not to a closed cyclonic circulation at the surface. When upper-level conditions are ideal, the entire life cycle from incipient cyclone to bent-back occlusion can occur in less than 36 hrs.

CYCLONE BOMB

Cyclones sometimes undergo explosive development. A rapidly intensifying extratropical cyclone, called a *bomb*, is defined as a cyclone whose central pressure drops by at least 24 mb in 24-hrs. Few cyclones meet this criterion, and most of those that do satisfy it occur in winter over warm ocean currents such as the Kuroshio Current off the coast of Japan and the Gulf Stream off the East Coast of the United States.

Because these oceanic cyclones are not closely monitored on a day-to-day basis, field experiments have investigated them using specially instrumented research aircraft. The *Genesis of Atlantic Lows Experiment (GALE)*, sponsored by a number of agencies, including the National Science Foundation (NSF), the U.S. Office of Naval Research (ONR), and the National Oceanic and Atmospheric Administration (NOAA), took place from January to March 1986; its primary goal was to observe the initial formation of oceanic cyclones in a region off the coast of North Carolina. The *Experiment on Rapidly Intensifying Cyclones over the Atlantic (ERICA)* was sponsored primarily by ONR and was conducted from December 1988 to February 1989 over the North Atlantic. The goal of ERICA was to monitor rapidly intensifying cyclones throughout their life cycle.

An extreme example of a developing cyclone was observed during ERICA on 4-5 January 1989. An incipient cyclone with a central pressure of 996 mb was first identified off Cape Hatteras, NC at approximately 7:00 p.m. (EST) on 4 January. Twenty-four hours later, the storm was located 700 km (430 mi) south of Newfoundland with a central pressure of 936 mb. The storm deepened by 60 mb in 24 hrs, 2.5 times the criterion for a *cyclonic bomb*!

CONVEYOR BELT MODEL

The Norwegian cyclone model provides a reasonably good description of the characteristics and life cycle of a typical midlatitude cyclone. A better understanding of upper-air circulation in the years since the Norwegian model was first developed led to a new three-dimensional cyclone model, combining horizontal and vertical air motions. This so-called **conveyor belt model** depicts the circulation within a mature cyclone in terms of three broad interacting airstreams, referred to as *conveyor belts*. Just as mechanical conveyor belts transport goods (or even people) from one location to another, atmospheric conveyor belts transport air with certain properties from one location to another and are named for the type of air they transport. The three conveyor belts are the warm, humid air stream, cold air stream, and dry air stream. A schematic of this model is shown in Figure 10.11.

In the conveyor-belt model, a warm conveyor belt originates in the cyclone's warm sector and follows the warm side of the cold front near the Earth's surface, south and east of the storm center. Typically, this warm and humid airstream ascends slightly as it flows northward in the cyclone's warm sector at low levels.

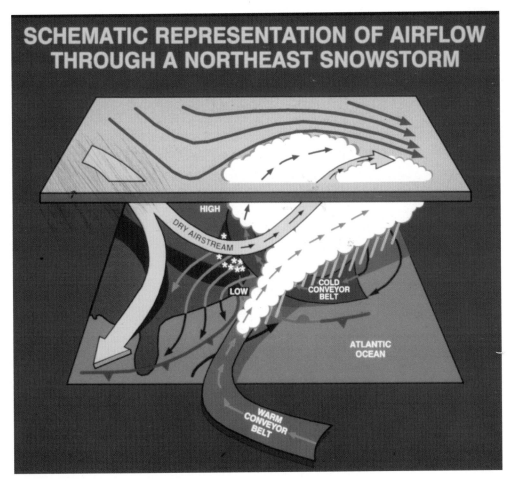

SCHEMATIC REPRESENTATION OF AIRFLOW THROUGH A NORTHEAST SNOWSTORM

FIGURE 10.11
Schematic representation of the three-dimensional circulation within an intense coastal cyclone showing warm, cold, and dry airstreams. [From P.J. Kocin and L.W. Uccellini, *Snowstorms Along the Northeastern Coast of the United States: 1955–1985.* Boston, MA: American Meteorological Society, 1990, p. 71.]

over the warm frontal surface, winds in the lower troposphere are directed from the east or northeast and are relatively cold. This flow forms the cold conveyor belt and, at low levels, is located just to the cold (north) side of the warm front. Like the warm conveyor belt, this airstream ascends as it progresses toward the west. Warm frontal precipitation falls through the cold airstream, increasing its relative humidity to saturation so that clouds and precipitation develop in the cold conveyor belt. As the cold conveyor belt ascends, it comes under the influence of mid- and upper-level tropospheric winds and turns clockwise, following the southwesterly or westerly flow aloft. The ascending saturated cold conveyor belt produces the *comma cloud* to the northwest of the cyclone center.

The warm conveyor belt then ascends more rapidly as it glides northward over the sloping warm frontal surface north of the surface warm front. As the air ascends, it cools adiabatically, water vapor condenses forming clouds, with subsequent development of rain or snow. Latent heat released during this process adds to the buoyancy of air in the warm conveyor belt.

Recall that at this stage in the cyclone's life cycle, winds aloft blow more or less from the southwest, with the surface cyclone located between the upper-level trough and ridge. The warm conveyor belt thus turns from a southerly to a southwesterly or even westerly direction as it ascends over the warm front and follows the upper-level flow. This airstream therefore helps explain the broad region of stratiform cloudiness and steady precipitation north of the warm front.

While the warm conveyor belt is gliding up and

The third conveyor belt is the dry airstream. Whereas air ascends at the cyclone center and along associated fronts, air descends just west of the cold front. Descending air originates aloft in the upper troposphere and lower stratosphere upstream of the upper-level trough and is very dry, especially compared to the humid air in the warm conveyor belt. As the dry conveyor belt descends, one branch turns southward behind the cold front, causing clearing skies. The other branch of the dry airstream first descends as it moves northward and toward the surface low center, and then ascends as it comes under the influence of the winds aloft east of the upper-level trough. As the dry airstream moves toward the surface low center, it forms the *dry slot* that separates the head and tail of the comma cloud and is especially prominent in the later phases of the cyclone's life cycle (Figure 10.10).

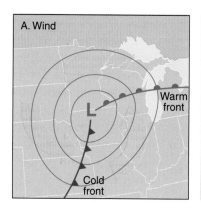
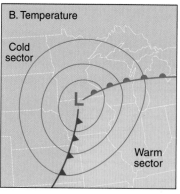
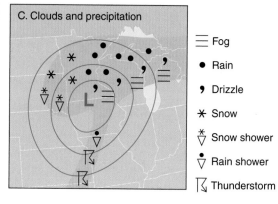

FIGURE 10.12
A mature cyclone centered on the border between Minnesota and Iowa, showing typical patterns of (A) surface winds, (B) surface air temperatures, and (C) areas of cloudiness and precipitation.

CYCLONE WEATHER

As an illustration of typical cyclone weather, consider a winter wave cyclone that developed a well-defined warm sector as it moved into the upper Midwest. Although the storm is still intensifying, its circulation, clouds, and precipitation already affect weather over a broad region. The typical diameter of such a cyclone is between 1000 and 2000 km (about 600 to 1200 mi). Figure 10.12 is a schematic representation of the winds, air temperatures, cloud shield, and precipitation pattern associated with this mature low.

Ideally, based on surface weather, a mature wave cyclone can be divided into four sectors about the low center. The lowest air temperatures occur to the northwest of the storm center, where strong and gusty northwest winds advect *cP* air or *A* air southward and eastward. Strong winds make the air feel colder than the temperature might suggest (wind-chill effect). Stratiform clouds and non-convective precipitation in this sector is associated with the head of the comma cloud and may be either rain or snow, depending upon air temperature. The cold front is south of the low center and is accompanied by a relatively narrow band of cumuliform clouds and showers and thunderstorms. Sinking air and generally clear skies characterize the southwest sector of the storm system. The mildest air is in the southeast (warm) sector of the cyclone, where south and southeast winds advect *mT* air northward from over the Gulf of Mexico. Skies are generally partly cloudy, dewpoints are high, and scattered convective showers are possible, especially during the afternoon. To the north and northeast of the storm center is an extensive zone of overrunning as *mT* air surges over a wedge of cool air maintained by east and northeast winds at the surface. Skies in the northeast sector are cloudy, and precipitation is steady and substantial.

Weather conditions in the various sectors of a mature wave cyclone are consistent with our earlier description of cold and warm frontal weather. Although the flow in and around cyclones is mostly horizontal, uplift of air along frontal surfaces is responsible for the cloud and precipitation pattern illustrated in Figure 10.12.

PRINCIPAL CYCLONE TRACKS

The specific track taken by any extratropical cyclone depends on the pattern of the upper-level westerlies in which the storm is embedded. The storm center tends to move in the direction of the wind blowing directly above the system at the 500-mb level. As a general rule, the cyclone center moves forward at about one-half the speed of the 500-mb winds. Keep in mind, however, that the upper-level circulation pattern also changes with time, and the cyclone's path shifts accordingly.

Principal storm tracks across the lower 48 states of the United States are plotted in Figure 10.13. All storms tend to converge toward the northeast; their ultimate destination is usually the Icelandic low of the North Atlantic or Western Europe. Although many storm tracks appear to begin just east of the Rocky Mountains, in reality they originate over the Pacific Ocean, near the Aleutians or in the Gulf of Alaska. As a cyclone travels through the mountains, it often loses its identity temporarily but redevelops on the Great Plains just east of the Front Range of the Rockies. For more on what happens, see the Essay, "The Case of the Missing Storm."

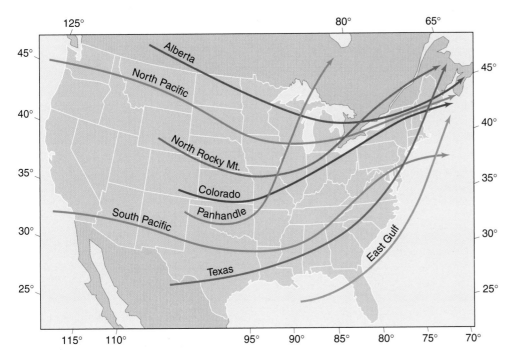

FIGURE 10.13
Principal cyclone tracks across the lower 48-states of the United States.

Cyclones that consistently follow some of the principal tracks have nicknames. The *Panhandle hook* forms to the lee of the Rockies and *hooks* southeastward and then turns almost due north, whereas the *Alberta clipper* develops to the lee of the Canadian Rockies and travels rapidly from west to east across southern Canada or the northern tier of states. *Nor'easters* track toward the northeast along the East Coast. It may be confusing to think of a storm as moving *toward* the northeast, while the winds in the northeast sector of the storm blow *from* the northeast. Actually, nor'easters usually intensify off the North Carolina coast and then track in a northeasterly direction up the coast. In effect, two motions exist: (1) movement of the storm center along the coast and (2) the counterclockwise circulation of winds about the storm center. The circulation of a storm is for the most part independent of the storm's path, much as the spin of a Frisbee is independent of its trajectory.

Some nor'easters intensify into powerful systems, such as the ERICA storm described earlier or the *Blizzard of '88* described in the Case-in-Point of Chapter 1, and bring heavy precipitation and strong winds to the East Coast. The potential impact of a nor'easter is well illustrated by the East Coast storm of 12-15 March 1993. That storm's path enabled the system to draw into its circulation copious amounts of water vapor and dump extraordinary amounts of snow over an unusually broad area stretching from Alabama to Maine. For more on this storm, see the Essay, "Nor'easters."

Generally, cyclones that form in the south yield more precipitation than those that develop in the north because the southerly flow in the warm sector of southern cyclones is better positioned to draw moisture-rich *mT* air from source regions such as the Gulf of Mexico. For this reason, Alberta cyclones typically yield only light amounts of rain or snow, whereas Colorado and Gulf-track storms often produce heavy accumulations of rain or snow. The forward speed of the cyclone also affects the total precipitation received at a particular location. For places in the path of a cyclone, a fast-moving system may yield rain or snow for only a few hours whereas a slower moving storm may precipitate for 12 or more hours.

We have examined the linkage between cyclones and winds aloft. The circulation in the mid- and upper-troposphere supplies *upper-air support* (horizontal divergence) for cyclones and determines their paths. Just as the planetary-scale circulation undergoes seasonal changes, so too do cyclones. In summer, when the mean position of the polar front and jet stream is across southern Canada, few well-organized cyclones occur in the United States, and the Alberta storm track shifts northward across Canada. In winter, however, when the mean position of the polar front and jet stream shifts southward, cyclogenesis is more frequent in the United States. Alberta-track storms are the most common because they occur year-round, whereas storms with more southerly tracks develop primarily in winter. In fact, one sign of the beginning of winter circulation patterns is the appearance of Colorado lows.

FIGURE 10.14
A winter cyclone develops over eastern Colorado and tracks northeastward toward the western Great Lakes. Track *A* takes the storm to the west and north of Chicago so that the city is on the warm side of the system. Track *B* takes the storm to the south and east of Chicago so that the city is on the cold side of the system.

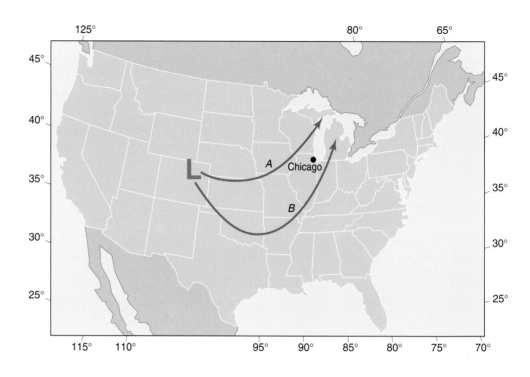

COLD SIDE/WARM SIDE

A midlatitude wave cyclone has a cold side and a warm side. In Figure 10.12, the coldest air is located to the northwest of the low center (where surface winds are blowing from the northwest), and the warmest air is to the southeast of the low center (where surface winds are blowing from the south). Hence, as a cyclone travels eastward across the continent, the weather to the left (cold side) of the storm track is quite different from the weather to the right (warm side) of the track.

Consider, for example, the two storm tracks plotted in Figure 10.14. A winter cyclone develops over eastern Colorado. As the storm matures, it moves northeastward toward the Great Lakes region and may take either track *A* or track *B*, depending on the direction of winds aloft. In both cases, the center of the cyclone passes within 150 km (95 mi) of Chicago. With track *A*, the storm moves west and then north of Chicago whereas track *B* takes the storm to the south and then east of the city. The storm's influence on Chicago's weather depends on its track, as summarized in Table 10.2.

If the storm follows track *A*, Chicago residents experience the warm side of the storm. Stratiform clouds thicken and steady rain (or perhaps snow briefly at the

TABLE 10.2
General Sequence of Weather Conditions at Chicago as Winter Cyclone Tracks West of the City (Track *A*) or East of the City (Track *B*)

		Track *A*						Track *B*		
Wind direction	E	SE	S	SW	W	NW	E	NE	N	NW
Frontal passage	-	WF	-	CF	-	-	-	-	-	-
Advection	-	Wm	Wm	Cld	Cld	Cld	-	-	Cld	Cld
Air pressure tendency	F	F	F	R	R	R	F	F	R	R
			Time	⇒				Time ⇒		

WF = Warm front, CF = Cold front, Wm = Warm air advection, Cld = Cold air advection
F = Pressure falling, R = Pressure rising

onset) give way to drizzle and fog after 12 to 24 hrs. As the warm front passes over the city, skies partially clear and winds shift abruptly from east to southeast, advecting warm and humid (*mT*) air at the surface. Clearing is short-lived, however, as convective clouds develop and are accompanied by scattered showers and thunderstorms that herald the arrival of colder air. As the surface cold front passes through the city and then on toward the east and northeast, winds *veer* (turn clockwise with time), blowing first from the southwest, then west, and finally northwest. Skies clear again, and the air temperature and dewpoint fall.

In contrast, if the storm takes track *B*, Chicago residents experience the cold side of the storm and no frontal passages. Lowering and thickening stratiform clouds accompany gusty east and northeast winds that drive steady snow or rain (depending on air temperatures) for 12 hrs or longer. Winds gradually *back* (turn counterclockwise with time) to a northerly direction, precipitation tapers off to snow flurries or showers, and air temperatures drop. Finally, winds shift to northwest, the sky begins to clear, and air temperatures continue to fall in response to strong cold air advection.

In summary, if you are located on the warm side of a cyclone's path, the wind direction veers with time and a cold front follows a warm front. If you are on the cold side of a cyclone's path, the wind direction backs with time without the passage of fronts. In winter in a northern location, substantial snowfall is much more likely on the cold side than on the warm side of a cyclone's track, with the axis of heaviest snowfall running parallel about 240 km (150 mi) to the northwest of the storm track.

COLD- AND WARM-CORE SYSTEMS

An occluded cyclone is a cold-core system; that is, the lowest temperatures occur throughout the column of air above the low-pressure center. In vertical cross section, constant pressure (isobaric) surfaces within an occluded **cold-core cyclone** dip downward in the center of the cyclone (concave upward) (Figure 10.15). Furthermore, the depth of the low increases with altitude implying that a cyclonic circulation prevails throughout the troposphere and is most intense at high altitudes. The thickness of an air layer defined above and below by pressure surfaces is directly proportional to the mean temperature of the air layer. The requirement that the thickness (mean temperature) be lowest at the center of the low produces the characteristic isobaric pattern.

On the other hand, the lowest air temperatures in

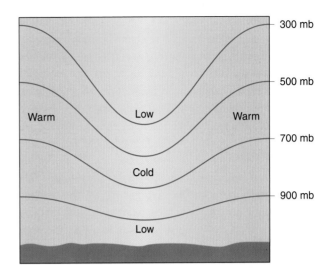

FIGURE 10.15
Vertical cross-section through an occluded cold-core cyclone showing isobaric (constant pressure) surfaces. The vertical scale is greatly exaggerated.

a non-occluded wave cyclone are northwest of the system's center and the highest temperatures are to the southeast. Figure 10.16 is a vertical cross-section through the low from northwest to southeast. Thickness (mean temperature) arguments indicate that until occlusion the low center aloft is not located above the low center near the surface, but rather is displaced to the cold side of the storm, implying that the system tilts with altitude. This structure is consistent with Figure 10.9B, which shows the upper-level trough lagging behind the surface cyclone.

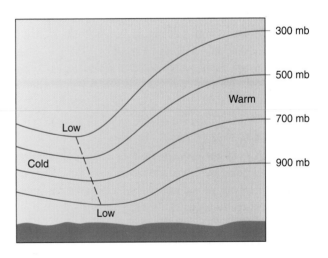

FIGURE 10.16
Vertical cross-section through a non-occluded wave cyclone showing isobaric (constant pressure) surfaces. The vertical scale is greatly exaggerated.

A different type of low that sometimes is plotted on a surface weather map has characteristics markedly different from those of cold-core or tilted lows. Cyclones of this type are stationary, have no fronts, and are associated with fair weather. They form over a broad expanse of arid or semiarid land, including the interior of Mexico and the southwestern United States, in response to intense solar heating of the ground. The hot surface heats the overlying air, lowering the density of the air column over an area broad enough for a synoptic-scale low to appear on a surface weather map. This **warm-core cyclone** (or *thermal low*) is very shallow, and its circulation weakens rapidly with altitude, that is, away from the heat source (the ground). The surface counterclockwise circulation frequently reverses at some altitude, and an anticyclone overlies the thermal low in the mid and upper troposphere.

In this section, we have focused on the life cycle, tracks, and general weather conditions associated with extratropical cyclones. In the daily march of weather in midlatitudes, anticyclones follow cyclones.

Anticyclones

An **anticyclone** (or *high*) is, in many ways, the opposite of a cyclone (or *low*). In midlatitudes, cyclonic circulation favors the convergence (coming together) of contrasting air masses and development of fronts, clouds, and precipitation. In anticyclones, by contrast, subsiding air and diverging surface winds favor formation of a uniform air mass and generally fair skies. Air modifies as it moves away from a center of high pressure so that a semi-permanent anticyclone of Earth's planetary-scale circulation can be the source of different types of air masses. Like cyclones, anticyclones have either cold or warm cores.

ARCTIC AND POLAR HIGHS

A **cold-core anticyclone** is actually a dome of continental polar (*cP*) or arctic (*A*) air and, depending on the specific type of air mass, is labeled either a **polar high** or an **arctic high**. Cold anticyclones are products of extreme radiational cooling over the often snow-covered continental interior of North America well north of the polar front. They are shallow systems in which the clockwise circulation weakens with altitude and often reverses aloft. A cold trough typically overlies a cold anticyclone.

Cold-core anticyclones are most intense (that is, they exert highest surface pressures) in winter, when the associated air mass is coldest. The air is extremely stable in these systems, and soundings indicate a temperature inversion in the lowest kilometer or two, associated with strong subsidence and adiabatic heating above the inversion. Massive arctic or polar anticyclones with very high central pressures tend to remain stationary over their source regions. The *Siberian high*, for example, is centered near Lake Baikal and from late November to early March has an average sea level pressure greater than 1030 mb. Lobes of cold air (smaller cold highs) often break away from arctic or polar highs. In North America, for example, cold air masses move out of source regions in Alaska and northwest Canada and slide southeasterly across southern Canada into the lower United States. These cold anticyclones interact with the circulation of migrating cyclones and help to maintain and strengthen the temperature contrast across the cyclone's cold front. Hence, clearing skies and sharply lower temperatures usually follow winter storms.

On some occasions in winter, a particularly strong arctic high brings a surge of bitterly cold air that sweeps as far south as Florida, and rarely can even traverse the Gulf of Mexico into Central America. The resulting subfreezing temperatures can spell disaster for citrus growers. In the 1980s, for example, three exceptional cold waves invaded the Florida peninsula. For three days, beginning on Christmas Day 1983, an arctic air mass gripped Florida, dropping temperatures well below freezing through much of the state. Citrus trees covering 93,000 hectares (230,000 acres) were either killed or damaged for a total loss in excess of $1 billion. Again, on 21 January 1985, a cold wave of even greater severity than the one in 1983 further damaged surviving citrus trees. Subfreezing temperatures were reported as far south as Miami on the morning of 22 January. Losses from this double blow reduced citrus-producing acreage by almost 90% in Lake County, formerly Florida's second largest citrus-producing county. Some growers replanted, and the groves were beginning to recover when yet another deep freeze over Christmas weekend 1989 ruined much of central Florida's citrus crop.

WARM HIGHS

A **warm-core anticyclone** forms south of the polar front and consists of extensive areas of subsiding warm, dry air. Thickness arguments show that, like cold-core cyclones, warm-core anticyclones strengthen with altitude. They are massive systems with a circulation extending from Earth's surface up to the tropopause. The semipermanent subtropical anticyclones, such as the *Bermuda-Azores high*, are examples of warm-core highs, but other warm-core anticyclones may develop over the interior of North America, especially in summer.

A cold high coincides with a shallow mass of cold, dense air and produces high surface air pressures compared to surrounding areas. How does a warm anticyclone produce high surface pressures? Whereas the central surface air pressure may be the same in both cold and warm anticyclones, an important difference is found in their vertical structures. The column of cold air in a cold-core anticyclone does not extend vertically nearly as far as the column of less dense warm air in a warm-core anticyclone. The greater mass of air over the center of a warm-core anticyclone, related to a higher tropopause, is responsible for the warm anticyclone's high surface pressure.

Cold-core anticyclones modify as they travel and may eventually become warm-core systems. As noted earlier, a cold-core anticyclone is actually a dome of relatively cold air, and as that air mass traverses land that is snow free, it is heated from below and moderates considerably, its pressure decreasing significantly. As a cold-core high drifts southeastward over the United States, air mass modification may be sufficient so that the pressure system eventually merges with the warm-core subtropical high over the Atlantic.

ANTICYCLONE WEATHER

An anticyclone is a fair-weather system because surface winds blowing in a clockwise (in the Northern Hemisphere) and outward direction induce subsidence of air over a broad area. Because subsiding air is compressionally warmed, the relative humidity drops and clouds usually dissipate or fail to develop. Although anticyclones are fair-weather systems, they can produce extremes in weather. The lowest air temperatures of the winter are usually associated with an arctic high, whereas drought and excessive heat are associated with a warm anticyclone that stalls in summer.

In addition, the horizontal air pressure gradient is weak over a broad region about the center of an anticyclone so that prevailing winds are light or the air is calm. At night, clear skies and light winds favor intense radiational cooling and perhaps dew, frost, or fog. Away from the broad central region of an anticyclone, the horizontal air pressure gradient strengthens and so does the wind. With stronger winds, significant advection oc-

curs. Typically, well to the east of the high center, northwest winds advect cold air southeastward, whereas to the west of the high center, southeast winds advect warm air northwestward. Air mass advection helps to increase the temperature contrast across the trough that separates highs, thereby favoring frontogenesis.

An understanding of the basic circulation characteristics of an anticyclone helps us to anticipate the sequence of weather events as an anticyclone travels into and out of a midlatitude location. Consider what happens in winter as a cold anticyclone slides southeastward out of southern Canada and into the northeastern United States. The following sequence may take place over several days to a week, depending on the anticyclone's forward speed.

Ahead of the anticyclone, strong northwest winds bring a surge of cold continental polar or arctic air. Strong winds and falling temperatures produce low windchill temperatures. To the lee of the Great Lakes, heavy lake-effect snow showers break out (Chapter 8). Even hundreds of kilometers downwind of the lakes, instability showers bring light accumulations of snow (e.g., over the hills of West Virginia and western Pennsylvania). However, as the anticyclone drifts closer, winds slacken, skies clear, and nocturnal radiational cooling produces very low surface temperatures. Under these conditions, air temperatures dip to their lowest readings, especially if the ground is snow-covered. Then, as the anticyclone moves away toward the southeast, winds again strengthen, but this time from a southerly direction, and warm air advection begins. The first sign of warm air advection is the appearance of high, thin cirrus clouds in the western sky.

In summer, a Canadian high-pressure system causes the same advection patterns as in winter except that the temperature contrast between air masses is considerably less at that time of year. Air advected ahead of the high on northwesterly winds may not be much cooler than the air advected behind the high on southerly winds. Often, the most noticeable difference between northerly and southerly winds at midlatitudes in summer is a contrast in humidity. Air advected ahead of the high is often less humid, and therefore more comfortable, than air advected behind the high.

At times in summer, however, an anticyclone becomes established east of the Rocky Mountains and remains in place for weeks. Such anticyclones are warm-core systems with a very deep circulation that is not readily displaced. Aloft, over the high is a warm ridge that may become cut-off from the prevailing westerly flow. With a *block* in place (Chapter 9), subsiding air associated with the anticyclone suppresses cloudiness and precipitation. Persistence of this pattern causes unusually high temperatures and drought.

The pattern of air mass advection in an anticyclone also applies to a ridge. Cold air advection usually occurs ahead (to the east) of a ridge, and warm air advection occurs behind (to the west of) a ridge. The circulation of an anticyclone (or ridge) does not occur in isolation from that of a cyclone (or trough). The atmosphere is, after all, a continuous fluid, with anti-cyclones following cyclones and cyclones following anticyclones. Northwest winds develop ahead of the high and on the back (west) side of a retreating low. Winds are caused by horizontal pressure gradients that develop between migrating anticyclones and cyclones.

Local and Regional Circulation Systems

Air masses, fronts, cyclones, and anticyclones are synoptic-scale systems that dominate the weather of middle latitudes. Many other weather systems operating at smaller spatial and temporal scales also contribute to the variable weather of midlatitudes. In some cases, synoptic-scale systems set the boundary conditions that make smaller scale weather systems possible. We open our discussion of meso- and microscale systems with land and sea (or lake) breezes, chinook winds, desert winds, and mountain and valley breezes.

LAND AND SEA (OR LAKE) BREEZES

For people lucky enough to live near the ocean or a large lake, sea or lake breezes bring welcome respite from the oppressive heat of a summer afternoon. On warm days, if the synoptic-scale air pressure gradient is relatively weak, a cool wind sweeps inland from over the sea or large lake. Depending on the source, this refreshing wind is called either a **sea breeze** or a **lake breeze**. Both breezes owe their existence to differential heating of land and water.

When land and water are exposed to the same intensity of solar radiation, the land surface warms up more than the water surface (Chapter 4). The relatively warm land heats the overlying air, thereby lowering air density. Compared to the land, the water is relatively cool, as is the air overlying the water. A local horizontal

air pressure gradient develops between land and water, with the highest pressure over the water surface (Figure 10.17 Top). In response to this horizontal air pressure gradient, cool air sweeps inland. Aloft, continuity requires a return flow of air directed from land to water, with air rising over the relatively warm land and sinking over the relatively cool water.

Sea (or lake) breezes are shallow circulation systems, generally confined to the lowest 1000 m (3300 ft) of the troposphere. Typically, the breeze begins near the shoreline several hours after sunrise and gradually expands both inland and out over the body of water, attaining maximum strength by mid afternoon. The

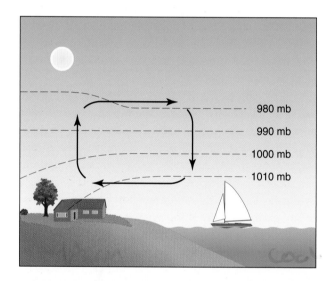

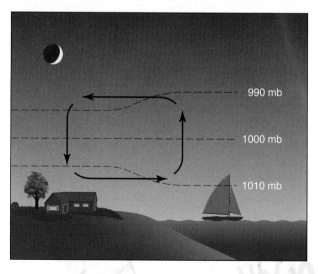

FIGURE 10.17
Vertical cross-sections showing isobaric (constant pressure) surfaces and circulation in a (Top) sea (or lake) breeze, and (Bottom) land breeze.

inland extent of the breeze varies from only a few hundred meters to tens of kilometers, depending in part on local topography.

Near sunset, the sea (or lake) breeze dies down. By late evening, however, surface winds begin to blow offshore as a **land breeze**. A reversal in heat differential between land and water is responsible for this wind shift. At night, radiational cooling chills the land surface more than the water surface. Air over the land surface thus becomes cooler (denser) than air over the water surface. A horizontal gradient in air density gives rise to a horizontal air pressure gradient with higher pressure over the land and lower pressure over the sea (or lake). A cool offshore breeze develops, along with a return airflow aloft; air sinks over the relatively cool land and rises over the relatively warm water (Figure 10.17 Bottom). A land breeze attains maximum strength around sunrise but is generally weaker than a sea (or lake) breeze.

Earth's rotation typically does not significantly influence the direction of sea (or lake) and land breezes because the duration of these regimes is too short and the distances too small. In some places, however, the Coriolis effect is responsible for a gradual shift in the direction of a sea breeze through the course of a day. Furthermore, in some localities, uplift produced along a sea-breeze front spurs the development of convective clouds and perhaps thunderstorms (Chapter 11).

CHINOOK WINDS

A **chinook wind** is a relatively warm and dry wind, which develops when air aloft is adiabatically compressed as it descends the leeward slopes of a mountain range. For every 1000 m of descent, the air temperature rises about 9.8 Celsius degrees (*dry adiabatic lapse rate*). Air that flows down the slopes of high mountain ranges such as the eastern side of the Rocky Mountains in the U.S. and Canada can undergo considerable warming. As this unsaturated air warms, its relative humidity decreases. During a chinook, a band of clouds, known as the *chinook arch*, overhangs the Rocky Mountain crest marking the location where clouds forming on the windward slopes vaporize.

Typically, a chinook develops when strong winds force a layer of stable air in the lower troposphere to ascend the windward slopes of a mountain range. When the air reaches the leeward slopes, its stability causes it to return (descend) to its original altitude. The larger-scale circulation causes further descent of the air. For example, chinook winds descending the leeward slopes of the Rocky Mountain Front Range are directed downslope by strong west winds associated with cyclones and anticyclones centered over the Great Plains well east of the mountains.

At the onset of a chinook, surface air temperatures often climb abruptly tens of degrees in response to the arrival of compressionally warmed air. On 6 January 1966, at Pincher Creek, Alberta, a chinook sent the temperature soaring 21 Celsius degrees (38 Fahrenheit degrees) in only 4 minutes! The sudden spring-like warmth may just as quickly give way to bitter cold. For example, along the foot of the Rockies, a shift of synoptic-scale winds from west to north brings an abrupt end to the chinook, the return of polar or arctic air, and plunging temperatures.

Chinook is a Native American word that, according to tradition, means *snoweater*. The term stems from the warm, dry wind's catastrophic effect on a snow cover. As noted in Chapter 6, air ascending the windward slopes of a mountain range loses much of its water vapor to condensation and deposition (cloud formation) and precipitation. As air descends the leeward slopes and is compressionally warmed, the relative humidity drops dramatically. Because the chinook is both warm and very dry, a snow cover sublimates and melts rapidly. It is not unusual for a half meter of snow to disappear in this way in only a few hours.

Chinook winds are strong and gusty and, locally, may reach destructive speeds, especially along the foothills of the Front Range of the Rocky Mountains. At Boulder, Colorado, in the foothills just northwest of Denver (Figure 10.18), violent downslope winds, sometimes gusting to 160 km (100 mi) per hr or higher,

FIGURE 10.18
Boulder, CO is situated in the foothills of the Rockies and experiences some particularly strong and destructive downslope winds.

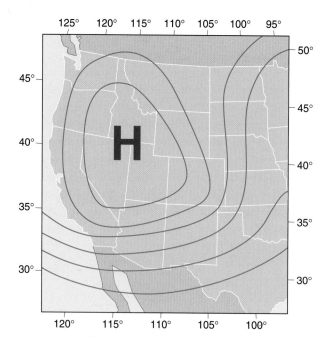

FIGURE 10.19
Schematic representation of the surface weather pattern that
favors development of Santa Ana winds over Southern
California. Solid lines are isobars.

unroof buildings and topple power poles. On average,
the community sustains about $1 million in property
damage each year because of these destructive winds.

Researchers do not agree on the precise
mechanism that triggers destructive chinook winds, but
according to one view, strong downslope winds are
linked to the disturbance of the planetary-scale westerlies
as they pass over the Rockies. Recall from Chapter 7
that a prominent north-south oriented mountain range
such as the Rockies deflects a westerly air flow into a
standing wave, a stationary pattern of crests and troughs
extending downwind of the mountain range and giving
rise to mountain-wave clouds. A chinook is actually a
segment of the wave that dips down the leeward slopes of
the mountain range. Apparently, in a violent chinook,
very energetic turbulent eddies that are generated aloft
are transported downward into the foothills so that
surface winds become very strong and gusty.

Chinook-type winds are not restricted to the
leeward slopes of the Rockies. A similar wind, called the
foehn, blows into the Alpine valleys of Austria and
Germany. The same type of warm, dry wind is drawn
down the eastern slopes of the Andes in Argentina, where
it is known as the *zonda*. In southern California, the
notorious **Santa Ana wind** is another chinook wind that

typically occurs in autumn and winter. A strong high
pressure system centered over the Great Basin sends
northeast winds over the southwestern United States,
driving air down slope from the desert plateaus of Utah
and Nevada, around the Sierra Nevada, and as far west as
coastal southern California (Figure 10.19). Adiabatic
compression produces hot, dry winds that desiccate
vegetation and contribute to outbreaks of forest and brush
fires. Santa Ana winds sometimes gust to 130 to 145 km
(80 to 90 mi) per hr and almost always cause some
property damage. For more on the Santa Ana wind and
the fire hazard in southern California, refer to the Essay,
"Santa Ana Winds and California Fire Weather."

DESERT WINDS

Deserts are windy places primarily because of
intense solar heating of the ground. In deserts, most of
the absorbed radiation goes into sensible heating because
water is scarce, vegetation is sparse, and relatively little
heat is used for evapotranspiration. In some spots, the
midday temperature of the surface can exceed 55 °C (131
°F). Such a hot surface generates a *superadiabatic* lapse
rate in the lowest air layer (that is, a lapse rate greater
than 9.8 Celsius degrees per 1000 m). A superadiabatic
lapse rate means great instability, vigorous convection,
and gusty surface winds. The strength and gustiness of
the wind vary with the intensity of solar radiation so that
wind speeds and gustiness usually peak in the early
afternoon and during the warmest months.

A **dust devil**, a whirling mass of dust-laden air,
is a common sight over flat, dry terrain. Dust devils
develop on sunny days in response to local variations in
surface characteristics (albedo, moisture supply,
topography) that give rise to localized hot spots. Air
over a hot spot is heated and forced to rise by cooler
surface winds that converge toward the hot spot. Shear
in the horizontal wind causes the column of rising hot air
to spin about a nearly vertical axis. The source of wind
shear may be nearby obstacles that disturb the horizontal
wind or the overturning of air induced by extreme
instability. In the process, dust is lifted off the ground
and whirled about, making the system visible. Unlike a
tornado, a dust devil is not linked to a cloud.

Dust devils are microscale systems. The most
common ones are small whirls less than 1.0 m (about 3
ft) across that typically last less than a minute. On rare
occasions, a dust devil exceeds 100 m (330 ft) in
diameter and whirls about for 20 minutes or longer. Such
intense dust devils may be visible to altitudes topping
900 m (3000 ft), but the invisible portion of the rising air

column may reach 4500 m (15,000 ft).

Most dust devils are too weak to cause serious property damage. The larger ones, however, are known to produce winds in excess of 75 km (45 mi) per hr and may cause some damage. According to the National Weather Service (NWS), every year in New Mexico, several large dust devils cause substantial damage to mobile homes, travel trailers, and buildings under construction. In the spring of 1991, a powerful dust devil that fortuitously passed over anemometers at the NWS office at Albuquerque produced a wind gust of 113 km (70 mi) per hr.

Thunderstorms or migrating cyclones produce larger scale winds in deserts. Surface winds associated with these weather systems can give rise to either dust storms or sandstorms, the difference between the two hinging on the size range of the loose surface sediments lifted by the wind. Dust consists of very small particles (diameters less than 0.06 mm or 0.002 in.) that can be carried by winds to great altitudes. Sand, typically covering only a small fraction of desert terrain, consists of larger particles (diameters of 0.06 to 2.0 mm or 0.002 to 0.08 in.) that are transported by the wind within about a meter of the ground.

The strong, gusty downdraft of a thunderstorm generates one of the most spectacular dust storms, known as a **haboob**. In a desert, rain falling from thunderstorm clouds often evaporates completely in the dry air beneath cloud base and does not reach the ground. The thunderstorm downdraft, however, exits the cloud base and strikes the ground as a surge of cool, gusty air. Dust lifted off the ground fills the air, severely restricting visibility, and the mass rolls along the ground as a huge ominous black cloud. A haboob may be more than 100 km (60 mi) wide and may reach altitudes of several thousand meters. These dust storms are most common in the northern and central Sudan especially near Khartoum. They also occur in the American southwest deserts.

MOUNTAIN AND VALLEY BREEZES

In summer, a localized circulation system may develop in wide, deep mountain valleys that face the sun (Figure 10.20). After winter snows have melted, bare valley walls strongly absorb solar radiation and sensible heating raises the temperature of air in contact with the valley walls. Air adjacent to the valley wall becomes warmer and less dense than air at the same altitude out over the valley floor. The cooler, denser air over the valley sinks as air adjacent to the valley walls blows upslope as a **valley breeze**. The ascending valley breeze

expands and cools and may trigger development of cumulus clouds near the summit.

A valley breeze is best developed between late morning and sunset. By midnight, the circulation reverses direction and persists until about sunrise. Under clear skies, nocturnal radiational cooling chills the valley walls and the air in contact with the walls also cools. Now air adjacent to the valley walls is colder and denser than the air at the same altitude above the valley. Air over the valley is forced to ascend as the cold, gusty **mountain breeze** flows down slope. Cold, dense air accumulates in the valley bottom where further radiational cooling may lead to formation of fog or low stratus clouds.

Valley breeze

Mountain breeze

FIGURE 10.20
Schematic representation of valley and mountain breeze circulations.

Mountain and valley breezes are most common during fair weather and when synoptic-scale winds are light or calm. Hence, these localized winds typically occur when mountainous regions are under the influence of a slow-moving anticyclone.

Conclusions

We have now examined the features of the principal weather makers of middle latitudes: air masses, fronts, cyclones, and anticyclones. We have seen how these synoptic-scale weather systems interact with one another and how they are linked to the planetary-scale circulation described in Chapter 9. Also recall from our discussion in Chapter 4 that synoptic-scale weather systems play an important role in poleward heat transport.

We also covered the characteristics of several local and regional weather systems. The domination of smaller scale weather systems by the larger scale circulation is apparent; that is, the planetary- and synoptic-scale patterns set boundary conditions for any smaller scale circulation system. In some cases, synoptic-scale winds reinforce mesoscale winds, as when regional winds blow in the same direction as a sea or lake breeze. In other cases, synoptic-scale winds overwhelm mesoscale winds, as when northerly winds sweep along the edge of the Rocky Mountains and eliminate the possibility of chinook winds. We continue our examination of weather systems in the next chapter with a look at thunderstorms and tornadoes.

Basic Understandings

- Air masses are classified on the basis of their temperature and humidity, characteristics that are acquired in their source regions. The four basic air mass types are cold and dry, cold and humid, warm and dry, and warm and humid.
- Air masses form over land (continental) and ocean (maritime), and at high latitudes (polar) and low latitudes (tropical). Arctic air is distinguished from polar air by its bitter cold.
- Air masses modify as they travel from one place to another. The degree of modification depends on air mass stability and the nature of the surface over which the air travels, specifically whether the surface is warmer or colder than the air mass.

- A front is a narrow zone of transition between air masses that contrast in temperature or water vapor concentration. The four types of fronts are: stationary, warm, cold, and occluded. A stationary front is just that, a front that exhibits essentially no lateral movement. A stationary front becomes a cold front if it begins to move in such a way that colder (more dense) air displaces warmer (less dense) air. If a stationary front begins to move in such a way that the warm (less dense) air advances while the cold (more dense) air retreats, the front changes in character and becomes a warm front.
- An occluded front forms late in the life cycle of a wave cyclone as the system moves into colder air. The three types of occluded fronts are cold, warm, or neutral based on the temperature contrast between the air behind the cold front and the air ahead of the warm front.
- Weather associated with stationary or warm fronts typically consists of a broad cloud and precipitation shield that may extend hundreds of kilometers on the cold side of the surface front. Weather along or ahead of a cold front usually consists of a narrow band of clouds and brief rain or snow showers, or thunderstorms.
- As a midlatitude cyclone progresses through its life cycle, it is supported and steered by the upper-level circulation toward the east and northeast. A cyclone typically begins as a wave along the polar front and deepens as the surface air pressure continues to drop. Winds strengthen and frontal weather develops. The cyclone finally occludes as the faster-moving cold front approaches the slower-moving warm front and the upper-level trough becomes vertically stacked over the surface cyclone.
- A midlatitude cyclone's life cycle can vary significantly in length. A typical cyclone takes about one day in each of the stages leading up to maximum intensity. However, when conditions are favorable, a cyclone can progress from birth to peak intensity in 36 hrs or less. In some cases, the cyclone never attains occlusion, instead spending most of its life span as a weak wave cyclone rippling along a stationary front.
- The conveyor-belt model describes the three dimensional structure of a mature cyclone in terms of three airstreams: warm and humid, cold, and dry.
- The track followed by a cyclone is key to the type of weather experienced at a given locality. On the cold side of the storm track, winds back with time (turn

counterclockwise) and there are no frontal passages. On the warm side of the storm track, winds veer (turn clockwise) with time and a cold front follows a warm front.

- Across eastern North America, Alberta-track cyclones are most frequent, but Colorado and coastal storm tracks are responsible for heavier precipitation, especially in winter.

- A warm-core cyclone (thermal low) is stationary, has no fronts, and is associated with hot, dry weather. These relatively shallow systems develop in the American Southwest.

- A cold-core anticyclone is a shallow system that coincides with a dome of continental polar or arctic air. Cold air advection occurs ahead of a high and warm air advection occurs behind a high.

- A warm-core anticyclone, such as a semipermanent subtropical high, extends from Earth's surface to the tropopause, and produces a broad area of subsiding warm, dry air.

- When synoptic-scale winds are weak, localized horizontal air pressure gradients can develop between land and sea (or lake). In response, winds blow onshore during the day (sea or lake breeze) and offshore at night (land breeze).

- A chinook wind consists of compressionally warmed, stable air that is forced down the leeward slopes of a mountain range by the circulation about cyclones or anticyclones.

- Intense solar heating in desert terrain produces a steep temperature lapse rate in the lowest air layer. Dust devils may develop in such unstable air. In addition, strong winds associated with thunderstorms or migrating cyclones may cause dust storms or sand storms.

- Mountain and valley breezes develop in summer in deep, wide mountain valleys that face the sun. A valley breeze is an upslope wind that forms during the day, and a mountain breeze is a downslope wind that forms at night.

ESSAY: The Case of the Missing Storm

A cyclone that sweeps ashore along the west coast of North America seems to disappear from the sea-level pressure pattern as it tracks inland over the mountainous West. A few days later, a low develops to the east of the Rocky Mountain Front Range, typically on the plains of Alberta or eastern Colorado. What actually happened to the storm as it crossed the mountains?

Visualize a cyclone moving ashore as a huge cylinder of air spinning about a vertical axis in a counterclockwise direction as viewed from above (see the Figure). The cylinder has Earth's surface as its lower boundary and the tropopause as its upper boundary. Because the tropopause remains at essentially the same altitude over the area, the column is forced to shrink in height as the cylinder moves up the windward slopes of a mountain range. The cylinder widens as it shrinks and the spin of the cylinder slows; that is, the storm's circulation weakens. As the cylinder of air then descends the leeward slopes of the mountain range, it stretches vertically and contracts horizontally, and the cyclonic spin strengthens. Weakening cyclonic circulation upslope followed by strengthening cyclonic circulation down slope account for the seeming disappearance and reappearance of a storm as it crosses mountainous terrain.

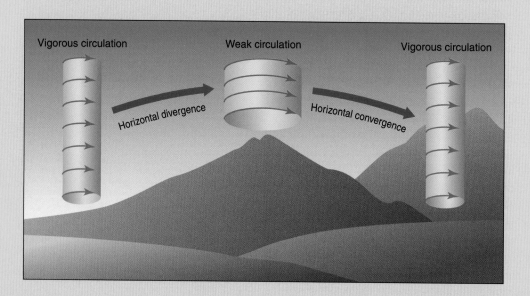

Figure
The cyclonic spin of a cylinder of air weakens as it ascends the windward slopes of a mountain range and strengthens as it descends the leeward slopes of the mountain range.

The changes in the storm's circulation are analogous to what happens to an ice skater performing a spin. The skater changes her spin rate by extending or drawing in her arms. When she extends her arms (analogous to horizontal divergence and the widening of our cylinder), her spin rate slows. When the skater brings her arms as close as possible to her body (analogous to horizontal convergence and the contracting of our cylinder), her spin rate increases. Both the spinning skater and the cyclone passing through the mountains conserve angular momentum. Simply put, *conservation of angular momentum* means that a change in the radius of a rotating mass is balanced by a change in its rotational speed. An increase in radius (horizontal divergence) due to shrinking of the vertical column is accompanied by a reduction in rotational rate whereas a decrease in radius (horizontal convergence) due to stretching of the column is accompanied by an increase in rotational rate.

ESSAY: Nor'easters

A *nor'easter* is an intense extratropical cyclone that tracks along the East Coast of North America and is named for the direction from which its powerful onshore winds blow. They are most frequent from October through April. The system moves toward the northeast and, if centered just offshore, strong onshore winds on the northern flank of the system (blowing from the east and northeast) can cause considerable coastal erosion, flooding, and property damage.

Although rarely attaining the strength of even a weak hurricane, a nor'easter can pack a powerful punch because it often impacts a much greater swath of coastline than a typical East Coast hurricane. A particularly destructive aspect of either storm system is the associated storm surge. A *storm surge* is a great dome of seawater, driven by strong onshore winds and low air pressure, that sweeps across the coastline. The diameter of an average nor'easter is about three times that of an average hurricane. Whereas a hurricane tends to track more directly from sea to land, impacting only about 100 to 150 km (60 to 90 mi) of coastline, a nor'easter often moves parallel to the coast so that its onshore winds may sweep over more than 1500 km (900 mi) of coastline. A less intense but slow-moving nor'easter can actually cause more damage than a more intense but fast-moving nor'easter. Winds in the slow-moving system blow in the same onshore direction for a longer period than the continually shifting wind in the fast-moving system. Persistent onshore winds increase the likelihood of a destructive storm surge.

One of the most noteworthy nor'easters of the 20th century in terms of its impact on virtually every sector of society occurred on 12-15 March 1993. A major storm system tracked from the Gulf of Mexico northeastward along the Eastern Seaboard. Although the storm system's size and central pressure (960 mb or 28.35 in., minimum) were not particularly unusual, its human impact was enormous, primarily because of its track along the heavily populated East Coast. This track enabled the storm to draw into its circulation a tremendous amount of moisture from the warm ocean waters just off the coast, while cold air inland helped the storm produce heavy snow over the region to the north and west of its track from Alabama to Maine.

Snow fell along the Gulf Coast with 7.6 cm (3 in.) reported at Mobile, AL and up to 13 cm (5 in.) measured in the Florida Panhandle, the greatest single snowfall in the state's history. The 33 cm (13 in.) that blanketed Birmingham, AL not only set a new 24-hr snowfall record for any month, but also set records for maximum snow depth, greatest snowfall in a single storm, and maximum snowfall for a single month. Total storm snowfall was tremendous in the Appalachians including 142 cm (56 in.) at Mt. LeConte TN and 127 cm (50 in.) at Mount Mitchell, NC. Just about every official weather station in West Virginia set a new 24-hr snowfall record with 76 cm (30 in.) reported at Beckley, WV. Further to the north, snow totaled 64 cm (25 in.) at Pittsburgh, PA, 69 cm (27 in.) at Albany, NY, and 109 cm (43 in.) at Syracuse, NY. Heavy snow was whipped by winds gusting in excess of hurricane force along the coast; Boston, MA recorded gusts to 131 km (81 mi) per hr. For the first time in history, snow forced the closing of every major East Coast airport. Thousands of people were isolated in the Appalachian Mountains; more than 3 million customers were without electricity at one time; hundreds of roofs collapsed under the weight of accumulated snow; and about 200 homes on North Carolina's Outer Banks were severely damaged by winds and flooding.

In addition to property damage caused by heavy snow and high winds, a squall line associated with the storm spawned 27 tornadoes in Florida, and a 3-m (10-ft) storm surge in the Gulf of Mexico flooded the Apalachicola area. All told, this so-called *storm of the century* claimed about 208 lives, more than three times the combined death toll of Hurricanes Hugo and Andrew, and caused damage estimated at $6 billion, the costliest extratropical storm in U.S. history.

ESSAY: Santa Ana Winds and California Fire Weather

Every autumn, Santa Ana winds descend the mountain slopes of interior southern California and sweep over the coastal plain to the sea. Named for the town of Santa Ana southeast of Los Angeles, these hot and dry winds blow over a landscape parched by the long dry summer, further desiccating the vegetation and making southern California particularly vulnerable to wildfire. Also contributing to the fire danger is a shrub community known as *chaparral* that dominates the hill slopes. The tissues of these plants contain oils that readily ignite; in fact, chaparral shrubs practically explode when exposed to fire. Periodic wildfires, whipped by Santa Ana winds roar down the canyons of southern California. One of the earliest accounts of such fires was in October 1542 when Juan Rodriguez Cabrillo, whose ship lay anchored off the Los Angeles Basin, observed hot desert winds blowing and fires burning.

In southern California, winter rains usually follow the summer dry season. (Dry summers and wet winters characterize a *Mediterranean-type* climate. For example, in Los Angeles, on average, almost 95% of annual rainfall occurs from November through April.) Falling on the nutrient-rich ashes of burnt vegetation, winter rains spur re-sprouting of shrubs. Furthermore, fire followed by rain also triggers germination of seeds that have been dormant in the soil. Renewal of vegetative cover helps restabilize hillsides denuded by wildfires. However, if heavy rains fall before vegetation has a chance to stabilize burnt-over slopes, soil and debris are washed from the bare ground and flushed into streambeds. Rainwater also percolates into the loosely consolidated upper soil layer. When this layer becomes saturated, it flows down steep hillsides as a river of mud, covering roadways and inundating homes. These mudflows are the greatest post-fire problem. Furthermore, even a single winter of below-average rainfall significantly elevates the potential for autumnal wildfires.

Native Americans adapted to the climate/fire regime of southern California by building temporary shelters that they could readily move out of harm's way. Furthermore, these people routinely set fires (*controlled burns*) that improved the habitat for game animals and thus made hunting easier. The arrival of European settlers in southern California marked the establishment of permanent dwellings and efforts to suppress wildfires. In subsequent centuries the population of southern California grew slowly, but since end of World War II, southern California has been one of the fastest growing regions of the nation. Hundreds of thousands of homes were constructed in the chaparral-covered foothills. A combination of autumnal Santa Ana winds, summer drought, fire-prone vegetation, and urban sprawl set the stage for highly destructive wildfires.

Such a holocaust erupted on 27 October 1993 when fifteen major fires broke out on hillsides from Ventura County south to San Diego County. Firefighters were largely unable to contain the wildfires until Santa Ana winds finally died down three days later. After several days of relatively weak winds, the regional air pressure gradient again strengthened and Santa Ana winds resumed, whipping flames through previously untouched areas of Malibu and Topanga. In the aftermath, state and federal emergency services estimated total property damage at more than $1 billion. Fires scorched more than 80,000 hectares (almost 200,000 acres) and destroyed more than 1200 structures. Only two years earlier, following several years of below average winter rainfall and normally dry summers, Santa Ana winds fanned one of the nation's worst-ever urban blazes. On 19 and 20 October 1991, fires raged across Oakland and Berkeley, California, killing 25 people and destroying nearly 3000 homes and apartments.

What can be done to reduce the Santa Ana fire hazard? Nothing can be done to alter the normal sequence of summer drought, autumnal Santa Ana winds, and variable winter rains. It may be possible to change where people live, however. Some people argue that zoning ordinances should be enacted and enforced that prohibit construction of houses in locales subject to natural hazards such as wildfires and mudflows. Alternatively, ordinances could require that dwellings be constructed of fireproof materials (particularly roofs and outside walls). Furthermore, chaparral vegetation could be cleared within a designated perimeter of buildings. Currently, many people permit shrubby vegetation to grow up to the edge of their homes for privacy and a sense of living with nature.

If fire is suppressed, chaparral shrubs grow larger and denser and annually add more leaves and twigs to the ground litter layer. Because of the buildup of fuel, the chaparral, once ignited, burns even more intensely. Periodic setting of controlled burns would reduce the frequency and severity of wildfires. With such a strategy, people work with and not against the forces of nature.

CHAPTER 11

THUNDERSTORMS AND TORNADOES

Focus should be to encourage and develop creativity in all children without the ultimate goal being to make all children inventors, but rather to develop a future generation of critical thinkers.

FARAQ MOUSA

Case-in-Point

On 3 May 1999, all the necessary ingredients for an outbreak of severe weather came together in the atmosphere over Oklahoma, northern Texas, and south central Kansas. Before that day was done, more than 70 tornadoes were reported in the region. Twenty-six tornadoes struck in or near Oklahoma City. The most intense of these tornadoes (rated F5 on the Fujita scale) took 38 lives as it tore through Moore and Bridge Creek, southern suburbs of Oklahoma City. Two hours later, an F4 tornado claimed 6 lives at Haysville, KS.

The rawinsonde observation at Norman, OK early on the afternoon of 3 May indicated a shallow layer of warm, humid air near the surface with dewpoints near 21 °C (70 °F). This warm, humid layer was capped by a temperature inversion at an altitude of about 1500 m (5000 ft) with much drier air aloft. As long as the capping inversion was in place, convection currents could surge no higher than about 1500 m but as the hours passed the temperature and humidity contrast between the low-level and upper-level air layers grew, increasing the potential for deep convection and severe thunderstorm development. The Norman rawinsonde sounding also indicated a change in wind direction with altitude, blowing from the south at the surface to west or west-southwest aloft. This vertical wind shear is another key ingredient for the development of severe weather.

The afternoon arrival of a jet streak triggered the round of tornadic thunderstorms by lifting the air column and erasing the capping inversion. Convection currents surged into the upper troposphere and even into the lower stratosphere. Massive supercell thunderstorms developed explosively and spawned violent tornadoes.

Driving Question:

What atmospheric conditions favor the development of severe weather?

Most of us are familiar with typical thunderstorm weather—the blackening sky and abrupt freshening of wind, followed by bursts of torrential rain, flashes of lightning, and rumbles of thunder. Often the system's cool breezes and rains bring welcome relief on a hot, muggy summer afternoon. For a farmer whose crops are wilting under the summer sun, the rains may be an economic lifesaver. Some thunderstorms become violent, however, and wreak havoc. Lightning starts fires, strong winds uproot trees and damage buildings, heavy rains cause flooding, and some thunderstorms spawn destructive hail or tornadoes.

Although less than 1% of all thunderstorms produce tornadoes, in some areas of North America, the possibility of a tornado is the principal reason why people fear thunderstorms. Tornadoes are the most intense of all weather systems. They are associated with certain severe thunderstorms and can take lives and cause considerable property damage. This chapter covers thunderstorms and tornadoes, their characteristics, life cycle, geographical and seasonal distribution, and associated hazards.

Thunderstorm Life Cycle

A **thunderstorm** is a mesoscale weather system; that is, it affects a relatively small area and is short-lived. A thunderstorm is the product of vigorous convection that extends high into the troposphere, sometimes reaching the tropopause or even higher. Upward surging air currents are made visible by billowing cauliflower-shaped cumuliform clouds. A thunderstorm consists of one or more convection cells, each of which progresses through a life cycle that is divided into three stages: cumulus, mature, and dissipating (Figure 11.1).

CUMULUS STAGE

If atmospheric conditions are favorable, cumulus clouds build both vertically and laterally. This is the initial or **cumulus stage** of thunderstorm formation. Over a period of perhaps 10 to 15 minutes, cumulus cloud tops surge upward to altitudes of 8000 to 10,000 m (26,000 to 33,000 ft). At the same time, neighboring cumulus clouds merge so that by the end of the cumulus stage, the storm's lateral dimension may be 10 to 15 km (6 to 9 mi).

Cumulus clouds are products of convection within the atmosphere (Chapter 4). Based on mode of origin, convection is of two types, free and forced. Intense solar heating of Earth's surface triggers **free convection**, but usually free convection is not sufficiently energetic to generate a thunderstorm. Frontal or orographic lifting or converging surface winds strengthen convection, a process known as **forced convection**. Most thunderstorms are products of forced convection.

Visualize the ascending branch of a convection current as a continuous stream of parcels of warm, unsaturated air. The rising parcels expand and cool at the dry adiabatic lapse rate (9.8 Celsius degrees per 1000 m) until they reach the *convective condensation level (CCL)*, where the relative humidity reaches 100%, water vapor condenses, and cumulus clouds start forming. The more humid the air is to begin with, the less the expansional cooling needed for parcels to achieve saturation and the lower is the base of cumulus clouds. The bases of cumulus clouds usually are lower in Florida, where the relative humidity is higher, than in New Mexico, where the relative humidity is lower.

Latent heat released during condensation adds to the buoyancy of the saturated (cloudy) parcels, and they surge upward while cooling at the moist adiabatic lapse rate (averaging 6 Celsius degrees per 1000 m).

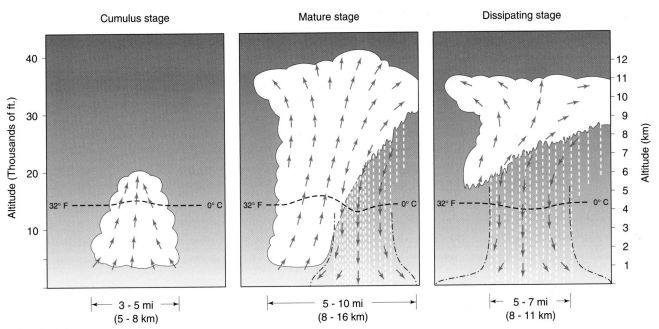

FIGURE 11.1
The life cycle of a thunderstorm cell consists of cumulus, mature, and dissipating stages. Temperatures are sub-freezing at altitudes above the dashed line.

Convective air parcels continue ascending as long as they are warmer, and thus less dense, than the surrounding air, that is, as long as the ambient air is unstable (Chapter 6). Some of the saturated parcels surge through the cloud top and evaporate in the relatively dry air above the cloud, increasing the water vapor concentration of that air. Because air above the cloud is now more humid, subsequent parcels are able to ascend higher before evaporating. As this process is repeated, the cumulus cloud billows upward. A cumuliform cloud that shows significant vertical growth and resembles a huge cauliflower is called **cumulus congestus**. If vertical growth continues, the cumulus congestus cloud builds into a **cumulonimbus** cloud, a thunderstorm cloud, producing precipitation, lightning, and thunder.

During the cumulus stage of the thunderstorm life cycle, saturated air streams upward throughout the cell as an *updraft*. The updraft is strong enough to keep water droplets and ice crystals suspended in the upper reaches of the cloud. For this reason, precipitation does not occur during the cumulus stage.

MATURE STAGE

By convention, the cumulus stage ends and the **mature stage** begins when precipitation reaches Earth's surface. Typically, this stage lasts for about 10 to 20 minutes. The cumulative weight of water droplets and

ice crystals eventually becomes so great that they can no longer be supported by the updraft. Rain, ice pellets, and snow descend through the cloud and drag the adjacent air downward, creating a strong *downdraft* alongside the updraft. At the same time, unsaturated air at the edge of the cloud is drawn into the cloud, a process known as *entrainment*.   (Actually,   entrainment   continues throughout the life cycle of a thunderstorm cell.) Entrained air mixes with the cloudy (saturated) air, causing some of the water droplets and ice crystals to vaporize. The consequent evaporative cooling weakens the buoyant uplift and strengthens the downdraft.

The downdraft exits the base of the cloud and spreads out along Earth's surface, well in advance of the parent thunderstorm cell, as a mass of cool, gusty air. The downdraft is relatively cool because of evaporative cooling beneath the cloud base that offsets to some extent compressional warming of the downdraft. At the surface, the arc-shaped leading edge of downdraft air resembles a miniature cold front and is called a **gust front**. Uplift along the gust front sometimes produces additional cumuliform clouds that may evolve into secondary thunderstorm cells tens of kilometers ahead of the parent cell (Figure 11.2).

Ominous-appearing low clouds are sometimes associated with thunderstorm gust fronts. A **roll cloud** is an elongated, tube-shaped cloud that appears to rotate

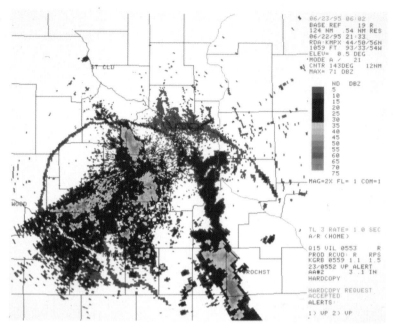

FIGURE 11.2
Thunderstorm cells can develop along the gust front ahead of the main thunderstorm. In this radar reflectivity image, the gust front shows up as an arc-shaped band to the north of the parent thunderstorm cells.

slowly about its horizontal axis. The roll cloud occurs behind the gust front and beneath, but detached from, the cumulonimbus cloud. How this cloud forms is not fully understood. Although appearances might suggest otherwise, roll clouds are seldom accompanied by severe weather. This is *not* the case for shelf clouds. A **shelf cloud**, also called an *arcus cloud*, is a low, elongated cloud that is wedge-shaped with a flat base. This cloud appears at the edge of a gust front and beneath and attached to the cumulonimbus cloud. A shelf cloud is

thought to develop as a consequence of uplift of stable warm and humid air along the gust front. Damaging surface winds may occur under a shelf cloud, and sometimes this cloud is associated with a severe thunderstorm.

A thunderstorm cell attains maximum intensity during its mature stage. This is when rain is heaviest (especially during the first 5 minutes or so), lightning is most frequent, and hail, strong surface winds, and even tornadoes may develop. Cloud tops can exceed an altitude of 18,000 m (about 60,000 ft). Strong winds at such altitudes distort the cloud top into an anvil shape (Figure 11.3). The flat top of the anvil indicates that convection currents have reached the extremely stable air of the tropopause. Only in severe thunderstorms will convection currents overshoot this altitude, causing clouds to billow into the lower stratosphere before collapsing back into the troposphere. Temperatures within the upper portion of the cloud are so low that the anvil is composed exclusively of ice crystals, giving it a fibrous appearance.

Viewed from space by satellite, clusters of mature thunderstorm cells appear as bright white blotches (Figure 11.4). The brightness of these clusters is due to

FIGURE 11.3
When the upward billowing cumulonimbus cloud reaches the tropopause, it spreads out forming a flat anvil top.

FIGURE 11.4
In this visible satellite image, clusters of intense thunderstorm cells appear as bright white blotches over Texas, Oklahoma, and Missouri..

the high albedo (reflectivity) of the cloud tops. Much of the solar radiation that penetrates cumulonimbus clouds is absorbed, so that the sunlight emerging at the cloud base is weakened considerably. For this reason, from our perspective on Earth's surface, the daytime sky darkens with the approach of a thunderstorm.

DISSIPATING STAGE

As precipitation spreads throughout the thunderstorm cell, so does the downdraft, heralding the demise of the cell. During the **dissipating stage**, subsiding air replaces the updraft throughout the cloud, effectively cutting off the supply of moisture provided by the updraft. Adiabatic compression warms the subsiding air, the relative humidity drops, precipitation tapers off and ends, and convective clouds gradually vaporize.

Thunderstorm Classification

Thunderstorms are classified on the basis of the number, organization, and intensity of their constituent cells. Thunderstorms occur as single cells, multicellular clusters, and supercells.

Single-cell thunderstorms are usually relatively weak weather systems that appear to pop up randomly almost anywhere within a warm, humid air mass. In reality, a single-cell thunderstorm is not a random phenomenon and almost always develops along some boundary within an air mass. The boundary, for example, may be the leading edge of the outflow from a distant thunderstorm cell. Or it may be the edge of a bubble of cool, dense air that is the remnant of another thunderstorm cell that has dissipated. At least in middle latitudes, solar-driven free convection alone usually is not sufficiently strong to generate a thunderstorm cell.

Typically, a thunderstorm cell completes its life cycle in 30 minutes or so, but sometimes lightning, thunder, and bursts of heavy rain persist for many hours. This is because most thunderstorms are *multicellular*; that is, a thunderstorm usually consists of more than one cell. Each cell may be at a different stage of its life cycle, and new cells form and old cells dissipate continually. A succession of many cells is thus responsible for a prolonged period of thunderstorm weather. Although a locality may be in the direct path of a distant, intense thunderstorm cell, the relatively brief life span of an individual cell means that severe weather may dissipate before reaching that locality. The multicellular nature of

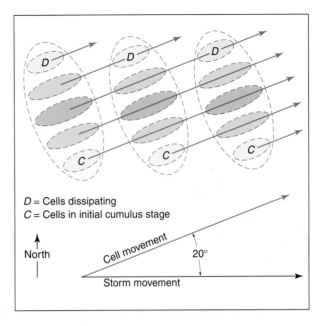

FIGURE 11.5
In this idealized situation viewed from above, the component cells of a multicellular thunderstorm travel at about 20 degrees to the eastward movement of the thunderstorm. As they move toward the northeast, the individual cells progress through their life cycle.

most thunderstorms complicates the motion of the weather system in that a thunderstorm may track at some angle to the paths of its constituent cells. For example, a thunderstorm may track from west to east, while its component cells head off toward the northeast (Figure 11.5). In terms of organization, two types of multicellular thunderstorms are the squall line and the mesoscale convective complex. Cells in either of these systems can produce severe weather, which is most likely to occur near the interface between the updraft and the downdraft of a mature cell.

A **squall line** is an elongated cluster of thunderstorm cells that is accompanied by a continuous gust front at the line's leading edge (Figure 11.6). A squall line is most likely to develop in the warm southeast sector of a mature cyclone ahead of and parallel to the cold front. (A more general name for a thunderstorm formed by uplift along cold or warm fronts is *frontal thunderstorm*.) Squall line thunderstorm cells are usually more intense than isolated single-cell thunderstorms because of the contribution of the synoptic-scale circulation to vertical motion. That is, the circulation within the associated low pressure system strengthens the updraft. As the squall-line gust front surges forward, warm, humid air is lifted and forced into

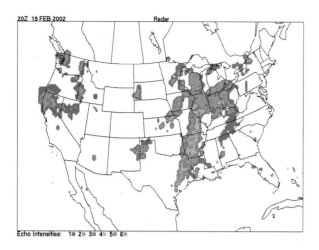

FIGURE 11.6
Radar image of a squall line from Texas to Illinois.

the updraft located at the leading edge of the squall line. The heaviest rain or hail typically occurs just behind (to the west of) this updraft. Light to moderate precipitation is produced by the older cells situated behind the mature cells at the leading edge.

A **mesoscale convective complex (MCC)** is a nearly circular cluster of many interacting thunderstorm cells covering an area that may be a thousand times larger than that of an isolated thunderstorm cell. In fact, it is not unusual for a single MCC to cover an area equal to that of the state of Iowa (Figure 11.7). MCCs are primarily warm-season (March through September) phenomena that generally develop at night and occur chiefly over the eastern two-thirds of the United States, where more than 50 may be expected in a single season.

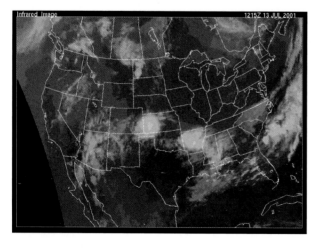

FIGURE 11.7
Infrared satellite image showing mesoscale convective complexes over western Kansas and most of Arkansas.

An MCC is not associated with a front and usually develops during weak synoptic-scale flow, often near a ridge of high pressure and on the cool side of a stationary front. A low-level jet feeds warm humid air into the developing system while a low pressure system forms at mid-levels of the troposphere. Rising temperatures at low levels and radiational cooling at upper levels destabilizes the troposphere. New cells develop and old cells die continually within an MCC, so the life expectancy of the system is at least 6 hrs and often 12 to 24 hrs. The longevity of an MCC, coupled with its typically slow movement (15 to 30 km, or 9 to 18 mi, per hr), means that rainfall is widespread and substantial. MCCs account for perhaps 80% of growing season rainfall in the Great Plains and Midwest. An MCC also has the potential of producing severe weather, including weak tornadoes, moderate-sized hail, and flash flooding.

A **supercell thunderstorm** is a relatively long-lived, large, and intense system. It consists of a single cell with an exceptionally strong updraft, in some cases estimated at 240 to 280 km (150 to 175 mi) per hr. A distinguishing characteristic of a supercell is the tendency of the updraft to develop a rotational circulation that may evolve into a tornado. Because of their link to strong to violent tornadoes, supercell thunderstorms are covered in more detail later in this chapter.

Geographical and Seasonal Distribution of Thunderstorms

Atmospheric conditions that favor thunderstorm development vary with latitude, season, and time of day. Key to understanding the spatial and temporal variability of thunderstorms is the conditions required for their formation: (1) humid air in the low- to mid-troposphere, (2) atmospheric instability, and (3) a source of uplift.

Most thunderstorms develop within masses of warm, humid air, that is, maritime tropical (*mT*) air, when that air mass is destabilized. Uplift is the key to destabilizing *mT* air. Usually, maritime tropical air is conditionally stable and becomes unstable (buoyant) only when lifted to the condensation level. Recall from Chapter 6 that conditional stability means that the ambient air is stable for unsaturated (clear) air parcels but unstable for saturated (cloudy) air parcels. The more humid the air, the less the ascent (expansional cooling) needed for destabilization.

Most thunderstorms develop when maritime tropical air is lifted (1) along fronts, (2) up mountain slopes, or (3) via horizontal convergence of surface winds. Furthermore, cold air advection aloft and/or warm air advection at the surface enhances the potential instability of *mT* air. Either of these processes increases the air temperature lapse rate and thereby reduces ambient air stability.

Solar heating drives atmospheric convection, so it is not surprising that thunderstorms tend to be most frequent when and where solar radiation is most intense. In most places, thunderstorms are most frequent during the warmest hours of the day, but there are lots of exceptions to this rule. Mesoscale convective complexes and squall lines develop both night and day. In the Missouri River Valley and adjacent portions of the upper Mississippi River Valley, even single-cell thunderstorms are more frequent at night than during the day. One possible explanation for the nocturnal thunderstorm maximum in the upper Missouri/Mississippi valleys centers on the role of a *low-level jet stream* of maritime tropical air that flows from off the Gulf of Mexico and northward up the Mississippi River Valley. This jet stream strengthens at night and causes warm air advection at low levels that destabilizes the air and spurs the buildup of cumuliform clouds.

Thunderstorm frequency usually is expressed in numbers of thunderstorm days per year, where a *thunderstorm day* is defined as a day when thunder is heard. This conventional method of expressing thunderstorm frequency likely underestimates the actual number of thunderstorms, particularly if more than one line of thunderstorms passes over a weather station on the same day. With this limitation in mind, thunderstorms occur with greatest frequency over the continental interiors of tropical latitudes. The steamy Amazon Basin of Brazil, the Congo Basin of equatorial Africa, and the islands of Indonesia have the highest frequency of thunderstorms of anywhere in the world, experiencing at least 100 thunderstorm days a year. Because the surfaces of large bodies of water do not warm as much as land surfaces in response to the same intensity of solar radiation, thunderstorms are less frequent over adjacent bodies of water.

In the subtropics and tropics, intense solar heating may combine with converging surface winds to trigger thunderstorm development. As noted in Chapter 9, this combination characterizes the intertropical convergence zone (ITCZ), a discontinuous band of thunderstorms more or less paralleling the equator that moves north and south seasonally with the sun.

In North America, thunderstorm frequency generally increases from north to south with the highest frequency over central Florida (Figure 11.8). Some interior localities of that state can expect an average of 100 thunderstorm days a year. The Florida thunderstorm maximum is due to convergence of sea breezes that

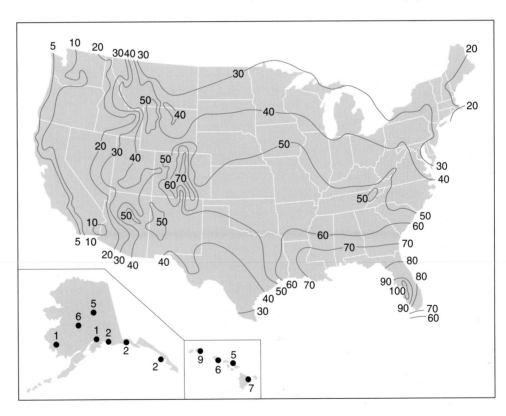

FIGURE 11.8
Average annual number of thunderstorm-days in the United States. [NOAA data]

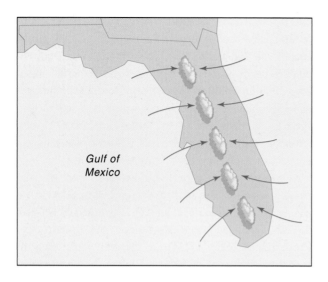

FIGURE 11.9
The relatively high frequency of thunderstorms over peninsular Florida is linked to the convergence of sea breezes blowing inland from both the Gulf and East Coasts.

develop along both the east and west coasts of the Florida peninsula (Figure 11.9). Sea breeze convergence over the interior induces ascent of maritime tropical air and formation of cumulonimbus clouds.

Portions of the Rocky Mountain Front Range rank second to interior Florida in thunderstorm frequency over North America. On average, more than 60 thunderstorm days occur per year in a band from southeastern Wyoming southward through central Colorado and into north central New Mexico. This high thunderstorm frequency is linked to topographically related differences in heating. Mountain slopes facing the sun absorb direct solar radiation and become relatively warm. The warm slopes, in turn, heat the air in immediate contact with the slopes, and that air rises. At the same time, the air at the same altitude, but located to the east of the mountains out over the relatively flat terrain of the western Great Plains, is much cooler. As warm air rises over the mountain slopes, it is replaced by cooler air sweeping westward from over the Plains. Updrafts over the mountain slopes produce cumuliform clouds that often evolve into thunderstorm cells, particularly during the warmest hours of the day. Thunderstorm development is enhanced whenever the synoptic-scale air pressure pattern favors winds blowing from the east over the western Great Plains.

To this point, our discussion has focused on conditions conducive to thunderstorm development. Under other conditions, convection and ascent of air is inhibited and thunderstorms do not form. This is the case when air masses reside or travel over relatively cold surfaces and are thereby stabilized. For example, snow-covered ground cools and stabilizes the overlying air. Convection is suppressed, so thunderstorms are relatively rare in middle and high latitudes in winter. Nonetheless, thunderstorms can be associated with winter cyclones, usually developing ahead of the surface warm front. In that case, *thundersnows* can be locally heavy.

Thunderstorms are unusual over coastal areas that are downwind from relatively cold ocean water. For example, thunderstorms are infrequent along coastal California, where prevailing winds are onshore and a shallow layer of maritime polar air often flows inland from off the relatively cold California Current. Cool *mP* air at low levels suppresses deep convection and thunderstorm development. Thus, the average annual number of thunderstorm-days is only 2 at San Francisco, 6 at Los Angeles, and 5 at San Diego.

Thunderstorms are also relatively rare in Hawaii, primarily because of the trade wind inversion (Chapter 9). At first-order weather stations such as Hilo and Honolulu, the average annual number of thunderstorm-days is under 10. The trade wind inversion, at an average altitude of about 2000 m (6600 ft), restricts vertical development of towering cumulus clouds. Cumulus clouds normally do not attain the altitude needed to develop into cumulonimbus clouds.

Severe Thunderstorms

By convention, a **severe thunderstorm** is accompanied by locally damaging winds, frequent lightning, or large hail. The official National Weather Service (NWS) criterion for designating a thunderstorm as *severe* includes any one or a combination of the following: hailstones 0.75 in. (1.9 cm) or larger in diameter; tornadoes or funnel clouds; surface winds stronger than 58 mi (93 km) per hr.

As a general rule, the greater the altitude of the top of a thunderstorm, the more likely it is that the system will produce severe weather. Why do some thunderstorm cells surge to great altitudes and trigger severe weather, whereas others do not? The key factor appears to be vertical wind shear. **Vertical wind shear** is the change in horizontal wind speed or direction with increasing altitude. Weak vertical wind shear (little change in wind with altitude) favors short-lived updrafts, low cloud tops, and weak thunderstorms, whereas strong vertical wind shear favors vigorous updrafts, great vertical cloud development, and severe thunderstorms.

With weak vertical wind shear, the flow of warm humid air into the cell is relatively weak. The downdraft's outflow pushes the gust front well ahead of the system, eventually cutting off the supply of warm humid air entirely. Also, precipitation falls through and thereby weakens the updraft. With increasing vertical wind shear, however, the gust front is held closer to the cell and the inflow of warm humid air is sustained for a longer period. Furthermore, most precipitation falls alongside rather than through (and against) the updraft so that the updraft maintains its strength and the thunderstorm surges to great altitudes. On the other hand, if the vertical wind shear is too strong, a developing cumulonimbus cloud is torn apart.

In the United States and Canada, most severe thunderstorms break out over the Great Plains and are associated with mature synoptic-scale cyclones. Severe thunderstorm cells usually form as part of a squall line within the cyclone's warm sector, ahead of and parallel to a fast-moving and well-defined cold front. The squall line appears as an ominous, twisting mass of low, dark clouds, often hundreds of kilometers long. Severe cells can produce large hail, heavy rain, and downbursts (discussed later in this chapter). Most tornadoes that do form are weak, although some squall line cells evolve into supercells that can spawn strong to violent tornadoes.

A synoptic situation that favors severe thunderstorm development is shown schematically in Figure 11.10. A mature cyclone is centered over western Kansas with a well-defined cold front trailing across the Oklahoma and Texas panhandles into West Texas and a warm front stretching southeastward from the low center. The midlatitude (polar front) jet and a low-level (*mT*) jet, which are also plotted on the map, cross to the southeast of the cyclone center. The most intense thunderstorm cells are likely to develop near the intersection of these two jets. The *subtropical jet stream* (Chapter 9) is sometimes also present at high levels in the troposphere, blowing from west to east over the warm sector of the cyclone. In that event, intense squall lines are likely to form between the midlatitude and subtropical jet streams.

The polar front jet stream produces strong vertical wind shear that maintains a vigorous updraft and great vertical development of thunderstorm cells. In addition, the jet contributes to a stratification of air that increases the potential instability of the troposphere. As

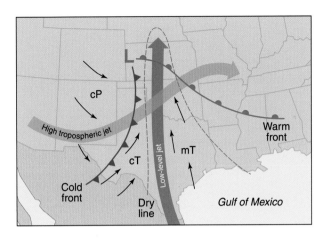

FIGURE 11.10
A synoptic weather pattern that favors development of severe thunderstorms.

described in Chapter 9, a jet streak induces both horizontal divergence and horizontal convergence of air aloft. Recall that diverging horizontal winds trigger ascent of air and cyclone development under the left-front quadrant of a jet streak. Meanwhile, air converges in the right-front quadrant of a jet streak, causing weak subsidence of air over the warm sector of the cyclone. The subsiding air is compressionally warmed and its relative humidity decreases. The subsiding, warming air is prevented from reaching Earth's surface by a shallow layer of maritime tropical air that surges northward as a low-level jet from over the Gulf of Mexico. The *mT* jet occurs on the western flank of the Bermuda-Azores subtropical high and is particularly strong at about 3000 m (9800 ft).

As a consequence of compressional warming, the air subsiding from aloft becomes warmer than the underlying layer of *mT* air. A zone of transition, that is, a temperature inversion, develops between the two layers of air (Figure 11.11). As noted in Chapter 6, an air layer characterized by a temperature inversion is extremely stable, so the two air layers do not mix and convection is confined to the surface *mT* air layer. In this circumstance, the inversion is known as a *capping inversion*. As long as this stratification persists, the contrast between air layers mounts; that is, subsiding air becomes drier and the underlying *mT* air becomes more humid. The potential for severe weather continues to grow, and all that is needed is a trigger that will enable updrafts to penetrate the capping inversion.

The necessary upward impetus may be supplied by a combination of the intense solar heating of mid-

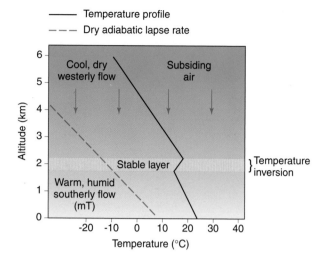

FIGURE 11.11
A temperature sounding that favors the development of severe thunderstorm cells. A capping temperature inversion separates subsiding dry air aloft from warm, humid air near the surface.

FIGURE 11.12
Pouch-like mammatus clouds occur on the underside of a thunderstorm anvil and sometimes indicate a severe storm system.

afternoon and the lifting of air caused by an approaching cold front or jet streak. Updrafts eventually break through the capping inversion, and cumulus clouds billow upward at explosive speeds that may exceed 100 km (62 mi) per hr. Such updrafts can even penetrate the tropopause and surge into the lower stratosphere. A severe thunderstorm is born.

With the synoptic pattern shown in Figure 11.10, air ahead of the cold front and west of the warm, humid tongue (enclosed by a dashed line) is usually quite dry. For this reason, the western boundary of the *mT* air mass is called the **dryline**. West of the dryline and east of the cold front, hot and dry (*cT*) air flows eastward and downhill from the Southwest desert and Mexican Plateau. A dryline commonly extends hundreds of kilometers from Texas northward onto the Western Plains and is often the site of squall line and severe thunderstorm development. Typically, the dryline advances eastward during the day and retreats westward at night.

Sometimes, ominous pouch-like mammatus clouds appear at the base of the spreading anvil top of a cumulonimbus cloud (Figure 11.12). (They are also associated with a variety of other cloud types including cirrus, altocumulus, and stratocumulus.) Contrary to popular belief, **mammatus clouds** do not always indicate a severe thunderstorm. In fact, in the mountainous western United States, they often are associated with relatively weak convective showers. Their unusual appearance is attributed to blobs of cold, cloudy air that

descend from the anvil into the unsaturated (clear) air beneath the anvil. Ice crystals at the margin of a blob sublimate, causing further cooling of the blob. This cooling offsets compressional warming of the descending air so that the blob sinks, bulging downward into the dry air. Ultimately, the size of the bulge is limited by the increasing rate of vaporization of the constituent ice crystals.

Thunderstorm Hazards

Thunderstorm hazards include lightning, downbursts, flash floods, hail, and tornadoes. Because tornadoes are an especially severe weather hazard, they are covered in a later section of this chapter.

LIGHTNING

A convective rain or snow shower is a thunderstorm if accompanied by lightning. **Lightning** is a brilliant flash of light produced by an electrical discharge within a cumulonimbus cloud or between the cloud and Earth's surface (Figure 11.13). Lightning is a weather phenomenon that can be directly hazardous to human life. Over a recent 30-year period (1971-2000), lightning annually killed an average of 73 people in the United States. This death toll may seem surprisingly high, perhaps because fatal lightning bolts are typically isolated events that seldom make headlines. A single, fatal lightning bolt is not as newsworthy as a single disastrous tornado that takes many lives, but so many

FIGURE 11.13
Lightning is a brilliant flash of light associated with an electrical discharge between clouds and Earth's surface, within a cloud, or between clouds. [NOAA photograph]

lightning strikes occur daily that fatalities and injuries add up. For some tips on lightning safety, refer to this chapter's first Essay.

Lightning not only kills and injures people, but also it ignites forest and brush fires. In the Rocky Mountain region, for example, lightning is the most common cause of forest fires, starting more than 9000 each year. One might think that heavy rain associated with a thunderstorm would quickly quench a lightning-ignited fire. In the western basins, however, the base of a cumulonimbus cloud is usually so far above the ground (perhaps 2400 m or 8000 ft), and the air below the cloud so dry, that much if not all the rain evaporates before reaching the fire.

Lightning is very costly to electrical utilities, each year causing tens of millions of dollars in damage to equipment (power lines, transformers) and in service restoration expenses. A useful tool in utilities' efforts to cope with the lightning hazard is the **lightning detection network (LDN)**, a system that provides real-time information on the location and severity of lightning strokes. Each monitoring station in the network consists of direction-finding equipment that can sense electro-magnetic fields associated with cloud-to-ground lightning. A computer superimposes locations of lightning strokes on a map of the region. An electrical utility that subscribes to a LDN service accesses lightning data through a computer link. Such information enables the utility to mobilize repair crews more efficiently and speeds up restoration of disrupted service.

What causes lightning? The potential for an electrical discharge exists whenever a charge difference develops between two objects. A normally neutral object becomes negatively charged when it gains electrons (negatively charged subatomic particles) and positively charged when it loses electrons. That is, the object is *ionized*. When large differences in electrical charge develop within a cloud or between clouds or a cloud and the ground, the stage is set for lightning.

On a clear day, Earth's surface is negatively charged with respect to the upper atmosphere (ionosphere). This charge distribution changes as a cumulonimbus cloud develops. Within the cloud, charges separate so that the upper portion and a much smaller region near the cloud base become positively charged. In between, a disk-shaped zone of negative charge forms that is a few hundred meters thick and several kilometers in diameter. At the same time, the region of negative charge in the developing cumulonimbus induces a positive charge on the ground directly under the cloud.

Air is a very good electrical insulator, so that as a thunderstorm cell forms and electrical charges separate and build, a tremendous potential soon develops for an electrical discharge. By about the time the thunderstorm cell enters its mature stage, the electrical field has strengthened to the point that the electrical resistance of air breaks down and electrons flow as a lightning discharge, thereby neutralizing the electrical charges. Lightning may forge a path between oppositely charged regions of a cloud, or between clouds or a cloud and the ground.

The cause of charge separation within cumulonimbus clouds is not well understood, but field studies and laboratory simulations offer a promising ex-planation focusing on graupel and the convective circulation within cumulonimbus clouds. **Graupel** (German for *soft hail*) consists of millimeter- to centimeter-size ice pellets formed when supercooled water droplets collide and freeze on impact. Within the cloud, as graupel descend and strike smaller ice crystals in their path, opposite charges develop on the graupel and the ice crystals. For collisions that take place at temperatures below about -15 °C (5 °F), graupel become negatively charged and ice crystals acquire a positive charge. Vigorous updrafts separate the particles, carrying the smaller positively charged ice crystals to the upper portion of the cloud, whereas the larger negatively charged graupel concentrate mostly in the lower portion of the cloud. This mechanism can explain the positive charge of the upper cloud region and the negative charge of the disk-shaped zone of the lower cloud region.

What accounts for the positive charge near the cloud base? Typically, temperatures near the cloud base are higher than -15 °C (5 °F). At those temperatures, collisions between graupel and ice crystals induce a positive charge on graupel and a negative charge on ice crystals. Ice crystals are swept up in the updraft, leaving heavier graupel to accumulate near the cloud base, giving that region a positive charge.

Lightning discharges between a cloud and the ground pose the greatest hazard for people, although these discharges represent only about 20% of all lightning bolts. Using high-speed photography, scientists have determined that a lightning flash involves a regular sequence of events. Initially, streams of electrons surge from the cloud base toward the ground in discrete steps, each of which is about 20 to 100 m (65 to 330 ft) long. These so-called *stepped leaders* describe a branching path and produce a narrow ionized channel. When a branch of the stepped leaders comes within about 100 m (330 ft) of the ground, it is met by a positively charged *return stroke* from the ground. The return stroke follows the path of least resistance and often emanates from tall, pointed structures such as a metal flagpole or tower. Now an ionized channel, only a few centimeters in diameter, links Earth's surface to the cloud. Electrons flow, neutralization occurs, and the channel is illuminated.

Following this initial electrical discharge, subsequent surges of electrons from the cloud, called *dart leaders*, follow the same conducting path. Each dart leader is met by a return stroke (from the ground), and the conducting path is again illuminated. Typically, a single lightning discharge consists of two to four dart leaders plus return strokes. Sometimes, a dart leader is met by a return stroke that forges a new conducting path from the ground. The result is a forked lightning bolt that strikes the ground at more than one place.

The sequence just described is the most common occurrence of cloud-to-ground lightning. In less than 10% of cases, a positively charged leader emanates from the cloud and initiates a lightning discharge. Much more rarely, positive or negative stepped leaders propagate upward from the ground and meet a return stroke surging downward from a cloud. This ground-to-cloud lightning usually is initiated from mountaintops or tall structures such as antenna towers.

Electricity flows at the astonishing rate of nearly 50,000 km (31,000 mi) per second; hence, the entire lightning sequence takes place in less than two-tenths of a second. The human eye has difficulty separating the individual flashes of light that constitute a single lightning bolt, so that we perceive a lightning flash as a flickering light.

Where there's lightning, there's thunder, although sometimes we see distant lightning but do not hear the thunder. Lightning heats the air along the narrow conducting path to temperatures that may exceed 25,000 °C (45,000 °F). Such intense heating occurs so rapidly that air density cannot respond, at least initially. The rapid rise in air temperature is accompanied by a tremendous increase in air pressure locally that generates a shock wave. The shock wave propagates outward, producing sound waves that are heard as **thunder**. The first thunder we hear is generated by the nearest part of the lightning bolt. Subsequent sound waves reach us from portions of the bolt that are progressively further away; hence, the rumble of thunder. Thunderstorm cells that are more than 20 km (12 mi) away are too distant for thunder to be heard although lightning can be seen. This is called *heat lightning*.

Light travels about a million times faster than sound so that we see lightning almost instantaneously, but we hear thunder later. The closer we are to a thunderstorm cell, the shorter is the time interval between the lightning flash and thunder. As a rule, thunder takes about 3 seconds to travel 1 km (or 5 seconds to travel 1 mi). If you must wait 9 seconds between lightning flash and thunderclap, the thunderstorm cell is about 3 km (1.8 mi) away. By noting the time difference between light and thunder of successive lightning discharges, you can deduce whether the thunderstorm cell is approaching or moving away, or even predict when it is dangerously close.

DOWNBURSTS

Severe, and sometimes not so severe, thunderstorms can produce a **downburst**, an exceptionally strong downdraft that, upon striking Earth's surface, diverges horizontally as a surge of potentially destructive winds. Downbursts occur with or without rain; if without rain, the downburst usually occurs under a curtain of virga (Chapter 7). The late T. Theodore Fujita of The University of Chicago is credited with discovering downbursts, and he coined the term. Fujita's discovery stemmed from his airborne survey of property damage near Beckley, WV shortly after the supertornado outbreak of 3-4 April 1974. He observed debris spread

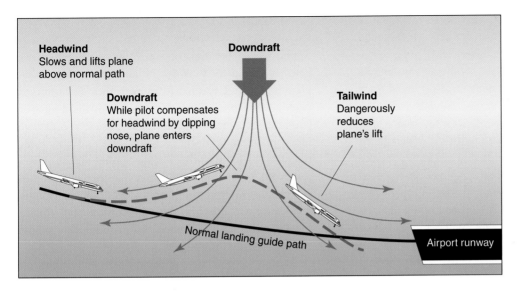

FIGURE 11.14
A microburst causes a landing aircraft to depart from its normal glide path. [Adapted from Michael Shibao, National Center for Atmospheric Research]

over the countryside in a starburst pattern—distinct from the swirling pattern that is characteristic of tornado damage. Downbursts blow down trees, flatten crops, and wreck buildings. Based on size of affected area, a downburst is classified as either a macroburst or a microburst.

A **macroburst** cuts a swath of destruction over a distance of more than 4 km (2.5 mi) with surface winds that may top 210 km (130 mi) per hr. The leading edge of a macroburst may be marked by a gust front, and the system lasts up to 30 minutes. A **microburst** is smaller and shorter-lived than a macroburst. By convention, its path of destruction is 4 km (2.5 mi) or less, top surface winds may be as high as 270 km (170 mi) per hr, and its life expectancy is less than 10 minutes.

Microbursts are particularly dangerous at airports, where they can play havoc with aircraft. Microbursts trigger *wind shear*, a change in wind speed or direction with distance, that can disturb the lift forces acting on the wings of an aircraft—perhaps causing an abrupt change in its altitude (Figure 11.14). An aircraft that flies into a microburst first encounters a strong headwind and then a strong tailwind. In a matter of seconds, the aircraft's speed relative to air can drop by more than 160 km (100 mi) per hr. If this happens during takeoff or landing, the aircraft's speed may drop below the minimum required for flight. T.T. Fujita underscored the microburst hazard for aircraft when he demonstrated that microbursts probably contributed to

two commercial aircraft accidents in 1975. Apparently, in both cases, the pilots unwittingly flew their jet planes through the center of a microburst, one on takeoff and the other while attempting to land. Both planes abruptly lost altitude and crashed.

In response to the microburst hazard to aviation, the FAA required airlines to install an approved microburst detection system on their aircraft by the end of 1995. As of this writing, the most effective systems employ sensors that can detect wind shear using the same principle as Doppler radar (Chapter 7). The goal is to provide pilots with at least 20 to 40 seconds advance warning of microburst-induced wind shear. Such advance warning is considered sufficient for pilots to take evasive action.

A squall line or mesoscale convective complex sometimes produces a family of straight-line downburst winds that impacts a path that may be hundreds of kilometers long. This severe weather system is known as a **derecho** (Spanish for *straight ahead*), a term that was coined in the 1880s by Gustavus Hinrichs, founder of the Iowa State Weather Service. The criterion for a derecho, as opposed to ordinary gusty thunderstorm winds, is sustained winds in excess of 94 km (58 mi) per hr; they generally track from northwest to southeast. On 4 July 1977, a derecho consisting of 25 individual downbursts struck several counties in northern Wisconsin, felling trees and buildings along a path that was 268 km (166 mi) long and 27 km (17 mi) wide. One person was killed, 35 were injured, and damage to buildings and timber totaled in the millions of dollars. Based on the extent of damage, surface winds may have been as high as 250 km (155 mi) per hr.

FLASH FLOODS

A **flash flood** is a short-term, localized, and often unexpected rise in stream level usually in response to torrential rain falling over a relatively small geographical area. Typically, stream level rises and falls within 6 hrs of the rain event. Excessive rainfall may occur when successive thunderstorm cells (in a squall line or an MCC) mature over the same area. Alternatively, a stationary or slow-moving intense thunderstorm cell may produce flooding rains. A thunderstorm is stationary or slow-moving when the system is (1) embedded in weak steering winds aloft and/or (2) maintained by a persistent flow of humid air up the slopes of a mountain range.

Atmospheric conditions that favor flash floods differ somewhat from those that give rise to other types of severe weather (e.g., hail and tornadoes). Flash-flood producing thunderstorms are more common at night and form in an atmosphere with weak vertical wind shear and abundant moisture through great depths. Flash flooding is most likely in an atmosphere that is *precipitation efficient,* that is, high values of precipitable water and relative humidity, and a thunderstorm cloud base having temperatures above freezing. High precipitable water and relative humidity reduces the amount of falling precipitation that vaporizes. A relatively warm cloud base favors the collision-coalescence process that leads to exceptionally heavy rainfall.

A combination of atmospheric conditions and topography contributed to the flash flood that claimed 139 lives (many of them campers) and caused $35.5 million in property damage in the Big Thompson Canyon of Colorado on 31 July 1976. Big Thompson Canyon is located in the Colorado Front Range about 80 km (50 mi) northwest of Denver. A persistent flow of humid air from the east up the mountain slopes triggered development of a succession of thunderstorm cells and heavy rainfall. With weak winds aloft, thunderstorm cells remained nearly stationary over essentially the same geographical area for at least 4 hrs. Rainfall in the headwaters of the Big Thompson River was estimated at 25 to 30 cm (10 to 12 in.) with perhaps 20 cm (8 in.) falling in only 2 hrs. Runoff cascaded down the steep mountain slopes and into the river that winds along the narrow canyon floor. The river rose abruptly and overflowed its banks as a flash flood. At one place along the river, the discharge (volume of water flowing per second) was more than 200 times greater than the long-term average. A wall of water almost 6 m (20 ft) high destroyed 418 houses and washed away 197 motor vehicles.

Some 21 years later, on 28 July 1997, Colorado residents were reminded of the Big Thompson Canyon disaster when a thunderstorm complex deluged nearby Fort Collins with record rainfall. Between 5:30 and 11:00 p.m. (MDT) up to 25 cm (10 in.) of rain drenched the southwest side of Fort Collins. A flash flood on Spring Creek, just downstream from where rainfall was heaviest, claimed 5 lives. Floodwaters also caused property damage (estimated at more than $200 million) over a large part of the city; especially hard-hit were buildings on the Colorado State University campus. Flood plain management measures, initiated well before the flood, particularly along Spring Creek, likely prevented greater loss of life and residential property damage. Management strategies included moving residents out of the floodway and constructing retention basins along the floodplain that could accommodate excess water during a major rainfall event.

A combination of atmospheric conditions and topography was responsible for the extreme rainfall of 28 July 1997 (just as at Big Thompson Canyon). Fort Collins is located just east of the Rocky Mountain Front Range. A persistent westward flow of unusually humid air over eastern Colorado toward higher terrain plus weak steering winds aloft set the stage for a succession of slow moving thunderstorm cells over the Front Range, dumping heavy rain for hours over essentially the same geographical area.

Flash flooding is especially hazardous in mountainous terrain, and motorists and campers are well advised to head for higher ground in the event of a flash flood warning (Figure 11.15). But even where the topography is relatively flat, a prolonged period of heavy

FIGURE 11.15
Road sign in the Colorado Rockies warns visitors to climb to higher ground in the event of a flash flood.

rain (more than 7.6 mm, or 0.3 in., per hr) can greatly exceed the infiltration capacity of the ground. Simply put, the ground cannot absorb all of the rainwater. Excess water runs off to creeks, streams, rivers, or sewers or collects in other low-lying areas. If a drainage system cannot accommodate the sudden input of huge quantities of water, a flash flood is the consequence.

Because of their design and composition, urban areas are prone to flash floods during intense downpours. Concrete and asphalt make the surfaces of a city virtually impervious to water, so elaborate storm sewer systems are required to transport runoff to nearby natural drainageways. Storm sewer systems have a limited capacity for water, however, and may be unable to handle the excess runoff produced during a torrential rainfall. Water collects in underpasses, at intersections, in dips in the roadway, and in other low-lying areas. Sometimes water levels rise so fast in these areas that motorists are trapped in their vehicles. In many cases, motorists misjudge the depth of the water and don't realize the power of even shallow water in motion. They unwittingly drive into water that sweeps their vehicle downstream and puts their lives in extreme danger.

Severe thunderstorms are not the only culprits in triggering flash flooding. Breaching of a dam or levee, or the sudden release of water during breakup of a river ice jam, can also cause an abrupt rise in water level. In any event, when in flood-prone areas, people are well advised to heed the precautions listed in Table 11.1.

HAIL

Hail is precipitation in the form of balls or lumps of ice more than 5 mm (0.2 in.) in diameter, called *hailstones* (Figure 11.16). Hail almost always falls from cumulonimbus clouds that are characterized by strong updrafts, great vertical development, and an abundant supply of supercooled water droplets. Hailstones range from pea size to the size of an orange or even larger. In the United States, the largest hailstone on record was collected at Coffeyville, KS on 3 September 1970. It weighed 766 g (1.7 lb) and measured 44.5 cm (17.5 in.) in circumference and 14.2 cm (5.5 in.) in diameter, about the size of a softball.

A hailstone develops when an ice pellet is transported vertically through portions of a cumulonimbus cloud containing varying concentrations of supercooled water droplets. The ice pellet may descend slowly through the entire cloud, or it may follow a more complex pattern of ascent and descent as it is caught alternately in updrafts and downdrafts. In the

TABLE 11.1
Some Flash Flood Safety Tips[a]

- Avoid driving on flooded roadways or bridges. Floodwaters only 0.5 m (1.5 ft) deep can carry away most automobiles. Also, flooded roads may be undermined.
- If your vehicle stalls in high water, abandon it and seek high ground.
- On foot never attempt to cross a stream if the water level is above your knees.
- Keep children away from drainage ditches and culverts.
- Find out the elevation of your property and the flood history of your community. If there is a risk of flooding, prepare an evacuation plan.
- Exercise caution when hiking or camping in remote areas. Avoid camping in mountain valleys or near dry streambeds. Take along a battery-powered radio to monitor changing weather conditions.

[a]Based on NOAA recommendations.

process, the ice pellet grows by accretion (addition) of freezing water droplets. In general, the stronger the updraft, the greater is the size of the ice pellet. Eventually, when it becomes too large and heavy for updrafts to support it, the ice pellet descends and falls out of the cloud base. If the ice does not melt completely during its journey through the above freezing air beneath the cloud, it reaches Earth's surface as a hailstone.

FIGURE 11.16
Hailstones are lumps of ice that fall from intense thunderstorm cells.

When an ice pellet enters a portion of the cloud containing a relatively high concentration of supercooled water droplets, water collects on the ice pellet as a liquid film, which freezes slowly to form a transparent layer, or *glaze*. When the ice pellet travels through a portion of the cloud where the concentration of supercooled water droplets is relatively low, droplets freeze immediately on contact with the ice pellet. As droplets freeze, many tiny air bubbles are trapped within the ice, producing an opaque whitish layer of granular ice, or *rime*. Hence, a hailstone is composed of alternating lamina of clear (glaze) and opaque (rime) ice, which, in cross section, resembles the internal structure of an onion. Scientists have counted as many as 25 layers in a single large hailstone.

Often hailstones cover the landscape in a long, narrow stripe known as a **hailstreak** (or *hail swath*). A typical hailstreak may be 2 km (1.2 mi) wide and 10 km (6.2 mi) long, and a single large thunderstorm may produce several hailstreaks. From a study of numerous hailstorms and hailstreaks, scientists at the Illinois State Water Survey devised a model to explain hailstreak development (Figure 11.17). Hail forms in the upper portion of a cumulonimbus cloud. After several minutes of hail formation, the updraft weakens, allowing the hail to descend within the cloud; about 4 minutes later, the first hailstones reach the ground. In the ensuing 10 minutes, as the thunderstorm continues to move along, the entire volume of hail is deposited along the ground as a hailstreak.

On rare occasions, the fall of hail is so great that snowplows must be called out to clear highways. This happened in Milwaukee, WI on 4 September 1988, when pea-sized hail formed drifts to 46 cm (18 in.) in some north-side neighborhoods. On 6 August 1980, at Orient, IA drifts of hail were reported to be 1.8 m (6 ft) deep. On the afternoon of 13 June 1984, hailstones as large as golf balls fell for up to 1.5 hrs in the western suburbs of Denver, CO producing an accumulation of 25 cm (10 in.), with drifts greater than a meter.

Hailstones may be large enough to smash windows and dent automobiles, but the most costly damage is to crops. In the U.S. each year, hail causes an average $1 billion in damage, primarily to crops, livestock, and roofs. Hail usually falls during the growing season, and, in a matter of minutes, large hail can wipe out the fruits of a farmer's year of labor. Traditionally, farmers cope with the hail hazard by purchasing insurance. Illinois farmers, for example, lead the United States in crop-hail insurance, purchasing an average annual liability coverage that tops $600 million. On 11 July 1990, softball-sized hailstones in Denver caused $625 million in property damage, mostly to automobiles and roofs. On 5 May 1995, the most costly hailstorm in U.S. history struck the Fort Worth, TX area. Golf ball- to baseball-size hailstones injured scores of people who were caught in the open during the storm. Total property damage was estimated at more than $2 billion. Refer to the Essay for information on efforts to suppress hail formation.

Perhaps surprisingly, hail frequency is not necessarily related to thunderstorm frequency. Although Florida experiences the greatest frequency of thunderstorms in the United States, hail is unusual in that state.

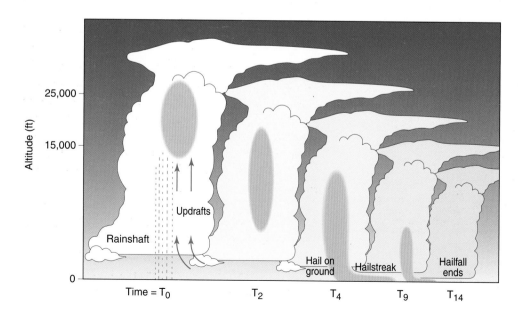

FIGURE 11.17
Model of hailstreak development. [Adapted from S.A. Changnon and J.L. Ivens, *Hail in Illinois, Public Information Brochure* 13, Illinois State Water Survey, 1987, p. 5.]

In North America, hail is most likely on the High Plains just east of the Rocky Mountains where Colorado, Nebraska, and Wyoming meet. In this so-called *hail alley*, hail can be expected to fall from about 10% of all thunderstorms and the average number of hail days a year is between seven and nine.

Tornadoes

Of the roughly 10,000 severe thunderstorms that occur in the United States in an average year, about 10% produce tornadoes. A **tornado** is a violently rotating column of air in contact with the ground that is almost always produced by a thunderstorm and often (but not always) made visible by water droplets formed by condensation and/or by dust and debris that are drawn into the system (Figure 11.18).

Tornadoes are the most violent of all weather systems. Fortunately, most are small and short-lived and

FIGURE 11.18
A tornado is a small-scale weather system that has the potential of taking many lives and causing considerable property damage. [Photo by Joe Willey, Plainview, TX}

often strike sparsely populated regions. Occasionally, however, a major tornado outbreak causes incredible devastation, death, and injury. Over a 16-hr period on 3-4 April 1974, 148 tornadoes struck 13 states in the east central United States, the most widespread and costly outbreak of tornadoes in the nation's history. This so-called *super outbreak* left 315 people dead and 6142 people injured and caused property damage in excess of $600 million. In early May 1999, an outbreak of particularly intense tornadoes hit portions of Oklahoma, Kansas, Texas, and Tennessee. As described in this chapter's Case-in-Point, Oklahoma City, OK was hardest hit. Fifty-five people lost their lives and damage topped $1.1 billion.

Tornado Characteristics

Probably the most striking characteristic of a tornado is its **funnel cloud**, a localized lowering of cloud base into a tapered column of tiny water droplets. Although often funnel-shaped, in fact, tornadoes take a variety of forms, ranging from cylindrical masses of roughly uniform dimensions in cross-section to long and slender, ropelike pendants. A funnel cloud forms in response to the steep air pressure gradient that is directed from the tornado's outer edge toward its center. Humid air expands and cools as it is drawn inward toward the center of the system. Cooling of air below its dewpoint causes water vapor to condense into cloud droplets. However, if the air is exceptionally dry, a funnel cloud may not form, and the tornado may be made visible by a whirl of dust and debris lifted off the ground.

A funnel cloud may not reach the ground. If there is no corresponding whirl of dust or debris on the ground, the system is reported as a *funnel aloft*. The system may or may not develop into a tornado. In some cases, a whirl of dust or debris appears on the ground even before a funnel cloud appears. Furthermore, the actual tornadic circulation covers a much wider area than is suggested by the funnel cloud. Typically, the diameter of a funnel cloud is only about one-tenth the diameter of the associated tornadic circulation.

A weak tornado's path on the ground typically is less than 1.6 km (1 mi) long and 100 m (330 ft) wide, and the system has a life expectancy of only a few minutes. Wind speeds are less than 180 km (110 mi) per hr. They account for less than 5% of all tornado fatalities. At the other extreme, a violent tornado can cause damage along a path more than 160 km (100 mi) long and 1.0 km (3000 ft) wide, and the lifetime of the system may be 10 minutes to more than 2 hrs. Based on indirect measurements, wind speeds in violent tornadoes range up to 500 km (300 mi) per hr.

The deadliest tornado in North American history was the Tri-State tornado of 18 March 1925. Traveling at a peak forward speed of 118 km (73 mi) per hr, the system lasted 3.5 hrs and produced a 353-km (219-mi) path of devastation from southeastern Missouri through the southern tip of Illinois and into southwest Indiana. The tornado caused 695 fatalities and 2000 injuries, and 11,000 people were made homeless. (There are some indications that more than one tornado contributed to the Tri-State disaster.) That same day, seven other tornadoes claimed an additional 97 lives.

Most tornadoes are spawned by, and travel with, intense thunderstorm cells. Tornadoes and their parent cells usually track from southwest to northeast, but any direction is possible. Tornado trajectories are often erratic with many tornadoes causing a hopscotch pattern of destruction as they alternately touch down and lift off the ground. Tornadoes have been known to move in circles and even to describe figure eights. Average forward speed is around 48 km (30 mi) per hr, although there are reports of tornadoes racing along at speeds approaching 120 km (75 mi) per hr (e.g., the 1925 Tri-State tornado).

An exceptionally steep horizontal air pressure gradient is ultimately responsible for a tornado's vigorous circulation. The air pressure drop over a horizontal distance of only 100 m (330 ft) may be equivalent to the normal air pressure drop between sea level and an altitude of 1000 m (3300 ft), that is, about a 10% reduction. The continually changing direction of the inward-directed pressure gradient force produces the

centripetal force that maintains the rotation of the air column about a vertical axis.

Most Northern Hemisphere tornadoes rotate in a counterclockwise direction (viewed from above); only about 5% rotate clockwise. While the Coriolis effect has a negligible influence on the scale of the funnel cloud, the counterclockwise bias is inherited from the larger parent thunderstorm.

Geographical and Seasonal Distribution of Tornadoes

The central United States is one of only a few places in the world where synoptic weather conditions and terrain are ideal for tornado development; interior Australia is another. Although tornadoes have been reported in all 50 states and throughout southern Canada, most occur in **tornado alley**, a north-south corridor stretching from eastern Texas and the Texas Panhandle northward through Oklahoma, Kansas, Nebraska, and into southeastern South Dakota (Figure 11.19). Central

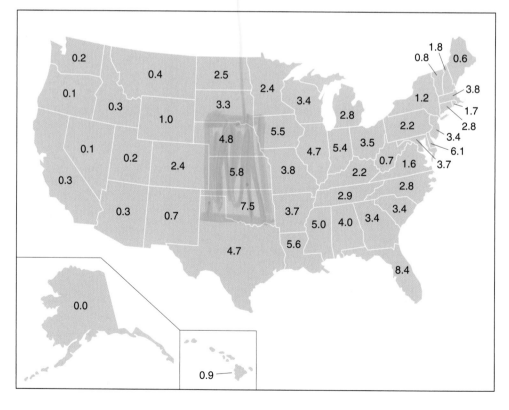

FIGURE 11.19
Average annual number of tornadoes per 10,000 square miles by state, 1950-1995.

Oklahoma has the highest annual incidence of tornadoes per unit area, whereas local tornado frequency maxima occur from central Iowa eastward to central Indiana, along the Gulf Coast, and in Florida.

Concentration of tornadoes over the central Plains is sometimes taken to imply that tornadoes do not occur in regions of great topographical relief. Actually, strong to violent tornadoes are largely unaffected by terrain; there are numerous examples of such tornadoes traversing rugged and even mountainous topography. On the other hand, weak tornadoes are more likely over flat than rough terrain.

Each year, the United States can anticipate between 700 and 1100 tornadoes. A new national record was set in 1998, with 1389 tornadoes reported, 33 (2.4%) of which were responsible for 129 fatalities. Slightly more than half of all tornadoes develop during the warmest hours of the day (10 a.m. to 6 p.m., local time), and almost three-quarters of tornadoes in the United States occur from March to July. The months of peak tornado activity are May and June when atmospheric conditions are optimal for spurring vigorous convection and the severe thunderstorms that spawn tornadoes.

One factor that contributes to the spring peak in tornado frequency is the relative instability of the lower atmosphere at that time of year. During the transition from winter to summer, daylight lengthens and solar radiation incident on Earth's surface becomes more intense. Heat is transported from the relatively warm ground into the troposphere, but it takes time for the entire troposphere to adjust to heating from below. The upper troposphere, in fact, usually retains its winter-like cold well into spring. The steep air temperature lapse rate (warm at low levels and cold aloft) favors supercell thunderstorm development.

Another factor that contributes to the spring tornado maximum is the greater likelihood that ideal synoptic weather conditions will occur at that time of year. Recall from earlier in this chapter that severe thunderstorms typically develop in the warm southeast sector of a strong extratropical cyclone. Such cyclones achieve their greatest intensity when sharp north-south temperature contrasts develop across the nation, that is, in spring when the polar front is well defined.

During late winter and spring, tornado occurrences progress northward. In effect, the center of maximum tornado frequency follows the sun, as do the midlatitude jet stream, the principal storm tracks, and northward incursions of maritime tropical air. By late February, maximum tornado frequency, on average, is along the central Gulf States. In April, the maximum frequency shifts to the southeast Atlantic States; in May and June, the highest tornado incidence is usually over the southern Plains, and by early summer, it moves into the northern Plains, the Prairie Provinces east of the Rockies, and the Great Lakes region. From late summer through autumn, tornado occurrence shifts southward.

What are your chances of experiencing a tornado? Very slim. Less than 1% of all thunderstorms produce tornadoes, and even in the most tornado-prone regions of North America, a tornado is likely to strike a given locale only once every 250 years. There are, of course, exceptions to the rule. Tornadoes have hit Oklahoma City no less than 27 times since 1892. To be prepared in the unlikely event that a tornado should strike your area, study the recommendations in Table 11.2.

TABLE 11.2
Some Tornado Safety Tips[a]

- Seek shelter in a tornado (storm) cellar, an underground excavation, or a steel-framed or substantial reinforced concrete building.

- Avoid auditoriums, gymnasiums, supermarkets, or other structures that have wide, free-span roofs.

- In an office building or school, go to an interior hallway on the lowest floor or a designated shelter area. Lie flat on the floor with your head covered.

- At home, go to the basement. If there is no basement, go to a small room (closet, bathroom, or interior hallway) in the center of the house on the lowest floor. Seek shelter under a mattress or a sturdy piece of furniture.

- Stay clear of all windows and outside walls.

- In open country, do not try to outrun a tornado in a motor vehicle. Tornadoes often move too fast and their paths are too erratic to avoid even if you drive at right angles to the apparent track of the storm. Instead stop and seek shelter indoors, and if this is not possible, lie flat in a ravine, dry creek bed, or open ditch.

- Do not seek shelter under a highway overpass, or in mobile homes or motor vehicles.

[a]Based on NOAA recommendations.

Tornado Hazards and the F-Scale

Tornadoes threaten people and property because of (1) extremely high winds, (2) a strong updraft, (3) subsidiary vortices, and (4) an abrupt drop in air pressure. Winds that may reach hundreds of kilometers per hr blow down trees, power poles, buildings, and other structures. Flying debris causes much of the death and injury associated with tornadoes. Broken glass, splintered lumber, and even vehicles become lethal projectiles. In violent tornadoes, the updraft near the center of the storm's funnel may top 160 km (100 mi) per hr and is sometimes strong enough to lift a railroad car off its tracks or a house off its foundation. Some tornadoes consist of two or more subsidiary vortices that orbit about each other or about a common center within a massive tornado. These multi-vortex tornadoes are usually the most destructive of all tornadoes.

It was once widely believed that a tornado caused buildings to explode, presumably because the air pressure within the building could not adjust rapidly enough to the abrupt pressure drop associated with the tornado. In the event of a tornado sighting, people were advised to open windows to help equalize the internal and external air pressure. In fact, most buildings have sufficient air leaks that a potentially explosive pressure differential never really develops. Hence, people should close windows and stay away from them. Structural damage to a building is caused either by winds slamming debris against the walls or by very strong currents of air that stream over the roof, causing it to lift. The same lifting happens as air flows over the curved upper surface of an airplane wing. As the roof is lifted, the walls collapse or are blown in, and the building disintegrates.

T.T. Fujita devised a six-point intensity scale for rating tornado strength and damage to structures (Table 11.3). The so-called **F-scale** is based on rotational wind speeds estimated from property damage and categorizes tornadoes as *weak* (F0, F1), *strong* (F2, F3), or *violent* (F4, F5). An F0 tornado produces minor damage, snapping twigs and small branches and breaking some windows. F1 and F2 tornadoes can cause moderate to considerable property damage and even take lives. F1 tornadoes can down trees and shift mobile homes off their foundations, and a F2 tornado can rip roofs off frame houses, demolish mobile homes, and uproot large trees. A F3 tornado can partially destroy even well constructed buildings and lift motor vehicles off the ground. At the violent end of Fujita's scale, destruction is

TABLE 11.3
The Fujita Tornado Intensity Scale

| | | Estimated wind speed | |
F-Scale	Category	km/hr	mi/hr
0	Weak	65-118	40-73
1		119-181	74-112
2	Strong	182-253	113-157
3		254-332	158-206
4	Violent	333-419	207-260
5		420-513	261-318

described as devastating to incredible, with the potential for many fatalities. A F4 tornado can level sturdy buildings and other structures and toss automobiles about like toys. In a F5 tornado, sturdy frame houses are lifted and transported some distance before disintegrating.

Fortunately, F5 tornadoes are rare. Of the 800 or so tornadoes that strike the United States in an average year, perhaps only one will be rated F5. When one does occur, however, the impact can be catastrophic. In about 1 minute, a F5 tornado leveled the village of Barneveld, WI on 8 June 1984; 100 homes were totally destroyed and nine lives were lost. The Xenia, OH tornado, which was part of the tornado outbreak of 3-4 April 1974, rated F5 over a portion of its 51-km (32-mi) path and claimed 34 lives.

In 1997, the American Meteorological Society reported that, in a typical year, 86% of all tornadoes are weak, 13% are strong, and only about 1% are violent. The few violent systems are responsible for the majority of all fatalities, however. Between 1961 and 1990, the average annual fatalities from tornadoes was 82, with 42% in April and 17% in May. In an average year, tornadoes injure about 1500 people.

As a tornado travels along the ground and progresses through its life cycle, its rating on the F-scale can fluctuate as the system's intensity changes. The Wichita-Andover, KS tornado of 26 April 1991 is an example. As shown in Figure 11.20, the tornado first touched down near Clearwater, KS about 5:57 p.m. (CDT) and then tracked northeastward to just north of El

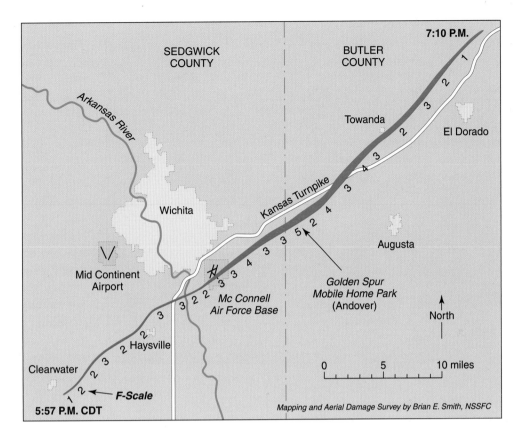

FIGURE 11.20
Path of the Wichita-Andover, Kansas tornado of 26 April 1991. Numbers along the path indicate rating on the Fujita tornado intensity scale which is based on property damage. [NOAA]

Dorado, KS where it dissipated about 7:10 p.m. The tornado rated F3 or F4 along most of its 77-km (48-mi) path. However, the rating rose to F5 as the tornado demolished all 241 mobile homes at the Golden Spur Mobile Home Park in Andover, KS. This tornado was responsible for 17 deaths.

The Tornado-Thunderstorm Connection

Most strong to violent tornadoes are spawned by a highly organized system known as a **supercell thunderstorm** (or just *supercell*) (Figure 11.21). Supercells are very energetic with updraft speeds sometimes in excess of 240 km (150 mi) per hr. They can last for several hours and produce more than one tornado.

Tornado development in a supercell begins with an interaction between the updraft and the larger-scale horizontal wind. In a tornadic supercell, the horizontal

wind exhibits strong vertical shear in both speed and direction. That is, the wind strengthens and veers (turns clockwise) with altitude, from south or southeast at the surface to southwest or west aloft. This shear in wind speed causes the air to rotate about a horizontal axis in a rolling motion. When this rotation interacts with the updraft, the tube of rotating air is tilted from a horizontal to nearly vertical orientation. The shear in the horizontal wind direction also adds to this rotation of air about a vertical axis. Consequently, the updraft spins in a counterclockwise direction (viewed from above), forming a **mesocyclone**, 3 to 10 km (2 to 6 mi) across. In only about 10% of all cases, a mesocyclone evolves into a tornado.

A **wall cloud**, a roughly circular lowered portion of the rain-free base of a thunderstorm, often accompanies a mesocyclone (Figure 11.22). A wall cloud is typically about 3 km (2 mi) in diameter and forms in the region of strongest updraft, usually to the rear (south or southwest) of the main shaft of precipitation. The updraft draws humid, rain-cooled air

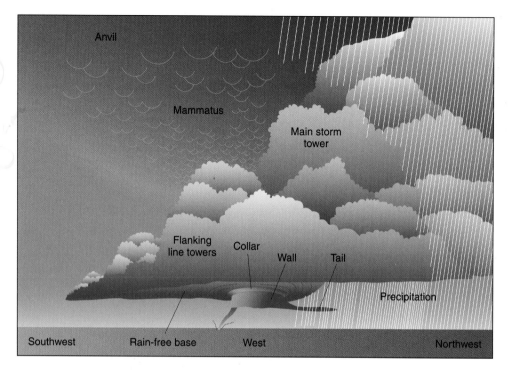

FIGURE 11.21
A tornadic supercell thunderstorm. [Adapted from NOAA, *Advanced Spotter's Field Guide*]

into the system, and as that air ascends, expands and cools, water vapor begins to condense at an altitude that is lower than cloud base. Hence, the base of the thunderstorm descends as a wall cloud. As foreboding as they may appear, most wall clouds do not produce tornadoes. Tornadic wall clouds typically are relatively long-lived, preceding the tornado by perhaps 10 to 20 minutes. Rotation of the wall cloud is obvious and persistent, becoming violent just prior to appearance of

FIGURE 11.22
A thunderstorm wall cloud.

the tornado. Also, surface winds blowing in toward a tornadic wall cloud are strong (in the range of 40 to 55 km, or 25 to 35 mi, per hr).

A mesocyclone circulation is most intense at an altitude of about 6100 m (20,000 ft), and from there it builds both upward and downward. Meanwhile, the updraft strengthens as more air converges toward the base of the supercell. The updraft sometimes becomes so strong that it overshoots the top of the thunderstorm and produces a dome-like cloud bulge on top of the spreading anvil cloud. In a tornadic supercell, the mesocyclone narrows and builds downward toward the ground. Relatively low air pressure in the mesocyclone causes water vapor to condense so that a funnel cloud forms. As the spinning column of air narrows, its circulation strengthens, sometimes to extremely high speeds. (This increase in wind speed is analogous to what happens when an ice skater performs a spin: As the skater pulls her arms closer to her body, her spin rate increases.) A tornado typically appears near the updraft and toward the rear of a supercell. Hence, in some instances, a shaft of heavy rain falling between the tornado and an observer makes it difficult to detect visually an approaching tornado.

At about the same time that the tornadic circulation begins to descend towards Earth's surface, a downdraft develops near the rear edge of the supercell. The downdraft strikes the Earth's surface within minutes of the tornado touchdown and begins to wrap around the tornado and mesocyclone. Eventually, the downdraft completely surrounds the tornado and mesocyclone and cuts off the inflow of warm, humid air. Consequently, the tornado weakens. As the tornado dissipates, its funnel shrinks and tilts and develops a ropelike appearance.

Although supercells typically produce the most devastating tornadoes, potentially destructive tornadoes can develop in multicellular squall lines and clusters. Squall-line tornadoes are most common in cells located just north of a break in the line or in the southernmost (*anchor*) cell. Sometimes very weak tornadoes spin off a thunderstorm gust front. Recall that a *gust front* is the leading edge of a downdraft after it strikes the ground and begins spreading out ahead of a thunderstorm cell. These weak, short-lived tornadoes are sometimes called *gustnadoes*.

Perhaps 90% of all North American tornadoes are spawned by thunderstorms that are associated with mature cyclones. Most of the others are products of convective instability triggered by hurricanes. In fact, many hurricanes that strike the southeastern United States are accompanied by tornadoes. Tornadoes often develop on the northeast flank of an Atlantic hurricane, after the system has turned toward the north and northeast (Chapter 12).

Monitoring Tornadic Thunderstorms

Scientific understanding of tornadoes and their genesis is somewhat tentative because direct monitoring of tornadoes generally is not feasible; traditional weather instruments are not durable enough to withstand tornadic winds. Between 1981 and 1983, *storm chasers* from the University of Oklahoma tried to deploy a 180-kg (400-lb) tornado-resistant instrument package in the path of tornadoes without much success. Today, storm chasers rely primarily on photography, balloon-borne instruments that monitor atmospheric conditions surrounding severe storms, and portable Doppler radar that can resolve the circulation within supercells (Figure 11.23).

FIGURE 11.23
Truck-mounted portable Doppler weather radar unit.

In communities where severe weather is a major concern, volunteer *storm spotters*, trained by the National Weather Service, monitor atmospheric conditions for signs of severe thunderstorms. Spotters provide a valuable service that complements weather radar, satellites, and networks of weather instruments and lightning detectors. But even trained storm spotters must be cautious of phenomena that may resemble tornadoes or funnel clouds. For more on this topic, refer to the Essay, "Tornado Look-Alikes."

As noted in Chapter 7, weather radar is an indispensable tool in monitoring severe weather systems. Operating in the reflectivity mode, weather radar cannot detect a tornado directly, but in some cases when the parent mesocyclone is present, a hook-shaped echo appears on the radar screen on the southeast side of a severe thunderstorm cell (Figure 11.24). A **hook echo** is produced by rainfall as it is drawn around the mesocyclone within a severe thunderstorm.

Weather radar operating in the velocity (Doppler) mode monitors the circulation *within* a weather system rather than just the intensity, location, and general displacement of an area of precipitation. Doppler radar *sees* inside severe thunderstorms and detects mesocyclones, developing tornadoes, gust fronts, and strong wind shear associated with microbursts. Mesocyclone signatures are sometimes identified up to 30 minutes prior to formation of a tornado and at distances up to 230 km (140 mi) from the storm system. Doppler radar can monitor a tornado as it evolves from a mesocyclone and before it descends to Earth's surface. A tornadic circulation within several tens of kilometers of the radar may show up on the radar image as a *tornado vortex signature (TVS)*, a small region of rapidly changing wind direction within a mesocyclone (Figure 11.25). Doppler radar provides the public with more advance warning of severe weather than was possible with reflectivity-only radars. Advance warning of up to 10 minutes or more (compared to less than 5 minutes with the conventional radars in use in the late 1980s) is feasible and has already saved many lives.

Although the advent of Doppler weather radar has greatly improved the ability of meteorologists to forecast tornadoes, storm spotters and visual surveillance of thunderstorms are still needed for several reasons: (1) Not all tornadoes are associated with mesocyclones and these tornadoes are particularly difficult to forecast. (2) Most mesocyclone signatures and some tornado vortex signatures are not associated with tornadoes. (3) Many tornado-bearing thunderstorms do not produce distinctive

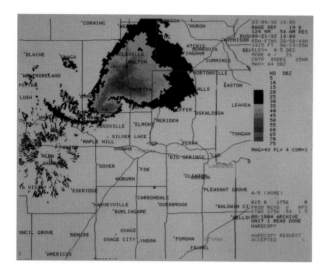

FIGURE 11.24
Weather radar image in the reflectivity mode showing a hook echo which may indicate air circulation in a developing tornado. [NOAA, National Weather Service]

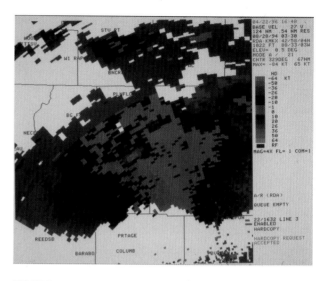

FIGURE 11.25
A tornado vortex signature (TVS) is visible near the center of this Doppler weather radar image in the velocity mode. [NOAA, National Weather Service]

radar signatures and some storms give false signatures. (4) The resolution of radar imagery is not fine enough to detect some severe weather phenomena.

Conclusions

Thunderstorms are the products of convection currents that surge to great altitudes within the troposphere. As such, thunderstorms channel excess heat at Earth's surface into the atmosphere via a combination of sensible and latent heating. Thunderstorms are thus most frequent in the warmest regions of the Earth, especially over the continental interiors of tropical latitudes. Most thunderstorms consist of more than one cell, each of which completes its life cycle in well under an hour. A thunderstorm cell reaches its maximum intensity during its mature stage and produces lightning, heavy rain, and strong gusty surface winds.

Although less than 1% of all thunderstorms produce tornadoes, in some areas of North America, the possibility of a tornado is the principal reason why people are fearful of thunderstorms. A tornado is a short-lived, small-scale weather system that is usually spawned by a severe thunderstorm. Tornado development requires a special combination of atmospheric conditions and terrain, so tornadoes are most frequent in spring over the central United States. Hazards of tornadoes are

extremely high winds, a powerful updraft, subsidiary vortices, and an abrupt drop in air pressure. The intensity of a tornado is rated from F0 (weak) to F5 (violent) based on wind speed estimated from property damage. The most intense tornadoes develop in the strong updraft of supercells.

In the next chapter, our discussion of weather systems continues with a focus on tropical cyclones (e.g., tropical storms and hurricanes).

Basic Understandings

- A thunderstorm is a mesoscale weather system produced by strong convection currents that surge high into the troposphere. The life cycle of a thunderstorm cell consists of a three-stage sequence: cumulus, mature, and dissipating.

- During the cumulus stage, cumulus clouds build vertically and laterally, updrafts characterize the entire system, and there is no precipitation. The mature stage begins when precipitation reaches Earth's surface. At that stage, updrafts occur alongside downdrafts and the system attains peak intensity. During the dissipating stage, subsiding air spreads through the entire cell and clouds vaporize.

- A thunderstorm usually consists of more than one cell, each of which may be at a different stage of its life cycle. A cluster of thunderstorm cells may track in a different direction than the individual constituent cells.

- Most thunderstorms develop in maritime tropical air as a consequence of uplift (1) along fronts, (2) on mountain slopes, (3) via convergence of surface winds, or (4) through intense solar heating of Earth's surface.

- Thunderstorms develop along boundaries within a mass of maritime tropical air, as a squall line along or ahead of a cold front, or in a mesoscale convective complex (MCC). MCCs account for a substantial portion of growing season rainfall over the Great Plains and Midwest.

- Worldwide, thunderstorms are most common over the continental interiors of tropical latitudes. In North America, thunderstorm days are most frequent in central Florida where converging sea breezes induce uplift of maritime tropical air. Convection is inhibited and thunderstorms are unlikely to develop in air masses that reside or travel over relatively cold surfaces and are thereby stabilized.

- Severe thunderstorm cells typically form along a squall line ahead of a fast-moving, well-defined cold front associated with a mature midlatitude cyclone. The polar front jet stream causes dry air to subside over a surface layer of maritime tropical air. This produces a layering of air that favors explosive convection and development of severe thunderstorms.

- Lightning is a brilliant flash of light produced by an electrical discharge within a cloud, between clouds, or between clouds and the ground. Cloud-to-ground lightning consists of a very rapid sequence of events involving stepped leaders, return strokes, and dart leaders.

- The reason for electrical charge separation in a cumulonimbus cloud is not well understood, but collisions between graupel and ice crystals, plus updrafts and downdrafts within the cloud likely play important roles.

- Some thunderstorm cells produce downbursts, intense downdrafts that spread out (diverge) at Earth's surface as potentially destructive winds. Based on size, a downburst is classified as either a macroburst or microburst. A squall line or mesoscale convective complex sometimes produces a family of straight-line downburst winds, known as a derecho.

- A flash flood is a short-term, localized, and often unexpected rise in stream level. Flash flooding is a special hazard in mountainous terrain, where steep slopes channel excess runoff into narrow stream and river valleys, and in urban areas, where impervious surfaces cause excess runoff to collect in low-lying areas.

- Hail develops in intense thunderstorm cells characterized by strong updrafts, great vertical development, and an abundant supply of supercooled water droplets.

- A tornado is a small mass of air that whirls rapidly about a nearly vertical axis and is usually made visible by condensed water vapor, and dust and debris drawn into the system.

- An exceptionally steep horizontal air pressure gradient between the tornado center and outer edge is the force ultimately responsible for the violence of a tornado.

- Most tornadoes occur in spring in a corridor stretching from Texas northward to southeastern South Dakota, from central Iowa eastward to central Indiana, and along the Gulf Coast states.

- Synoptic weather conditions favorable for the outbreak of tornadoes progress northward (with the sun), from the Gulf Coast in early spring to southern Canada by early summer.

- When a tornado strikes, very high winds, a strong updraft, subsidiary vorticies, and an abrupt air pressure drop are responsible for considerable property damage. Based on rotational wind speed estimated from damage to structures, tornadoes are classified as weak, strong, or violent on the F-scale. Most tornadoes are weak, but most fatalities are caused by rare violent tornadoes.

- Most intense tornadoes develop out of a mesocyclone that forms in the strong updraft of a supercell thunderstorm. A tornadic circulation is generated by an interaction between a thunderstorm updraft and strong vertical shear in the horizontal wind.

- Doppler radar monitors the circulation within a severe thunderstorm and can provide advance warning of tornado development.

ESSAY: Lightning Safety

The odds of being struck and killed by lightning are actually very slim, about 1 in 600,000. By comparison, the odds of being struck and killed by an auto are more than 50 times greater. Although the danger of lightning cannot be ignored, some simple precautions will minimize the risk of injury during a thunderstorm. No place is absolutely safe, but the risk can be significantly reduced in even the most lightning-prone area of the nation (south and central Florida, where an estimated 10 lightning bolts strike every square kilometer of land yearly).

When a thunderstorm threatens, the wisest strategy is to seek shelter in a house or other building, avoiding contact with conductors of electricity that provide pathways for lightning. These include pipes (don't shower), stoves (don't cook), and wires (don't use the telephone or computer). Electrical appliances pose no hazard if grounded, but why tempt fate by assuming they are properly wired?

Some confusion surrounds the safety of motor vehicles during a thunderstorm. Contrary to popular belief, the rubber tires of a motor vehicle provide occupants with little or no protection from lightning. When lightning strikes a motor vehicle, the electrical current is conducted over the metal surface of the vehicle, through the steel frame to the tires, and then to the ground. Without direct contact with the metal frame, passengers usually escape injury. All things considered, a person is much safer in a car (not a convertible) than outdoors during a thunderstorm.

What about the lightning threat to aircraft? Pilots steer clear of thunderstorm cells not only because of lightning but also because of the associated strong wind shear and the possibility of hail. All large aircraft (and many small ones) carry so-called static wicks on various trailing edges of the craft that are designed to bleed electrical charges back into the atmosphere. This prevents a build-up of electrical charge and lessens the chance of a lightning discharge between clouds and aircraft.

If a building or an auto is not accessible when a thunderstorm approaches, find shelter under a cliff, in a cave, or in a low area, such as a ravine, a valley, or even a roadside ditch (not subject to flooding). Avoid (1) tall, isolated structures, such as trees, telephone poles, and flagpoles, (2) metallic objects, such as wire fencing, rails, wire clotheslines, bicycles, and golf clubs, (3) high areas, such as the tops of mountains, hills, and roofs, and (4) bodies of water such as swimming pools and lakes. Stay off riding lawn mowers and farm tractors (unless equipped with a metal cab). While isolated trees in open spaces are hazardous, a thick grove of trees may offer safe haven.

If caught out in the open and your hair stands on end, lightning may be about to strike nearby. Immediately crouch down with feet together, placing your hands on your knees, and bend forward. In this way, you present the smallest possible target for a return stroke. Do not lie down on the ground, however, because if lightning does strike, the electrical charge can run along the ground. Such a charge traveling through your body between your head and feet will deliver much more electrical energy to you than if the charge runs along the ground between your feet. Furthermore, unless they are wet, shoes will reduce the charge conducted to your body should a current move along the ground. A group of people in the open should spread out, staying several meters apart.

Even as a thunderstorm approaches, some people believe that they are safe as long as it is not raining. They continue doing what they have been doing, even if it is an inherently dangerous activity during a thunderstorm (e.g., getting in that one last hole of golf or inning of baseball). Unfortunately, lightning can strike well outside of the area of rainfall. In fact, lightning has been reported as far as 16 km (10 mi) beyond the main shaft of thunderstorm precipitation. And, as pointed out elsewhere in this chapter, in arid or semi-arid regions, thunderstorm precipitation may completely vaporize in the dry air below cloud base. Such thunderstorms still produce lightning.

Contrary to popular belief, lightning can strike the same structure more than once (as long as the structure was not destroyed). The top of the Empire State Building in New York City typically is struck more than 20 times a year, and on one occasion was hit 15 times in only 15 minutes.

Two of every three people *struck* by lightning recover fully. Most survivors are jolted by a nearby lightning bolt and are not hit directly. Such near misses, however, can produce severe burns because of the intense heat generated by lightning. If an electrical current from a lightning strike passes through the body, however, it can disrupt a special system in the heart that generates rhythmical electrical impulses that control the regular contractions of the heart muscle. If such a disruption is severe, heart muscles are unable to work together and cardiac output declines. Blood flow can slow to the point that body tissues are damaged by an inadequate supply of oxygen and nutrients, a condition known as *circulatory shock*. However, the body possesses a number of mechanisms that attempt to return cardiac output to normal.

Emergency medical personnel should be summoned immediately for a lightning-strike victim. Meanwhile, cardiopulmonary resuscitation (CPR) or mouth-to-mouth resuscitation may revive the victim. Covering the victim with a blanket or additional clothing will reduce heat loss from the body. It is a myth that victims of lightning strikes carry an electrical charge; they should be attended to without delay. Treatment can aid bodily processes that counteract shock. Hence, if circulatory shock is not severe, the victim usually recovers.

With severe electrical disruption of the heart, circulatory shock can reach a critical stage in which shock breeds more shock. That is, reduced cardiac output causes the circulatory system (including the heart) to deteriorate, which in turn, causes a further reduction in blood flow. This condition, known as *progressive shock*, can trigger a vicious cycle of cardiovascular deterioration. Without immediate medical intervention, progressive shock can rapidly reach a critical stage where the person cannot be saved.

ESSAY: Hail Suppression

Humankind's efforts to suppress hail have deep historical roots. In 14th century Europe, church bells were rung and cannons fired in the belief that the accompanying noise would ward off hail. A period of particularly intense hail suppression activity took place in the grape-growing regions of Austria, France, and Italy during the late 19th century. M. Albert Stiger, a wine grower and burgomaster of Windisch-Feistritz, Austria, designed and built a special funnel-shaped hail suppression cannon. Stiger believed that the smoke particles in the cannon fire would inhibit hailstone development. Amazingly, in experimental firings in 1896-97 at Windisch-Feistritz, Stiger reported no hail, although severe hail damage occurred in neighboring areas.

Word of Stiger's apparent success spread throughout the vineyard regions of Europe, and hail cannons soon became commonplace. There were so many cannons that accidental shooting of people became a serious problem in some localities. After Stiger's much heralded success was not duplicated elsewhere, interest in hail cannons rapidly waned, and by 1905 this early attempt at weather modification ended.

The modern era of hail suppression experimentation began after World War II. Although founded on a much better, albeit incomplete, understanding of cloud physics, the new techniques shared some similarities with earlier efforts. For example, until the practice was outlawed in the early 1970s, farmers in Italy routinely fired explosive rockets into threatening clouds in an attempt to shatter developing hailstones. In the former Soviet Union, scientists fired silver iodide (AgI) crystals, a cloud-seeding agent, into cumulonimbus clouds. They hypothesized that AgI crystals would stimulate the formation of large numbers of small ice pellets (graupel), which would melt long before reaching the ground, instead of the normal development of small numbers of larger ice pellets that reach the ground as hailstones.

U.S. agricultural losses to hail amount to hundreds of millions of dollars every year. It is not surprising then that U.S. scientists set out to test the Soviet hypothesis of hail suppression and to learn more about hail-producing thunderstorms. To these ends, the *National Hail Research Experiment* was launched over northeastern Colorado in 1972. Although much was learned about hailstorms, three years of seeding potential hail-producing thunderstorms failed to confirm the Soviet hypothesis. Some scientists, however, questioned the experimental design and whether it was a viable test of the Soviet technique. In an interesting parallel with events in 19th century Europe, declining public confidence in the effectiveness of modern hail-suppression efforts brought an end to federal funding of hail suppression research in 1979.[*]

[*] Much of this discussion is based on S.A. Changnon, Jr., and J.L. Ivens, "History Repeated: The Forgotten Hail Cannons of Europe," *Bulletin of the American Meteorological Society* 62(1981):368-375.

ESSAY: Tornado Look-Alikes

Experienced watchers of the sky are well aware that a number of atmospheric features resemble funnel clouds or tornadoes. These include waterspouts, virga, scud clouds, and dust devils. Even distant smoke plumes are sometimes mistaken for tornadoes.

A *waterspout* is a tornado-like disturbance that occurs over the ocean or over a large inland lake. It is so named because it consists of a whirling mass of water that appears to stream out of the base of its parent cloud, which can be either cumulus congestus or cumulonimbus. A waterspout is usually considerably less energetic, smaller, and shorter-lived than a tornado. The rare intense waterspout may well be a tornado that formed over land and then traveled out over a body of water. In any event, boaters should steer clear of waterspouts.

A curtain of *virga* (rain or snow that vaporizes before reaching the ground) or *scud clouds* (ragged-appearing low clouds) are sometimes mistaken for a tornado or funnel cloud because their occurrence might coincidently exhibit a cylindrical or funnel-shaped profile. Observing these features for a minute or two reveals that they lack organized rotation about a vertical or near-vertical axis, clearly distinguishing them from tornadoes or funnel clouds. In deserts, or wherever the soil dries out, intense solar heating often gives rise to a swirling mass of dust, known as a *dust devil* (Chapter 10). A dust devil resembles a tornado, but forms near the ground, is not attached to any clouds, and causes little if any property damage.

CHAPTER 12

TROPICAL WEATHER SYSTEMS

> Wheeling, the careening winds
> arrive with lariats
> and tambourines of rain.
> Torn-to-pieces, mud-dark
> flounces of Caribbean
> cumulus keep passing,
> keep passing. By afternoon
> rinsed transparencies begin
> to open overhead.
> Mediterranean
> windowpanes of clearness.
>
> AMY CLAMPITT, *The Kingfisher,*
> *"The Edge of the Hurricane"*

Case-in-Point

In the three hours it took Hurricane Andrew to cross extreme southern Florida on the morning of 24 August 1992, its peak sustained winds of 234 km (145 mi) per hr with gusts in excess of 264 km (165 mi) per hr, contributed to the deaths of 15 people and left 180,000 homeless. With property damage estimated at $27 billion (in 1992 dollars), Andrew was the most costly hurricane in U.S. history. To make matters worse, Andrew continued to track west and northwestward across the Gulf of Mexico and, two days later, struck the Louisiana coast where it claimed four more lives and caused an additional $400 million in property damage.

Although Andrew was the most costly hurricane in terms of property damage in the United States, its impact pales in comparison to that of Hurricane Mitch on Central America (especially Honduras and Nicaragua). Mitch was the deadliest hurricane to strike the Western Hemisphere since the "Great Hurricane of 1780" (estimated to have killed up to 22,000 people in the Caribbean). The final death toll from Mitch likely will never be known, but the official loss of life was listed at 11,000 with thousands more missing. More than 3 million people were made homeless or had their lives and livelihoods severely affected and the estimated total property damage was at least $5 billion.

Mitch began as a weak tropical depression on 22 October 1998. By 2100 UTC on 26 October, the system's sea-level central pressure had dropped to 905 mb with sustained winds of 290 km (180 mi) per hr and gusts well over 320 km (200 mi) per hr, placing the storm in the category of most powerful hurricanes (category 5 on the Saffir-Simpson Hurricane Intensity Scale). Mitch stayed at this high level of intensity for 33 consecutive hours. After threatening Jamaica, Mitch tracked westward and by 2100 UTC on 27 October, was centered about 95 km (60 mi) north of Trujilo on the north coast of Honduras. Over the next two days, Mitch slowly drifted southward and finally made landfall, devastating offshore islands and coastal areas of Honduras. Mitch then drifted westward through the mountainous interior of Honduras, finally reaching the border with Guatemala on 31 October. Although winds slowly abated, orographically induced rainfall was torrential, falling at the rate of 30 to 60 cm (12 to 24 in.) per day in many mountainous regions and totaling in some places as much as 190 cm (75 in.) for the entire storm. Flooding and mudflows swept away whole villages and their inhabitants, virtually destroyed the entire infrastructure of Honduras, and devastated parts of Nicaragua, Guatemala, Belize, and El Salvador.

Driving Question:

What conditions are required for the development of tropical cyclones?

This chapter focuses on tropical weather systems, primarily tropical cyclones. **Tropical cyclones** are synoptic-scale low-pressure systems that originate over the tropical ocean and include tropical depressions, tropical storms, hurricanes, and typhoons. At maturity, a tropical cyclone is one of the most intense and potentially most destructive storms in the world. We describe the characteristics, geographical and seasonal distribution, associated hazards, and life cycle of tropical cyclones. We also consider in separate sections, the hurricane threat to the southeastern United States, efforts to forecast hurricanes based on weather conditions over West Africa, and unsuccessful experiments to modify hurricanes. We set the stage by summarizing some of the basic characteristics of weather in the tropics.

Weather in the Tropics

Weather in the tropics, the belt between the Tropics of Cancer and Capricorn (from 23.5 degrees N to 23.5 degrees S) exhibits very little seasonal variation in temperature. Temperatures are uniformly high in response to year-round high solar altitude and nearly uniform day length (Chapter 3). In fact, in much of the tropics, the diurnal (day to night) temperature variation is greater than the annual temperature variation.

In the tropics, broad expanses of air are uniformly warm and humid so that fronts and frontal weather are not present. Intense solar radiation heats Earth's surface and spurs deep convection and isolated afternoon thunderstorms that produce brief periods of heavy rain. For reasons presented in Chapters 4 and 11, thunderstorms are more frequent over continents and large islands than the adjacent ocean. Sometimes thunderstorm cells are aligned as narrow bands known as *tropical non-squall clusters*. On other occasions, a band of more intense cells forms, similar to a midlatitude squall line. In these so-called *tropical squall clusters*, winds can be strong and gusty and rainfall may persist for several hours and can be very heavy especially at the onset. The ITCZ (Intertropical Convergence Zone) stimulates thunderstorm activity and its north-south seasonal shifts are responsible for seasonal variations in precipitation in portions of the tropics (Chapter 9). The ITCZ follows the sun, moving northward in the Northern Hemisphere spring and southward in the Northern Hemisphere autumn. The rainy season is summer (high sun) and the dry season is winter (low sun).

Little variation in the thermal and moisture properties of tropical air means that there is little horizontal variation in surface air pressure. Hence, isobaric analysis is of little value in routine weather analysis and forecasting in the tropics. Instead, meteorologists rely upon streamline analysis. A **streamline** is a line that is everywhere parallel to the wind direction and portrays graphically the horizontal flow of air. Streamline analysis can be used to identify regions of divergence and convergence such as associated with an easterly wave. As described later in this chapter, an easterly wave is an important weather-maker in the tropics and sometimes is the precursor of a tropical cyclone.

Lengthy episodes of tranquil warm weather are typical of the tropics, but occasionally tropical cyclones develop that bring torrential rains and strong winds. The most feared and potentially most destructive of these tropical cyclones is the hurricane.

Hurricane Characteristics

Hurricane is likely derived from *Haracan*, the name of the storm god of the Taino people who inhabited Caribbean islands at the time of Spanish exploration of the New World. A **hurricane** is a violent tropical cyclone that originates over tropical ocean waters, usually in late summer or early fall. By definition, a hurricane has a maximum sustained wind speed of 119 km (74 mi) per hr or higher, although winds in a particularly intense hurricane may top 250 km (155 mi) per hr. By convention in the United States, a *sustained wind speed* is a one-minute average measured at the standard anemometer height of 10 m (33 ft).

A convenient way to describe a hurricane is to contrast it with the *extratropical cyclone* examined in Chapter 10. A hurricane develops in a uniform mass of very warm and humid air, so the system has no associated fronts or frontal weather (Figure 12.1). Air pressure is distributed symmetrically about the system's low-pressure center so that isobars form a pattern of closely spaced nearly concentric circles. Typically, the central pressure at sea level is considerably lower and the horizontal air pressure gradient much steeper in a hurricane than in an extratropical cyclone. Reconnaissance aircraft

extrapolated a surface air pressure of 888 mb (26.22 in.) at the center of Hurricane Gilbert while the storm was over the northwest Caribbean Sea on 13 September 1988; this is the lowest sea-level air pressure ever recorded in the Western Hemisphere. A hurricane is usually a much smaller system, averaging a third the diameter of a typical midlatitude cyclone. Rarely do hurricane-force winds extend much more than 120 km (75 mi) beyond the system's center. And structurally, a mature hurricane is a warm-core low that weakens rapidly with altitude, especially above 3000 m (9800 ft). In the upper troposphere, at altitudes above about 12,000 m (40,000 ft), the circulation usually becomes anticyclonic (clockwise when viewed from above in the Northern Hemisphere). Recall that an extratropical cyclone is a cold-core system whose circulation strengthens with altitude.

At the center of a hurricane is an area of almost cloudless skies, subsiding air, and light winds (less than 25 km or 16 mi per hr), called the **eye** of the storm (Figure 12.2). The eye generally ranges from 10 to 65 km (6 to 40 mi) across, typically shrinking in diameter as the hurricane intensifies and winds strengthen. At a hurricane's typical rate of forward motion, the eye may take up to an hour to pass over a given locality. People are sometimes lulled into thinking the storm has ended when skies clear and winds abruptly slacken following a hurricane's initial blow. They may be experiencing passage of the hurricane's eye; heavy rains and ferocious winds will soon resume but blow from the opposite direction.

Encircling the eye of a mature hurricane is the **eye wall**, a ring of thunderstorm (cumulonimbus) clouds that produce heavy rains and very strong winds. The most dangerous and potentially most destructive part of a hurricane is the portion of the eye wall on the side of the advancing system where the wind blows in the same direction as the storm's forward motion.

FIGURE 12.1
Schematic diagram of the internal structure of a hurricane. Note that the vertical dimension is greatly exaggerated. [From NOAA, *Hurricane*. Washington, DC: Superintendent of Documents, 1977]

FIGURE 12.2
The central eye of Hurricane Felix is apparent in this visible satellite image.

On that side, hurricane winds combine with the storm's forward motion producing the system's strongest surface winds. In the Northern Hemisphere, this dangerous semicircle of high winds (and high ocean waves) occurs on the right side of the storm when facing in the direction of the storm's forward movement. Cloud bands producing heavy convective showers and areas of hurricane-force winds spiral inward toward the eye wall. And at high altitudes cirrus or cirrostratus clouds spiral outward from the system.

Whereas hurricanes differ substantially from extratropical cyclones, they share some characteristics with intense storms that sometimes develop over the ocean at high latitudes. For more on these systems, refer to this chapter's first Essay, "Polar Lows With Hurricane Characteristics."

Geographical and Seasonal Distribution

Three conditions are required for a tropical cyclone to form: (1) relatively high sea-surface temperature, (2) adequate Coriolis deflection, and (3) weak winds aloft. To a large extent, these requirements dictate the geographical and seasonal distribution of hurricanes and other tropical cyclones.

Tropical cyclone formation requires a sea-surface temperature of at least 26.5 °C (80 °F) through an ocean depth of 60 m (200 ft) or more. Such exceptionally warm ocean water sustains the system's

circulation by the latent heat released when water, evaporated from the ocean surface, is conveyed upward and condenses within the storm system. Temperature largely governs the rate of evaporation of water, so the higher the sea-surface temperature, the greater is the supply of latent heat for the storm system. Furthermore, ocean spray readily evaporates and adds to the supply of latent heat. As a tropical cyclone moves over colder water or land, however, it loses its warm-water energy source and weakens. Also, the strong winds of a tropical cyclone stirs up the surface ocean waters, bringing cold water to the surface. The lower sea-surface temperatures can inhibit the development of subsequent tropical cyclones over the same region of the ocean.

In recent years, meteorologists discovered that tropical cyclone development may be influenced by changes in sea-surface temperature caused by warm and cold rings, large turbulent eddies that break off from ocean surface currents. A ring forms when a meander in a current forms a loop that pinches off, separating from the main current and moving as an individual eddy. Ocean water to the north of the Gulf Stream is relatively cold with temperatures averaging less than 10 °C (50 °F). Rings that break off from the north side of the Gulf Stream entrain relatively warm water from the Sargasso Sea whereas rings that break off from the south side of the Gulf Stream entrain relatively cold water. *Warm rings* rotate in a clockwise direction and have diameters to 200 km (125 mi) and a pool of warm water that may reach a depth of 1500 m (4900 ft). *Cold rings* rotate in a counterclockwise direction, are larger (diameters to 300 km or 185 mi) and their associated pool of cold water may reach the ocean floor (to a depth of more than 4000 m or 13,000 ft). Tropical cyclones may intensify over warm rings and weaken over cold rings.

The second prerequisite for tropical cyclone formation is a significant Coriolis deflection; that is, the influence of Earth's rotation must be sufficiently strong to initiate a cyclonic circulation. As noted in Chapter 8, the Coriolis force weakens toward lower latitudes and is zero at the equator. The minimum latitude where the Coriolis force is adequate for formation of tropical cyclones appears to be about 4 degrees.

The first two conditions favorable for tropical cyclone formation (i.e., relatively high sea-surface temperature and sufficient Coriolis effect) occur only over certain portions of the ocean. Plotted in Figure 12.3 are the main ocean breeding grounds for hurricanes and other tropical cyclones along with average storm trajectories. Most hurricanes form in the 8- to 20-degree

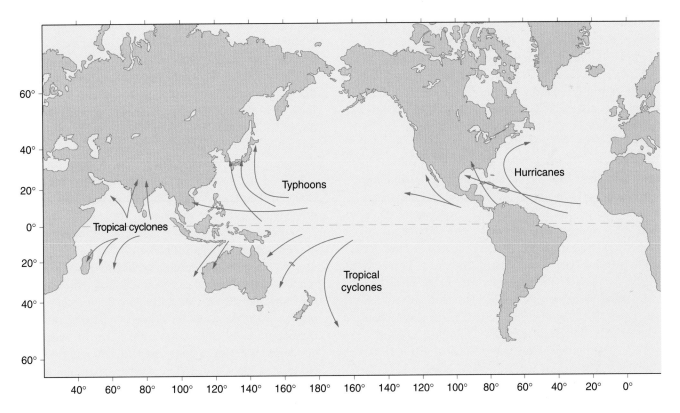

FIGURE 12.3
Tropical cyclone breeding grounds are located only over certain regions of the world ocean. Arrows indicate average hurricane trajectories.

latitude belt. Major hurricane breeding grounds are: (1) the tropical North Atlantic west of Africa (including the Caribbean Sea and Gulf of Mexico), (2) the North Pacific Ocean west of Mexico, (3) the western tropical North Pacific and China Sea, where a hurricane is called a *typhoon* (from the Cantonese, *tai-fung*, meaning great wind), (4) the South Indian Ocean east of Madagascar, (5) the North Indian Ocean (including the Arabian Sea and the Bay of Bengal), and (6) the South Pacific Ocean from the east coast of Australia eastward to about 140 degrees W. In the Indian Ocean and near Australia, hurricanes are simply called *cyclones*.

The requirement of relatively high sea-surface temperature makes hurricane occurrence distinctly seasonal. The temperature change of surface ocean waters lags the regular seasonal variations in incoming solar radiation. Sea-surface temperatures reach a seasonal maximum long after the date of most intense solar radiation. Most Atlantic hurricanes develop when surface waters are warmest, that is, in late summer and early autumn; the official hurricane season runs from 1 June to 30 November, with the peak hurricane threat for the U.S. coastline between mid-August and late October.

The third requirement for tropical cyclone development is relatively weak winds aloft over oceanic breeding grounds. Weak winds aloft allow a cluster of cumulonimbus clouds to organize over tropical seas, the first step in the evolution of a hurricane. Strong west to east winds in the middle and upper troposphere shear off the tops of westward tracking thunderstorms, preventing the systems from building vertically and organizing. Wind shear is the principal reason no hurricanes form off the east or west coasts of South America (although Caribbean hurricanes occasionally impact the north coast of Venezuela). During El Niño, winds are unusually strong at altitudes near 12,000 m (40,000 ft) over the Atlantic hurricane breeding ground. Hence, Atlantic hurricanes are infrequent during El Niño. For example, the Atlantic hurricane season of 1997 occurred during an intense El Niño and produced 3 hurricanes, only one of which made landfall on the U.S. coast.

A worldwide average of about 80 tropical cyclones develops each year and approximately one-half strengthen to hurricanes. The western Pacific Ocean, with its vast expanse of warm surface water, is the most active area for tropical cyclones, with roughly 32 systems each season, of which about 18 intensify into typhoons.

Only hurricanes spawned over the tropical

Atlantic, Caribbean Sea, and Gulf of Mexico pose a serious threat to coastal North America. A seasonal average 10 named tropical storms (precursors to hurricanes) form over these waters. Of these systems, on average about 6 intensify into hurricanes and 2 of these hurricanes strike the U.S. coast. The 1995 Atlantic hurricane season was the second most active in recorded history with 19 tropical storms, 11 of which attained hurricane strength. But the annual number of hurricanes may have little bearing on the number of landfalling hurricanes and their impact. Whereas only 4 hurricanes occurred during the 1992 Atlantic season, one of them, Hurricane Andrew, was the most costly in terms of property damage in U.S. history. Hurricanes have hit every Atlantic and Gulf Coast State from Texas to Maine. Florida is the most hurricane-prone of all the states, with 57 hurricanes crossing its coastline between 1900 and 1996. In the same period, Texas was second with 36 and Louisiana and North Carolina had 25 each.

The Pacific Ocean off Mexico and Central America ranks second to the western North Pacific in average annual number of tropical cyclones (13), the majority of which develop into hurricanes. And yet, the Pacific Coast of North America is rarely a target of hurricanes. Prevailing winds (northeast trades) are directed offshore and usually steer tropical cyclones that form west of Central America away from the coast. Also, the southward flowing cold California Current plus upwelling just off the southern California (and Baja California) coast produce sea-surface temperatures that normally are too low to sustain hurricanes that travel toward the northeast. During unusual circulation regimes, however, tropical cyclones have struck coastal Southern California and even traveled over the Desert Southwest. For example, on 9–10 September 1976, Hurricane Kathleen crossed Baja California and tracked near Yuma, AZ bringing torrential rains and floods to the Imperial and Lower Colorado River valleys. A more recent example was Hurricane Nora in September 1997. Nora made landfall on 25 September near Punta Eugenia on the Baja California coast with maximum sustained winds of 137 km (85 mi) per hour. The system weakened to a tropical storm and crossed into the United States near the California/Arizona border. Rainfall up to 7.5 cm (3 in.) was common in the Desert Southwest and some mountain locations in Arizona reported almost 30 cm (12 in.).

The Hawaiian Islands are sometimes threatened by tropical cyclones that develop over the central tropical Pacific or track into that region from the hurricane breeding grounds west of Mexico. Fortunately, in an average year, only 3 tropical storms affect the central Pacific and since 1957 only 4 hurricanes have impacted the Islands. However, the most recent one, Hurricane Iniki, in September 1992, was devastating. Iniki tracked directly over the island of Kauai with estimated maximum sustained winds of 185 km (115 mi) per hr with gusts to 258 km (160 mi) per hr. A storm surge of 1.5 m (5 ft) topped by wind-driven waves caused extensive flooding. Seven people lost their lives. Total property damage was estimated at $2.3 billion (in year 2000 dollars), making this the most costly natural disaster in the history of the State of Hawaii.

The primary breeding ground for Atlantic hurricanes exhibits a general east/west seasonal shift. Early in the hurricane season (May and June), hurricanes form mostly over the Gulf of Mexico and the western Caribbean. By July, the main area of hurricane development begins to shift eastward across the tropical North Atlantic. By mid-September, most hurricanes develop in a belt stretching from the Lesser Antilles (in the eastern Caribbean Sea) eastward to south of the Cape Verde Islands (off Africa's west coast). After mid-September, hurricanes again originate mostly over the Gulf of Mexico and the western Caribbean.

Hurricane Hazards

Hazards of hurricanes are (1) heavy rains and floods, (2) strong winds, (3) tornadoes, and (4) storm surge. According to research conducted by scientists at the Tropical Prediction Center/National Hurricane Center, in the 30-year period from 1970 to 1999, freshwater floods were responsible for almost 60% of the 600 U.S. deaths attributed to tropical cyclones or their remnants. One hundred thirty-eight people drowned either in or while attempting to abandon their motor vehicles. In those three decades, far more people (351) died from inland flooding than from coastal storm surge flooding (only 6 deaths). For example, of the 56 people who perished during Hurricane Floyd (1999), 50 drowned due to inland flooding. During the same 30-year period, about 12% of fatalities were wind-related.

WATER AND WIND

Hurricanes produce very heavy rainfall with amounts typically in the range of 13 to 25 cm (5 to 10 in.). Even if the system tracks well inland, heavy rains often persist and may trigger costly flooding (Figure 12.4).

In June 1972, rains from the remnants of Hurricane Agnes accounted for much of the storm's $8.4 billion (in year 2000 dollars) in property damage. Agnes tracked up the Eastern Seaboard from southwest of the Florida Keys to North Carolina and then northward into Pennsylvania (Figure 12.5). The storm brought beneficial rains to the Southeast, which had been experiencing a dry spell, but heavy rains in the mid-Atlantic region were anything but beneficial. During the week prior to Agnes' arrival, frontal showers and thunderstorms produced soaking rains from Virginia to New England. Rainfall averaged 3 to 8 cm (1 to 3 in.) and locally topped 15 cm (6 in.). Agnes' torrential rains falling on already saturated soils produced runoff that triggered many record-breaking river crests and devastating floods.

Heaviest rains from Agnes fell in a band about 350 km (215 mi) wide from western North Carolina northeastward into central New York State. In this hilly terrain, rainfall totals were quite variable but generally ranged from 20 to 30 cm (8 to 12 in.) for the period 18-25 June. Some localities reported total rainfall in excess of 46 cm (18 in.) with more than 25 cm (10 in.) falling in only 24 hrs. Flooding was particularly severe in central Pennsylvania where rising waters forced more than

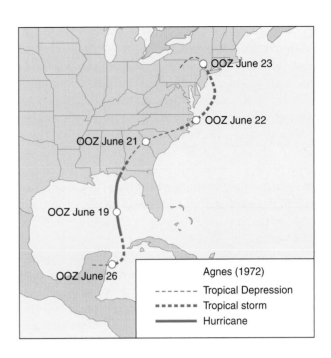

FIGURE 12.5
The track of Hurricane Agnes in June 1972.

FIGURE 12.4
This roadside marker in central Virginia commemorates the fatalities and property damage caused by the torrential rains associated with the remnants of Hurricane Camille in August 1969.

250,000 people from their homes. The Susquehanna River crested at 4 to 5.5 m (13 to 18 ft) above flood stage. Of the 122 fatalities attributed to Agnes, 50 were in Pennsylvania, and flood damage in that state accounted for almost two-thirds of total storm damage.

Alberto and Allison are other, more recent examples of how devastating rains can be produced by the remnants of a tropical cyclone. The first tropical storm of the 1994 season, Alberto never achieved hurricane status, yet its rains took a terrible toll in a band from southeastern Alabama to central Georgia. Alberto developed in the western Caribbean Sea and strengthened to a tropical storm on 1 July. After drifting slowly west/northwestward, Alberto turned northward, and on the morning of 3 July, came ashore near Fort Walton Beach, FL with peak sustained wind speeds near 97 km (60 mi) per hour. The storm then weakened rapidly as it drifted across southeastern Alabama and into central Georgia.

Alberto likely would have received little notoriety except that its moisture-rich remnants slowly meandered over central Georgia and eastern Alabama for three days. Many area weather stations reported 3-day (3-5 July) rainfall totals in excess of 25 cm (10 in.). Americus, GA set an all-time state record for 24-hr rainfall with 53.6 cm (21.1 in.). Rivers rose to record high crests, and flooding was extensive. The Ocmulgee

River at Macon, GA broke its old record by 1.7 m (5.5 ft) when it crested at more than 5.2 m (17 ft) above flood stage. Most of the storm's 32 fatalities were victims of floodwaters. Property damage was estimated at $1 billion.

With sustained winds of 97 km (60 mi) per hr, Allison was a minimal tropical storm before moving ashore at the eastern end of Galveston Island, TX on 5 June 2001 (with sustained winds of 80 km, or 50 mi, per hr). Yet this weak system produced near-record rainfall and massive flooding from eastern Texas eastward across the Gulf States and along the mid-Atlantic coast and ranks as the most deadly and costly tropical storm ever to strike the U.S. mainland.

With weak steering winds aloft, from the 4th through the 9th, Allison followed an erratic path, describing a meandering counterclockwise loop over eastern Texas, weakening to a tropical depression on the 7th, and twice soaking the Houston metropolitan area (Figure 12.6). From the 4th through the 9th, rainfall totals in the Houston area included 94 cm (36.99 in.) at Port of Houston and 90.1 cm (35.67 in.) at Greens Bayou. The system's remnant circulation then drifted southward into the northwestern Gulf of Mexico where it intensified into a *subtropical cyclone* (a weather system having characteristics of both tropical and midlatitude cyclones). On the 11th, Allison moved inland over Louisiana where

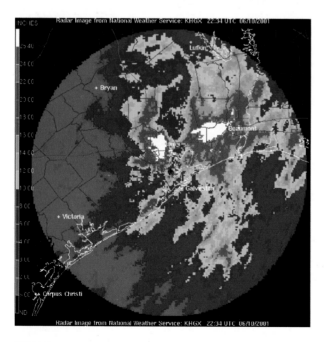

FIGURE 12.6
Radar-determined cumulative rainfall over southeast Texas (Houston and Galveston near center) caused by the remnants of tropical storm Allison for the period ending on 10 June 2001.

torrential rains totaled 75.8 cm (29.86 in.) at Thibodaux and 70 cm (27.55 in.) at Salt Point. The system slowly drifted over southeastern Mississippi and then moved northeastward to eastern North Carolina where it again stalled on the 14th. Tallahassee, FL recorded a 24-hr record rainfall of 25.7 cm (10.13 in.) from the 11th to the 12th. On 17 June, Allison tracked toward the northeast off the mid-Atlantic coast and eventually dissipated on the 19th over the North Atlantic southeast of Nova Scotia. All told, Allison took the lives of 41 people (including 23 in Texas and 8 in Florida) and caused at least $5 billion in property damage.

Winds pushing on the outside walls of a building exert a pressure that increases dramatically as the winds strengthen. *Wind pressure*, the force per unit area caused by air in motion, increases with the square of the wind speed. Hence, doubling the wind speed increases the wind pressure on an external wall by a factor of four. Furthermore, debris transported by the wind and hurled against structures exacerbates the damage potential of strong winds.

Meteorologists have assumed that strong gusts are responsible for the most serious hurricane wind-related property damage. After detailed analysis of the aftermath of Hurricane Andrew, however, T.T. Fujita argued that small but powerful whirlwinds embedded in the hurricane circulation were responsible for the most severe property damage. Whirlwinds combined with the larger-scale hurricane circulation to produce winds estimated at 320 km (200 mi) per hr. Based on mode of origin, Fujita described these whirlwinds as *spin-up vortices*. Apparently, small eddies spin off the eye wall and are stretched vertically by powerful convection currents. Stretching accelerates the eddy circulation.

Hurricane winds diminish rapidly once the system makes landfall, so that most wind damage is confined to within about 200 km (125 mi) of the coastline. In fact the database of hurricane tracks maintained by the National Hurricane Center shows no hurricane centered more than about 325 km (200 mi) inland in the United States. Two factors account for the abrupt drop in wind speed once a hurricane makes landfall. A hurricane over land is no longer in contact with its energy source, warm ocean water. In addition, the frictional resistance offered by the rougher land surface slows the wind and shifts the wind direction toward the low-pressure center of the system (Chapter 8). This wind shift causes the storm to begin to fill; that is, its central pressure rises, the horizontal pressure gradient weakens, and winds slacken.

As noted in Chapter 11, tornadoes can accompany a hurricane once it makes landfall. As a hurricane moves over land its surface winds weaken rapidly while winds aloft remain strong. The wind shear that is created between horizontal winds at the surface and aloft contributes to tornado development. Usually only a few tornadoes occur with a hurricane but in 1967, Hurricane Beulah reportedly spawned as many as 115 tornadoes across southern Texas. Tornadoes are most probable after the hurricane enters the westerly steering current and curves toward the north and northeast; they form mostly to the northeast of the storm center, often outside the region of hurricane-force winds.

STORM SURGE

A **storm surge** is a dome of ocean water, 80 to 160 km (50 to 100 mi) wide that sweeps over the coastline near the hurricane's landfall (Figure 12.7). A storm surge can demolish marinas, piers, cottages, and other shoreline structures, erode beaches, and wash out coastal roads and railway beds.

Prior to 1970, storm surges were responsible for the majority of hurricane-related fatalities. In fact, in

October 1972, the American Meteorological Society adopted a policy statement that attributed 90% of all hurricane-related fatalities near the U.S. coast to the combination of storm surge and accompanying waves. Since then, a better-informed population, early storm warnings, and evacuation of populations from vulnerable coastal areas spurred a sharp decline in storm surge related deaths. As pointed out earlier, during the 30-year period from 1970 to 1999, the majority of hurricane-related deaths in the United States were from drowning in freshwater floods. Only 6 deaths were due to storm surges. But the recent decline in storm surge related deaths should not lessen the vigilance of coastal residence should a hurricane threaten for history recounts many catastrophic storm surges.

A storm surge accounted for much of the considerable loss of life in a hurricane that struck the southeast United States on 27 August 1893, one of six hurricanes to strike the U.S. mainland that year (equaled in number by only 1916 and 1985). The unnamed hurricane came ashore at night just north of Savannah, GA with little warning. (U.S. Weather Bureau meteorologists predicted that the hurricane, first detected near Nassau the previous day, would make landfall somewhere along the East Coast south of New York.) The actual track placed the system's *dangerous semicircle*, the band of strongest onshore winds, over the Sea Islands of Beaufort County, SC. Most of the inhabitants of the Sea Islands were freed slaves, their children and grandchildren. They were mostly farmers who raised a highly valued variety of cotton. No telegraph lines connected the islands to the mainland so there was no warning of the hurricane's approach. With no means of evacuation, residents were completely at the mercy of 175 to 210 km (110 to 130 mi) per hr winds and a 3.7 to 4.6 m (12 to 15 ft) storm surge

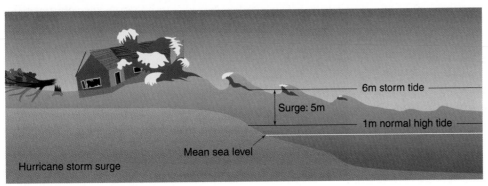

FIGURE 12.7
Onshore winds and low air pressure associated with a hurricane cause a dome of water to move shoreward, causing considerable coastal erosion and flooding. Adding to the destructive potential of a storm surge are wind-driven waves.

that swept over the islands. Some 20,000 to 30,000 people were left homeless, their crops were destroyed, and the death toll was estimated at 2000 (and possibly as high as 2500). The Sea Islands disaster came at a time of national recession and prior to the time of organized federal or state natural disaster assistance. Intervention by the Red Cross was all that saved the survivors from almost certain famine.

In the most deadly natural disaster in U.S. history, more than 8000 people perished, mostly by drowning, when a hurricane storm surge flooded Galveston, TX on 8 September 1900. Floodwaters inundated the entire Gulf Coast island city. Some 3600 homes were destroyed; a dam of wreckage that built up some six blocks inland from the beach probably spared the business district total destruction. To protect the city from future storm surges, the U.S. Army Corps of Engineers elevated Galveston some 3 m (10 ft) and erected a massive concrete seawall along the city's Gulf shore.

Hurricane Camille, an exceptionally intense system with winds to 300 km (186 mi) per hr, produced a maximum storm surge of 7.3 m (24 ft) at Pass Christian, MS on 17 August 1969. The total death toll from the storm was 256, about half from coastal flooding. In 1970, one of the most disastrous storm surges ever flooded the Bay of Bengal coast of East Pakistan (now Bangladesh). A storm surge of nearly 7 m (23 ft) inundated the vast low-lying coastal plain of the Ganges Delta, claiming an estimated 300,000 lives mostly by drowning. There was no high ground for people to flee to.

Strong onshore winds plus relatively low air pressure are responsible for a hurricane storm surge. Hurricane-force winds drive a mound of ocean water toward coastal areas and relatively low air pressure in a hurricane causes sea level to rise (about 0.5 m for every 50 mb drop in pressure or about 1 ft for every 1 in. of mercury drop in pressure). The storm surge is superimposed on normal ocean tidal oscillations so that the potential impact is greatest when the storm surge coincides with astronomical high tide. This timing is most important where the tidal range is relatively great. *Tidal range* is the difference in water level between high and low tides and is greater along the Eastern Seaboard (typically several meters) than along the Gulf Coast (less than 1 m). In addition, wind-driven waves having heights of at least 1.5 to 3 m (5 to 10 ft) top the storm tide and add to the property damage potential. These waves are likely to be highest where the water offshore is relatively deep.

The intensity, track, and forward speed of the hurricane plus coastal configuration and changes in water depth along the coast and seaward play important roles in a storm surge's damage potential. In general, a storm surge of 1 to 2 m (3 to 6.5 ft) can be expected with a weak hurricane, whereas the storm surge accompanying a violent hurricane may top 5 m (16.4 ft). A storm surge is most likely on the side of the approaching hurricane where winds are onshore. Low-lying coastal plains are vulnerable to a hurricane storm surge especially if no protective offshore islands are present that would dissipate much of the energy of the surge.

A special numerical model developed in 1979 accurately predicts the location and height of a hurricane storm surge. Weather forecasters report considerable success with the *SLOSH* (*Sea, Lake, and Overland Surges from Hurricanes*) model. Using the output from SLOSH and consulting local topographic maps, forecasters can identify areas most likely to be inundated by floodwaters when a hurricane threatens. Hypothetical hurricanes are simulated using various combinations of storm intensity, forward speed, and expected location of landfall to predict the maximum high water value for various SLOSH basins along the coast. This information is very useful for public safety officials in developing evacuation plans. SLOSH model coverage includes the entire East and Gulf coastline, as well as parts of Hawaii, Guam, Puerto Rico, and the Virgin Islands.

SAFFIR-SIMPSON INTENSITY SCALE

Like tornadoes, hurricanes have an intensity scale, known as the **Saffir-Simpson Hurricane Intensity Scale** after its designers, H. S. Saffir, a consulting engineer, and R. H. Simpson, former director of the National Hurricane Center. Developed in the early 1970s and first included in hurricane advisories in 1975, the scale rates hurricanes from 1 to 5 corresponding to increasing intensity (Table 12.1). The scale provides an estimate of potential coastal flooding and property damage from a hurricane landfall. Wind speed is the primary determining factor for a hurricane's rating on the Saffir-Simpson Scale as storm surge heights are highly dependent on the underwater topography and other factors in the region of landfall. Each intensity category specifies (1) a range of central air pressure, (2) a range of maximum sustained wind speed, (3) storm surge potential, and (4) the potential for property damage.

Property damage potential rises rapidly with ranking on the Saffir-Simpson Scale. Destruction from a category 4 or 5 hurricane is 100 to 300 times greater than

TABLE 12.1
Saffir-Simpson Hurricane Intensity Scale

Category	Central pressure mb (in.)	Wind speed km/hr (mi/hr)	Storm surge m (ft)	Damage potential
1	≥ 980 (≥ 28.94)	119-154 (74-95)	1-2 (4-5)	Minimal
2	965-979 (28.50-28.91)	155-178 (96-110)	2-3 (6-8)	Moderate
3	945-964 (27.91-28.47)	179-210 (111-130)	3-4 (9-12)	Extensive
4	920-944 (27.17-27.88)	211-250 (131-155)	4-6 (13-18)	Extreme
5	< 920 (< 27.17)	> 250 (> 155)	> 6 (> 18)	Catastrophic

that caused by a category 1 hurricane. A category 1 (weak) hurricane causes no real damage to buildings but may affect unanchored mobile homes, shrubbery, and trees and cause minor damage to piers. With a category 2 (moderate) hurricane, we can expect some damage to the roofing materials, doors, and windows of buildings. Damage to trees, mobile homes, and piers can be considerable and small craft in unprotected anchorages may break their moorings. A category 3 (strong) hurricane can destroy mobile homes, cause some structural damage to small residences and utility buildings, and blow down large trees. Flooding is likely along the coast and some evacuation may be necessary. With a category 4 (very strong) hurricane, the roof structure of small residences may fail. Beaches experience major erosion and considerable damage is done to structures near the shore. Major coastal flooding may occur requiring massive evacuation of residential areas. In a category 5 (devastating) hurricane, we can expect complete roof failure on many residences and industrial buildings. In some instances, buildings experience complete failure. Major coastal flooding is likely requiring massive evacuation of residential areas.

Of the 162 hurricanes that struck the U.S. Atlantic or Gulf coasts between 1900 and 1998, 65 (about 40%) were classified as *major*; that is, they rated 3 or higher on the Saffir-Simpson scale. The 25 major hurricanes that made landfall along the Gulf or Atlantic coast between 1949 and 1990 accounted for three-quarters of all property damage from all landfalling tropical storms and hurricanes during the same period. Tables 12.2 and 12.3 list the 10 most intense and 10

TABLE 12.2
Ten Most Intense U.S. Hurricanes at Landfall, 1900-2000[a]

Rank	Hurricane	Year	Category	Pressure (mb)
1	Florida Keys	1935	5	892
2	Camille (MS, LA, VA)	1969	5	909
3	Andrew (FL, LA)	1992	4	922
4	FL Keys, south TX	1919	4	927
5	FL (Lake Okeechobee)	1928	4	929
6	Donna (FL, eastern US)	1960	4	930
7	TX (Galveston)	1900	4	931
7	LA (Grand Isle)	1909	4	931
7	LA (New Orleans)	1915	4	931
7	Carla (Texas)	1961	4	931

[a]From NOAA, National Hurricane Center

TABLE 12.3
The Ten Deadliest U.S. Hurricanes, 1900-1998[a]

Rank	Hurricane	Year	Category	Deaths
1	Unnamed, Galveston, TX	1900	4	8000[b]
2	Unnamed, Lake Okeechobee, FL	1928	4	1836
3	Unnamed, Florida Keys/south TX	1919	4	600[c]
4	"New England"	1938	3	600
5	"Labor Day," Florida Keys	1935	5	408
6	Audrey, SW LA and N TX	1957	4	390
7	Unnamed, NE United States	1944	3	390[d]
8	Unnamed, Grand Isle, LA	1909	4	350
9	Unnamed, New Orleans, LA	1915	4	275
10	Unnamed, Galveston, TX	1915	4	275

[a] From NOAA, National Hurricane Center
[b] May actually have been as high as 10,000 to 12,000.
[c] Over 500 lost on ships at sea; 600-900 estimated deaths.
[d] Some 344 of these lost on ships at sea.

deadliest hurricanes to strike the United States since 1900.

The last category 5 Atlantic basin hurricane of the 20th century to make landfall was Gilbert in September 1988. (Camille in August 1969 was the next previous one and Mitch in 1998 was a category 5 hurricane at peak intensity over the western Caribbean.) Gilbert originated as a tropical wave off the West African coast on 3 September. The wave followed a steady west-northwest track and seven days later strengthened to a hurricane while south of Puerto Rico. Gilbert then passed over the length of Jamaica on the 12th, intensified over the northwest Caribbean, and on 14 September made landfall as a category 5 hurricane near Cozumel on Mexico's Yucatan Peninsula. At landfall, sustained winds were estimated at 275 km (171 mi) per hr. The hurricane then continued northwestward over the Gulf of Mexico and made its final landfall (as category 3) on the coast of Mexico about 200 km (125 mi) south of the Texas border. Gilbert brought much death and destruction to both Mexico and Jamaica. The death toll was 202 in Mexico and 45 in Jamaica, and combined property damage in the two countries topped $4 billion.

Life Cycle of a Hurricane

The first sign that a hurricane may be in the making is the appearance of an organized cluster of cumulonimbus clouds over tropical seas. This region of convective activity is labeled a **tropical disturbance** if a center of low pressure is detectable at the surface. Chances are that the tropical disturbance was triggered by the ITCZ, by a trough in the westerlies that intrudes into the tropics from midlatitudes, or by a wave (or ripple) in the easterly trade winds (an *easterly wave*).

The most intense hurricanes that threaten North America usually develop out of convective cloud clusters associated with easterly waves that travel westward off the West African coast. An **easterly wave** is a ripple in the tropical easterlies (trade winds) featuring a weak trough of low pressure (Figure 12.8). The system likely owes its origin the disturbance of the prevailing easterlies by the Ethiopian Highlands of East Africa. They are most common from August through October. An easterly wave propagates from east to west across the tropical Atlantic generally more slowly than the current in which it is embedded. Some waves pass over Central America and into the Pacific where they can trigger tropical cyclone development.

Over the ocean, to the west of the trough line

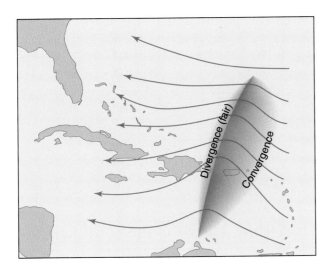

FIGURE 12.8
An easterly wave moving westward across the tropical Atlantic
may develop into a tropical cyclone.

within an easterly wave is an area of lower-tropospheric divergence, a shallow moist layer, and fair weather. To the east of the trough line is an area of intense lower-tropospheric convergence, a deep moist layer, and heavy rain showers. Convergence on the east side of an easterly wave helps to organize convective activity into a developing tropical cyclone. Storms originating in this way over the Atlantic are referred to as *Cape Verde-type* hurricanes.

Only a small percentage of convective cloud clusters in the tropical Atlantic actually evolve into full-blown hurricanes for the following reasons: (1) Strong subsidence of air on the eastern flank of the Bermuda-Azores anticyclone and the associated *trade-wind inversion* inhibit deep convection (Chapter 9). (2) Vertical wind shear (change of horizontal wind with altitude) over the tropical Atlantic is usually too great for hurricane formation. As noted earlier, strong horizontal winds aloft disrupt deep convection. (3) The middle troposphere is usually too dry; low vapor pressure at these levels inhibits intensification of the system.

If atmospheric/oceanic conditions favor hurricane development and if those conditions persist, the surface air pressure falls and a cyclonic circulation develops. Water vapor condenses within the storm, releasing latent heat of vaporization, and the heated air rises. Expansional cooling of the rising air triggers more condensation, release of even more latent heat, and a further increase in buoyancy. Rising temperatures in the core of the storm, coupled with anticyclonic outflow of

air aloft, cause a sharp drop in surface air pressure, which in turn, induces more rapid convergence of air at the surface. The consequent uplift surrounding the developing eye leads to additional condensation and release of latent heat, an example of positive feedback.

Through these processes, a tropical disturbance intensifies and its winds strengthen. When sustained wind speeds top 37 km (23 mi) per hr, the developing storm is called a **tropical depression**. When wind speeds reach 63 km (39 mi) per hr, the system is classified as a **tropical storm** and assigned a name. (See the Essay, "Naming Hurricanes.") Once sustained winds exceed 118 km (73 mi) per hr, the storm is officially designated a hurricane. As a hurricane weakens and decays, the system is downgraded by reversing this classification scheme.

Cape Verde-type hurricanes usually drift slowly westward with the trade winds (along the southern flank of the Bermuda-Azores subtropical high) across the tropical North Atlantic and into the Caribbean. At this stage in the storm's trajectory, it is not unusual for the system to travel a mere 10 to 20 km (6 to 12 mi) per hr and take a week to cross the Atlantic. Once over the western Atlantic, however, the storm usually picks up speed and begins to curve northward along the western flank of the Bermuda-Azores high, and northeastward as the system is caught in the westerlies of middle latitudes. Precisely where this curvature takes place determines whether the hurricane enters the Gulf of Mexico (perhaps then tracking up the lower Mississippi River Valley or over the Southeastern States), moves up the Eastern Seaboard, or curves back out to sea.

When an Atlantic coastal hurricane reaches about 30 degrees N, it may begin to acquire extratropical characteristics, as colder air is drawn into the circulation of the system and fronts develop. From then on, the storm follows a life cycle similar to that of any extratropical cyclone, often occluding over the North Atlantic. Many hurricanes, however, depart significantly from the track just described. Some of the hurricane tracks plotted in Figure 12.9, for example, are very erratic. A hurricane can describe a complete circle or reverse direction. In November 1999, Hurricane Lenny (only the fifth Atlantic basin storm in November on record to achieve category 3 or 4 status) tracked toward the east and northeast across the Caribbean Sea—opposite the direction of movement of most Atlantic basin hurricanes. According to the National Hurricane Center, this unusual track caused unprecedented wave and storm surge damage to westward facing harbors on

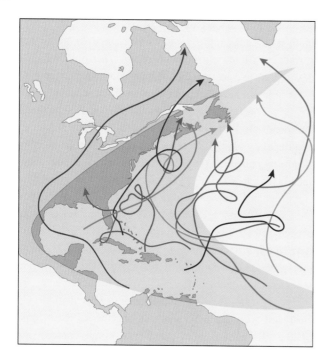

FIGURE 12.9
Tropical cyclone trajectories are often erratic, as shown by these samples. As indicated by the shaded area, however, Atlantic tropical cyclones initially drift westward and curve toward the north and northeast when they reach the western Atlantic. [From NOAA, *Hurricane*. Washington, DC: Superintendent of Documents, 1977

Caribbean islands. During the 20th century only three other hurricanes moved mostly east or northeastward through the Caribbean.

In addition, some hurricanes maintain their tropical characteristics even after traveling far north along the Atlantic Coast where they are fueled by warm Gulf Stream waters. The eye of Hurricane Hazel, for example, was still discernible when the system passed over Toronto in October 1954, after making landfall about 1100 km (700 mi) to the south near Wilmington, NC. New England, situated more than 25 degrees of latitude north of the usual hurricane breeding grounds, has been the target of many full-blown hurricanes. The most deadly of these was the unnamed hurricane of 21 September 1938 (category 3) that raced through New England and into Canada at an estimated 80 km (50 mi) per hr. A storm surge ravaged the New England coast, with a 4-m (16-ft) wall of water sweeping into downtown Providence, RI. Winds gusting over 200 km (125 mi) per hr severely damaged forests, and torrential rains caused flash flooding on rivers and streams. Fatalities were estimated at 600 and property damage totaled $400

million ($4.7 billion in year 2000 dollars). A more recent New England hurricane was Bob (category 2) in August 1991. Hurricane Bob moved swiftly up the East Coast from its origin over the northern Bahamas. The storm center passed just east of Cape Hatteras, NC and then remained offshore until striking Rhode Island's south coast. The eye then passed just east of Boston, MA and Portland, ME. Top wind speed was 185 km (115 mi) per hr, the death toll was 18, and property damage totaled $2.0 billion (in year 2000 dollars).

Hurricane Threat to the Southeast

From 1970 through 1989, only 10 major hurricanes (category 3 or higher) struck the U.S. mainland—down from 15 during 1950-69 and 16 in 1930-49. Between 10 August 1980 and 17 August 1983, no hurricanes struck the United States. Fewer major hurricanes during the 1970s and 1980s lulled many coastal residents of the southeast United States into a false sense of security and encouraged development and population growth in areas that could be devastated by a major hurricane. Development of the U.S. coastal zone is proceeding at a pace that is 2 to 3 times the national average. More and more resort hotels, high-rise condominiums, and expensive homes are being constructed perilously close to the shoreline and even among coastal sand dunes (Figure 12.10).

About 45 million permanent residents inhabit hurricane-prone portions of the nation's coastline, and

FIGURE 12.10
The relative infrequency of hurricanes and tropical storms in recent decades has lulled some residents of coastal areas into a false sense of security and inspired construction of hotels, cottages, and condominiums within meters of high-tide level.

the population continues to climb. Population growth is most rapid from Texas through the Carolinas with Florida leading the nation in both population growth and hurricane potential. Today, perhaps 80% to 90% of Atlantic and Gulf Coast residents have never experienced the full impact of a major hurricane. Some of them may have weathered with relative ease a weak hurricane or the fringes of a strong system. But such an experience may only lull them into complacency so that they are less likely to prepare adequately in the event of a major hurricane threat. Compounding the problem of the growth of the resident population in the Southeast is the arrival of holiday, weekend, and seasonal visitors to seaside resorts. During vacation periods, the population of some of these locales swells ten- to one-hundred-fold. Many of these resorts are in low-lying coastal areas or on exposed beaches that are subject to rapid inundation by a storm surge.

The hurricane danger is particularly acute for people living on or visiting the nearly 300 barrier islands that fringe portions of the Atlantic and Gulf coasts. A **barrier island** is a long, narrow strip of sand parallel to the coast and separated from the mainland by a lagoon or backwater bay. A barrier island is a continually changing system. Sea waves breaking on the shores of barrier islands dissipate their energy by shifting sands and modifying the shape of the island. A barrier island gradually grows toward the mainland as sediments settle in the lagoon while waves erode the ocean side of the island.

A barrier island faces the open ocean and absorbs the brunt of powerful storm-driven sea waves and storm surges thereby providing some protection for coastal beaches, wetlands, and shoreline structures. But many barrier islands have been developed, especially for cottages and resorts, and their sands are temporarily stabilized under a layer of asphalt or concrete. Some coastal cities, including Atlantic City, NJ, Miami Beach, FL, and Virginia Beach, VA are built entirely on barrier islands. Such exposed locations are particularly vulnerable to the ravages of tropical cyclones. On developed barrier islands, much of the energy of storm waves would be expended in demolishing buildings and roads as well as shifting sands.

Evacuation of people from barrier islands, as well as other low-lying coastal areas, is the traditional strategy in the event of a major storm threat (Figure 12.11). The effectiveness of coastal evacuation plans was tested in the late summer of 1985 when Hurricane Elena (category 3) menaced the Gulf of Mexico coast

FIGURE 12.11
In hurricane-prone areas, evacuation routes are marked by special road signs.

(Figure 12.12). For four days, Elena followed an erratic path over the Gulf of Mexico, first taking aim at southern Louisiana, then the Florida panhandle, and later central Florida before reversing direction and finally coming ashore near Biloxi, MS on 2 September. Nearly a million people from Sarasota, FL to New Orleans, LA were forced to leave coastal communities and flee to inland shelters. Some returned home only to evacuate again as Elena changed course. Although property damage was considerable ($2 billion in year 2000 dollars) because of extensive flooding and winds that exceeded 160 km (100

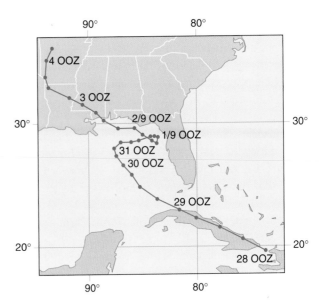

FIGURE 12.12
The erratic track of Hurricane Elena from 28 August through 4 September 1985. The hurricane's position is reported every 6 hrs.

mi) per hr, only 4 fatalities were attributed to Elena and none in the area of hurricane landfall. Timely evacuation of residents of low-lying coastal areas also saved many lives when Hugo (September 1989) and Andrew (August 1992) threatened; they were both category 4 hurricanes.

The potential downside of evacuation was illustrated in September 1999 when Floyd, an unusually massive hurricane, threatened much of the U.S. east coast. In the greatest evacuation in U.S history, more than 2 million residents of the coastal area from South Florida to South Carolina took to the roads and fled inland. In many areas the result was gridlock—too many vehicles on too few highways. As it turned out, Floyd spared most of the evacuated area and made landfall near Cape Fear, NC on 16 September 1999 as a category 2 storm. The gradually weakening system crossed eastern North Carolina and Virginia and then tracked northeastward along the New Jersey coast and became an extratropical system over New England. Torrential rainfall, totaling as much as 38 to 50 cm (15 to 20 in.) over portions of eastern North Carolina and Virginia caused extensive inland flooding, 56 deaths, and property damage of at least $6 billion. Floyd's track along the U.S. east coast required hurricane warnings from South Florida to Massachusetts.

Successful evacuation of coastal communities hinges on sufficient advance warning of a hurricane's approach, but as noted earlier in this chapter, hurricanes are notorious for sudden changes in direction, forward speed, and intensity. (In retrospect, about 75% of all coastal storm evacuations prove to be unnecessary.) A hurricane's erratic behavior is especially troublesome for people living in isolated localities and congested cities where highway systems may not have kept pace with population growth. In such places the time required for evacuation may be lengthy. The Federal Emergency Management Agency (FEMA) estimates evacuation times during the peak tourist season as 50-60 hrs for New Orleans, LA and Fort Myers, FL, 40-49 hrs for Ocean City, MD, 30-39 hrs for the Florida Keys and Cape May County, NJ, and 20-29 hrs for Long Island, NY, Atlantic City, NJ, the Outer Banks of North Carolina, and Galveston, TX.

As coastal communities continue to grow, the time required for evacuation of their population inevitably lengthens. Evacuation must begin earlier when a storm is farther away and greater uncertainty surrounds its likely track. Such uncertainty necessitates a broader zone of evacuation that translates into greater economic losses associated with evacuation (e.g., closed businesses). Cognizant of the problems posed by lengthy evacuation times, some coastal communities are looking into alternative strategies such as establishing shelters in public and commercial buildings and growth control ordinances.

Other strategies adopted or advocated by public safety and other government officials to minimize loss of life and property to hurricanes are (1) stringent building codes, (2) preservation of mangrove swamps, and (3) elimination of federal floodplain insurance. Building codes may call for new homes to be more wind resistant by requiring bolts that anchor the floor to the foundation and steel brackets that attach the roof to the walls. In addition, in many areas, all new buildings must be elevated above the once-in-a-century flood level (Figure 12.13). Some buildings are constructed so that the lower floor gives way to flood waters, thus reducing damage to main structural supports. Unfortunately, many structures were built prior to these tough codes, and as the aftermath of Hurricane Andrew revealed, stringent building codes are not always enforced.

Mangrove swamps along tropical coastlines are the final line of natural defense against hurricanes and other ocean storms. Their extensive root systems anchor soil and vegetation and dissipate the energy of a storm surge. In too many cases, however, shoreline development has destroyed these swamps, often because the vegetation obstructed the ocean view for residents of seaside dwellings. Preservation of mangrove swamps and other natural buffers, such as coastal sand dunes, has been given renewed emphasis in hurricane-prone areas.

A controversial strategy is to eliminate federal

FIGURE 12.13
This home, located on North Carolina's Outer Banks, is designed so that the first floor will give way to storm-surge floodwaters. The second and third floor living area of the home is supported by wooden beams driven deeply into the sand.

flood insurance for flood-prone coastal areas as has been done already on undeveloped portions of barrier islands. Some people argue that by providing policies at low cost, federal flood insurance programs actually encourage development in flood-prone areas. Furthermore, this insurance enables homeowners to rebuild structures destroyed by flooding in the same hazardous location.

Atlantic Hurricanes and West African Rainfall

Professor William M. Gray of Colorado State University discovered an interesting relationship between the frequency of major hurricanes in the tropical Atlantic and rainfall in West Africa. Each year, Gray uses this relationship along with other factors to forecast the number of tropical cyclones (tropical storms, hurricanes, and major hurricanes) in the upcoming Atlantic hurricane season. Gray discovered that major Atlantic hurricanes (categories 3-5 on the Saffir-Simpson Scale) are much more numerous when West Africa is relatively wet than when West Africa is relatively dry. This finding may prove valuable to long-range weather forecasters and, based on historical trends in West African rainfall, suggests that over the long term Atlantic hurricane activity may be on the upswing in the decades to come.

Gray analyzed rainfall records from 38 climate stations in the Western Sahel (Senegal, southern Mauritania, western Mali, Gambia, and Guinea Bissau) and 24 stations in the Gulf of Guinea for the 42-year period, 1949-1990. Recall from Chapter 9 that monsoon circulation dominates this region's climate such that summer is the rainy season and winter is dry. Recall also that this region is subject to prolonged drought.

In West Africa, the 1950s and 1960s were relatively wet. During that period, 13 major hurricanes made landfall along the U.S. east coast. On the other hand, during the Sahelian drought of 1970-1987, only one major hurricane (Gloria, 1985, category 3) struck the East Coast. In fact, Gray found that when rainfall in West Africa is above average prior to 1 August, the likelihood of a major East Coast hurricane after 1 August is 10 to 20 times greater than when West Africa experiences below average rainfall prior to 1 August. On the other hand, Gray found no clear relationship between West Africa rainfall and hurricane intensity along the Gulf Coast (including the Florida Panhandle).

Why the apparent link between Atlantic hurricane intensity and West African rainfall? As noted elsewhere in this chapter, convergence associated with easterly waves that travel westward from West Africa is a key to Atlantic hurricane development. According to Gray, such waves are stronger and better organized when rainfall in West Africa is relatively abundant. As long as all other atmospheric and sea-surface conditions are favorable, stronger and better organized easterly waves are more likely to develop into hurricanes.

Experts on the climate of the African Sahel, such as Professor Sharon Nicholson of Florida State University, point out that for at least several hundred years the Sahel has experienced alternating multi-decadal episodes of dry and wet conditions. Beginning in the late 1980s, signs appeared that the latest Sahelian drought was coming to an end. If this is the case, and if past weather is any indicator of the future, then West Africa may soon enter an extended wet episode. If Gray's correlation holds, an upswing in intense hurricane activity in the Atlantic may also be in the offing.

Hurricane Modification

Spurred by a succession of six destructive hurricanes that lashed the U.S. east coast during the mid-1950s, atmospheric scientists stepped up efforts to learn more about hurricanes and to devise hurricane modification techniques. To these ends, the U.S. government funded *Project STORMFURY* beginning in 1961. Until the project's termination in 1983, staff scientists advanced our understanding of the formation, structure, and dynamics of hurricanes; improved hurricane forecasting; and upgraded hurricane reconnaissance methods. The original goal of hurricane modification, however, was never realized.

The working hypothesis of *STORMFURY* was that seeding hurricanes with silver iodide (AgI) crystals would reduce wind strength. The argument went as follows: Seeding the band of convective clouds just beyond the eye wall would trigger additional latent heat release (as supercooled water droplets converted to ice crystals) and enhance convection. This artificially invigorated convection would then dominate convection in the eye wall, and a new eye wall would form at a greater distance from the storm center. Thereby, the band of strongest winds would be displaced farther out from the eye center and the hurricane's circulation would weaken (just as a skater's spin rate slows when she extends her arms).

The 1960s and 1970s presented few opportunities for hurricane seeding experiments primarily because few hurricanes met the selection criteria. Systems selected for seeding had to be mature, relatively intense, expected to remain far from land for at least 24-hrs following seeding, and have a well-developed eye wall. Some success was reported, however, in reducing winds (between 10% and 30%) in four hurricanes that were seeded. Later, these apparent successes were dismissed when new data called into question the original *STORMFURY* hypothesis. For one, it was discovered that convective clouds in hurricanes contain too little supercooled water for seeding to be effective. For another, monitoring of unmodified hurricanes found that changes in eye-wall diameter occur as part of the storm's natural evolution. It is likely, then, that *STORMFURY's* apparent success at hurricane modification was by chance rather than by seeding.

Conclusions

Hurricanes are intense, long-lived, tropical cyclones that develop over the tropical ocean but can track into midlatitudes. The requirements of relatively high sea-surface temperature plus significant Coriolis deflection restrict hurricane formation to certain regions of the globe. In the Atlantic Basin, these systems initially drift westward in the trade winds and then turn north and northeastward as they are caught up in the midlatitude westerlies. Torrential rains, strong winds, and storm surge associated with a hurricane can take many lives and cause considerable property damage. Based on intensity and potential for storm surge and property damage, a hurricane is assigned a rating on the Saffir-Simpson Scale.

With this chapter, we have completed our description of the genesis, life cycle, and characteristics of the major atmospheric circulation systems that affect the weather in middle latitudes.

Basic Understandings

- Weather in the tropics exhibits very little seasonal temperature variation because of year-round high solar altitudes and nearly uniform day length. Rainfall is primarily the result of convective activity driven by intense solar heating. Principal weather systems are the ITCZ, easterly waves, and tropical cyclones.

- A hurricane is a violent tropical cyclone that originates over the tropical ocean, usually in late summer or early fall. It develops in a mass of warm and humid air, has no fronts, and is about one-third the diameter of an average extratropical cyclone.

- The most dangerous and potentially most destructive part of a hurricane is the portion of the eye wall on the side of the advancing system where the wind blows in the same direction as the storm's forward motion.

- For a tropical cyclone to develop, sea-surface temperatures must be 26.5 °C (80 °F) or higher through a depth of at least 60 m (200 ft), the Coriolis force must be sufficient (latitude of at least 4 degrees north or south of the equator) to initiate a cyclonic circulation, and winds aloft must be relatively weak. The first two requirements mean that tropical cyclone development is limited to certain regions of the tropical ocean. Weak winds aloft allow a cluster of cumulonimbus clouds to organize over tropical seas, the first step in the evolution of a hurricane.

- The annual number of hurricanes can have little bearing on the number of hurricanes that make landfall and their impact.

- Hazards of hurricanes include heavy rains and flooding, strong winds, tornadoes, and storm surge.

- Once a hurricane makes landfall, it loses its warm-water energy source and experiences the frictional resistance offered by greater surface roughness. Its circulation weakens rapidly so that most wind damage is confined to within about 200 km (125 mi) of the coastline. Torrential rains often continue well inland, however, and can cause severe flooding.

- A storm surge is a dome of ocean water driven onshore by storm winds. The greatest potential for coastal flooding and erosion occurs when a storm surge coincides with high tide and where the tidal range is relatively great.

- Based on intensity and potential for flooding and

property damage, a hurricane is rated from 1 (weak) to 5 (very intense) on the Saffir-Simpson Hurricane Intensity Scale.

- Most major hurricanes that impact North America originate as convective cloud clusters associated with easterly waves in the trade winds. With certain atmospheric/oceanic conditions, some convective cloud clusters evolve into tropical disturbances that can develop into tropical depressions, tropical storms, and finally hurricanes.

- Rapid population growth in coastal areas coupled with the lull in intense hurricane activity during the 1970s and 1980s has heightened the hurricane hazard in the southeastern United States.

- Attempts to modify hurricanes through cloud seeding have been unsuccessful. However, out of such efforts has come a greater understanding of the dynamics of hurricanes.

ESSAY: Polar Lows With Hurricane Characteristics

Some intense storms that form over the ocean off the coasts of Alaska and Norway have characteristics in common with hurricanes and tropical storms. For this reason, these polar lows are sometimes called *Arctic hurricanes*.

These polar lows are similar to tropical hurricanes in several respects. Both are relatively small-scale cyclones that are highly symmetric and feature a relatively warm core. Most clouds are convective with a ring of cumulonimbus clouds surrounding a clear calm eye, and strongest winds are associated with those cumulonimbus clouds. Both systems are triggered by other pre-existing atmospheric disturbances and both derive their energy from the ocean surface.

Arctic hurricanes differ from tropical hurricanes in the following ways: Arctic hurricanes are shallower systems that develop in cold air north of the polar front and evolve to full maturity in only 12 to 24 hrs. On the other hand, tropical hurricanes form in warm, humid air and may require 3 to 7 days to attain full strength. In addition, Arctic hurricanes travel nearly twice as fast as tropical hurricanes. Sustained wind speeds are somewhat higher in a tropical hurricane than in an Arctic hurricane. Minimum sustained wind speeds are about 110 km (68 mi) per hr in an Arctic hurricane and, by convention, 119 km (74 mi) per hr in a tropical hurricane.

In a tropical hurricane, the flux of heat from the sea-surface to the atmosphere is the principal source of energy that sustains the system. Most of this energy flux consists of latent heating (that is, evaporation of water at the sea-surface and subsequent condensation and release of latent heat within the storm system). Much less important is sensible heating (that is, conduction and convection of heat from the sea-surface to the atmosphere). Although the total heat flux from sea to air is roughly the same for both Arctic and tropical hurricanes, sensible heating plays a proportionally greater role in Arctic hurricanes.

As noted elsewhere in this chapter, a tropical hurricane is triggered by the ITCZ, a trough in the westerlies, or an easterly wave. The trigger for an Arctic hurricane may be a polar trough or jet streak. Development is most likely along a front that marks the transition zone between modified and unmodified arctic air. Arctic air is modified (becoming milder and more humid) by virtue of its trajectory over ocean water.

ESSAY: Naming Hurricanes

When sustained winds in a tropical depression reach 63 km (39 mi) per hr, the intensifying system is designated a *tropical storm* and assigned a name. This practice is intended to improve communication between forecasters and the general public especially since tropical storms or hurricanes are relatively long-lived weather systems and more than one may occur in the same ocean basin at the same time.

Originally, hurricanes were identified by their latitude and longitude and later by letters of the alphabet. Late in the 19th century, an Australian forecaster Clement Wragge was apparently the first to apply women's names to tropical cyclones; he also named them for politicians he disliked. During World War II, U.S. Army Air Corps and Navy meteorologists informally applied women's names to tropical cyclones. From 1950 to 1952, tropical cyclones in the North Atlantic were labeled using the phonetic alphabet (e.g., Able, Baker, etc.). In 1953, U.S. Weather Bureau meteorologists began using an alphabetical sequence of female names for tropical storms and hurricanes in the Atlantic basin. Women's names were first used for tropical cyclones near Hawaii in 1959 and for the rest of the northeast Pacific basin the next year. Starting in 1978, names consisted of alternating male and female names in alphabetical order, in English, Spanish, and French. The same practice began in the Atlantic basin in 1979. Six lists of names are repeated every 6 years, although the names of very destructive hurricanes have been removed from the list (e.g., Betsy, 1965; Agnes, 1972; David and Frederic, 1979; Allen, 1980; Elena and Gloria, 1985; Gilbert and Joan, 1988; Hugo, 1989; Bob, 1991; Andrew, 1992; Alicia, 1993; Opal and Roxanne, 1995; Mitch 1998).

In the northwest Pacific Basin, women's names were used for tropical cyclones from 1945 to 1979. Then, alternating male and female names were used until 1 January 2000. On that date, the World Meteorological Organization's Typhoon Committee adopted a unique list of Asian names, most of which are not personal names but rather the names of flowers, animals, birds, and trees. Furthermore, the list of names is organized by the contributing nation (in alphabetical order). The Tokyo Typhoon Centre of the Japanese Meteorological Agency is responsible for assigning names from this list to developing tropical storms.

Separate lists of names are maintained for the southwest Indian Ocean (west of 90 degrees E), western Australia (90 degrees to 125 degrees E), northern Australia (125 degrees to 137 degrees E), and eastern Australia (137 degrees to 160 degrees E). Lists of names are also available for the regions near Fiji and Papua New Guinea. Tropical cyclones in the north Indian Ocean are not named.

CHAPTER A

WEATHER ANALYSIS AND FORECASTING

Probable nor'east to sou'west wind, varying to the southard and westard and eastard and points between;
high and low barometer sweeping round from place to place;
probable areas of rain, snow, hail, and drought, succeeded or preceded by earthquakes, with thunder and lightning.

MARK TWAIN, *New England Weather*

Most people readily recall occasions when an erroneous weather forecast upset their plans. It may have been an unexpected thundershower that brought an abrupt end to a softball game, or a raging blizzard that appeared instead of the anticipated clearing skies, or the promised spring-like weekend that turned out to be anything but spring-like. People seem to remember missed weather forecasts all too clearly and overlook the vast majority of times when the forecast was on target.

Weather forecasts, especially for the next day or so, are quite accurate and forecasters achieve some skill with forecasts out to 7 days. Nonetheless, most people readily recall occasions when the weather pulled unpleasant, and in some cases, hazardous surprises on them. In many situations, these surprises were not from the lack of a weather forecast, watch, or warning. The

fact is that changes in weather or the potential for change, including the more dramatic and hazardous changes, are routinely predicted to a high degree of accuracy. Unfortunately, forecasts might not reach as far into the future as we would like or need. The quality of weather forecasts, especially short-term forecasts, is such that it has become more important than ever to know the latest forecast. Simply stated, people who know what the weather is and what it is likely to be, benefit in many ways.

Of course, not all weather is predicted nor is it currently predicted with the accuracy one would like beyond a few days. Our present understanding of the workings of the atmosphere and our ability to monitor it are neither perfect nor complete. Also, weather forecasts are not always prepared to the detail we might desire for our particular situation. Weather prediction will never be perfect because forecasters will always be working with incomplete information on the state of the atmosphere and with some scientific questions not yet answered. However, when viewed with the objectivity of statistical analysis, short-range weather forecasting is surprisingly accurate. The U.S. **National Weather Service (NWS)**, an agency of the **National Oceanic and Atmospheric Administration (NOAA)**, issues 24-hr weather forecasts that are correct nearly 85% of the time. The popular notion that weather forecasting is not very accurate stems from the simple fact that a missed forecast is more memorable because of the inconvenience a person might experience.

How are weather forecasts made? What are the limits of forecast accuracy? On the basis of what you learn in this course, how can you make your own weather forecasts? These are the principal questions addressed in this chapter.

World Meteorological Organization

The atmosphere is a continuous fluid that envelops the globe, so that weather observation, analysis, and forecasting require international cooperation. To this end, the *International Meteorological Organization* (*IMO*) was founded in 1878. In 1951, the IMO became the **World Meteorological Organization (WMO)**, an agency of the United Nations. Today, the WMO, headquartered in Geneva, Switzerland, coordinates the efforts of 185 nations and territories in a standardized global weather-monitoring program known as **World Weather Watch (WWW)**.

At standard observation times, the state of the atmosphere is monitored worldwide. This *Global Observing System* consists of almost 10,000 land stations, about 4000 of which contribute data to regional synoptic weather networks. Additional observational data are gathered by more than 7000 ships at sea, reconnaissance and commercial aircraft, satellites, weather radar, and almost 1000 radiosonde stations. Observations are transmitted to three *World Meteorological Centers* located near Washington, DC, Moscow, Russia, and Melbourne, Australia, where maps and charts representing the current state of the atmosphere are prepared. From human and computer analysis of this information, weather forecasts are prepared. Maps, charts, and forecasts are then sent to 34

Regional Specialized Meteorological Centers (*RSMCs*) and to *National Centers for Environmental Prediction* (*NCEP*) located in WMO member nations. At RSMCs and NCEPs, weather information and forecasts are generated and interpreted for each center's area of responsibility and distributed first to local weather service forecast offices and then to the public. In the United States, the **National Centers for Environmental Prediction (NCEP)** are headquartered in Camp Springs, MD.

The above description implies that weather forecasting entails (1) acquisition of present weather data, (2) graphical depiction of the state of the atmosphere by plotting those data on maps and charts, (3) analysis of data and maps, (4) prediction of the future state of the atmosphere, and (5) dissemination of weather information and forecasts to the public.

Acquisition of Weather Data

Since invention of the first weather instruments in the 17th century, weather observation has undergone considerable refinement (Chapter 2). Denser monitoring networks, more sophisticated calibrated instruments and communications systems, and better trained weather observers have produced an increasingly detailed, reliable, and representative record of weather and climate.

SURFACE WEATHER OBSERVATIONS

Nearly 1000 stations across the United States routinely monitor surface weather. These stations may be operated by any of the following: (1) National Weather Service personnel, (2) the staff of other government agencies, including the Federal Aviation Administration (FAA), or (3) private citizens or businesses in cooperation with the NWS. At sea and on the Great Lakes, more than 2000 ships also voluntarily gather weather data.

The NWS also maintains networks of automated weather stations in locations where manned observations are not feasible. These include, for example, more than 110 automated weather stations operated by the National Data Buoy Center (NDBC). Some NDBC stations are attached to moored buoys in offshore locations (including the Great Lakes), and others are at lighthouses, fishing piers, and offshore oil platforms. NDBC stations provide data on storm intensity and track and are considered an essential part of the hurricane warning system. Satellites relay data from the buoy network to users.

Weather stations gather data for (1) preparation of weather maps and forecasts, (2) exchange with other nations, and (3) use by aviation. Stations report cloud height and sky cover, wind speed and direction, visibility, precipitation type and amount, air temperature, dewpoint, and air pressure. Also, airport weather stations report altimeter setting for guidance to aviation.

Weather observations must be taken at the same time everywhere so as to accurately represent the state of the atmosphere. To this end, weather observers adhere to **Coordinated Universal Time (UTC)** (or *Universel Temps Coordinné*), previously known as *Greenwich Mean Time* (*GMT*), the time along the prime meridian, 0 degree longitude. (The prime meridian passes through the Old Royal Observatory, Greenwich, England.) For reference, at 0600 UTC, it is midnight Central Standard Time (CST) in Chicago, IL and 10 p.m. Pacific Standard Time (PST) in San Francisco, CA. For more on time keeping, see the Essay in Chapter 1.

During the 1990s, weather-observing facilities of the National Weather Service underwent extensive re-structuring and modernization with the goal of upgrading the quality and reliability of weather observation and forecasting. NWS Weather Forecast Offices were consolidated at 115 sites across the United States, each with a designated area of responsibility (Figure A.1). In addition, automated weather stations replaced the old system of manual hourly observations. This **Automated Surface Observing System (ASOS)** consists of modern sensors, computers, and fully automated communications ports (Figure 2.7). A network of about 1700 ASOS units operate continuously, feeding observational data to NWS Forecast Offices and local airport control towers. Also as part of its modernization program, the National Weather Service replaced old weather radars with new Doppler radars at 113 sites nationwide. The FAA and the U.S. Department of Defense operate another 39 Doppler units. As noted in Chapters 7 and 11, Doppler radar offers significant (possibly life saving) advantages over the old weather radar in providing more advance warning of the development of severe weather systems.

Besides the numerous land-based weather stations that provide information of potential use for weather forecasting and aviation, another 11,000 cooperative weather stations are scattered across the United States (Figure 2.8). These stations are cooperative in that people volunteer their time and labor to monitor instruments and the NWS provides

FIGURE A.1
A National Weather Service Weather Forecast Office.

instruments and data management. The principal function of the **NWS Cooperative Observer Network** is to record daily precipitation and temperatures for hydrologic, agricultural, and climatic purposes. Observers report 24-hr precipitation totals and maximum/minimum temperatures based on observations made daily by 8 a.m., local time; some observers also report river levels. Traditionally, observers telephoned reports to the local NWS Weather Forecast Office but a new program that automates cooperative stations enables the observer to enter data into a microcomputer that formats and transmits data to computer workstations in the NWS Advanced Weather Interactive Processing System (AWIPS) (described later).

UPPER-AIR WEATHER OBSERVATIONS

As noted in Chapter 2, meteorologists still rely on radiosondes to monitor the upper atmosphere (Figure 2.9). A *radiosonde* is a radio-equipped instrument package carried aloft by a balloon that transmits to a ground station temperature, pressure, and humidity profiles (*soundings*) to a maximum altitude of about 30,000 m (100,000 ft). In addition, winds at various levels in the atmosphere are computed by tracking the balloon's drift using a radio direction-finding antenna (a *rawinsonde observation*). Worldwide, readings are made twice each day, at 0000 UTC and 1200 UTC.

As of this writing, the U.S. National Weather Service is developing the Radiosonde Replacement System (RRS) to take the place of the current generation of radiosondes. The RRS consists of a new global positioning system (GPS) tracking antenna, new GPS radiosondes that operate at a radio frequency of 1680

FIGURE A.2
Map showing locations of radiosonde stations.

MHz, plus a new NT-based workstation. The new system will provide more detailed and accurate upper-air data. Readings will be available at 1-second intervals, corresponding to altitude levels about 5 m (16 ft) apart.

The U.S. operates almost 90 radiosonde (*RAOB*) stations, a quarter of which use special high-altitude balloons that obtain data from altitudes above 30,000 m (100,000 ft). There are about 70 RAOB stations in the continental U.S., 14 in Alaska, 2 in Hawaii, and one each in Guam and Puerto Rico (Figure A.2). Occasionally, meteorological rockets probe much higher altitudes (up to 100 km or 62 mi) but data from these probes are used primarily for research. In some urban areas, low-level soundings (up to 3000 m or 9800 ft) monitor atmospheric conditions to assess air pollution potential. Satellites, weather radar, aircraft, and dropwindsondes (similar to a radiosonde but dropped

from an aircraft) also supply upper-air weather data. In fact, satellite-derived observational data currently account for 84% of all data assimilated in NCEP numerical models.

Data Depiction on Weather Maps

Computers at the *Hydrometeorological Prediction Center (HPC)*, one of the National Centers for Environmental Prediction (NCEP), use special symbols for plotting weather observations on synoptic and hemispheric weather maps. The weather reported by each observation station is depicted on a map by following a conventional **station model**. The station model in Figure A.3 shows symbols for surface weather conditions. By

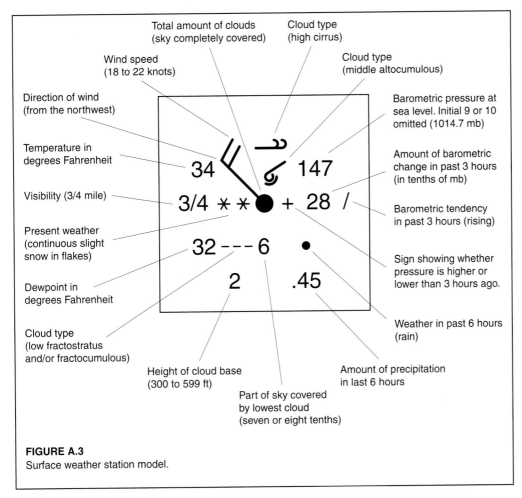

Total amount of clouds (sky completely covered)
Cloud type (high cirrus)
Wind speed (18 to 22 knots)
Cloud type (middle altocumulous)
Direction of wind (from the northwest)
Barometric pressure at sea level. Initial 9 or 10 omitted (1014.7 mb)
Temperature in degrees Fahrenheit
Amount of barometric change in past 3 hours (in tenths of mb)
Visibility (3/4 mile)
Barometric tendency in past 3 hours (rising)
Present weather (continuous slight snow in flakes)
Sign showing whether pressure is higher or lower than 3 hours ago.
Dewpoint in degrees Fahrenheit
Weather in past 6 hours (rain)
Cloud type (low fractostratus and/or fractocumulous)
Height of cloud base (300 to 599 ft)
Amount of precipitation in last 6 hours
Part of sky covered by lowest cloud (seven or eight tenths)

FIGURE A.3
Surface weather station model.

international agreement, the same station model and symbols are used throughout the world.

Weather systems are three-dimensional so that both surface and upper-air weather maps are needed to depict the state of the atmosphere. A very different approach is used for the two types of maps, however. Surface weather data are plotted on a constant-altitude surface (usually sea-level), whereas upper-air weather data are plotted on constant-pressure (*isobaric*) surfaces.

SURFACE WEATHER MAPS

It is standard practice for meteorologists to adjust surface air pressure readings to sea level (Chapter 5), thereby eliminating the influence of station elevation on air pressure readings. This procedure enables meteorologists to compare surface air pressure readings at weather stations that are located at different elevations above sea level. The adjusted air pressure readings are plotted on surface weather maps. *Isobars*, lines of equal air pressure, are then constructed at 4-mb intervals, a procedure that requires interpolation between stations.

An isobaric analysis reveals the location of such features as anticyclones (*Highs*) and cyclones (*Lows*), troughs and ridges, as well as horizontal air pressure gradients (Figure A.4).

On surface weather maps, a cyclone center, where air pressure is lowest, is indicated by the symbol *L* or *LOW*. Closely spaced isobars surrounding the storm center indicate relatively strong horizontal pressure gradient and strong surface winds. Fronts originate at the storm center and isobars bend (*kink*) as they cross fronts; hence, fronts often are found in troughs. Because we can infer surface wind direction from the isobar pattern, the bending of isobars at fronts tells us that the wind changes direction as we cross a front. As noted in Chapter 10, a wind shift is one characteristic of a frontal passage. Relatively weak horizontal air pressure

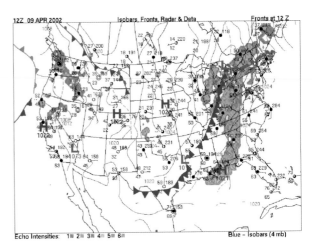

FIGURE A.4
Sample surface isobaric analysis.

gradient (widely spaced isobars) and weak winds generally occur over a broad area about the center of an anticyclone (mapped as *H* or *HIGH*).

Surface synoptic weather maps are drawn every 3 hrs for North America and at 6-hr intervals for the Northern Hemisphere. Special maps and charts are also constructed that summarize a variety of surface weather elements including, for example, (1) maximum and minimum temperatures for 24-hr periods, (2) precipitation amounts for 6 and 24 hrs, and (3) observed snow cover. A national composite radar summary is also issued hourly. You can access these weather maps and charts on the course Homepage.

UPPER-AIR WEATHER MAPS

Upper-air weather data acquired by radiosondes are plotted on constant-pressure surfaces using a modified station model. Applying basic laws of atmospheric physics, meteorologists compute the altitudes corresponding to these pressure values. For example, meteorologists determine the altitude of the 500-mb surface, that is, the altitude where the air pressure is about one-half its average sea-level value. By plotting altitudes of the 500-mb level as monitored simultaneously by all radiosonde stations across North America, meteorologists construct a map representing the *topography* of the 500-mb surface (Figure A.5). The 500-mb map is particularly useful for weather forecasting because circulation at that level closely approximates the upper-level steering winds and the trough and ridge patterns responsible for the development and displacement of surface weather systems such as cyclones.

Meteorologists (or their computers) draw contour lines where the 500-mb level is at the same altitude, interpolating between weather stations. By convention, height contours are labeled in meters above sea level and the contour interval (difference between successive contours) is 60 m. Dashed lines, also plotted on the map, are *isotherms*, lines of equal temperature, labeled in degrees Celsius. Arrows and barbs represent wind direction and speed at the 500-mb level.

The altitude of a pressure surface (such as the 500-mb surface) varies from one place to another, primarily because of differences in mean temperature of the air below the pressure surface. Air pressure drops with altitude more rapidly in cold air than in warm air. Raising the temperature of air reduces its density so that greater altitudes are required for warm air to exhibit the same drop in pressure as cold air. This means, for example, that the 500-mb level is at a lower altitude where the air below is relatively cold and at a higher altitude where the air below is relatively warm. Height contours on an isobaric surface therefore always show a gradual slope downward from the warm tropics to the colder polar latitudes.

Upper-air observations indicate that horizontal winds generally parallel the height contour lines at the 500-mb level. Where contours are closely spaced (a relatively steep height gradient), winds are strong, and where contours are far apart (a relatively weak height gradient), winds are light. Why are winds associated with a gradient in the height of a pressure surface? A height gradient develops in response to a horizontal gradient in air temperature, and where there is a horizontal air temperature gradient, there are also horizontal gradients in air density and air pressure. As discussed in Chapter 8, a horizontal air pressure gradient generates wind. The 500-mb surface is so far above the atmospheric boundary layer that the planetary- and synoptic-scale winds are essentially *geostrophic* where contours are straight, and *gradient* where contours are curved (Chapter 8). Hence, the large-scale wind at the 500-mb level is the product of interactions among a horizontal height gradient, the Coriolis force, and the centripetal force.

On upper-air weather maps, contours exhibit both cyclonic (counterclockwise) and anticyclonic (clockwise) curvature. These are the *troughs* and *ridges*, respectively, in the westerlies described in Chapter 9. Contours sometimes define a series of nearly concentric circles, perhaps indicating a cutoff or blocking circulation pattern. At the center of a ridge, the air

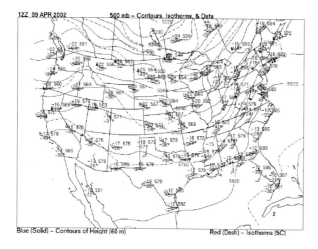

FIGURE A.5
Sample 500-mb analysis.

column is relatively warm and contour heights are high, so we label the ridge with an *H*. An upper-air ridge often is linked to a warm-core anticyclone at the surface. In contrast, near the center of a trough, the air column is relatively cold and contour heights are low, so the trough is labeled with an *L*. An upper-air trough may be linked to a cold-core cyclone at the surface. Warm-core cyclones (thermal lows) and cold-core anticyclones (polar and arctic highs) are too shallow to appear on 500-mb weather maps.

Winds blowing across regional isotherms produce cold or warm air advection (Chapter 4). **Cold air advection** takes place where winds blow from colder localities toward warmer localities. **Warm air advection** occurs where winds blow in the opposite sense. Warm air advection causes the 500-mb surface to rise, whereas cold air advection lowers the 500-mb surface. Cold air advection between the surface and 500-mb thus deepens troughs and weakens ridges, whereas warm air advection strengthens ridges and weakens troughs. When cold air advects into troughs at the same time that warm air advects into ridges, the circulation becomes more meridional. On the other hand, if warm air advects into troughs while cold air advects into ridges, the circulation becomes more zonal. As described in Chapter 9, shifts in the upper-air circulation pattern between meridional and zonal flow have important implications for north/south air mass exchange and the development and track of cyclones.

NCEP's Hydrometeorological Prediction Center issues 500-mb maps twice each day based on upper-air observations at 0000 UTC and 1200 UTC. One set of maps covers North America and another the Northern Hemisphere. Although we have focused our discussion of upper-air weather maps on the 500-mb level, similar analyses are routinely constructed twice daily for the 850-, 700-, 300-, 250-, 200-, and 100-mb levels. The 300-, 250-, and 200-mb charts are useful in locating the polar front jet stream. The altitude of the jet stream depends on the mean temperature of the underlying air column and is lower in winter than in summer. Hence, for jet stream analysis, meteorologists rely upon the 300-mb chart during the coldest time of year and the 200-mb chart during the warmest time of year.

The deluge of real-time weather information (e.g., weather maps, satellite images, soundings, radar maps) has spurred development of computerized data management systems. An example is **McIDAS** *(Man-computer Interactive Data Access System),* developed by scientists at the Space Science and Engineering Center at

FIGURE A.6
Workstations in the National Weather Service AWIPS. [Photo by R.S. Weinbeck]

the University of Wisconsin-Madison. McIDAS ingests weather data and integrates and organizes those data into guidance products for potential users. Products include two- and three-dimensional composite images. Users access McIDAS on a computer terminal, select some satellite image or other data display, and may choose, for example, to overlay different fields of data.

McIDAS inspired the new computer workstations that have been operating at NWS Weather Forecast Offices since 2000. This **Advanced Weather Interactive Processing System (AWIPS)** receives and organizes ASOS data plus analysis and guidance products from the National Centers for Environmental Prediction and enables the meteorologist to display, process, and overlay images, graphics, and other data (Figure A.6).

Weather Prediction

Meteorologists at the Hydrometeorological Prediction Center analyze satellite observations and weather data plotted on surface and upper-air weather maps and computer model output. From these analyses, they prepare general short- to medium-range weather forecasts for the United States. They issue 12-, 24-, 36-, and 48-hr forecasts plus 3- to 7-day extended outlooks each day, and weather outlooks up to 10 days three times weekly. Regional NWS Weather Forecast Offices prepare detailed short-term forecasts for their areas of responsibility. In addition, the Climate Prediction Center generates monthly (30-day), seasonal (90-day), and multi-seasonal outlooks for the globe.

Weather forecasting is an extremely challenging endeavor, primarily because it involves many variables and a vast quantity of weather data. For these reasons, computerized numerical models of the atmosphere have been developed to assist forecasters.

NUMERICAL WEATHER FORECASTING

In the early 1950s, John Von Neuman and Jule Charney pioneered the use of electronic computers for weather forecasting at the Institute for Advanced Study at Princeton University. A primitive computer, less powerful than today's personal computer, successfully forecast the horizontal air pressure pattern at 5000 m (16,400 ft) altitude over North America 24 hrs in advance. By 1955, computers were routinely generating weather forecasts from surface and upper-air weather observations.

An electronic computer is programmed with a numerical model of the atmosphere, that is, a model consisting of computational versions of the mathematical equations that relate winds, temperature, pressure, and water vapor concentration. Beginning with present (real-time) weather data, numerical models predict the values of quantities on a uniform grid of points on pressure-surfaces for some future time, say a few minutes from now. With these predicted conditions as a new starting point, another forecast is then computed for the subsequent few minutes. The computer repeats this procedure again and again until weather maps are generated for the next 12, 24, 36, and 48 hrs (or longer). In this process, tens of millions of computations must be performed each second on a vast array of observational data; hence the need for a high-speed electronic computer (*supercomputer*) that can accommodate huge quantities of data. (In fact, meteorology pioneered the application of supercomputers.) Each day, NCEP's latest high performance supercomputer (IBM SP at the Bowie Computer Center, Bowie, MD) ingests over 3.9 million observations, and produces 100 gigabytes of information that is the basis of 174,314 guidance products issued daily by NCEP.

NCEP computers are programmed with several different numerical models of the Earth-atmosphere system. Currently, among the basic regional forecast models is the *Nested Grid Model (NGM)* which divides the troposphere into 16 vertical levels with data points at intervals of 80 km (50 mi) and issues forecasts twice a day, at 0000 UTC and 1200 UTC. Since its introduction

in 1984, the NGM has proved its value especially in predicting East Coast winter storms. The *Eta* model, a more recent NCEP numerical model, provides greater resolution than the NGM; it consists of 60 levels with 12 km (8 mi) between data points. The Eta model is run four times per day.

Some computerized numerical models of the atmosphere are designed to operate over different spatial scales depending on the forecast range. For medium-range forecasts (up to 10 days), observational data are fed into the computer from all over the globe, since within that forecast range a weather system may travel thousands of kilometers. On the other hand, for short-range forecasts (up to 3 days), the model utilizes data drawn from a more restricted region of the globe. Compared to a global model, a regional model offers the advantage of greater resolution of data over a smaller area of interest.

SPECIAL FORECAST CENTERS

Not all weather forecasts are prepared by NCEP's Hydrometeorological Prediction Center. Special forecast centers are responsible for predicting tropical cyclones (hurricanes and tropical storms), severe weather, river flow and flooding, and weather for mariners, and providing guidance information for domestic and international aviation. Also scientists at a special forecast center (*Space Environment Center* at Boulder, CO) predict the "weather" of Earth's near-space environment for satellite and space shuttle operations.

In 1873, the U.S. Army Signal Corps (then in charge of the national system of weather observation and forecasting) began gathering weather reports from Havana and Santiago, Cuba to help detect tropical cyclones in the Caribbean. A hurricane was plotted for the first time on a surface weather map on 28 September 1874 (centered offshore near Savannah, GA). In 1890, with the transfer of weather observation and forecasting from military to civilian hands (the U.S. Weather Bureau in the Department of Agriculture), primary emphasis was on weather over the continent and little attention was given to tropical cyclones. Although six land-falling hurricanes claimed more than 4000 U.S. lives in 1893, the primary impetus for expanding the number of weather stations in the Caribbean was the Spanish-American War of 1898 and the fear of what a hurricane near Cuba could do to the U.S. fleet. A U.S. Weather Bureau office was even established in Havana.

In subsequent decades, monitoring (and study) of tropical cyclones profited greatly from technological advances in both communications and remote sensing. Invention of the radio made possible ship-to-shore reports on the state of the atmosphere and seas near tropical cyclones; the first such report was filed in August 1909. By the late 1930s, the upper-air was regularly monitored providing insights on the relationship between tropical cyclones and steering winds aloft. Beginning in the late 1950s, weather radar at coastal stations observed rain bands and other structures of an approaching hurricane. Since the 1960s, weather satellites have proved valuable in tracking and monitoring changes in tropical cyclones as they progress through their life cycles. The value of satellites in data gathering is underscored by the fact that there is only one upper-air station for 10 million square km (4 million square mi) of the tropical and subtropical Atlantic. Furthermore, in recent decades, coastal and offshore buoys provide information on sea-surface conditions.

During the 1940s, responsibility for forecasting Atlantic hurricanes was split among U.S. Weather Bureau offices in New Orleans, Miami, Washington, DC, Boston, and San Juan. In 1967, the Miami Weather Bureau office was officially designated as the **National Hurricane Center (NHC)**. And in 1984, the NHC was separated from the Miami National Weather Service Weather Forecast Office. Today, responsibility for forecasting tropical cyclones is divided between the National Hurricane Center (NHC) on the campus of Florida International University in Miami, FL and the Central Pacific Hurricane Center in Honolulu, HI.

At the NHC, a branch of the *Tropical Prediction Center*, meteorologists watch for development of tropical cyclones over the Atlantic Basin, including the Caribbean and Gulf of Mexico, plus the Eastern Pacific. The Center operates the SLOSH model for prediction of storm surges, prepares and distributes hurricane watches and warnings for the public, conducts research on hurricane forecasting techniques, and sponsors public awareness programs. Local NWS Weather Forecast Offices transmit information from the NHC as advisories, warnings, or special weather statements to people in their areas of responsibility. The goal is to provide at least 12 hrs of daylight warning for coastal residents so that they can prepare for a hurricane and evacuate if necessary.

Specially instrumented reconnaissance aircraft complement satellite, radar, and buoy surveillance of tropical cyclones when they threaten any land in the Atlantic, Caribbean, Gulf of Mexico, Hawaii, or California. Aircraft fly directly through and over the storm and determine the location of the eye, measure wind speeds, and obtain soundings by deploying GPS (global positioning system) dropwindsondes. By extrapolating from air pressure readings at flight altitude (usually 1500 to 3000 m or 5000 to 10,000 ft), meteorologists also determine the sea-level air pressure at the storm center. Reconnaissance aircraft do a better job than satellite sensors in determining wind speed in the storm system and locating its eye. These data are immediately communicated to the NHC.

Reconnaissance flights into tropical cyclones began during World War II and originally involved aircraft from both the U.S. Navy and Army Air Forces (later the U.S. Air Force). Today, the 53d Weather Reconnaissance Squadron, a U.S. Air Force reserve unit, conducts most hurricane reconnaissance from their base at Keesler Air Force Base in Biloxi, MS. In addition, NOAA supports special aircraft designed for hurricane research and reconnaissance. Based at MacDill Air Force Base in Tampa, FL, these flights gather data for input into computer models at the National Centers for Environmental Prediction with products sent on to the NHC.

The principal challenge for NHC forecasters is prediction of the track of a hurricane. Normally, such forecasts are issued every 6 hrs and cover periods up to 72 hrs in the future. The basis for hurricane track forecasts is a blend of climatology (records of tracks of similar hurricanes in the past), special numerical models, and the experience of the forecaster. Although the advent of satellite monitoring significantly improved the accuracy of tropical cyclone track forecasts during the 1960s, skill in track forecasting has improved little since then. Beyond 36 hrs, accuracy declines rapidly and approaches zero for 72-hr forecasts. Apparently, the major problem is the poor quality of observational data initially fed into the numerical models. Also, models do not do well in predicting the change in intensity of tropical cyclones. On the other hand, if the hurricane track forecast is on target, the SLOSH model can accurately predict the location and height of the *storm surge* (a dome of water pushed ashore by storm winds) along coastal areas. By consulting local topographic maps, public safety officials use SLOSH predictions to identify areas likely to be inundated by floodwaters and order evacuation accordingly.

Since the 1983 hurricane season, the NHC has included a probability forecast as part of its advisory statements. Probability is defined as the percent chance that the center of a hurricane or tropical storm will pass within 105 km (65 mi) of any one of 46 designated Gulf

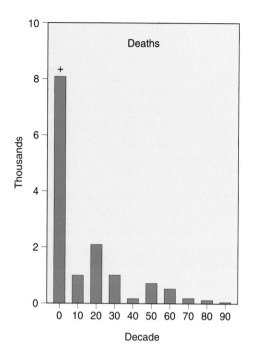

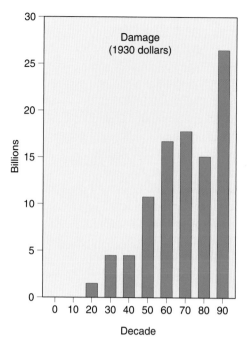

FIGURE A.7
During the 20th century, tropical cyclone fatalities in the United States trended downward, whereas property damage trended upward.

lull in hurricane activity coupled with more effective hurricane forecasting and response programs is responsible for an abrupt decline in hurricane related fatalities since the early part of the 20th century (Figure A.7). During the same period, however, the dollar cost of property damage has skyrocketed, presumably because of greater coastal development.

As of 1 October 1995, the *National Severe Storms Forecast Center* (*NSSFC*) in Kansas City, MO, became the *Aviation Weather Center* (*AWC*) and the *Storm Prediction Center* (*SPC*). The AWC remains in Kansas City where it shares facilities with the National Weather Service Training Center (Figure A.8). The SPC moved to Norman, OK in 1997 and is co-located with the *National Severe Storms Laboratory* (*NSSL*), a center for severe weather research.

Meteorologists at the Aviation Weather Center

and East Coast communities from Brownsville, TX to Eastport, ME. The first probability forecast is usually issued 72 hrs in advance of the storm's anticipated landfall. At that time, by convention, the probability of a hurricane landfall is set no higher than 10% for any community. Probabilities increase to 13% to 18% at 48 hrs, 20% to 25% at 36 hrs, 35% to 50% at 24 hrs, and 60% to 80% at 12 hrs, prior to the storm's expected landfall.

A *hurricane watch* is issued for a coastal area when winds of 119 km (74 mi) per hr or higher are possible within the next 36 hrs; a *hurricane warning* means that hurricane conditions are expected in 24 hrs or less. The uncertainty in hurricane track forecasts and the considerable threat posed by a hurricane requires forecasters to *over-warn* the public. That is, hurricane watches and warnings are issued for a stretch of coastline that is much longer than the diameter of the approaching hurricane. Over-warning is costly for people who prepare for a hurricane that never shows up. Millions of dollars are spent on preparations, such as boarding up windows and closing businesses, and additional millions of dollars are lost due to disruption of commerce and industry. And the cost of evacuation can be considerable. Nonetheless, these costs are outweighed by the benefits of being prepared just in case. The recent

FIGURE A.8
The National Weather Service Training Center co-located with the Aviation Weather Center at Kansas City, MO. [Photo by R.S. Weinbeck.]

issue warnings, forecasts, and analysis of weather conditions for aviation interests. These functions support the Federal Aviation Administration (FAA) Air Traffic Control responsibility to safely and efficiently manage the nation's airspace. AWC forecasters identify weather hazards to aircraft in flight and generate warnings that are made available to the aviation community immediately. In addition, forecasts are made for weather conditions that may impact domestic and international airspace over the next 24 hrs. Efficient use of aviation forecasts has proven economically beneficial to airlines in fuel savings. These forecasts complement airport terminal and route forecasts provided by local NWS Weather Forecast Offices.

Meteorologists at the Storm Prediction Center monitor atmospheric conditions for potential development of severe local storms, including winter blizzards, and wild fires. Several times daily, the SPC issues *severe thunderstorm outlooks* identifying areas expected to experience severe local storms over the next 1 to 3 days. *Outlooks* specify the type, coverage, and intensity of the severe weather expected. The SPC also issues watches for severe thunderstorms and tornadoes. Watches essentially fine-tune forecast areas identified in the *severe thunderstorm outlook*. A typical watch

encompasses about 65,000 square km (25,000 square mi) and is valid for 4 to 6 hrs.

The U.S. Army Signal Corps began the first river and flood forecast service in 1872. A disastrous flood on the Kansas River in 1903 prompted Congress to make the river and flood forecast service a special division of the U.S. Weather Bureau (now the National Weather Service). By 1913, the service had expanded to 483 river gauging stations (many operated by the U.S. Geological Survey). In 1946, the Weather Bureau established the nation's first River Forecast Centers (RFCs) at Cincinnati, OH and Kansas City, MO. Currently, 13 RFCs serve the nation with the primary goal of minimizing the loss of life and property damage caused by floods by providing the public with timely river and flood forecasts (Figure A.9).

At RFCs, staff hydrologists and hydrometeorologists develop river, reservoir, and flood forecasts using special numerical models that simulate the response of river and stream flow to rainfall and snowmelt. The objective is to predict the volume of water that runs off the land surface into a river or stream, the discharge at specific locations along major rivers or small streams near urban areas, and variations in river discharge and stage (height of the river) as the water

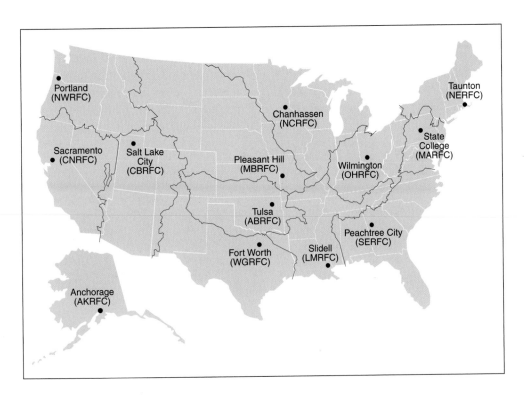

FIGURE A.9
Locations of the River Forecast Centers (RFCs) and their areas of responsibility.

volume moves downstream. In late winter and early spring, RFC staff prepares spring snowmelt outlooks to identify the potential for spring flooding, assuming a normal rate of snowmelt. Specialized services include low flow forecasts during dry periods, ice advisories for navigable rivers in winter, and reservoir inflow forecasts. Forecasts are forwarded to personnel at local NWS Weather Forecast Offices who in turn issue forecasts and outlooks for water resource managers, emergency management officials, and the public. Flood forecasts specify the expected height of the flood crest, the date and time when the river is expected to flood, and the date and time when the river is expected to return to its banks.

NCEP's *Marine Prediction Center* issues forecasts, warnings, and other guidance information for marine interests in the Northern Hemisphere. Predictions cover a period up to 5 days.

FORECAST SKILL

Just how accurate are today's computer-guided weather forecasts? This question can be answered by comparing the skill of modern weather forecasting with that of predictions based on *persistence* (forecasting no change in present weather) or *climatology* (forecasts derived from past weather records). On this basis, weather forecasting skill declines rapidly for periods longer than 48 hrs and is minimal beyond 10 days.

Several factors contribute to the drop in forecasting skill. Computerized forecasts are only as accurate as the input (observational) data and predictive equations (numerical models) allow them to be. Errors are introduced by (1) missing or inaccurate observational data, (2) failure of weather stations to detect all mesoscale and microscale circulation systems, and (3) imprecise equations in numerical models that include assumptions and first approximations. Unfortunately, the adverse effects of errors in initial conditions grow as the forecast period lengthens. In today's numerical models of the atmosphere, the impact of even small errors doubles about every 2 to 2.5 days over the forecast period. When computerized numerical models are run out to 10 days, some elements of daily forecasts are useful for some localities to about 6 or 7 days. Beyond that, *chaos*, the term for sensitivity to initial conditions, sets in.

The accuracy of numerical weather forecasting, particularly in the 1- to 5-day range, has shown slow but steady improvement over the past few decades, and prospects for continued progress are good. This optimistic outlook stems from: (1) improved understanding of the atmospheric processes that makes

possible the development of more realistic numerical models of the Earth-atmosphere system, (2) larger and faster computers, (3) more reliable and sophisticated observational tools, including Doppler radar and remote sensing by satellite, and (4) denser weather observational networks worldwide.

Today, meteorologists routinely employ two techniques to optimize the accuracy of forecasts based on numerical models: (1) *ensemble forecasting* and (2) *model comparison*. In the first, a numerical model generates several forecasts, each based on slightly different sets of initial conditions. If the forecasts are consistent, then the forecast is considered reliable. In the second technique, a comparison is made among forecasts generated by several different models. If they all agree, then the forecast is issued with relatively high confidence. Using either technique, if forecasts are inconsistent, any one forecast is considered unreliable.

Although computerized numerical models have revolutionized weather forecasting, computers are never likely to replace human weather forecasters (Figure A.10). The products of computerized numerical models function as guidance materials for the forecaster. He or she must understand the basic atmospheric processes, and the characteristic errors and biases of the various numerical models and weather-observing systems. As much as numerical weather prediction has advanced in recent years, the best forecasters still rely heavily on personal experience and intuition, tempered by their knowledge of how the atmosphere works. A good forecaster does not start with the model output in preparing a forecast, but rather he or she begins with previous and current observations, forming a view of

FIGURE A.10
A computer is never likely to replace a meteorologist.

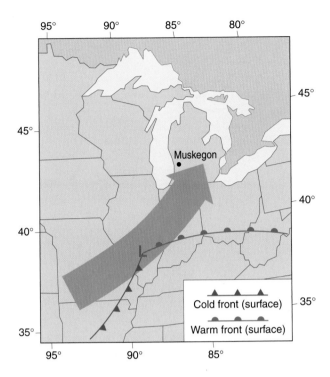

FIGURE A.11
An early winter cyclone tracks northeastward through the Midwest, causing heavy snowfall at Muskegon, MI.

what the atmosphere has been doing and ideas on what is likely to happen in the future. Furthermore, special local or regional conditions may call for significant modification of a computer-generated weather forecast. Forecasters must analyze and interpret computerized predictions and adapt those forecasts as necessary to regional and local circumstances. Consider an example.

Suppose that an intense early winter cyclone tracks northeastward through the Midwest, across central Illinois, and into eastern Lower Michigan, as shown in Figure A.11. Muskegon, MI is on the cold, snowy side of the storm's path, so residents experience snow and strong northeast winds. After the storm passes, winds shift to the north and northwest, and cold air advection begins. A computerized numerical model might predict clearing skies for Muskegon, but local conditions dictate a different result. Strong winds from the northwest advect cold, dry air across the relatively warm waters of Lake Michigan, giving rise to localized lake-effect snows on the lake's downwind eastern shore (Chapter 8). Because of this local effect, northwest winds may bring more snow to Muskegon after the storm has passed than northeast winds did when the storm was nearby.

LONG-RANGE FORECASTING

Meteorologists at NCEP's *Climate Prediction Center* in Camp Springs, MD prepare 30-day (monthly), 90-day (seasonal), and multi-seasonal generalized weather *outlooks* that identify areas of expected positive and negative **anomalies**, departures from long-term averages, in temperature and precipitation (Figure A.12).

Forecasting the prevailing circulation pattern at the 700-mb level is the first step in forecasting monthly temperature and precipitation anomalies at the surface. The present circulation pattern is extrapolated into the future, although an effort is also made, based on historical data, to identify features of the present pattern that are most likely to persist. The prevailing westerly flow at the 700-mb level permits identification of areas of strong warm and cold air advection as well as principal storm tracks. Predictions of surface temperature and precipitation anomalies are derived from these data.

A somewhat different approach is utilized for 90-day outlooks. Forecasters rely more on long-term trends and recurring events, attempting to isolate persistent circulation features from prior months and seasons. In the *analog technique*, for example, a computer searches its 40-year archive (memory) of past seasonal weather patterns for the closest match with the present season's weather pattern. The 90-day outlook is then based on whatever follows the best historical match.

In January 1995, meteorologists at the Climate Prediction Center began issuing 15-month (multi-seasonal) outlooks for regions of expected positive and negative anomalies in temperature and precipitation. Each month 13 forecasts are issued, each one covering a 3-month period. The first forecast covers the three months beginning two weeks from the date of the forecast's release. Each subsequent 3-month forecast overlaps the previous one by two months.

A promising area of research in long-range weather forecasting relies on identification of teleconnections. A **teleconnection** is a linkage between changes in atmospheric circulation occurring in widely separated regions of the globe, often many thousands of kilometers apart. For example, Jerome Namias and his colleagues at the Scripps Oceanographic Institution argued that a coupling between ocean and atmosphere triggered the 1988 Midwestern drought. Anomalously cold surface water in the central and eastern tropical Pacific (La Niña) plus unusually warm surface water to the north (near Hawaii) interacted with the atmosphere to produce a long-wave pattern that was more meridional than usual. As described in Chapter 9, this circulation

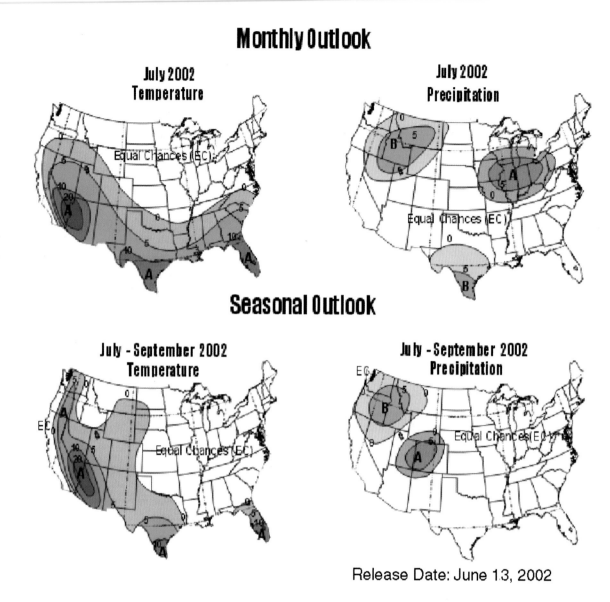

FIGURE A.12
A sample long-range monthly and seasonal outlook from the Climate Prediction Center.

pattern featured a warm anticyclone that persisted over the central contiguous United States through much of the spring and summer of 1988.

Identification of teleconnections associated with El Niño and La Niña is the primary basis for extension of long-range forecasting to multi-seasons. As described in detail in Chapter 9, meteorologists report success in predicting the onset of El Niño a year or so in advance and have identified linkages between El Niño and weather anomalies in midlatitudes—especially during winter.

Statistics on seasonal forecast skill can be a bit confusing. For more than two decades the National Weather Service has reported average skills of only about 8% for temperature outlooks and 4% for precipitation outlooks. To put these statistics in perspective, forecast skill would be 0% if based on chance alone and 100% for a perfect forecast. The greatest success so far has been with winter temperature outlooks, with a forecast skill of 16% to 18%.

SINGLE-STATION FORECASTING

Even without access to an official NWS forecast, short-term weather forecasts based on weather observations at one location, known as **single-station forecasts**, may be derived from the principles of weather behavior examined in this course. Because such forecasts are based on rules applied at only one location, they tend to be generalized and tentative, and complications often crop up as local conditions are modified by changes elsewhere. Table A.1 is a sample list of rules of thumb applicable to midlatitude weather. You may wish to add to this list. Other approaches base forecasts on fair-weather bias, persistence, or climatology.

Analysis of records of past weather events may aid single-station weather forecasting. Such records reveal a **fair-weather bias**. That is, fair-weather days outnumber stormy days almost everywhere. In fact, if we boldly predict that all days will be fair, we probably will be correct more than half the time. The only merit of this exercise is to establish a baseline for evaluating the skill of more sophisticated weather forecasting techniques. That is, we would expect traditional forecasting methods to score higher than forecasting methods based solely on fair-weather bias.

Another characteristic behavior of weather is **persistence**, that is, the tendency for weather episodes to persist for some period of time. For example, if the weather has been cold and stormy for several days, the weather may well continue that way for many more days. Weather records show, however, that in midlatitudes one weather type typically gives way to another weather type very abruptly, usually in a day or less. Weather forecasts based on persistence alone are therefore prone to error.

A third approach is based on **climatology**; that is, we prepare a weather forecast for a particular day based on the type of weather that occurred on the same day in years past. Suppose, for example, that the climate record of your area indicates that it has rained on 7 August only 12 times in the past 100 years. Accordingly, you predict that the probability of rain next 7 August is only 12%, and you confidently plan a picnic for that day. The problem with the climatology approach is that, statistics aside, there is no guarantee that it will not rain next 7 August.

PRIVATE SECTOR FORECASTING

Our description has focused mainly on the role of the federal government in weather forecasting. In addition, many private sector meteorologists analyze weather maps and other guidance materials supplied by the National Centers for Environmental Prediction. Many television and radio stations and some newspapers employ their own meteorologists or contract with private forecast services. Some private meteorologists tailor their forecasts to the specific needs of their commercial, agricultural, or industrial clients. For example, a private weather forecaster retained by an appliance store chain might alert store executives to a pending heat wave so that stores might be stocked with an adequate supply of fans and air conditioners. Another private forecaster might advise an electric power utility of expected summer temperatures so the energy supplier can better anticipate customer demand for air conditioning. A private forecaster might provide frost and freeze warning service for orchards or cranberry bogs. In this way, private sector forecasters supplement the efforts of government weather forecasters.

TABLE A.1
Some Rules of Thumb for Single-Station Weather Forecasting

- At night, air temperatures are lower if the sky is clear than if the sky is cloud-covered.
- Clear skies, light winds, and snow-covered ground favor extreme nocturnal radiational cooling and very low air temperatures by sunrise.
- Steadily falling air pressure often signals the approach of stormy weather whereas rising air pressure suggests that continued fair weather or improving weather is in the offing.
- Appearance of cirrus, cirrostratus, and altostratus clouds, in that order, indicates overrunning ahead of a warm front and the likelihood of precipitation.
- A counterclockwise wind shift from northeast to north to northwest (called *backing*) is usually accompanied by clearing skies and cold air advection.
- A clockwise wind shift from east to southeast to south (called *veering*) is often accompanied by clearing skies and warm air advection.
- A wind shift from northwest to west to southwest is usually accompanied by warm air advection and increasing clouds.
- If radiation fog lifts by late morning, a fair afternoon is likely.
- With west to northwest winds, a steady or rising barometer, and scattered cumulus clouds, fair weather is likely to persist.
- Towering cumulus clouds by mid-morning forewarn of afternoon thunderstorms.

Communication and Dissemination

Weather maps, charts, and forecasts issued by the National Centers for Environmental Prediction are transmitted to local NWS Weather Forecast Offices to guide them in preparing specific forecasts for their areas of responsibility. Weather information is then distributed to public safety officials and the general public via a variety of communications systems.

When hazardous weather threatens, the National Weather Service issues outlooks, watches, warnings, and advisories. An *outlook* is provided for people who require considerable advance notice so that they might adequately prepare for a specific weather event. For example, the outlook for flooding from spring snowmelt in the Upper Midwest may be issued many weeks in advance.

When hazardous weather is occurring or appears possible or probable, the National Weather Service issues watches, warnings, and advisories. These statements cover severe local storms, winter storms, floods, hurricanes, and non-precipitation hazards. A **weather watch** is indicated when hazardous weather is possible based on current or anticipated atmospheric conditions. The actual occurrence or timing of hazardous weather can be uncertain. People in the designated area need not interrupt their normal activities except to remain alert for threatening weather and to keep the television or radio on for further information. A **weather warning** is issued when hazardous weather is taking place somewhere in the region or is imminent. People are advised to take all necessary safety precautions. *Advisories* refer to anticipated weather hazards that are less serious than those covered by a warning. Examples include a winter weather advisory, a heavy surf advisory, or a heat advisory.

The local NWS Weather Forecast Office issues a *tornado warning* only after detection of a thunderstorm that is known or likely to include a tornado. Indication of a tornadic circulation is based on Doppler radar and/or reports of funnel clouds or cloud-base rotation by specially trained storm spotters. A warning covers a much smaller area than a watch, usually a county or portion thereof, and specifies the location of the tornado, its anticipated path, and the time when the tornado is expected in the warning area. Tornado warnings usually are valid for 30 to 60 minutes.

Winter storm warnings may specify heavy snow, blizzard conditions, or an ice storm. Usually, a *heavy snow warning* is issued if snowfall is expected to total at least 10 to 15 cm (4 to 6 in.) in less than 12 hrs, or at least 15 to 20 cm (6 to 8 in.) in less than 24 hrs. A **blizzard warning** means that falling or blowing snow is expected to be accompanied by sustained winds (or frequent gusts) of 56 km (35 mi) per hr or higher, reducing visibility to less than 400 m (1300 ft). Conditions are expected to persist for at least 3 hrs. A *severe blizzard* produces winds stronger than about 73 km (45 mi) per hr, visibility near zero, air temperature below -12 °C (10 °F), and dangerously low wind chill equivalent temperatures. Ice storm warnings mean that potentially dangerous accumulations of freezing rain or sleet are expected on the ground and other exposed surfaces.

To alert the public to the flash flood hazard, NWS Weather Forecast Offices issue watches, warnings, and advisories. A *flash flood watch* indicates that flash flooding is possible within or close to the designated watch area based on current or anticipated weather conditions. Residents of the watch area are advised to prepare to take action in the event that a flash flood warning is issued or flooding is observed. A *flash flood warning* is issued when a dangerously rapid rise in river level is imminent in the warning area or is currently occurring due to heavy rain (or dam failure). The public is advised to take appropriate action immediately. An *urban and small stream flood advisory* alerts the public to potential flooding that may cause inconvenience but is not life threatening for those living in the affected area. These advisories are issued when heavy rain is anticipated, which could flood streets and other low-lying urban areas, or if small streams are expected to reach bank full.

Because weather is changeable, weather observations and guidance information such as weather maps, forecasts, watches, and warnings must be communicated as rapidly as possible both nationally and internationally. For this reason, the World Meteorological Organization and weather services of its member nations maintain elaborate communications networks consisting of a variety of systems. Weather information is relayed by satellite, teletypewriter, radio, Internet, and facsimile systems (which reproduce maps, charts, and satellite images). Modernization of National Weather Service facilities is making the weather communications network more efficient and reliable. Earlier in this chapter, we described the Automated Surface Observing System (ASOS) and the Advanced Weather Interactive Processing System (AWIPS).

The public receives regular weather reports and forecasts via commercial and public radio, the NOAA weather radio, commercial television, cable-TV (*The Weather Channel* and others), the Internet, and newspapers (Chapter 1). In many cities, weather information is also available via recorded telephone announcement systems. In addition, during hurricane season, the National Hurricane Center (in cooperation with NOAA, NBC News, and *USA Today)* operates a 24-hr telephone hotline. A recorded message provides the latest information on an approaching hurricane or tropical storm.

Conclusions

Weather forecasting is a complex and challenging science that depends on the efficient interplay of weather observation, data analysis by meteorologists and computers, and rapid communication systems. Meteorologists have achieved a very respectable level of skill for short-range weather forecasting. Further improvement is expected with denser surface and upper-air observational networks, more precise numerical models of the atmosphere, larger and faster computers, and more sophisticated techniques of remote sensing by satellite. If these advances are to be realized, however, continued international cooperation is essential, for the atmosphere is a continuous fluid that knows no political boundaries.

Basic Understandings

- The fluid nature of the atmosphere means that international cooperation is required in gathering and interpreting surface and upper-air weather data. To this end, the World Meteorological Organization (WMO) coordinates an international effort of weather observation, analysis, and forecasting.
- Weather forecasting entails acquisition of present weather data, graphical depiction of the state of the atmosphere on weather maps and charts, computer-aided analysis of data and maps, prediction of the future state of the atmosphere, and dissemination of weather information and predictions to the public.
- Surface weather is monitored at land stations and by automated weather stations and ships at sea. Rawinsondes (profiling temperature, pressure,

humidity, and wind), radar, aircraft, and satellites monitor the atmosphere above the Earth's surface. Modernization of the National Weather Service during the 1990s upgraded NWS Weather Forecast Offices, and introduced the Automated Surface Observing System (ASOS) and Doppler radar.

- Using standard station models, surface weather data are plotted on constant altitude (sea-level) maps whereas upper-air weather observations are plotted on maps of constant pressure (isobaric) surfaces (e.g., 500-mb map). The altitude of a specific pressure surface varies from one place to another primarily because of differences in mean temperature of the air below the pressure surface. For example, the 500-mb surface is at a lower altitude where the air below is relatively cold and at a higher altitude where the air below is relatively warm.
- The NWS Advanced Weather Interactive Processing System (AWIPS) consists of computer workstations that ingest and organize ASOS, satellite, and radar data plus analysis and guidance products from the National Centers for Environmental Prediction (NCEP). AWIPS enables meteorologists to display, process, and overlay images, graphics, and other data.
- With numerical weather forecasting, an electronic computer is programmed with a mathematical model of the Earth-atmosphere system that relates winds, temperature, pressure, and humidity. Current weather observational data are used to initialize the model and the model then predicts a future state of the atmosphere. With the predicted conditions serving as a new starting point, another forecast is computed for a subsequent period of time. Repeated iterations yield weather forecasts for the next 12, 24, 36, and 48 hrs or longer.
- Special forecast centers are responsible for predicting tropical cyclones (hurricanes and tropical storms), severe weather (e.g., tornadoes), weather guidance for domestic and international aviation, river flow and flooding, and weather for mariners.
- Although the skill of short- and medium-range weather forecasting has improved steadily in recent decades, forecasting skill declines rapidly for periods longer than 48 hrs and is minimal for periods beyond 10 days. Errors are introduced by missing or inaccurate observational data, failure of weather stations to detect all microscale and mesoscale weather systems, and imprecise equations in

numerical models that include assumptions and first approximations. These errors in initial state grow with the forecast period.

- NCEP's Climate Prediction Center prepares 30-day, 90-day, and multi-seasonal generalized weather outlooks that identify areas of expected positive and negative anomalies in temperature and precipitation. Teleconnections, linkages between changes in atmospheric circulation occurring in widely separated regions of the globe, are particularly useful in long-range weather forecasting. Examples are anomalies in midlatitude winter weather during El Niño and La Niña.

- Single-station weather forecasting is based upon principles of meteorology, fair-weather bias, climatology, or persistence of weather episodes.

CHAPTER B

ATMOSPHERIC OPTICS

- Halo
- Rainbow
- Corona
- Glory
- Mirage
- Blue Sky and Red Sun
- Twilight
- Conclusions
- Basic Understandings

A rainbow in the morning,
Is the sailor's warning.
A rainbow at night
Is the sailor's delight.

Weather proverb.

As the sun's rays travel through the atmosphere, they are reflected or refracted by cloud droplets or ice crystals, or by raindrops. The consequence is a variety of optical phenomena, including halos, rainbows, coronae, and glories. In these cases, refraction (bending of light) occurs when solar radiation travels from one transparent medium into another (i.e., from air into water or ice). Solar rays are also refracted as they pass through the atmosphere because of variations in the density of air. This type of refraction gives rise to mirages. In addition, scattering of sunlight by particles within the atmosphere is responsible for the blue of the daytime sky, the white of clouds, red horizon at sunset, and twilight. The characteristics and origins of these and other atmospheric optical effects are subjects of this chapter.

Halo

A **halo** is a whitish (sometimes slightly colored) ring of light surrounding the sun or the moon (Figure B.1). It forms when the tiny ice crystals that compose high, thin clouds such as cirrus or cirrostratus refract the sun's rays. **Refraction** is the bending of light as it passes from one transparent medium (such as air) into another transparent medium (such as ice or water). The light rays bend because the speed of light is greater in air than in ice or

water. Refraction occurs whenever light rays strike the interface between two different transparent media at an angle other than 90 degrees (Figure B.2). Beams of light that strike a water or ice surface at 90 degrees are not refracted.

An analogy helps explain refraction. Suppose you are driving your auto down a highway and suddenly encounter a patch of ice along the right side of the road. The right wheels travel over the ice, while the left wheels remain on dry pavement. You slam on the brakes. The

FIGURE B.1
This halo about the sun is caused by refraction of sunlight by tiny ice crystals composing high thin clouds. [Photo by E.J. Hopkins.]

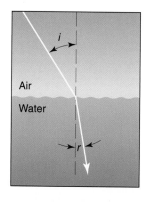

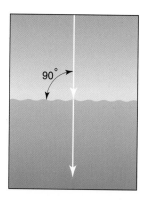

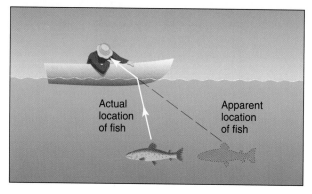

FIGURE B.2

Light rays may be refracted (bent) as they travel from one transparent medium into another as in (A) where a light ray is refracted as it travels from air to water. The speed of light is less in water than in air, so the light ray is bent toward a line drawn perpendicular to the water surface, and angle *r* is less than angle *i*. (B) Light rays that enter the water at a 90 degree angle are not refracted. (C) Refracted light can be deceptive. The fish appears to be farther away from the boat than it really is because of refraction and because human perception assumes that light always travels in a straight line.

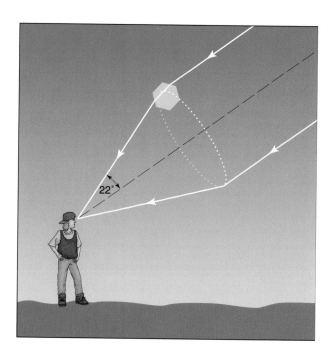

FIGURE B.3

A ray of sunlight is refracted as it passes through an hexagonal ice crystal from one side (top side in the drawing) to another side (bottom side in the drawing). The angle between the two sides of the ice crystal is 60 degrees. This type of refraction produces a 22-degree halo centered on the sun or moon.

left side of the car slows while the right side of the car slips on the ice; hence, the auto swerves to the left. The swerving of the auto is analogous to a light ray bending toward the medium in which light travels more slowly.

In clouds, ice crystals occur as hexagonal (six-sided) plates, columns, or dendrites (stars). Most halos are formed when light is refracted upon entering and exiting either plates or columns. A light ray entering an ice crystal through one rectangular side and exiting through another rectangular side is refracted by at least 22 degrees (Figure B.3). If ice crystals are relatively small and randomly oriented, most rays are refracted at 22 degrees so that light is focused in a circle having a radius of about 22 degrees. Suppose, for example, that you see a halo around the moon. Visualize two lines: one joining you with the moon's center, and the other line joining you with any point on the halo. The angle

between the two imaginary lines is 22 degrees. For reference, the halo's radius appears to be the same as the width of this book when held at arm's length.

A less common halo has a radius of about 46 degrees about the sun (or moon). In this case, columnar ice crystals with diameters in the range of 15 to 25 micrometers refract the light. Light rays travel through an ice crystal from one rectangular side to the top or from side to base rather than from side to side (Figure B.4).

The Zuni tribe of New Mexico has a weather proverb that considers a halo to be a harbinger of stormy weather: "*When the sun is in his house, it will rain soon.*" The phrase "*in his house*" refers to a halo surrounding the sun. Cirrus and cirrostratus clouds may signal the beginning of overrunning as warm air is advected aloft by the circulation associated with an approaching low pressure system (cyclone). However, the appearance of a halo around the sun (or moon) is no guarantee that precipitation will follow. Cirrus or cirrostratus clouds may spread as much as 1000 km (620 mi) ahead of the storm center, and the storm could change direction or even weaken and die out prior to reaching your location.

Sometimes light is concentrated on either side of

FIGURE B.4
In a less common situation than shown in Figure B.3, a ray of sunlight is refracted from side to top or base, or vice versa, as it passes through a columnar ice crystal. The angle between the side and top or base is 90 degrees. This type of refraction produces a 46-degree halo centered on the sun or moon.

the sun as two brilliant spots, known as **sundogs** (because they appear to follow the sun around the sky), also called *parhelia* (from the Greek, meaning "beside the sun") (Figure B.5). Cirrus clouds composed of relatively large

plate-like ice crystals are responsible for sundogs. Because of air resistance, the top and bottom surfaces of these crystals remain nearly horizontal as they fall through the atmosphere. The rectangular sides of the ice crystals are vertically oriented so that light is refracted either to the right or left so that sundogs always appear at the same altitude as the sun. Side-to-side refraction concentrates light in two spots about 22 degrees on either side of the sun. In some cases, sundogs are part of a 22-degree halo.

An ice crystal acts like a glass prism and disperses sunlight into its component colors. The more energetic short-wavelength end of the visible spectrum (violet light) is refracted the most, so that the violet portion of sundogs is farthest from the sun. The less energetic long-wavelength end of the visible spectrum (red) is refracted the least, so the red portion of sundogs is closest to the sun. Because of the optics involved, halos and sundogs are brightest at low solar altitudes and when ice clouds are relatively thin.

Rainbow

A **rainbow** is a circular arc of concentric colored bands, caused by a combination of refraction and reflection of sunlight (or rarely of moonlight) by raindrops (Figure B.6). Sunlight striking a shaft of falling raindrops is refracted twice and internally reflected by each drop of rain. As

FIGURE B.5
Bright colored spots appearing on either side of the sun and at the same elevation, called parhelia or sundogs, are caused by refraction of sunlight by plate-like ice crystals falling through the atmosphere.

FIGURE B.6
A rainbow is a nearly circular arc of color caused by refraction and internal reflection of sunlight by falling raindrops. Refraction disperses visible light into its component colors.

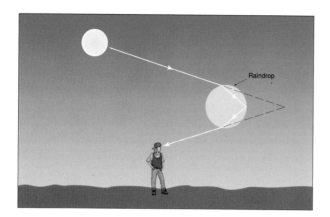

FIGURE B.7
With a primary rainbow, a solar ray is refracted upon entering a raindrop, reflected internally, and refracted again upon exiting the raindrop.

shown in Figure B.7, a solar ray is refracted as it enters a raindrop; then the ray is reflected by the inside back of the drop before being refracted again as it exits the drop.

Because of reflection, a rainbow appears to an observer who has his or her back to the sun and is facing a distant rain shower (Figure B.8). A rainbow never forms when the sky is completely cloud covered; the sun must be shining. You can create your own rainbow on a sunny day by directing the spray from a garden hose so that you observe the spray with the sun at your back. Midlatitude weather systems usually progress from west to east so that appearance of a rainbow to the east in the evening usually signals improving weather. Rain showers to the east are moving away, and clearing skies are approaching from the west, where the sun is setting. Conversely, a morning rainbow to the west signal approaching rain.

Like a glass prism, raindrop refraction disperses sunlight into its component colors forming the concentric bands of color of a *primary rainbow*. From outer to innermost band, the colors are red, orange, yellow, green, blue, and violet. In many cases, a much dimmer, *secondary rainbow* appears about 8 degrees above the

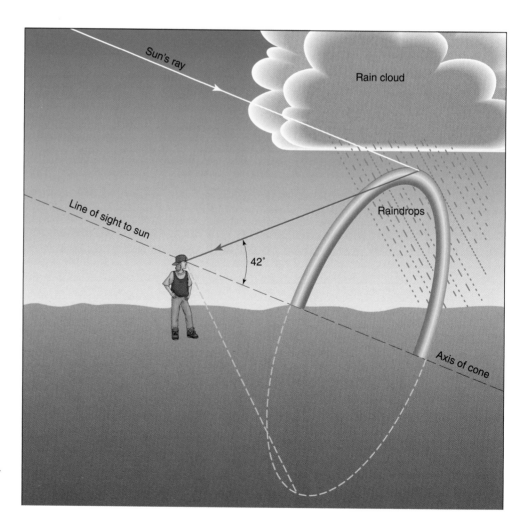

FIGURE B.8
A rainbow appears to an observer who has his or her back to the sun and faces a distant rain shower.

FIGURE B.9
Refraction of solar rays by raindrops plus double reflection within raindrops produces a dimmer secondary rainbow just above the primary rainbow. A secondary rainbow is visible in Figure B.6.

FIGURE B.10
Iridescent clouds exhibit brilliant spots or borders of colors, usually red and green. [Photo by E.J. Hopkins.]

primary rainbow (visible in Figure B.6).   Double reflection within raindrops produces the secondary rainbow, with the order of colors reverse that of the primary rainbow (Figure B.9).

Because ice crystals also refract sunlight, you may wonder why halos are only weakly colored at best. In fact, ice crystals disperse light into its component colors, but because the size and shape of ice crystals vary more than the size and shape of raindrops, the colors produced by an assemblage of ice crystals tend to overlap one another rather than form discrete bands. In halos, colors therefore wash out, although occasionally a reddish tinge is visible on the inside of a halo (as in Figure B.1)

Corona

At times, a series of alternating light and dark rings surround the moon or, less often, the sun. This optical phenomenon is known as a **corona**. Typically, a corona is only a few degrees in radius and is far smaller than a halo. It is caused by diffraction of light around water droplets that compose a thin, translucent veil of altocumulus, altocumulus lenticularis, or stratocumulus clouds. Often only one ring is actually observed.

Diffraction is the slight bending of a light wave as it moves along the boundary of an object such as a water droplet. As light waves bend, they interfere with one another. Where the crests of one light wave coincide with the crests of another wave, interference is constructive and a larger wave results. On the other hand, where the crests of one wave coincide with the troughs of another wave, the interference is destructive, and the waves cancel each other. Where light waves interfere constructively, we see a ring of bright light, and where they interfere destructively, we perceive darkness. If cloud droplets are of uniform size, then a corona is colored with blue-violet on the inside and red on the outside of each ring. Dependency of diffraction on wavelength causes this separation of color: the longer wavelength red light is diffracted more than the shorter wavelength blue-violet light.

Diffraction is also responsible for **iridescent clouds**.   These are clouds (usually altocumulus, cirrostratus, or cirrocumulus) having bright spots, bands, or borders of color, usually red and green (Figure B.10). Iridescent clouds typically appear up to about 30 degrees from the sun.

Glory

Prior to the age of aircraft travel, about the only way a person could view a glory was from the top of a lofty mountain peak. To see a glory, an observer must be in bright sunshine above a cloud or fog layer, and the sun must be situated so as to cast the observer's shadow on the clouds below. The observer then sees a **glory** as concentric rings of color centered about the shadow of his or her head. Although less distinct, the colors of a glory are the same as in a primary rainbow, with the innermost band being violet and the outermost band being red. Today, aircraft pilots and observant passengers often see a glory around the shadow of their aircraft on a cloud deck below.

A glory depends on much the same optics as a primary rainbow with two important differences, that is, the size of the reflecting and refracting particles, and the direction of reflected and refracted light. Whereas rainbows occur when sunlight strikes a mass of falling raindrops, glories are the consequence of sunlight interacting with a mass of much smaller suspended water droplets of uniform size (radii less than about 25 micrometers) that compose a cloud. In both cases, the sun's rays undergo refraction upon entering the droplet, followed by a single internal reflection, and another refraction upon exiting the droplet. In a glory, sunlight is refracted and reflected *directly* back toward the sun due to the small droplet size. A spherical cloud droplet diffracts light rays ever so slightly toward the droplet, so that light rays incident on a cloud droplet parallel those returning from the cloud droplet (Figure B.11).

The special optics of a glory explains why it appears about the shadow of the observer who is situated in the direct path of both the incident solar rays and the returning (refracted and reflected) solar rays. This also explains an observation that must have fascinated ancient mountain mystics. Suppose that you and a friend are standing side by side high on a mountain slope at a site favorable for viewing a glory (cloud deck or fog bank below and the sun behind your shoulder). On the cloud deck below, you see your own shadow next to that of your friend, but a glory appears about your head and not about your friend's head. Lest you presume that you have been singled out, note that your friend has the opposite observation, for your friend sees a glory only about his or her head. The fact is that each observer is in a position to view only one glory.

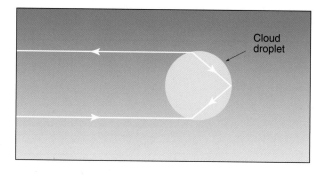

FIGURE B.11
In the special optics that produce a glory, both the incident and returning solar rays are diffracted slightly toward the surface of the cloud droplet. Consequently, the incident and returning rays follow parallel paths.

Mirage

Appearances can be deceiving, especially in the case of mirages. A **mirage** is an optical phenomenon in which an image of some distant object appears to be displaced from its normal view. Distant buildings or hills can appear higher or lower than they really are. A nonexistent pool of water may suddenly appear on the highway ahead, or a sailboat viewed from shore may appear upside down. Mirages are caused by the refraction of light rays within the lower atmosphere.

Light travels at different speeds through different transparent substances. Light changes speed, for example, as it travels from air into an ice crystal. For reasons discussed earlier in this chapter, a light ray is refracted at the interface between two different transparent media. The speed of light also varies within a single medium if the density of that medium is not uniform. Hence, light rays bend as they pass through a substance of varying density, such as the atmosphere.

If the density of the atmosphere were the same throughout, then light rays would always travel along straight paths at constant speed, and no refraction would occur. As described in Chapter 5, however, air density varies with changes in temperature and pressure. Because our concern here is with optical phenomena that occur within the lower troposphere and involve relatively short viewing distances, we need not be concerned with the influence of horizontal air pressure gradients on air density. For the same reason, we also ignore the effects of horizontal temperature gradients on air density. But we cannot ignore the effect of vertical temperature profiles on the change of air density with altitude.

As a rule, light rays traveling through the atmosphere are refracted such that denser air is on the inside (concave side) of the bend, and less dense air is on the outside (convex side) of the bend. Air density almost always decreases with altitude so that light reflected from a distant object follows a downward curved path to the observer. Because human perception is based on the assumption that light travels in a straight path, to the observer the object *appears* to be higher than it really is.

Because the atmosphere refracts sunlight, the image of the setting or rising sun that we see is slightly higher in the sky than it would be without an atmosphere. One implication of this has to do with the length of daylight on the equinox. As noted in Chapter 3, on the equinox, day and night are assumed to be of equal length

(12 hrs) everywhere (except right at the poles). Actually, days are slightly longer than nights on the equinox. At Washington, DC, for example, day length on the autumnal equinox of 23 September 2002 is 12 hrs and 6 minutes. The length of daylight is defined to be the period between sunrise and sunset, and sunrise and sunset occur when the upper edge of the solar disc is just visible on the unobstructed level horizon. Because the sun appears higher than it actually is at both sunrise and sunset, there is slightly more than 12 hours between sunrise and sunset on the equinox.

If the air temperature decreases with altitude at less than the usual rate, or if there is a temperature inversion (an increase of temperature with altitude), air density decreases with altitude faster than normal. Hence, light rays reflected from a distant object bend more sharply than usual before reaching the viewer, and objects appear higher that normal. This is called a *superior mirage*. On the other hand, if the lowest air layer features a greater than usual temperature lapse rate, rays are refracted less than normal and objects appear lower than we usually see them. This is called an *inferior mirage*.

These are a few examples of the many possible types of mirages. As the vertical air temperature profile becomes more complex, so too do the types of mirages that can appear. For example, the familiar oasis mirage in a desert is an inverted image of the sky seen below the horizon. All mirages, then, are displacements or distortions of something real.

Blue Sky and Red Sun

Visible light is composed of the entire spectrum of colors: red, orange, yellow, green, blue, and violet. This can be demonstrated by passing a beam of sunlight through a glass prism and onto a screen. A spectrum of colors appears because the glass refracts the light beam. The more energetic violet wavelengths of visible light are bent the most, and the less energetic red wavelengths are bent the least.

Refraction is not the only way that visible light is dispersed into its component colors. Scattering of sunlight has a similar effect and is responsible for the blue color of the daytime sky. Scattering occurs when tiny particles in the atmosphere interact with light waves and send those light waves in different directions. If the radius of the scattering particles is much smaller than the

wavelength of the scattered light, then the amount of scattering varies with wavelength. This happens, for example, when visible light is scattered by the gas molecules composing air. In a now classic experiment performed in 1881, the English physicist and mathematician Lord Rayleigh demonstrated that this type of scattering is inversely proportional to the fourth power of the wavelength. Hence, violet light, at the short-wavelength end of the visible spectrum, is scattered much more efficiently than red light, at the long-wavelength end of the visible spectrum. This optical effect is known as **Rayleigh scattering**.

As sunlight travels through the atmosphere, component colors are scattered selectively out of the solar beam: violet is scattered more than blue, blue more than green, green more than yellow, and so forth. Dependence of scattering on wavelength predicts that scattered sunlight should be mostly violet. But the daytime sky appears blue rather than violet. One reason is the human eye's greater sensitivity to blue light than violet light, so that the sky appears bluer than it really is. Another factor contributing to the sky's blueness is dilution of violet light by all the other scattered colors; although the other colors are scattered less than violet, they tend to wash violet into blue.

We can now understand why the setting sun turns the horizon red on a clear evening. When the sun is on the horizon, its path through the atmosphere is about 40 times longer than when the sun is directly overhead. Consequently, at sunset (or sunrise), the interaction (e.g., scattering) between incoming solar radiation and the atmosphere's component gas molecules is considerable. Arriving at sea level, the horizontal solar beam has about 23% of its initial red light but only 0.000006% of its initial violet light. Just before sunrise or just after sunset (during twilight), the clear sky horizon turns red. On the moon, which has a highly rarefied atmosphere, the sun appears as a white disk in a black sky and stars are visible even in bright sunshine; scattering of sunlight is inconsequential in the lunar atmosphere.

The result is different when the particles doing the scattering are much larger than the wavelength of light. In such instances, scattering is not wavelength dependent; that is, visible radiation is scattered more or less equally at all wavelengths. The particles composing clouds (i.e., water droplets and/or ice crystals) are sufficiently large to scatter sunlight in this way. For this reason, sunlit clouds have the same color as the sun, that is, they are white during the day and red or pink at sunrise and sunset.

Particles that are about the same size as the wavelength of light are responsible for **Mie scattering**, named for the German physicist Gustav Mie. In 1908, Mie demonstrated that particles having diameters slightly less than 1 micrometer scatter the most light for their size (scatter most efficiently). Aerosols (solid and liquid particles suspended in the atmosphere) between about 1 and 2 micrometers in diameter scatter red light more efficiently than violet light. Sulfurous aerosols injected into the stratosphere by violent volcanic eruptions are in this size range and are responsible for the enhanced red and orange sunsets that may persist for more than a year following such an eruption. Most aerosols have diameters of 0.02 to 0.5 micrometer and scatter sunlight with equal efficiency at all wavelengths. Hence, an abundance of these aerosols turn the sky a hazy white.

FIGURE B.12
Crepuscular rays consist of alternating light and dark bands that appear to diverge fanlike from the sun's position during twilight.

Twilight

Twilight, a period after sunset or before sunrise when the sky is illuminated, is also a product of scattering of sunlight. Multiple scattering of sunlight by constituents of the upper atmosphere illuminates the sky. The length of twilight varies with latitude and time of year. The annual range increases with latitude so that in polar regions, for example, the length of twilight ranges from 24-hrs to none at all. Twilight is divided into three sequential stages based on the level of illumination: civil twilight, nautical twilight, and astronomical twilight.

Illumination during *civil twilight* is just adequate for outdoor activities without the need for artificial lighting. It covers the period between sunrise/sunset and when the center of the sun's disc is 6 degrees below the horizon. On a clear day, a faint purple glow may develop over a large region of the western sky during the latter portion of civil twilight. During *nautical twilight*, light is sufficient to distinguish the outlines of objects on the ground. While the ocean horizon is visible at the beginning of nautical twilight, it is not at the end when the center of the sun's disc is 12 degrees below the horizon. *Astronomical twilight* occurs when the center of the sun's disc is between 12 and 18 degrees below the horizon. Sixth magnitude stars are visible directly overhead.

At the beginning of evening twilight, observant sky watchers may see crepuscular rays or the green flash. **Crepuscular rays** are alternating light and dark bands (solar rays and shadows) that appear to diverge in a fanlike pattern from the solar disc near twilight (Figure B.12). (Actually, solar rays are parallel and only appear to diverge because of the perspective of the observer.) Crepuscular rays can also form during the day as the sun's rays pass through holes in clouds or between clouds in a hazy sky.

The **green flash** is a thin green rim that appears briefly at the upper edge of the sun at sunrise or sunset and is best seen on a distant horizon when the atmosphere is very clear. The green flash is primarily the consequence of atmospheric refraction and scattering of light from a low sun. Refraction is most pronounced when the sun is on the horizon and, as noted earlier, violet/blue light at the short wavelength end of the visible spectrum is refracted (bent) more than red light at the long wavelength end. This explains the red rim at the bottom of the low sun and should produce a blue/violet rim at the top of the solar disc. However, Rayleigh scattering removes most of the blue/violet light leaving behind a sliver of green.

The sliver of green produced by atmospheric refraction alone is too small to be seen by the naked eye. With an ideal temperature profile (e.g., superadiabatic lapse rate near the surface), the atmosphere acts like a lens to magnify the green rim to visible proportions. The green flash is actually a mirage that appears as an island of green light floating above the setting or rising sun. At midlatitudes, the sun rises and sets so rapidly that the green flash lasts only about 1 second (hence the name green flash). But at polar latitudes, the sun rises and sets more slowly and the green flash can lasts upwards of 30 minutes.

Conclusions

We have seen that interactions of sunlight (or moonlight) with clouds or rainfall produce a variety of optical phenomena including halos, rainbows, coronae, and glories. These interactions involve reflection, refraction, and/or diffraction. In these cases, refraction takes place as light rays pass from one transparent medium into another (e.g., air to water or air to ice). In addition, because of variations in density within the atmosphere, light rays are refracted and give rise to mirages and the green flash. Scattering of sunlight by air molecules is wavelength-dependent and responsible for the blue of the daytime sky whereas scattering of sunlight by cloud particles (droplets and ice crystals) is uniform across wavelengths and accounts for the white of clouds.

Basic Understandings

- A halo is a whitish (or slightly colorized) ring of light surrounding the sun or moon, formed when the ice crystals in high thin clouds refract the sun's rays. Refraction is the bending of light as it passes from one transparent medium into another. Cirrus clouds composed of relatively large plate-like ice crystals are responsible for sundogs, brilliant spots of light located on either side of the sun.
- A rainbow is a circular arc of concentric colored bands, caused by a combination of refraction and internal reflection of sunlight by raindrops. In most cases a much dimmer secondary rainbow appears about 8 degrees above the primary rainbow. Double reflection within raindrops produces a secondary rainbow, with the order of colors reverse that of the primary rainbow.
- A corona consists of alternating light and dark rings that surround the moon (or less often the sun) caused by diffraction of light around water droplets that compose a thin, translucent veil of altocumulus or stratocumulus clouds. Diffraction is the slight bending of a light wave as it moves along the boundary of an object such as a water droplet, producing an interference pattern.
- A glory consists of concentric rings of color that appear around a shadow and is caused by interactions between sunlight and a uniform mass of water droplets composing a warm cloud. The optics of a glory differ from that of a rainbow in that spherical cloud droplets diffract light rays ever so slightly toward the droplet, so that light rays incident on a cloud droplet parallel those returning from the cloud droplet.
- A mirage is an optical phenomenon in which an image of some distant object appears to be displaced from its normal view. It is caused by refraction of light as it travels through the atmosphere. Changes in air density that accompany variations in the vertical temperature profile are responsible for the refraction of light.
- Rayleigh scattering of sunlight by the molecules composing air accounts for the blue of the daytime sky. If the radius of the scattering particles is much smaller than the wavelength of the scattered light, then the amount of scattering is inversely proportional to the fourth power of the wavelength of the light being scattered. The human eye's greater sensitivity to blue light than violet light plus the dilution of violet light by the other scattered colors also play a role in the blue of the sky.
- Scattering is not wavelength dependent when light is scattered by particles having a radius that is much greater than the wavelength of the radiation being scattered. This is the case for clouds and explains why clouds appear white during the day and pink or red near sunset or sunrise. Mie scattering applies to particles that have diameters about the same as the wavelength of visible light. Scattering of visible light is most efficient for particles having a diameter of 1 micrometer.
- Twilight is a period after sunset or before sunrise when the sky is illuminated and is a product of scattering of sunlight by components of the upper atmosphere. Based on decreasing level of illumination, twilight is designated civil, nautical, and astronomical.
- Twilight phenomena include crepuscular rays and the green flash. Crepuscular rays consist of alternating light and dark bands (solar rays and shadows) that appear to diverge in a fanlike pattern from the sun's position at about twilight. The green flash is a thin green rim that appears briefly at the upper edge of the sun at sunrise or sunset and is primarily the consequence of atmospheric refraction and scattering of light from a low sun.

CHAPTER C

CLIMATE AND CLIMATE CHANGE

I had a dream, which was not all a dream.
The bright sun was extinguish'd, and the stars
Did wander darkling in the eternal space,
Rayless, and pathless, and the icy Earth
Swung blind and blackening in the moonless air;
Morn came and went—and came, and brought
   no day,
And men forgot their passions in the dread
Of this their desolation; and all hearts
Were chilled into a selfish prayer for light …

LORD BYRON, 1816 *Darkness*

This chapter covers some of the basics of climatology, controls of climate, and global patterns of climate. We summarize the climate record, how climate varies geographically and with time, lessons of the climate past, and possible causes of climate change.

Describing Climate

Climate is defined as the weather of some locality averaged over some time period plus extremes in weather. Climate must be specified for a particular place and time period because, like weather, climate varies both spatially and temporally. Thus, for example, the climate of Minneapolis is different from that of Miami, and winters in Minneapolis were somewhat milder in the 1980s and 1990s than they were in the 1880s and 1890s. Extremes in weather are important aspects of the climate record. Hence, daily weather reports usually include the highest and lowest temperatures ever recorded for that date and climatic summaries typically identify such extremes as the coldest, warmest, driest, wettest, snowiest, or cloudiest month or year on record.

Knowledge of climate extremes has many practical applications. Farmers, for example, are interested in knowing not only the long-term average rainfall during the growing season but also the frequency of drought. Electric utilities are interested in the coldest and mildest winters and hottest and coolest summers on record so that they might assess the likely range in residential energy demand for space heating and cooling.

Climate is usually described in terms of normals, means, and extremes of a variety of weather elements including temperature, precipitation, and wind. Climatic summaries are available in tabular form for climatic divisions of each state and major cities along with a narrative description of local or regional climate usually accompanies these data. The U.S. National Weather Service is responsible for gathering the basic weather data used in climatological summaries. Data are processed, entered into archives, and made available for users at NOAA's **National Climatic Data Center (NCDC)** in Asheville, NC.

THE CLIMATIC NORM

Traditionally, the climatic *norm* or *normal* is equated to the average value of some climatic element such as temperature or snowfall. This tradition can sometimes cause misconceptions. For one, *normal* may be taken to imply that climate is static when, in fact, climate is inherently variable with time. Furthermore, normal may imply a Gaussian (bell-shaped) probability distribution, although many climatic elements are non-Gaussian. For our purposes, we can think of the climatic norm of some locality as encompassing the *total* variation in the climate record, that is, both averages and extremes. This implies, for example, that an exceptionally cold winter may not be *abnormal* because its mean temperature may fall within the expected range of variability of winter temperature.

By international convention, climatic norms are computed from averages of weather elements compiled over a 30-year period. Current climatic summaries are based on weather records from 1971 to 2000. Average July rainfall, for example, is the simple average of the total rainfall during each of 30 consecutive Julys from 1971 through 2000. The 30-year period is adjusted every 10 years to add the latest decade and drop the earliest one. In the United States, 30-year averages are computed for temperature, precipitation, and air pressure only. Averages of other climatic elements such as wind speed and cloudiness are derived from the entire period of record at a particular weather station. Extremes such as the highest temperature or driest month are also drawn from the entire period of record.

Selection of a 30-year period for averaging weather data may be inappropriate for many applications because climate varies over a wide range of time scales and can change significantly in periods much shorter than 30 years. For some purposes, a 30-year period provides a shortsighted view of the climate record. Compared with the long-term climate record, the current 1971-2000 *norm,* for example, was unusually mild over much of the United States. Nonetheless, people find averages and extremes of past climate to be a useful guide to future expectations.

Many people assume that the mean value of some climatic element is the same as the median (middle value); that is, 50% of all cases are above the mean and 50% of all cases are below the mean. This is a reasonable assumption for climatic elements such as temperature, which approximate a simple Gaussian-type probability distribution. Hence, for example, we might expect about half the Januarys will be warmer and half the Januarys will be colder than the 30-year mean January temperature. On the other hand, the distribution of some climatic elements, such as precipitation, is distinctly non-Gaussian, and the mean does not correspond to the median. For example, in a dry climate subject to infrequent deluges of rain during the summer, considerably fewer than half the Julys are wetter than the mean and many more than half of Julys are drier than the mean. In fact, for many purposes, the median value of precipitation is a more useful description of climate than the mean value.

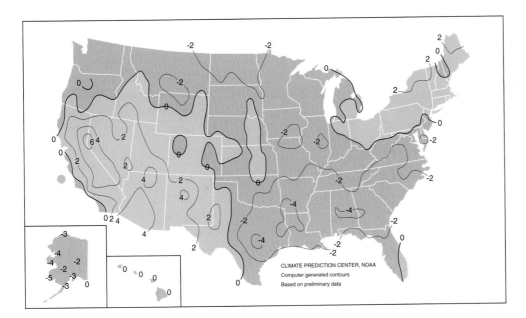

FIGURE C.1
Departure of average temperature from the long-term average (1971-2000) in Fahrenheit degrees for October 2001.

CLIMATIC ANOMALIES

Climatologists often compare the average weather of a specific week, month, or year with the past climate record. Such comparisons carried out over a broad geographical region show that departures from long-term climatic averages, called **anomalies**, do not occur with the same sign or magnitude everywhere. For example, the average temperature during October 2001 was above the long-term average (*positive anomalies*) through much of the western and northeastern U.S. but below the long-term averages (*negative anomalies*) in between (Figure C.1). Furthermore, the magnitude of the anomaly, positive or negative, varies from one place to another.

Precipitation anomalies typically form more complex patterns than do temperature anomalies (Figure C.2). This is due to greater spatial differences in precipitation arising from the variability of storm tracks

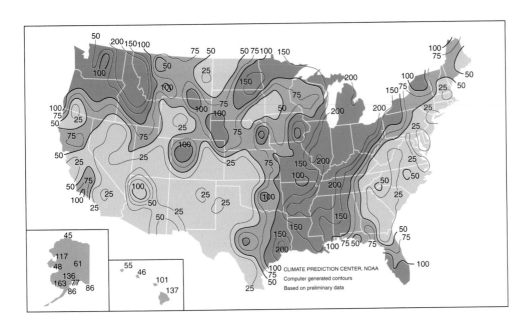

FIGURE C.2
Percent of long-term average (1971-2000) precipitation for October 2001.

and the almost random distribution of convective showers. For these reasons, in spring in midlatitudes, for instance, even adjoining counties may experience opposite rainfall anomalies (one having above-average rainfall and the other below-average rainfall). From an agricultural perspective, the geographic nonuniformity of climatic anomalies may be advantageous in that some compensation is implied. That is, poor growing weather and consequent low yields in one area may be compensated to some extent by better growing weather and increased yields elsewhere. This is known as *agroclimatic compensation* and generally applies to crops such as corn, soybeans, and other grains that are grown over broad geographical areas.

The geographic nonuniformity of climatic anomalies in midlatitudes is linked to the prevailing westerly wave pattern that ultimately governs cold and warm air advection, cyclogenesis, and storm tracks (Chapter 9). Hence, the pattern of the westerlies determines the location of weather extremes such as drought or very low temperatures. In view of the number of westerly waves that typically encircle the hemisphere, a single weather extreme never occurs over an area as large as the United States; that is, severe cold or drought never grips the entire nation at the same time.

Geographic nonuniformity also characterizes trends in climate. Hence, the trend in the average annual temperature of the Northern Hemisphere is not necessarily representative of all localities within the hemisphere. During the same period, some places experience cooling whereas other places experience warming regardless of the direction of the overall hemispheric (or global) temperature trend. Not only is it misleading to assume that the direction of large-scale climatic trends applies to all localities, but also it is erroneous to assume that the magnitude of climatic trends is the same everywhere. A small change in the average hemispheric temperature typically translates into a much greater change in some areas and into little or no change in other areas.

Climate Controls

Many factors working together shape the climate of any locality. Controls of climate consist of (1) latitude, (2) elevation, (3) topography, (4) proximity to large bodies of water, and (5) prevailing atmospheric circulation. For all practical purposes, the first four controls are fixed and

exert regular and predictable influences on climate. Seasonal changes in incoming solar radiation, as well as length of daylight, vary with latitude and air temperature responds to those regular variations (Chapters 3 and 4). Elevation influences air temperature and whether precipitation falls in the form of rain or snow. Topography can affect the distribution of clouds and precipitation so that the windward slopes of high mountain barriers usually are wetter than the leeward slopes (Chapter 6).

The relatively great thermal inertia of large bodies of water moderates the temperature of downwind localities, reducing the seasonal temperature contrast and lengthening the growing season. Furthermore, ocean currents strongly influence climate. Cold surface currents, such as the California Current, are heat sinks; they cool and stabilize the overlying air, thereby reducing the likelihood of thunderstorms. Relatively warm surface currents, such as the Gulf Stream, are heat sources; they supply heat and moisture to the overlying air, destabilizing the air and energizing storm systems (Figure C.3).

The final climate control, atmospheric circulation, encompasses the combined influence of all weather systems operating at all spatial scales (Chapters 8 through 12). Although strongly influenced by the other climate controls, atmospheric circulation is considerably less regular and less predictable than the others. This variability is especially evident in synoptic-, meso-, and micro-scale weather systems. Planetary-scale circulation systems (e.g., the prevailing wind belts, subtropical anticyclones, and the intertropical convergence zone), exert a more systematic influence on climate. How these and other controls interact to shape the climates of the continents will become clearer as we summarize the basic patterns of climate on Earth.

Global Climate Patterns

From a global perspective, climate exhibits some regular patterns. As we examine these patterns, keep in mind that they may be significantly altered by local and regional climate controls.

TEMPERATURE

Ignoring the influence of mountainous terrain on air temperature, mean annual isotherms roughly parallel latitude circles, underscoring the influence of

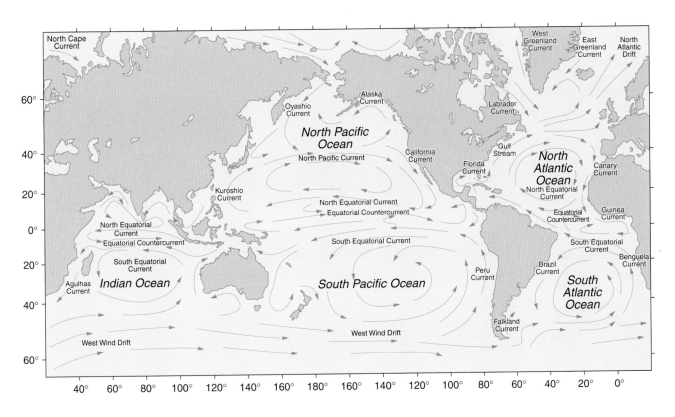

FIGURE C.3
Long-term average surface ocean currents exert an important influence on climate.

solar radiation and solar altitude on climate. The latitude of highest mean annual surface temperature, the so-called **heat equator**, is located about 10 degrees north of the geographical equator. Mean annual isotherms are symmetrical with respect to the heat equator, decreasing in magnitude toward the poles.

The heat equator is in the Northern Hemisphere because, overall, that hemisphere is warmer than the Southern Hemisphere for several reasons. For one, polar regions of the two hemispheres have different radiational characteristics so that the Arctic is warmer than the Antarctic. Most of the Antarctic continent is submerged under massive glacial ice sheets, so the surface has a very high albedo for solar radiation and is the site of intense radiational cooling, especially during the long polar night. By contrast, the Northern Hemisphere polar region is mostly ocean. Although the Arctic Ocean is usually ice covered, patches of open water develop in summer and lower the overall surface albedo.

A second factor contributing to the relative warmth of the Northern Hemisphere is that hemisphere's greater fraction of land in tropical latitudes. Because land surfaces warm up more than water surfaces in

response to the same incoming solar radiation, tropical latitudes are warmer in the Northern Hemisphere than in the Southern Hemisphere. A third contributing factor is ocean circulation which transports more warm water to the Northern Hemisphere than to the Southern Hemisphere.

Systematic patterns also appear when we consider the worldwide distribution of mean January temperature (Figure C.4) and mean July temperature (Figure C.5); January and July are usually the coldest/warmest months of the year in their respective hemisphere. Neglecting the influence of topography on air temperature, isotherms tend to parallel latitude circles. However, monthly isotherms exhibit some notable north-to-south bends, primarily because of land/sea contrasts and the influence of ocean currents. As the year progresses from January to July and to January again, isotherms in both hemispheres follow the sun and shift north and south in tandem. The latitudinal (north-south) shift in isotherms is greater over the continents than over the ocean; that is, the annual range in air temperature is greater over land than over the sea. Furthermore, the meridional temperature gradient is greater

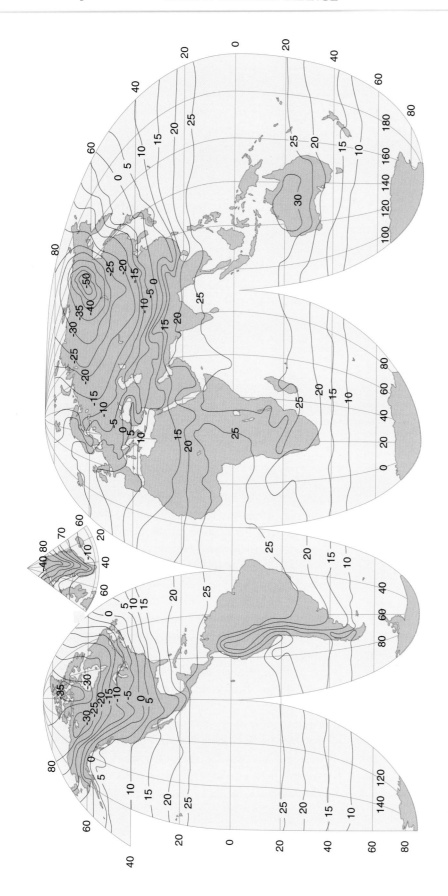

FIGURE C.4

Mean sea-level air temperature for January in degrees Celsius.

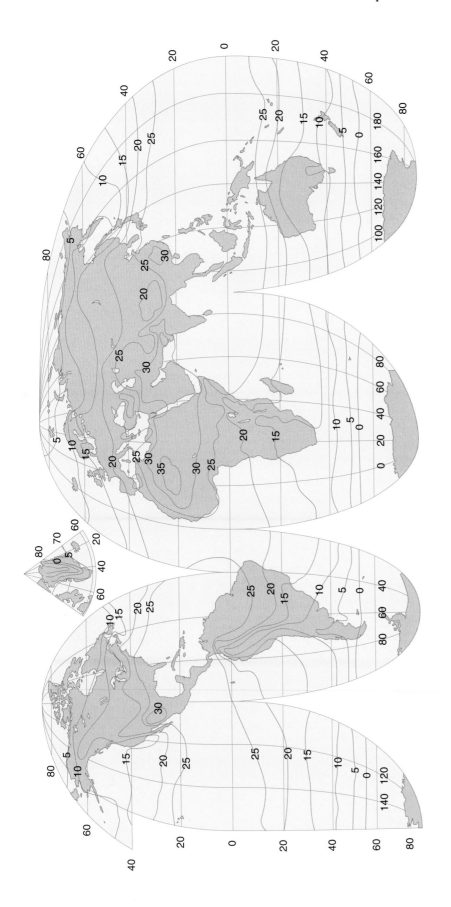

FIGURE C.5
Mean sea-level air temperature for July in degrees Celsius.

in the winter hemisphere than in the summer hemisphere. A steeper temperature gradient means a more vigorous circulation and stormier weather in the winter hemisphere (Chapter 9).

PRECIPITATION

The global pattern of mean annual precipitation (rain plus the liquid equivalent of snowfall) exhibits great spatial variability (Figure C.6). Some of this variability can be attributed to topography and the distribution of land and sea, but the planetary-scale circulation is also important (Chapter 9). The ITCZ, subtropical anticyclones, and the prevailing wind belts impose a roughly zonal pattern on precipitation distribution. In addition, regular shifts of these circulation features through the year are responsible for the seasonality of precipitation that typifies the climate of many localities.

In tropical latitudes, convective activity associated with intense solar heating and the trade wind convergence triggers abundant rainfall year-round. In the adjacent belt poleward to about 20 degrees latitude, rainfall depends on seasonal shifts of the ITCZ and the subtropical anticyclones. Poleward shift of the ITCZ causes summer rains, whereas equatorward shift of the subtropical highs brings a dry winter. This climatic zone includes the belt of tropical monsoon circulation, described in Chapter 9.

Poleward of this belt, from about 20 to 35 degrees N and S, subtropical anticyclones, centered over the ocean basins, dominate the climate all year. Subsiding dry air on the anticyclones' eastern flanks is responsible for Earth's major subtropical deserts (e.g., the Sahara). On the other hand, unstable humid air on the western flanks of subtropical anticyclones causes relatively moist conditions. Between about 35 and 40 degrees latitude, the prevailing westerlies and subtropical anticyclones govern precipitation. Typically, on the western side of continents, winter cyclones migrating with the westerlies bring moist weather, but in summer, westerlies shift poleward and the area lies under the dry eastern flank of a subtropical anticyclone. Hence, summers are dry. At the same latitudes, but on the eastern side of continents, the climate is dominated by westerlies in winter and the moist airflow on the western flank of a subtropical anticyclone in summer. Thus, rainfall is triggered by cyclonic activity in winter and by convection in summer and shows little seasonal variability.

Precipitation generally declines poleward of about 40 degrees latitude as lower temperatures reduce the amount of precipitable water (Chapter 6). Although precipitation is generally not seasonal, the tendency in the continental interiors is for more precipitation in summer. The summer precipitation maximum is due to higher air temperatures, greater precipitable water, and more vigorous convection at that time of year.

Our description of the global pattern of annual precipitation is somewhat idealistic and requires some qualification. Land/sea distribution and topography complicate the generally zonal distribution of precipitation. More rain falls over the ocean than over the continents, and mountain belts induce wet windward slopes and extensive leeward rain shadows. Furthermore, annual precipitation totals fail to convey some other important aspects of precipitation, including the average amount of rainfall per day and the season-to-season and year-to-year reliability of precipitation. As a rule, rainfall is most reliable in maritime climates, less reliable in continental localities, and least reliable in arid regions. However, drought is possible anywhere, even in maritime climates.

CLIMATE CLASSIFICATION

In response to the interaction of many controls, Earth's climates form a complex mosaic. For more than a century, climatologists have attempted to organize the myriad of climate types by devising classification schemes that group together climates having common characteristics. Classification schemes typically group climates according to (1) the meteorological basis of climate, or (2) the environmental effects of climate. The first is a *genetic* climate classification that asks why climate types occur where they do. The second is an *empirical* climate classification that infers the type of climate from such climatic impacts as the distribution of indigenous vegetation. In addition, the advent of electronic computers and databases has made possible *numerical* climate classification schemes that utilize sophisticated statistical techniques. Appendix III provides information on one of the more popular climate classification schemes.

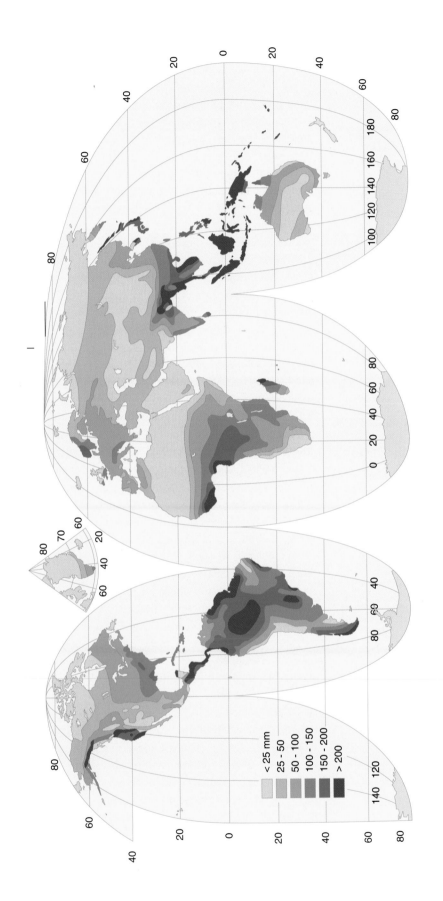

FIGURE C.6
Mean annual precipitation (rain plus melted snow) in millimeters (mm).

The Climate Record

Climate varies not only from one place to another but also with time. Examining records of past climate reveals some basic understandings regarding the nature of climate behavior and provides a useful perspective on current and future climate. In most places, however, the reliable instrument-based record of past weather and climate is limited to not much more than a century or so. For information on earlier variations in climate, scientists rely on reconstructions of climate based on historical documents (e.g., personal diaries, ship's logs, weather-sensitive economic data) and longer-term geological and biological evidence such as bedrock, fossil plants and animals, tree growth-rings, pollen, deep-sea sediment cores, and glacial ice cores. In this section, we summarize major features of Earth's climate record. We begin with a brief description of some of the techniques used to reconstruct climate prior to the instrument-based record.

CLIMATE RECONSTRUCTION

Tree growth-rings can yield a year-to-year record of past climate variations stretching back many thousands of years. Each spring/summer, living trees add a growth-ring whose thickness and density depend on growing season weather. A small hollow drill is used to extract a tree's growth-ring record. Although other environmental factors (e.g., soil type, drainage) can be important, tree growth-rings are especially sensitive to moisture stress and have been used to reconstruct the chronology of drought hundreds of years prior to the era of instrument-based records.

Pollen is a valuable source of information on past climates—especially over the past 15,000 years. *Pollen* is the tiny dust-like fertilizing component of a seed plant that is dispersed by the wind and accumulates at the bottom of lakes (and other depositional environments) along with other sediments. Scientists use a corer to extract a sediment column (core) that chronicles past changes in pollen (and therefore, vegetation). Back at the laboratory, pollen grains are separated from the other sediment in the core and identified as to source vegetation. Implications for past climate are drawn assuming that (1) the pollen is of local origin and (2) climate largely controls the type of vegetation. A change in dominant pollen type in a core signals a change in nearby vegetation, likely in response to a climate change.

A water molecule (H_2O) is composed of either of two stable isotopes of oxygen, ^{16}O or ^{18}O, and is the basis for the oxygen isotope technique of climate reconstruction. (Isotopes of a single element differ in atomic mass due to differences in the number of neutrons in the nucleus of the atom.) On average, water molecules containing the lighter ^{16}O isotope move faster than water molecules containing the heavier ^{18}O isotope and evaporate more readily from land and sea. During a glacial climatic episode, growing ice sheets sequester more and more light oxygen while ocean water ends up with proportionately less ^{16}O. With a shift to a non-glacial climate, ice sheets melt and water rich in ^{16}O drains into the ocean decreasing the ratio of ^{18}O to ^{16}O. Hence, the proportion of heavy to light oxygen in ocean water is directly proportional to the total volume of glacial ice on Earth.

Much of what we know about the climatic fluctuations of the Pleistocene Ice Age is based on oxygen isotope analysis of the shell remains of microscopic marine organisms extracted from deep-sea sediment cores. Marine organisms such as foraminifera build shells of calcium carbonate ($CaCO_3$), incorporating the same ratio of ^{18}O to ^{16}O as in the surrounding seawater. Hence, the ratio of ^{18}O to ^{16}O is relatively high in the shells of marine organisms that lived during glacial climatic episodes and relatively low in the shells of marine organisms that lived during non-glacial climatic episodes. When these organisms die, their shells settle to the ocean bottom and mix with other sediment. Oxygen isotope analysis of shells extracted from sediment cores taken from the deep ocean bottom yield a continuous record of glacial ice volume on the planet going back many hundreds of thousands of years.

Oxygen isotope analysis is also applied to cores extracted from glaciers. Glacial ice cores contain tiny bubbles of trapped air that provide information on past changes in atmospheric chemistry. Photosynthesis at the sea-surface transmits the oxygen isotope ratio of seawater to atmospheric oxygen (O_2). (Oxygen is a product of photosynthesis.) Atmospheric oxygen (with its inherited $^{18}O/^{16}O$ ratio) is trapped in the accumulating snow and glacial ice and yields a record of past variations in the planet's glacial ice cover (mirroring the record from deep-sea sediment cores). Analysis of trapped air bubbles are also useful for reconstructing long-term changes in the concentration of the greenhouse gases carbon dioxide (CO_2) and methane (CH_4). Furthermore, from analysis of the oxygen isotope ratio and deuterium (an isotope of hydrogen) concentration in layers of glacial ice, scientists obtain a more direct measure of past variations in local air temperature.

Cores extracted from the Greenland and Antarctic ice sheets provide the most lengthy climate record. During the summers of 1991-1993, two independent scientific teams, one American and the other European, drilled into the thickest portion of the Greenland ice sheet. The two drill sites were located within 30 km (19 mi) of each other, about 650 km (400 mi) north of the Arctic Circle. The extracted cores are about 3000 m (9850 ft) in length and spanned a period of roughly 200,000 years. The world's longest ice core was extracted from the East Antarctic ice sheet at Vostok station where, in the mid-1990s, drilling reached a record depth of 3100 m (10,170 ft), spanning two major glacial-interglacial transitions.

GEOLOGIC TIME

Attempts to reconstruct climate through the millions and billions of years that constitute **geologic time** become more challenging with increasing time before present. Descriptions of early climates are suspect because of lengthy gaps in the record and difficulty in determining the timing of events and correlating events that occurred in widely separated areas. Nonetheless, the available evidence supports some general conclusions regarding the climate of geologic time. For convenience of study, the climate record of the geologic past is subdivided using the *geologic time scale*, a standard division of Earth history into eons, eras, periods, and epochs based on large-scale geological events (Figure C.7).

Plate tectonics complicates climate reconstruction efforts that focus on periods spanning hundreds of million of years. The solid outer skin of the planet is divided into a dozen gigantic rigid plates that slowly drift over the face of the globe, moving continents and opening and closing ocean basins. Mountain building and most volcanic activity occur at plate boundaries. Alfred Wegener, a German geophysicist, was the first to formally propose the concept of plate tectonics in 1912 but the scientific community did not widely accept the theory until the 1960s when mounting geological evidence became convincing.

Through the span of human existence, topography and the geographical distribution of the ocean and continents were essentially fixed controls of climate. This was not so over the vast expanse of

geologic time. Mountain ranges rose and eroded away; seas invaded and withdrew from the land, and continents drifted over the face of the globe. Plate tectonics has probably operated on the planet for at least 2 billion years and explains such seemingly anomalous finds as glacial deposits in the Sahara Desert, fossil tropical plants in Greenland, and fossil coral in Wisconsin (Figure C.8). These discoveries reflect climatic conditions millions of years ago when landmasses were situated at different latitudes than they are today. Plate tectonics also impacted global climate by altering the course of heat-transporting surface ocean currents and conveyor belts (Chapter 4).

Geologic evidence points to an interval of major climatic fluctuations about 570 million years ago, corresponding to the transition between Precambrian and Phanerozoic Eons. Along Namibia's Skeleton Coast rock layers that were formed in tropical seas directly overlie glacial deposits. Some scientists interpret these rock sequences as indicating abrupt climatic shifts between extreme cold and tropical heat. During as many as four cold episodes, each lasting perhaps 10 million years, the continents were encased in glacial ice and the ocean froze to a depth of more than 1000 m (3300 ft). At the close of each cold episode, temperatures rose rapidly, and within only a few centuries, all the ice melted.

Global warming persisted through much of the Mesozoic Era, from about 245 to 66 million years ago. At the boundary between the Triassic and Jurassic Periods, the global mean temperature rose perhaps 3 to 4 Celsius degrees (5.5 to 7 Fahrenheit degrees), causing substantial changes in vegetation and a major extinction of animals. At peak warming during the Cretaceous Period, the global mean temperature was perhaps 6 to 8 Celsius degrees (11 to 14 Fahrenheit degrees) higher than now. Subtropical plants and animals lived as far north as 60 degrees N and dinosaurs roamed the North Slope of Alaska. Adding to the geological processes impacting climate, a great meteorite impact about 65 million years ago threw huge quantities of dust into the atmosphere, blocking sunlight, and causing cooling that apparently was responsible for the demise of the dinosaurs.

The Cenozoic Era was a time of great climatic fluctuations. About 55 million years ago, methane (CH_4) released by deep-sea sediments escaped into the ocean and then into the atmosphere where it enhanced the greenhouse effect, causing an already warm planet to become even warmer. But geologic evidence indicates

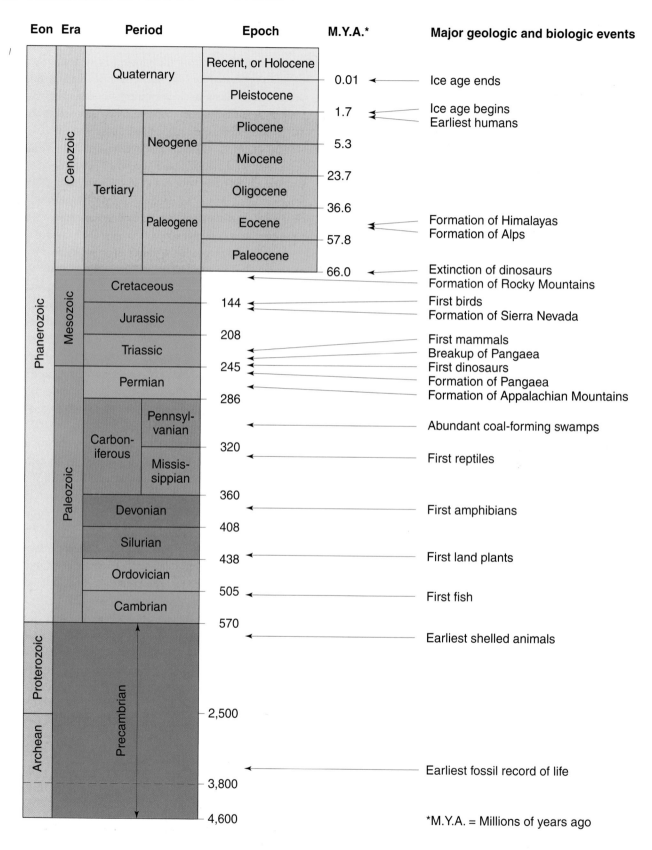

FIGURE C.7
Geologic time scale.

FIGURE C.8
This bedrock exposed in northeastern Wisconsin contains fossil coral that dates from nearly 400 million years ago. Based on the environmental requirements of modern coral, scientists conclude that 400 million years ago, Wisconsin's climate was tropical marine. Plate tectonics can help explain such a drastic change between ancient and modern conditions.

that more than 40 million years ago, Earth's climate began a change toward colder, drier, and more variable conditions setting the stage for the Pleistocene Ice Age. According to research findings of W.F. Ruddiman of Lamont-Doherty Geological Observatory of Columbia University and J.E. Kutzbach of the University of Wisconsin-Madison, mountain building may be the principal reason for this change in Earth's climate, specifically the rise of the Colorado Plateau, Tibetan Plateau, and Himalayan Mountains. Prominent mountain ranges not only influence the geographical distribution of clouds and precipitation, they also can alter the planetary-scale circulation. Furthermore, mountain building may disturb the global carbon cycle. Enhanced weathering of bedrock exposed in mountain ranges sequesters more atmospheric carbon dioxide in sediments thereby weakening the greenhouse effect.

In the American West, the region from the California Sierras to the Rockies, known as the Colorado Plateau, has an average elevation of 1500 to 2500 m (5000 to 8200 ft). Although mountain building began about 40 million years ago, about half of the total uplift took place between 10 and 5 million years ago. The Tibetan Plateau and Himalayan Mountains of southern Asia cover an area of more than 2 million square km (0.8 million square mi) and have an average elevation of more than 4500 m (14,700 ft). About half of total Himalayan uplift took place over the past 10 million years. These plateaus diverted the planetary-scale

westerlies into a more meridional pattern, increasing the north-south exchange of air masses and altering the climate over a broad region of the globe. Also, seasonal heating and cooling of the plateaus causes low pressure to develop in summer and high pressure in winter, contributing to an enhanced monsoon-type circulation.

PAST TWO MILLION YEARS

Reconstruction of the climate over the past 2 million or so years does not have to account for the effects of plate tectonics. For practical purposes, mountain ranges, continents, and the ocean were essentially as they are today. Climate varies over a wide range of time scales so that viewing the climate record of the past two million years in progressively narrower time frames is informative. Such an approach helps to resolve the complex oscillations of climate into more detailed fluctuations especially over the recent past.

Compared to the climate that prevailed through most of geologic time, the climate of the last two million years was anomalous in favoring the development of huge glacial ice sheets (although evidence also exists of ice ages earlier in geologic time). During much of Earth's history, the average global temperature may have been 10 Celsius degrees (18 Fahrenheit degrees) higher than it was over the past two million years. The most recent cooling set in about 40 million years ago and culminated in the Pleistocene Ice Age that began about 1.7 million years ago and ended about 10,500 years ago.

During the Pleistocene Ice Age the climate shifted numerous times between glacial climates and non-glacial climates (Figure C.9A). A **glacial climate** favors formation of glaciers or the expansion of existing glaciers whereas a **non-glacial climate** favors no glaciers or the shrinkage of existing glaciers. During major glacial climatic episodes, the Laurentide ice sheet developed over central Canada and spread westward to the Rocky Mountains, eastward to the Atlantic Ocean, and southward over the northern tier states of the United States. At the same time, a much smaller ice sheet formed over northwestern Europe including the British Isles. In mountainous regions, glaciers expanded down former river valleys and coalesced into large tongues of ice on the plains beyond. The vast quantity of water locked up in these huge ice sheets (and much smaller mountain glaciers) caused sea level to drop by about 120 m (400 ft), exposing portions of the continental shelf, including a land bridge linking Siberia and North America. The Laurentide and European ice sheets

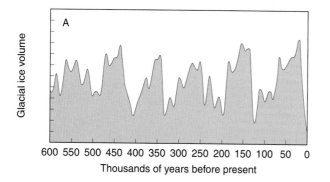

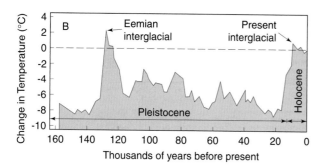

FIGURE C.9
(A) Variation in global glacial ice volume from the present back to about 600 million years ago based on analysis of oxygen isotope ratio in deep-sea sediment cores. (B) Temperature variation over the past 160,000 years based on oxygen isotope analysis of an ice core extracted from the Antarctic ice sheet at Vostok and expressed as a departure in Celsius degrees from the 1900 mean global temperature. [Compiled by R.S. Bradley and J.A. Eddy from J. Jousel et al., *Nature* 329(1987):403-408 and reported in *EarthQuest* 5, No. 1 (1991).]

Shifting focus to the past 160,000 years, resolution of the climate record improves. The temperature curve in Figure C.9B is derived from analysis of an ice core extracted from the Antarctic ice sheet at Vostok. A relatively mild interglacial episode, referred to as the *Eemian*, began about 127,000 years ago and persisted for about 7000 years. In some localities, temperatures may have been 1 to 2 Celsius degrees (2 to 4 Fahrenheit degrees) higher than during the warmest portion of the present interglacial. The Eemian interglacial was followed by numerous fluctuations between glacial and interglacial episodes. The last major glacial climatic episode began about 27,000 years ago and reached its peak about 18 to 20 thousand years ago when glacial ice cover over North America was about as extensive as it had ever been (Figure C.10).

The temperature curve in Figure C.11 covers the past 18,000 years and is based on a variety of climatic indicators. Global temperature 18,000 years ago averaged 4 to 6 Celsius degrees (7.2 to 10.8 Fahrenheit degrees) lower than at present. The general warming trend that followed was punctuated by relatively brief glacial climatic episodes. A notable example is the cold interval from about 12,900 to 11,600 years ago known as the *Younger Dryas* (named for the polar wildflower, *Dryas octopetala*, that reappeared in portions of Europe at the time). Return of glacial climatic conditions triggered short-lived re-advances of remnant ice sheets in

thinned and retreated, and may even have disappeared entirely, during relatively mild non-glacial climatic episodes, which typically lasted about 10,000 years. Throughout these non-glacial episodes, however, glacial ice persisted over most of Antarctica and Greenland as it still does today.

During glacial climatic episodes, temperatures were lower than they are today but the cooling was not geographically uniform. A variety of geologic evidence indicates that during the Pleistocene, temperature fluctuations between major glacial and non-glacial climatic episodes typically amounted to as much as 5 Celsius degrees (9 Fahrenheit degrees) in the tropics, 6 to 8 Celsius degrees (11 to 14 Fahrenheit degrees) at mid latitudes, and 10 Celsius degrees (18 Fahrenheit degrees) or more at high latitudes. An increase in the magnitude of a climatic change with increasing latitude is known as **polar amplification**, indicating that polar areas are subject to greater changes in climate.

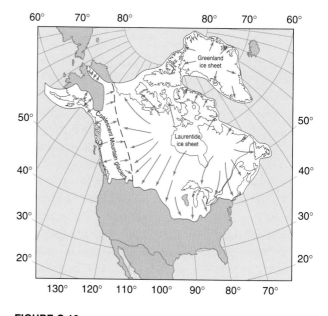

FIGURE C.10
The extent of glacial ice cover over North America about 20,000 to 18,000 years ago, the time of the last glacial maximum.

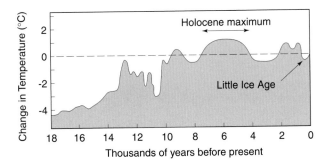

FIGURE C.11
Temperature variation over the past 18,000 years based on analysis of a variety of climatic indicators and expressed as a departure in Celsius degrees from the 1900 mean global temperature. [Compiled by R.S. Bradley and J.A. Eddy based on J.T. Houghton et al. (eds.), *Climate Change: The IPCC Assessment*, Cambridge University Press, U.K., 1990, and reported in *EarthQuest* 5, No. 1 (1991).]

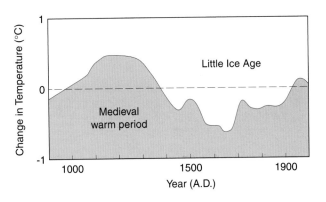

FIGURE C.12
Temperature variation over the past 1000 years based on analysis of historical documents and expressed as a departure in Celsius degrees from the 1900 mean global temperature. [Adapted from J.T. Houghton et al. (eds.), *Climate Change: The IPCC Assessment*, Cambridge University Press, U.K., 1990, p. 202.]

North America, Scotland, and Scandinavia. Glacial ice finally withdrew from the Great Lakes region about 10,500 years ago ushering in the present interglacial, the **Holocene**.

Although the Laurentide ice sheet was melting and would disappear almost entirely by about 5500 years ago, the Holocene has been an epoch of spatially and temporally variable temperature and precipitation. Cores extracted from the Greenland ice sheet and sediment cores taken from the bottom of the North Atlantic Ocean reveal that abrupt millennial-scale climate fluctuations interrupted the overall post-glacial warming trend. Warming after 10,000 years ago gave way to a cold episode about 8200 years ago; significant cooling also occurred between about 6100 and 5000 years ago and from about 3100 to 2400 years ago. On the other hand, at times during the mid-Holocene (classically known as the *Hypsithermal*) mean annual global temperature was perhaps 1 Celsius degree (2 Fahrenheit degrees) higher than it was in 1900, the warmest in more than 110,000 years, that is, since the Eemian interglacial. A pollen-based climate reconstruction indicates that 6000 years ago, July mean temperatures were about 2 Celsius degrees (3.6 Fahrenheit degrees) higher than now over most of Europe.

A generalized temperature curve for the past 1000 years, derived mostly from historical documents, is shown in Figure C.12. The most notable features of this record are the **Medieval Warm Period** from about 950 to 1250 AD and the cooling that followed, from about 1400 to 1850 AD, a period now known as the **Little Ice Age**. The Medieval Warm Period and the Little Ice Age

were not episodes of sustained warming and cooling, respectively. On the contrary, sediment and glacial ice core records plus historical evidence indicate decadal fluctuations in temperature and precipitation. During the Medieval Warm Period, the mean global temperature was about 0.5 Celsius degree (0.9 Fahrenheit degree) higher than in 1900. The first Norse settlements appeared along the southern coast of Greenland and vineyards thrived in the British Isles. Independent lines of evidence confirm that the Little Ice Age was a relatively cool period in many places, with mean annual global temperatures perhaps 0.5 Celsius degree (0.9 Fahrenheit degree) lower than it was in 1900. Sea ice cover expanded, mountain glaciers advanced, growing seasons shortened, and erratic harvests caused much hardship for many people, including abandonment of Norse settlements in Greenland.

INSTRUMENT-BASED TEMPERATURE TRENDS

Invention of weather instruments and establishment of world-wide weather observational networks made the climate record much more detailed and dependable. The most reliable temperature records date from the late 1800s when the birth of predecessors to today's World Meteorological Organization (WMO) and U.S. National Weather Service (NWS) ensured that weather observations were made and recorded using standardized methods. Examination of temperature trends over the past 100 years or so is instructive as to the short-term variability of climate.

FIGURE C.13
Instrument-derived trends in mean annual global, sea-surface, and land temperatures expressed as departures in degrees Celsius and Fahrenheit from the period average.

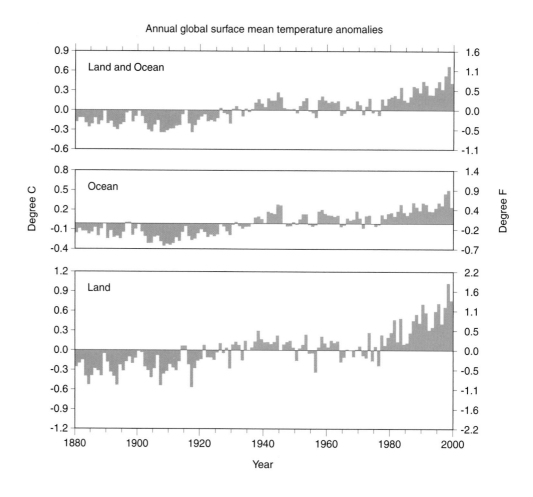

Annual global surface mean temperature anomalies

    Plotted in Figure C.13 are (1) global mean annual temperature (2) global mean sea surface temperature, and (3) global mean land surface temperature for the period since 1880, expressed as a departure (in degrees Celsius and Fahrenheit) from the period average. As expected, these temperature time series, assembled by scientists at the National Climatic Data Center (NCDC), indicate greater year-to-year variability over land than ocean. The trend in global mean temperature is generally upward from 1900 until about 1940, downward from 1940 to about 1970, and upward again through the 1990s. The total temperature fluctuation amounts to only about ±0.4 Celsius degrees (±0.7 Fahrenheit degrees) about the period average. This temperature record for the globe as a whole is not representative (in direction or magnitude) of all locations worldwide; that is, the trend was amplified or reversed, or both, in specific regions.

    In late 1999, the WMO reported that the global mean annual temperature was about 0.7 Celsius degree (1.3 Fahrenheit degrees) higher at the end of the 20th century than at the close of the 19th century. In Spring 2000, scientists at the NCDC reported that global warming accelerated during the last quarter of the 20th century. Since 1976, global mean annual temperature climbed at the rate of 2 to 3 Celsius degrees (4 to 5 Fahrenheit degrees) per 100 years. The seven warmest years in the instrument-based record occurred in the 1990s with 1998 being the warmest. During 1997 and 1998, record high global mean temperatures were set for 16 consecutive months. Furthermore, climate reconstructions (mostly in the Northern Hemisphere) indicate that the 20th century was the warmest in 1000 years. For example, analysis of ice cores extracted from a glacier at 7163 m (23,500 ft) in the Himalayan Mountains finds that the 1990s and the last half of the 20th century were the warmest of any equivalent periods in a millennium.

    Some critics question the integrity of large-scale (hemispheric or global) temperature trends. They cite as potential sources of error: (1) improved sophistication and reliability of weather instruments through the period of record, (2) changes in location and exposure of instruments at most long-term weather stations, (3) huge gaps in monitoring networks, especially over the ocean, and (4) the influence of urbanization and expansion of associated heat islands. By careful statistical treatment of available data, however, global-scale temperature trends have been confirmed with what is generally considered to be a reasonable degree of certainty.

Lessons of the Climate Past

What does the climate record tell us about how climate behaves through time? The following are some of the principal lessons of the climate record that are useful in assessing prospects for the climate future and the possible impacts of climate change.

1. *Climate is inherently variable over a broad spectrum of time scales ranging from years to decades, to centuries, to millennia.* Variability is an endemic characteristic of climate. The question for the future is not *whether* the climate will change but *how* the climate will change.

2. *Variations in climate are geographically nonuniform in both sign (direction) and magnitude.* Large-scale trends in climate are not necessarily duplicated at a particular location although the tendency is for temperature trends to amplify with increasing latitude (*polar amplification*). A large-scale change in climate is not likely to have the same impact everywhere.

3. *Climate change may consist of a long-term trend in mean temperature or mean precipitation or an increase or decrease in frequency of extreme weather events (e.g., drought, excessive cold).* Recall that climate encompasses mean values plus extremes. A trend toward warmer or cooler, wetter or drier conditions may or may not be accompanied by an increase or decrease in frequency of weather extremes. A climatic regime featuring relatively little change in mean temperature or mean precipitation through time may be accompanied by an increase or decrease in frequency of weather extremes.

4. *Climate change tends to be abrupt rather than gradual.* In the context of the climate record, *abrupt* is a relative term. If the time of transition between climatic episodes is much shorter than the duration of the episodes, then the transition is considered to be relatively abrupt. Analysis of cores extracted from the Greenland ice sheet indicates that cold and warm climatic episodes, each lasting about 1000 years, were punctuated by abrupt changes in periods as brief as a single decade. The abrupt-change nature of climate would test the resilience of society to respond effectively to climate change.

5. *Only a few cycles operate in the long-term climate record.* Reliable cycles include diurnal and seasonal variations in incoming solar radiation and temperature, meaning simply that days are usually warmer than nights and summers are warmer than winters. Quasi-regular variations in climate include El Niño (occurring about every 2 to 7 years), Holocene millennial-scale fluctuations identified in glacial ice cores, and the major glacial-interglacial climatic shifts of the Pleistocene Ice Age unlocked from deep-sea sediment cores and operating over tens of thousands to hundreds of thousands of years.

6. *Climate change can impact society.* History recounts numerous instances when climate change significantly impacted society. Consider some examples. Researchers have implicated prolonged drought in the decline and fall of Middle Eastern empires around 2000 BC, the Harappan civilization of the Indus Valley of India about 1700 BC, the Mycenaean civilization of Greece in 1200 BC, and the Anasazi abandonment of Mesa Verde (southwestern Colorado) around 1300 AD (Figure C.14). The cooling trend that heralded the Little Ice Age was likely a major factor in the disappearance of Norse settlements in Greenland about 1400 AD.

FIGURE C.14
Drought apparently forced the Anasazi Indians of Mesa Verde, in southwestern Colorado, to abandon their cliff dwellings around 1300 A.D.

Although modern societies are more capable of dealing with climate change than early peoples, a rapid and significant change in climate would seriously impact all sectors of modern society.

Possible Causes of Climatic Change

Climate varies over a broad spectrum of time scales, but no simple explanation exists for why climate varies. The complex spectrum of climate variability is a response to the interactions of many processes, both internal and external to the Earth-atmosphere system.

One way to organize our thinking on the many possible causes of climate variability is to match a possible cause (or *forcing*) with a specific climatic oscillation (or *response*), based on similar periods of oscillation. For example, plate tectonics might explain long-term climate changes operating over hundreds of millions of years. Systematic changes in Earth's orbit about the sun may account for climatic shifts of the order of 10,000 to 100,000 years. Changes in sunspot number and variations in the sun's energy output may be associated with climatic fluctuations of decades to centuries. Volcanic eruptions, El Niño, or La Niña may account for climatic fluctuations lasting several months to a few years. But matching some forcing mechanism with a climatic response based on similar periods of oscillation is no guarantee of a real physical relationship.

Another way to think about climatic variability is in terms of **global radiative equilibrium**, that is, energy entering the Earth-atmosphere system (i.e., absorbed solar radiation) must ultimately equal energy leaving the system (i.e., infrared radiation emitted to space). Any change in either energy input or energy output will shift the Earth-atmosphere system to a new equilibrium and change the planet's climate. A number of factors can alter global radiative equilibrium, including changes in solar energy output, volcanic eruptions, human activity, and changes in Earth's surface properties.

CLIMATE AND SOLAR VARIABILITY

Fluctuations in the sun's energy output, sunspots, or regular variations in Earth's orbital parameters (the Milankovitch cycles) are external factors that can alter the planet's climate.

Satellite measurements in the 1980s and 1990s confirmed long-held suspicions in the scientific community that the sun's total energy output at all wavelengths varies with time. Furthermore, numerical models predict that only a 1% change in the sun's energy output could significantly alter the mean temperature of the Earth-atmosphere system.

Changes in solar energy output are apparently related to sunspot number. A **sunspot** is a dark blotch on the face of the sun, typically thousands of kilometers across, that develops where an intense magnetic field suppresses the flow of gases transporting heat from the sun's interior (Figure C.15). A sunspot appears dark because its temperature is about 400 to 1800 Celsius degrees (720 to 3240 Fahrenheit degrees) lower than the surrounding surface of the Sun, the *photosphere*. In 1843, the German astronomer Samuel Heinrich Schwabe reported a regular variation in sunspot activity. A single sunspot typically lasts only a few days, but the rate of sunspot generation is such that the number of sunspots varies systematically (Figure C.16). The time between successive sunspot maxima or minima averages about 11 years with a range of 10 to 12 years. Also, the strong magnetic field associated with sunspots exhibits an approximate 22-year oscillation in polarity (the *double sunspot cycle*). Sunspot number reached a maximum in 1989, a minimum in 1996, and a maximum in 2000.

Satellite monitoring reveals that the sun's energy output varies directly with sunspot number; that

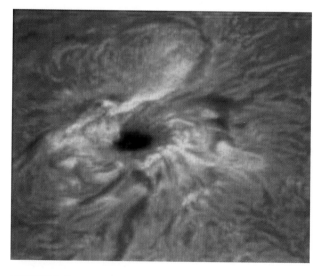

FIGURE C.15
A sunspot is a dark blotch that appears on the sun's photosphere, typically thousands of kilometers across, that develops where an intense magnetic field suppresses the flow of gases transporting heat from the sun's interior. [Courtesy of Dr. Donat G. Wentzel, University of Maryland and the National Optical Astronomical Observatories.]

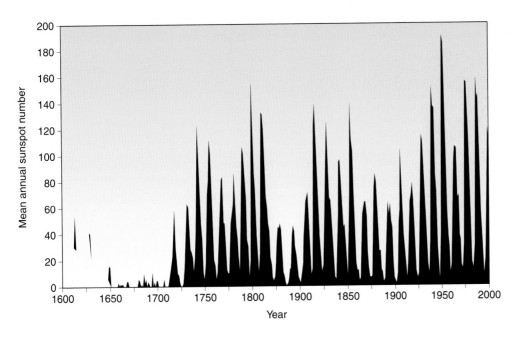

FIGURE C.16
Variation in mean annual sunspot number since the early 17th century. [Source: National Geophysical Data Center.]

is, a slightly brighter sun has more sunspots and a slightly dimmer sun has fewer sunspots. The variation in total solar energy output through one 11-year sunspot cycle amounts to less than 0.1%, with much of that change taking place in the ultraviolet (UV) portion of the solar spectrum. A brighter sun is associated with more sunspots because of a concurrent increase in bright areas, known as **faculae**, which appear near sunspots on the photosphere. Faculae dominate sunspots and the sun brightens. More sunspots may contribute to a warmer global climate and fewer sunspots may translate into a colder global climate.

How reasonable is the proposed link between global climate and sunspot number? In 1893, while studying sunspot records at the Old Royal Observatory at Greenwich, England, E. Walter Maunder discovered that sunspot activity was greatly diminished during the 70-year period between 1645 and 1715, now referred to as the *Maunder minimum*. The scientific community for the most part ignored Maunder's finding until the 1970s when John A. Eddy of the University Corporation for Atmospheric Research (UCAR) in Boulder, CO reinvestigated his work. Eddy pointed out that the Maunder minimum plus a prior period of reduced sunspot number, called the *Spörer minimum* (1400 to 1510 AD), occurred at about the same time as relatively cold phases of the Little Ice Age in Western Europe.

Furthermore, the Medieval Warm Period coincided with an interval of heightened sunspot activity between about 1100 and 1250 AD.

Skeptics dismiss the significance of the match between the Maunder minimum and a cooler climate. They argue that relatively cold episodes occurred in Europe just prior to and after the Maunder minimum (i.e., 1605-1615, 1805-1815), and relatively cool conditions did not persist throughout the Maunder minimum and were not global in extent. Furthermore, some scientists argue that a 0.1% variation in total solar energy output during the 11-year sunspot cycle is much too weak to significantly impact Earth's climate. Other scientists counter that certain mechanisms operating within the Earth-atmosphere system could amplify the effects of changes in total solar energy output, making the slight brightening and dimming of the sun an important player in Earth's climate variability.

Milankovitch cycles consist of regular variations in the precession and tilt of Earth's rotational axis and the eccentricity of its orbit about the sun (Figure C.17). Named for the Serbian astrophysicist Milutin Milankovitch (1879-1958) who studied them extensively in the 1920s and 1930s, these long-term, quasi-rhythmic changes are caused by gravitational influences exerted on Earth by other large planets, the moon, and the sun. Combined, Milankovitch cycles produce seasonal and latitudinal variations in the amount of solar radiation reaching the Earth-atmosphere system, driving climatic oscillations operating over periods of tens of thousands to hundreds of thousands of years. Milankovitch cycles were likely responsible for the major advances and recessions of the Laurentide ice sheet over North America during the Pleistocene.

Over a period of about 23,000 years, Earth's spin axis describes a complete circle, much like the

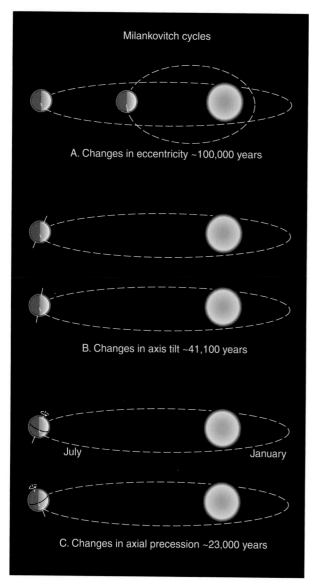

Milankovitch cycles

A. Changes in eccentricity ~100,000 years

B. Changes in axis tilt ~41,100 years

July   January

C. Changes in axial precession ~23,000 years

FIGURE C.17
Milankovitch cycles likely explain the large-scale fluctuations of
glacial ice cover during the Pleistocene. Note that diagrams
greatly exaggerate changes in Earth-sun geometry.

The tilt of Earth's spin axis varies from 22.1 degrees to 24.5 degrees and then back to 22.1 degrees over a period averaging about 41,000 years, the consequence of long-period changes in orientation of Earth's orbital plane with respect to its spin axis. As axial tilt increases, winters become colder and summers become warmer in both hemispheres. Earth's axial tilt has been slowly decreasing for about 10,000 years and will continue to do so for the next 10,000 years.

The shape of Earth's orbit about the sun changes from elliptical (high eccentricity) to nearly circular (low eccentricity) in an irregular cycle of 90,000 to 100,000 years. Variation in orbital eccentricity alters the distance between Earth and sun at aphelion and perihelion, thereby changing the amount of solar radiation received by the planet at those times of the year. When the orbit is highly elliptical, the amount of radiation received at perihelion would be 20% to 30% greater than at aphelion. Also, changes in eccentricity along with the precession cycle significantly influence the lengths of the individual astronomical seasons.

Milankovitch cycles do not alter appreciably the total amount of solar energy that enters the Earth-atmosphere system, but they do change significantly the latitudinal and seasonal distribution of incoming solar radiation. Milankovitch developed a numerical model based on the three cycles that calculated latitudinal differences in incoming solar radiation and the corresponding surface temperature for 600,000 years prior to the year 1800. He proposed that glacial climatic episodes began when Earth-sun geometry favored an extended period of increased solar radiation in winter and decreased solar radiation in summer at sub-polar latitudes. More intense winter radiation at these latitudes translates into somewhat higher temperatures, greater precipitable water, and more snowfall. Weaker solar radiation in summer meant that some of Canada's winter snow cover, especially north of 60 degrees N, would survive summer and repetition of such cool summers over many successive years would favor formation of glaciers. This process, according to Milankovitch, was the origin of the Laurentide ice sheet. At other times, Earth-sun geometry favored enhanced solar radiation in summer at sub-polar latitudes, triggering an interglacial (non-glacial) climate and recession of the Laurentide ice sheet.

Milankovitch's ideas on the cause of large-scale fluctuations in the planet's glacial ice cover were largely ignored for 50 years. Then in the mid-1970s, analysis of

wobble of a spinning top. This precession cycle changes the dates of *perihelion*, when Earth is closest to the sun, and *aphelion*, when Earth is farthest from the sun, increasing the summer-to-winter seasonal temperature contrast in one hemisphere and decreasing it in the other. Currently, perihelion is in early January and aphelion is in early July. In about 10,000 years, those dates will be reversed (perihelion in July and aphelion in January) and the seasonal contrast will be greater than it is now in the Northern Hemisphere and less in the Southern Hemisphere.

deep-sea sediment cores revealed regular shifts between glacial and interglacial climatic episodes over the past 450,000 years that closely correspond to periodic changes in Earth's three orbital parameters (i.e., precession, axial tilt, and eccentricity). This discovery strongly argues for Milankovitch cycles as the principal forcing of the regular large-scale waxing and waning of the planet's glacial ice cover during the Pleistocene. However, Milankovitch cycles do not account for the period of climate quiescence prior to the Pleistocene. Apparently, Milankovitch cycles (which likely operated throughout Earth history) were ineffective in initiating continental-scale glaciation until plate tectonics provided the appropriate boundary conditions, that is, landmasses at high latitudes and mountain ranges in place—conditions not achieved until onset of the Pleistocene.

Discovery of the possible relationship between systematic changes in Earth's orbital parameters and major glacial-interglacial climatic shifts prompted climatologists to search for other influences of Milankovitch cycles on climate variability. For example, J.E. Kutzbach of the University of Wisconsin-Madison has linked the Hypsithermal to Earth's precession cycle. Between 7000 and 5000 years ago, Earth was closest to the sun (perihelion) in June rather than in January as it is now. In the Northern Hemisphere, this meant about 5% more solar radiation during summer and 5% less solar radiation in winter. The consequent increase in air temperature contrast between summer and winter probably altered the planetary-scale circulation. For one, evidence suggests that the monsoon circulation of subtropical latitudes was more vigorous during the Hypsithermal than it is today.

CLIMATE AND VOLCANOES

The idea that volcanoes influence climate has been around for more than two centuries. Benjamin Franklin (1706-1790) proposed that eruption of Iceland's Laki volcano in the summer of 1783 was responsible for the severe winter of 1783-84. The unusually cool summer of 1816 (the so-called *year without a summer* in New England) followed the violent eruption of Tambora, an Indonesian volcano, in the spring of 1815. Several relatively cold years occurred on the heels of the 1883 eruption of Krakatau, also an Indonesian volcano. Is the relationship between volcanic eruptions and cooling real or coincidental?

Only explosive volcanic eruptions that are rich in sulfur oxide are likely to impact climate and then only

for a few years at most. A violent volcanic eruption can send sulfur oxide high into the stratosphere where it combines with moisture to form tiny droplets of sulfuric acid (H_2SO_4) and sulfate particles. Collectively, these droplets and particles are referred to as *sulfurous aerosols*. The small size of sulfurous aerosols (averaging about 0.1 micrometer in diameter), coupled with the absence of precipitation in the stratosphere, allow sulfurous aerosols to remain suspended in the stratosphere for many months to perhaps a year or longer before being cycled to Earth's surface. Successive volcanic eruptions has produced a permanent sulfurous aerosol veil in the stratosphere at altitudes of about 15 to 25 km (9 to 16 mi). Sulfur oxide emissions from clusters of volcanic eruptions temporarily thicken the stratospheric aerosol veil.

Research conducted after the 1991 eruption of Mount Pinatubo provided new insights on the relationship between sulfurous aerosols of volcanic origin and large-scale climate fluctuations. On 15-16 June 1991, Mount Pinatubo, located on Luzon Island in the Philippines, erupted violently injecting an estimated 20 megatons of sulfur dioxide into the stratosphere (Figure C.18). This was the most massive stratospheric volcanic aerosol cloud of the 20th century. By altering both incoming and outgoing radiation, these sulfurous aerosols affected temperatures in the stratosphere and troposphere, altered atmospheric circulation, and impacted surface air temperatures around the globe.

Sulfurous aerosols absorb both incoming solar radiation and outgoing infrared radiation causing warming of the lower stratosphere especially in the tropics. In the stratosphere, sulfurous aerosols also reflect sunlight to space. NASA scientists reported that in the months following the Mount Pinatubo eruption, satellite sensors measured a 3.8% increase in solar radiation reflected to space. Also, in the presence of chlorine, sulfurous aerosols destroy ozone (O_3) especially at high latitudes, allowing more solar UV radiation to reach Earth's surface. (This is essentially the same mechanism responsible for the Antarctic ozone hole described in Chapter 3.) In the two years following the Pinatubo eruption, mid-latitude ozone levels decreased by about 5%. The combination of stratospheric warming at low latitudes and ozone depletion at high latitudes strengthens the circumpolar vortex. And associated with this large-scale circulation change is a non-uniform change in surface temperatures; that is, some places cool while other places warm.

FIGURE C.18
Sulfurous aerosols produced in the spectacular eruption of Mount Pinatubo in the Philippines in June 1991 likely impacted global climate for a year or two.

Through its impacts on the flux of radiation and attendant changes in global circulation patterns, the Pinatubo eruption was likely responsible for the cool summer of 1992 over continental areas of the Northern Hemisphere (with surface temperatures up to 2 Celsius degrees or 3.6 Fahrenheit degrees lower than the long-term average). The Pinatubo eruption was also implicated in the temperature anomalies of the winters of 1991-92 and 1992-93, featuring higher than average temperatures over most of North America, Europe, and Siberia and lower than average temperatures over Alaska, Greenland, the Middle East, and China.

A violent sulfur-rich volcanic eruption is unlikely to lower the mean hemispheric or global surface temperature by more than about 1.0 Celsius degree (1.8 Fahrenheit degrees) although the magnitude of local and regional temperature change may be greater. The 1963 eruption of Agung in Bali, for example, decreased the mean temperature of the Northern Hemisphere an estimated 0.3 Celsius degree (0.5 Fahrenheit degree) for a year or two. The violent eruption of the Mexican volcano El Chichón on 4 April 1982 may have produced hemispheric cooling of about 0.2 Celsius degree (0.4 Fahrenheit degree). The Mount Pinatubo eruption temporarily interrupted the post-1970s global warming trend; from 1991 to 1992, the global mean annual temperature dropped 0.6 Celsius degree (1.1 Fahrenheit degrees).

CLIMATE AND HUMAN ACTIVITY

Human activity may contribute to climate variability. Humans modify the landscape (e.g., urbanization, clear-cutting of forests) and thereby alter radiational properties of Earth's surface. Urbanization elevates air temperatures in cities (Chapter 4). Combustion of fossil fuels (i.e., coal, oil, and natural gas) alters concentrations of certain key gaseous and aerosol components of the atmosphere. Of these human disturbances of the environment, the final one is most likely to impact climate on a global or hemispheric scale. Many atmospheric scientists as well as many public policy makers are concerned about the possible impact on global climate by the steadily rising concentrations of atmospheric carbon dioxide (CO_2) and other infrared-absorbing gases. Higher levels of these gases appear likely to enhance the so-called *greenhouse effect* and could contribute to warming on a global scale.

In 1957, systematic monitoring of atmospheric carbon dioxide began at NOAA's Mauna Loa Observatory in Hawaii under the direction of Charles D. Keeling of Scripps Institution of Oceanography. The observatory is situated on the slope of a volcano 3400 m (11,200 ft) above sea level in the middle of the Pacific Ocean—sufficiently distant from major industrial sources of air pollution that carbon dioxide levels are considered representative of at least the Northern Hemisphere. Also since 1957, atmospheric CO_2 has been monitored at the South Pole station of the U.S. Antarctic Program and that record closely parallels the one at Mauna Loa. The Mauna Loa (and South Pole) record shows a sustained increase in average annual atmospheric carbon dioxide concentration from 315.98 ppmv (parts per million by volume) in 1959 to 369.40 ppmv in 2000 (Figure 3.24). Superimposed on this upward trend is an annual carbon dioxide cycle due to seasonal changes in Northern Hemisphere vegetation.

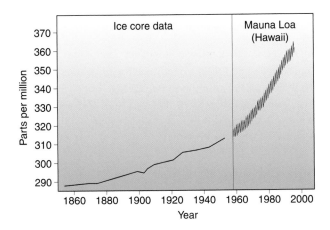

FIGURE C.19
Reconstructed and measured trend in atmospheric carbon dioxide concentration.

Levels of carbon dioxide fall during the growing season, reaching a minimum in October, and recover in winter, reaching a maximum in May.

The upward trend in atmospheric carbon dioxide was underway long before Keeling's monitoring and appears likely to continue well into the future (Figure C.19). Human contributions to the buildup of atmospheric CO_2 began roughly three centuries ago with the clearing of land for agriculture and settlement. Land clearing contributes CO_2 to the atmosphere via burning of vegetation, decay of wood residue, and reduced photosynthetic removal of carbon dioxide from the atmosphere. By the middle of the 19th century, growing dependency on coal burning associated with the beginnings of the Industrial Revolution triggered a more rapid rise in CO_2 concentration. Carbon dioxide is a byproduct of the burning of coal and other fossil fuels. The concentration of atmospheric CO_2 is now about 31% higher than it was in the pre-industrial era. Fossil fuel combustion accounts for roughly 75% of the increase in atmospheric carbon dioxide while deforestation (and other land clearing) is likely responsible for the balance. With continued growth in fossil fuel combustion, the atmospheric carbon dioxide concentration could top 550 ppmv (double the pre-industrial level) by the end of the 21st century.

Furthermore, rising levels of other infrared-absorbing gases (e.g., methane and nitrous oxide) could also enhance the greenhouse effect. Methane's concentration in the atmosphere has increased by an estimated 1060 ppb (parts per billion) (151%) since 1750. Methane (CH_4) is the product of organic decay in the absence of oxygen and its concentration may be rising because of more rice cultivation, cattle, landfills, and/or termites, all sources of methane. The atmospheric concentration of nitrous oxide (N_2O) has increased by an estimated 46 ppb (17%) since 1750 and the upward trend is likely linked to industrial air pollution. Although occurring in extremely low concentrations, methane and nitrous oxide are very efficient absorbers of infrared radiation and potentially could also contribute to global warming.

Atmospheric aerosols that are byproducts of human activity apparently have the opposite effect on temperatures at Earth's surface as greenhouse gases. These aerosols vary in size, shape, and chemical composition. Larger aerosols tend to settle out of the atmosphere rapidly whereas smaller ones may remain suspended for many days to weeks and can be transported thousands of kilometers by the wind, possibly impacting large-scale climate. Perhaps 90% of these aerosols are byproducts of fossil fuel burning in the Northern Hemisphere.

Sulfur oxides emitted from power plant smokestacks and boiler vent pipes combine with water vapor in air to produce tiny droplets of sulfuric acid and sulfate particles. These sulfurous aerosols appear to raise the atmosphere's reflectivity directly by reflecting sunlight to space and indirectly by acting as condensation nuclei that spur cloud development. Greater reflectivity cools the lower atmosphere. Sulfurous aerosols in the troposphere have a shorter-term impact on climate than carbon dioxide and other greenhouse gases. Rain and snow wash sulfurous aerosols from the atmosphere so that the residence time of these substances is typically only a few days. On the other hand, a CO_2 molecule typically resides in the atmosphere for upwards of three-quarters of a century before being cycled out by natural processes operating as part of the global carbon cycle.

CLIMATE AND EARTH'S SURFACE PROPERTIES

Earth's surface, which is mostly ocean water, is the prime absorber of solar radiation. Any change in the physical properties of Earth's water or land surfaces or in the relative distribution of ocean, ice, and land may affect Earth's radiation balance and climate.

Variations in mean regional snow cover may contribute to climate change because an extensive snow cover has a refrigerating effect on the atmosphere. Fresh-fallen snow typically reflects 80% or more of incident solar radiation, thereby substantially reducing the amount of solar heating and lowering the daily maximum temperature. Snow is also an excellent emitter of infrared radiation, so heat is efficiently radiated to space, especially on nights when the sky is clear. Because of this radiational feedback, a snow cover tends to be self-sustaining. This effect may be further enhanced by cyclone tracks that often follow the periphery of a regional snow cover, where horizontal air temperature gradients are relatively great. This places the snow-covered region on the cold, snowy side of migrating cyclones, thereby adding to the snow cover and reinforcing the chill. An unusually extensive winter snow cover favors persistence of an episode of cold weather. On the other hand, less than the usual extent of winter snow cover raises average air temperatures.

Whereas changes in regional snow cover might impact climate over the short-term (seasonal), changes in Earth's sea-ice or glacial-ice coverage are likely to have longer-lasting effects on climate. Global sea-ice (formed from the freezing of seawater) covers an average area of about 25 million square km (9.6 million square mi), about the area of the North American continent. Ice sheets, ice caps, and mountain glaciers cover a total area of about 15 million square km (5.8 million square mi), roughly 10% of the planet's land area. Sea-ice is distinct from floating ice shelves that form where glacial ice moves from land out over the ocean. Ice is much more reflective of incident solar radiation than either ocean or snow-free land so that any change in glacial or sea-ice cover would affect climate.

Polar climates are particularly sensitive to changes in sea-ice cover. Sea ice insulates the overlying air from warmer ocean water and a sea-ice surface reflects about five times more incident solar radiation than ocean water. If sea-ice cover shrinks, ice-free ocean expands and absorbs more solar radiation, sea-surface temperatures rise, and more ice melts. This runaway situation could rapidly reduce the sea-ice cover and greatly alter the flux of heat energy and moisture between the ocean and atmosphere.

Changes in ocean circulation and sea-surface temperatures contribute to large-scale climate change. As described in detail in Chapter 9, changes in sea-surface temperature patterns accompanying El Niño and La Niña have significant short-term influences on climate, typically impacting certain areas of the globe for periods of 12 to 18 months. Ocean circulation includes warm and cold surface currents and deep ocean currents that function as global-scale conveyor belts transporting heat throughout the world (Chapter 4). Regular changes in the strength of these conveyor belts may explain millennial-scale (1400 to 1500 year cycles) over the past 10,000 years. A strong conveyor belt brings relatively mild episodes to the North Atlantic and Western Europe whereas a weak conveyor belt triggers cooling.

The Climate Future

What does the climate future hold? Atmospheric scientists attempt to answer this question primarily by constructing global climate models that run on supercomputers.

GLOBAL CLIMATE MODELS

A **global climate model** simulates Earth's climate system. One type of global climate model consists of mathematical equations that together describe the physical interactions among the various components of the climate system, that is, the atmosphere, ocean, land, ice-cover, and biosphere. A global climate model differs from numerical models used for weather forecasting in that it predicts broad regions of expected positive and negative temperature and precipitation *anomalies* (departures from long-term averages), and the mean location of circulation features such as jet streams and principal storm tracks.

Global climate models, for example, are used to predict the potential climatic impacts of rising levels of atmospheric carbon dioxide (or other greenhouse gases). Inputting current boundary conditions (i.e., climate controls), a global climate model is first used to simulate present climates. Then, holding all other variables constant, the carbon dioxide concentration of the atmosphere is elevated (usually doubled) and the model is run to a new equilibrium state. By comparing the new climate state with the present climate, scientists deduce the probable impact of an enhanced greenhouse effect on patterns of temperature and precipitation.

Most modelers agree that global climate models are in need of considerable refinement. Today's models may not adequately simulate the role of small-scale weather systems (e.g., thunderstorms) or accurately portray local and regional conditions and may miss important feedback processes. A major uncertainty is the net feedback of clouds. Clouds cause both cooling (by reflecting sunlight to space) and warming (by absorbing and emitting to Earth's surface outgoing infrared radiation). The cooling effect prevails with an increase in low cloud cover whereas the warming effect prevails with an increase in high cloud cover.

Problems stem in part from the limited spatial resolution of global climate models. Today's models partition the global atmosphere into a three-dimensional grid of boxes with each box typically having an area of 250 square km (155 square mi) and a thickness of 1 km (0.6 mi). Limited spatial resolution in climate models stems from limited computational speed. Although today's supercomputers can perform 10 to 50 billion operations per second, the complexity of the climate system means that simulation of climate change over a century requires months of computing time. Much greater resolution of global climate models will come with development of faster supercomputers.

Another approach to predicting future climate is to identify the various factors that may have contributed to past fluctuations in climate and to extrapolate their influence into the future. Atmospheric scientists have probed climate records in search of (1) regular cycles that might be extended into the future and (2) analogs that might provide clues as to how the climate in specific regions responds to global-scale climate change. Few of the quasi-regular oscillations that appear in the climate record has practical value for climate forecasting at least over the next several centuries. Although the record of the climate past yields much useful information on how climate behaves through time, the search for analogs in the climate record for future global warming has been fruitless. Proposed analogs include relatively warm episodes of the mid-Holocene and the Eemian interglacial. But those analogs are inappropriate because the mid-Holocene and Eemian warming primarily affected seasonal temperatures with only a slight rise in global mean temperature. Furthermore, boundary conditions were different. During the mid-Holocene, sea level was lower, ice sheets were more extensive, and the dates of perihelion and aphelion were different than they are now or will be over the next several centuries. Although the level of atmospheric CO_2 trended upward during the mid-Holocene, the rate of increase (about 0.5 ppmv per century) was far lower than at present (more than 60 ppmv per century). Pre-Pleistocene analogs are also inappropriate because of the absence of ice sheets and significant differences in topography and land-ocean distribution.

ENHANCED GREENHOUSE EFFECT AND GLOBAL WARMING

Over the long term (the next 10 to 20 thousand years), Milankovitch cycles favor a return to Ice Age conditions. In the near term, that is, over the next century or so, scientists and public policy makers are concerned about a possible continuation of the present global warming trend. If all other controls of climate remain fixed, rising concentrations of atmospheric carbon dioxide and other greenhouse gases may cause global warming to persist throughout this century.

How much warming might accompany a doubling of atmospheric CO_2? Based on the most recent runs of global climate models, the *Intergovernmental Panel on Climate Change (IPCC)* in 2000 revised upward its earlier projections of the magnitude of global warming that could attend an enhanced greenhouse effect. Various models project that the globally averaged surface temperature will rise by 1.4 to 5.8 Celsius degrees (2.5 to 10.4 Fahrenheit degrees) over the period 1990 to 2100. Recall, however, that climate change is geographically non-uniform (in both magnitude and sign) so that this projected rise in global mean annual temperature is not necessarily representative of what might happen everywhere. For example, polar amplification suggests that global warming will be greater at higher latitudes. In any event, enhancement of the natural greenhouse effect could cause a climate change that would be greater in magnitude than any previous climate change in 10,000 years.

Global warming that arises from an enhanced greenhouse effect assumes that all other climate controls remain constant. To what extent that will actually happen is not known. Comparing the post-1957 trend in atmospheric CO_2 to the trend in mean annual global temperature strongly suggests that recent climate was shaped by many interacting factors. The rapid rise in CO_2 concentration was not accompanied by a consistent rise in global mean temperature over the same period. Recall, for example, that sulfurous aerosols from the June 1991 eruption of Mount Pinatubo apparently was responsible for significant global-scale cooling the following summer. Also as noted earlier, El Niño and La Niña influence climate in some areas of the globe over periods of 12 to 18 months. Furthermore, analysis of tiny air bubbles in cores extracted from the Greenland and Antarctic ice sheets indicate that during the Pleistocene Ice Age, atmospheric CO_2 varied between about 260 and 280 ppmv. Although CO_2 was consistently higher during milder interglacial climatic episodes than during colder glacial climatic episodes, fluctuations in CO_2-levels are out of phase with reconstructed variations in climate. For example, at the beginning of the last major glacial climatic episode, the decline in CO_2 concentration significantly lagged cooling in the Antarctic. Hence, fluctuations in atmospheric CO_2 may have been a *response* to large-scale climate oscillations rather than a *cause* of those oscillations.

POTENTIAL IMPACTS OF GLOBAL CLIMATE CHANGE

What are the potential impacts of a continued rise in global mean temperature? As noted earlier, global-scale trends in climate do not affect all regions in the same way. But if global warming persists, societal impacts would be extensive. Rising sea level, more frequent weather extremes, and disruption of agricultural systems are among the most serious potential consequences.

Higher temperatures would cause sea level to rise in response to melting of land-based polar ice sheets and mountain glaciers, and thermal expansion of seawater. (Melting of floating ice does not raise sea level.) Global scale warming is likely to be amplified at high latitudes (*polar amplification*) thereby threatening to partially melt the Greenland and Antarctic ice sheets. As noted in Chapter 6, the total amount of water in the global water cycle is essentially fixed so that a reduction in the glacial ice cover of the continents would be accompanied by a rise in sea level. (This inverse relationship between sea level and glacial ice has prevailed at least as far back as the beginning of the Pleistocene.) Also, seawater expands with rising temperature. As a general rule, mean sea level rises about 2 m (7 ft) for every 1 Celsius degree rise in water temperature. According to IPCC estimates, melting glaciers combined with thermal expansion of ocean water could raise mean sea level by 9 to 88 cm (4 to 35 in.) over the period 1990 to 2100. Higher sea level would accelerate coastal erosion, inundate some islands, and make low-lying coastal plains more vulnerable to storm surges.

According to a 1997 report by the U.S. Office of Science and Technology Policy, a 50-cm (20-in.) rise in sea level would cause a substantial loss of coastal land, especially along the Gulf and southern Atlantic coasts. Particularly vulnerable is South Florida where a third of the Everglades has an elevation of less than 30

cm (12 in.). Globally, a 50-cm (20-in.) rise in sea level would double the number of people at risk from storm surges from about 45 million at present to more than 90 million, not counting any population growth in the coastal zone.

While higher temperatures would mean higher sea level, the level of the Great Lakes is likely to fall. Higher summer temperatures coupled with less winter ice cover on the Great Lakes are likely to translate into greater evaporation. And less winter snowfall would reduce spring runoff. Depending on the model used, forecasts call for a drop in mean water level on Lake Michigan of up to 2 m (6.5 ft) by the year 2070.

Global warming may translate into more severe drought in some locations and floods in other places. Unfortunately, global climate models do not agree on the locations of precipitation extremes. A recent assessment of predictions generated by different state-of-the-art coupled atmosphere/ocean climate models underscores the rudimentary state of climate prediction at the regional scale. For example, the Canadian Climate Centre model predicts that global warming will be accompanied by drier summer soil conditions over the eastern two-thirds of the coterminous U.S., whereas the U.K. Hadley Center for Climate Research and Prediction model projects wetter soils in the same area.

Whereas regional response to global warming is uncertain, scientists and policy makers are confident that localities most vulnerable to water resource problems are places where the quality and quantity of fresh water are already a problem, that is, in semi-arid and arid regions. In the U.S., water supply problems are most likely in the drainage basins of the Missouri, Arkansas, Rio Grande, and lower Colorado Rivers. Water shortages in the Middle East and Africa may heighten political tensions especially in nations that depend on water supplies originating outside their borders.

Although global warming is not expected to substantially impact total global food production, higher temperatures and more frequent drought may severely affect certain regions of the world. Developed nations have the capacity to adapt their agricultural systems to climate change. While overall U.S. food production is expected to climb throughout the 21st century, there will be some losers. For example, greater demand for irrigation waters may further stress important reservoirs of groundwater (e.g., the High Plains aquifer). On the other hand, the agricultural systems of developing nations are much more vulnerable to the adverse effects of warming and are most likely to experience a significant decline in food production.

Are there signs that recent global warming might be already impacting the environment? Between 1980 and 1994, the frequency of weather extremes (e.g., drought, excessive heat, heavy rains) increased across the U.S. In tropical and subtropical latitudes, many mountain glaciers are retreating at accelerated rates. And data from submarines, ice-cores, and satellites indicate that since the 1950s, the Arctic sea-ice cover has thinned and diminished in area. On the other hand, in early 2002, scientists at the California Institute of Technology and the University of California-Santa Cruz reported that based upon their satellite analysis of flow measurements of the Ross ice streams, the West Antarctic ice sheet appears to be thickening.

Norwegian researchers reported that the area of multi-year ice cover in the Arctic decreased by about 7% per decade from 1978 to 1998. Their findings were derived from satellite monitoring of the spectrum of microwave energy emitted by the ice. Comparing ice-thickness measurements made by U.S. Navy submarines from 1958 to 1976 with those made during the *Scientific Ice Expedition* program in 1993, 1996, and 1997, University of Washington scientists found that ice had thinned from an average 3.1 m (10 ft) to an average 1.8 m (6 ft). Thinning at the rate of 15% per decade translates into a total volumetric ice loss of about 40% in three decades. (Upward-looking acoustic sounders were used to map ice depth.) Furthermore, scientists visiting the Arctic in Fall 2000 reported an ice-free patch more than a mile across. This large an ice-free area is very unusual in the Arctic and raises concern about a possible ice-albedo feedback mechanism (discussed earlier in this chapter) that would accelerate melting of the sea-ice cover and amplify warming.

Less sea ice in the Arctic may be the direct consequence of higher air temperatures or indirectly the result of changes in ocean circulation (i.e., greater input of warmer Atlantic water under the Arctic ice). Some scientists argue that higher air and ocean temperatures in the Arctic (leading to shrinking ice cover) are natural variations in climate associated with the so-called Arctic Oscillation (AO) and that the ice-cover will return to normal as the AO changes phase. Related to the North Atlantic Oscillation (Chapter 9), the AO consists of a seesaw variation in air pressure between the North Pole and the margins of the polar region. These pressure changes alter winds over the Arctic. During the present AO phase (operating since 1989) winds have been delivering warmer than usual air (and water) into the Arctic.

NASA research results released during Summer 2000 conclude that while the central interior of the Greenland ice sheet shows no sign of thinning, about 70% of the margin is thinning substantially. Two separate research teams—one using a global positioning system (GPS) to monitor ice flow and the other using an airborne laser altimeter to measure ice thickness—reached the same conclusion based on observations made between 1993 and 1999. The maximum melting rate at the margin is about 1 m per year. An estimated 50 cubic km (12 cubic mi) of Greenland's ice melts each year, enough to raise sea level by 0.13 mm annually.

In Spring 2000, NOAA scientists reported that the world ocean (Atlantic, Pacific, and Indian combined) warmed significantly between 1955 and 1995. The greatest warming occurred in the upper 300 m (984 ft) of the ocean and amounted to 0.31 Celsius degree (0.56 Fahrenheit degree). Water in the upper 3000 m (9840 ft) warmed by an average 0.06 Celsius degree (0.11 Fahrenheit degree). In February 2002, researchers at Scripps Institution of Oceanography reported that mid-depth (between 700 and 1100 m) temperatures in the Southern Ocean rose 0.17 Celsius degree between the 1950s and 1980s. Although these magnitudes of temperature change sound trivial, water has an unusually high specific heat so that even a very small change in temperature of such vast volumes of water represents a tremendous heat input into the ocean. Sequestering of vast quantities of heat in the ocean may help explain why global warming during the 20th century was less than some climate models predicted based on buildup of greenhouse gases. Heating of the ocean has partially offset warming of the lower atmosphere. This finding also underscores the moderating role of the ocean in global climate change.

Prospects of higher sea-surface temperatures prompted some scientists to predict an upturn in the number and intensity of tropical cyclones (i.e., hurricanes and tropical storms). However, as discussed in Chapter 12, relatively high sea-surface temperature is one of several factors required for tropical cyclone formation. In fact, some climate models predict stronger winds aloft accompanying a warmer climate that would inhibit tropical cyclone development. In January 2001, the IPCC concluded that in spite of the global warming trend, there is no evidence that top wind speeds or rainfall had increased in tropical cyclones during the second half of the 20th century. Resolving the likely impact of global warming on tropical cyclones awaits the development of more realistic climate models.

Conclusions

Climate encompasses both averages and extremes in weather for some locality over some specified time period (either 30 years or the entire period of record). Whereas average values of many climatic elements are useful, for some such as precipitation, the median value is a more appropriate descriptor of climate. Earth is a mosaic of many climate types that are shaped by a variety of many interacting controls, including latitude, elevation, topography, proximity to large bodies of water, and prevailing atmospheric circulation. Genetic, empirical, and numerical schemes have been devised to classify climates.

Climate varies not only spatially but also with time. In most places, the reliable instrument-based climatic record extends back only 100 years or so. For information on climate prior to this period, climatologists must rely on a variety of documentary, geological, and biological climatic indicators. Although the climate record loses detail, continuity, and reliability with increasing time before present, it is evident that climate is inherently variable and varies over a wide spectrum of time scales.

The interaction of many factors is responsible for the inherent variability of climate. Although we can isolate specific climatic controls that are internal or external to the Earth-atmosphere system, our understanding of how these controls interact is far from complete. This state of the art limits the ability of atmospheric scientists to forecast the climatic future. We should therefore be wary of simplistic scenarios of the climatic future, for not enough is known about the causes of climatic variations.

Continued research on climate is needed. It is reasonable to assume that physical laws govern climatic change; that is, variations in climate are not arbitrary, random events. As scientists more fully comprehend the laws regulating climatic variability, their ability to predict the climatic future will improve. Meanwhile, trends in climate must be monitored closely, especially in view of the strong dependence of our food and water supplies and energy demand on climate. In spite of, or perhaps because of, the scientific uncertainty, some experts argue that we should plan now for the worst possible future climate scenario.

Basic Understandings

- Climate is defined as the weather of some locality averaged over some time period plus extremes in weather. The climatic norm (or normal) at some place encompasses the total variability of the climate record, that is, averages plus extremes.

- By international convention, climatic norms are computed from averages of weather elements compiled over a specified 30-year period. Current climatic summaries are based on weather records from 1971 to 2000.

- Departures from long-term climatic averages, called anomalies, do not occur with the same sign or magnitude everywhere. Large-scale trends in climate are also geographically non-uniform.

- Controls of climate consist of latitude, elevation, topography, proximity to large bodies of water, and the prevailing atmospheric circulation.

- Mean annual isotherms roughly parallel latitude circles, underscoring the influence of solar radiation and solar altitude on climate. Through the course of a year, mean monthly isotherms in both hemispheres follow the sun, shifting north and south in tandem.

- The global pattern of mean annual precipitation (rain plus the liquid equivalent of snowfall) exhibits great spatial variability in response to topography, distribution of land and sea, and the planetary-scale circulation.

- For information on climates prior to the era of reliable instrument-based records, scientists rely on climatic inferences drawn from historical documents, and geological/biological evidence such as bedrock, fossils, pollen, tree growth-rings, glacial ice cores, and deep-sea sediment cores.

- Plate tectonics complicate climate reconstruction of periods spanning hundreds of millions of years. In the context of geologic time, topography and the geographical distribution of continents and the ocean are variable climate controls.

- More than 40 million years ago, global climate began shifting toward cooler, drier, and more variable conditions. Scientists have implicated tectonic forces and the building of the Colorado

Plateau, Tibetan Plateau, and Himalayan Mountains as the principal cause of this climatic change. Cooling after 40 million years ago culminated in the Pleistocene Ice Age that began about 1.7 million years ago. During the Pleistocene, the climate shifted numerous times between glacial climatic episodes (favoring expansion of glaciers) and non-glacial climatic episodes (favoring shrinkage of glaciers).

- Cooling during the Pleistocene was geographically non-uniform in magnitude with maximum cooling at high latitudes and minimum cooling in the tropics. This latitudinal variation in temperature change is known as *polar amplification.*

- Notable post-glacial climatic episodes were the relatively warm Hypsithermal, during the mid-Holocene, the Medieval Warm Period, from about 950 to 1250 AD, and the Little Ice Age, from about 1400 to 1850 AD. In all cases, the temperature change was geographically non-uniform in magnitude.

- The instrument-based record of global mean temperature indicates a gradual warming trend since the 1880s, interrupted by cooling at mid and high latitudes of the Northern Hemisphere from about 1940 to 1970.

- Analysis of the climate record reveals many useful observations regarding the temporal behavior of climate. Climate is inherently variable; climate change is geographically non-uniform; climate change may involve changes in mean values of temperature or precipitation as well as changes in the frequency of weather extremes; climate change tends to be abrupt rather than gradual; only a few reliable cycles operate in the climate record; and climate change can impact society.

- Matching some forcing mechanism with a climate response based on similar periods of oscillation is no guarantee of a real physical relationship. Factors that could alter the global radiative equilibrium and change Earth's climate include fluctuations in solar energy output, volcanic eruptions, certain human activities, and changes in Earth's surface properties.

- Changes in the sun's total energy output at all wavelengths apparently are related to sunspot activity. Solar output varies directly and minutely with sunspot number; that is, a slightly brighter sun has more sunspots (because of a concurrent increase in faculae, bright areas), and a slightly dimmer sun exhibits fewer sunspots.

- Milankovitch cycles apparently drive climatic oscillations operating over tens of thousands to hundreds of thousands of years and were likely responsible for the major advances and recessions of the Laurentide ice sheet over North America during the Pleistocene. They consist of regular variations in precession and tilt of Earth's rotational axis and the eccentricity of its orbit about the sun. These same cycles show up in deep-sea sediment cores that date from the Pleistocene Ice Age.

- Only violent volcanic eruptions rich in sulfur oxide gases are likely to impact global or hemispheric-scale climate. Such an eruption has a non-uniform effect on surface air temperature but is unlikely to lower the mean hemispheric temperature by more than about 1.0 Celsius degree (1.8 Fahrenheit degrees).

- Human activity may impact global-scale climate by increasing the concentration of greenhouse gases (causing warming) or sulfurous aerosols (causing cooling). The post-industrial upward trend in atmospheric carbon dioxide is primarily due to burning of fossil fuels and to a lesser extent the clearing of vegetation.

- Earth's surface, which is mostly ocean water, is the prime absorber of solar radiation so that any change in the physical properties of the water or land surfaces or in the relative distribution of ocean and land may impact the global radiation balance and climate.

- A global climate model consists of a series of mathematical equations that describes the physical laws that govern the interactions among the various components of the climate system. They are used to predict broad regions of expected positive and negative temperature and precipitation anomalies.

- The climate record does not contain reliable cycles that can be used to forecast the climate over a period of several centuries. Also, no satisfactory analogs exist in the climate record that would allow scientists to predict the regional response to global-scale climate change.

- Current global climate models predict that significant global warming will accompany a doubling of atmospheric CO_2 (possible by the close of this century). Warming may be greater in magnitude than any other prior climate change of the past 10,000 years.

- Projections of the magnitude of future global warming arising from CO_2-enhancement of the greenhouse effect assumes that all other controls of climate remain constant. To what extent this will happen is not known.

- Global warming is likely to cause sea level to rise in response to melting of polar ice sheets and expansion of seawater. On the other hand, warming is likely to cause levels of the Great Lakes to fall in response to increased evaporation and reduced winter snow pack.

- Global warming could alter the water cycle so that some areas experience more severe droughts whereas other areas have a greater frequency of floods. Arid and semi-arid localities already experiencing water supply problems are most vulnerable to the adverse effects of global warming.

APPENDIX I

CONVERSION FACTORS

	Multiply	By	To obtain
LENGTH	inches (in.)	2.54	centimeters (cm)
	centimeters	0.3937	inches
	feet (ft)	0.3048	meters (m)
	meters	3.281	feet
	statute miles (mi)	1.6093	kilometers (km)
	statute miles (mi)	0.869	nautical miles
	kilometers	0.6214	statute miles
	kilometers	0.54	nautical miles
	nautical miles	1.852	kilometers
	kilometers	3281	feet
	feet	0.0003048	kilometers
SPEED	miles per hour (mph)	1.6093	kilometers per hour (kph)
	miles per hour	1.1516	knots (kts)
	miles per hour	0.447	meters per second (m/s)
	knots	0.869	miles per hour
	knots	0.5148	meters per second
	kilometers per hour	0.6213	miles per hour
	kilometers per hour	0.540	knots
	meters per second	1.944	knots
	meters per second	2.237	miles per hour
WEIGHTS AND MASS	ounces (oz)	28.35	grams (g)
	grams	0.0353	ounces
	pounds (lb)	0.4536	kilograms (kg)
	kilograms	2.205	pounds
	tons	0.9072	metric tons
	metric tons	1.102	tons
LIQUID MEASURE	fluid ounces (fl oz)	0.0296	liters (L)
	gallons (gal)	3.785	liters
	liters	0.2642	gallons
	liters	33.815	fluid ounces

AREA	acres (A)	0.4047	hectares (ha)
	square yards (yd^2)	0.8361	square meters (m^2)
	square miles (mi^2)	2.590	square kilometers (km^2)
	hectares	0.010	square kilometers
	hectares	2.471	acres
PRESSURE/ FORCE	pounds force (lb)	4.448	newtons (N)
	newtons	0.2248	pounds
	millimeters of mercury at 0 °C	133.32	pascals (Pa; N per m^2)
	pounds per square inch (psi)	6.895	kilopascals (kPa; 1000 pascals)
	pascals	0.0075	millimeters of mercury at 0 °C
	kilopascals	0.1450	pounds per square inch
	bars	1000	millibars (mb)
	bars	100000	pascals
	bars	0.9869	atmospheres (atm)
ENERGY	joules (J)	0.2389	calories (cal)
	kilocalories (kcal)	1000	calories
	joules	1.0	watt-seconds (W-sec)
	kilojoules (kJ)	1000	joules
	calories	0.00397	Btu (British thermal units)
	Btu	252	calories
POWER	joules per second	1.0	watts (W)
	kilowatts (kW)	1000	watts
	megawatts (MW)	1000	kilowatts
	kilocalories per minute	69.93	watts
	watts	0.00134	horsepower (hp)
	kilowatts	56.87	Btu per minute
	Btu per minute	0.235	horsepower

APPENDIX II

MILESTONES IN THE HISTORY OF ATMOSPHERIC SCIENCE

ca. 525 B.C.	Greek philosopher Anaximenes of Miletus proposed that winds, clouds, rain and hail are formed by thickening of air, the primary substance.
ca. 500 B.C.	Parmenides classified world climates by latitude as torrid, temperate, or frigid.
ca. 400 B.C.	Rainfall measured in India.
334 B.C.	Aristotle produced his *Meteorologica*, the first work on the atmospheric sciences.
330 B.C.	Hippocrates writes *Air, Waters, and Places,* a treatise on climate and medicine.
ca. 135 B.C.	The Chinese writer Han Ying first notes the hexagonal shape of snowflakes in *Moral discourses illustrating the Han text of the "Book of Songs."*
61 A.D.	Seneca complained of the air pollution of Rome.
1304	Theodoric of Friebourg (now in Germany) writes *De iride* (On the Rainbow) describing his experiments with bulbs filled with water.
1442	Rain gauges used in Korea.
1450	Cardinal Cusa invented the balance hygrometer.
1591	The English mathematician Thomas Harriot (1560-1621) notes that snowflakes have six sides.
1592	Galileo Galilei invented the thermoscope, forerunner of the thermometer.
1621	Dutch physicist Willebrord Snell (1591-1626) discovers his law of refraction of light.
1637	French philosopher René Descartes publishes *Discours de la méthode,* which includes his ideas about the rainbow and cloud formation.
1638	Galileo Galilei determined that air has weight.
1641	Ferdinand II of Tuscany constructs a thermometer containing liquid.
1643	Torricelli invented the barometer.

1644	Rev. John Campanius Holm recorded the first weather data in America near the present site of Wilmington, DE.
1648	Florin Périer ascended the Puy-de-Dome, France, and demonstrated that air pressure falls with increasing altitude.
1660	Von Guericke noted that a severe storm followed a sudden drop in barometric pressure.
ca. 1670	Mercury was used in thermometers for the first time.
1683	Edmund Halley published the first comprehensive map of winds along with a partial explanation of the trade winds.
1686	Edmund Halley proposed an explanation for monsoons.
1687	Isaac Newton developed his three laws of motion.
1687	French physician Guillaume Amontons (1663-1705) invents a hygrometer.
1709	German physicist Gabriel Daniel Fahrenheit (1686-1736) constructs a thermometer with alcohol as the working fluid.
1714	Fahrenheit introduced the Fahrenheit temperature scale.
1730	French entomologist and physicist René Réamur (1683-1757) introduces a thermometer using a water/alcohol mixture.
1735	German explorer Johann Gmelin discovers permafrost in Siberia.
1735	George Hadley proposed the Hadley cell circulation.
1742	Swedish astronomer Anders Celsius introduced the Celsius temperature scale.
1743	Jean Pierre Christin inverted the fixed points on Celsius' scale, producing the scale used today.
1743	Benjamin Franklin deduced the progressive movement of a hurricane along the U.S. East Coast.
1747	Benjamin Franklin began a series of experiments on electricity that led to the use of the lightning rod.
1749	Franklin attached a lightning rod to his house in Philadelphia.
1749	In Glasgow, Scotland, Alexander Wilson used thermometers attached to a kite to obtain the first temperature profile of the lower atmosphere (up to perhaps 60 m or 200 ft).
1760	Joseph Black formulated the concept of specific heat.
1766	Cavendish discovered hydrogen.
1770	Rutherford discovered nitrogen.
1774	Scheele and Priestly discovered oxygen.
1781	Systematic weather recording began at New Haven, CT.
1800	Sir William Herschel discovered energy transfer in the infrared (IR).

1802	British pharmacist and amateur meteorologist Luke Howard (1722-1846) developed his classification of cloud types.
1802	J. L. Gay-Lussac published the *gas law*; Dalton defined *relative humidity*.
1804	French scientists J. L. Gay-Lussac and Jean Biot conducted the first manned balloon exploration of the atmosphere, reaching a maximum altitude of 7000 m (23,000 ft).
1805	British admiral Sir Francis Beaufort (1774-1857) proposed a scale of winds (now known as the *Beaufort scale*).
1820	English chemist John Frederic Daniell (1790-1845) invented the dew-point hygrometer.
1821	Swiss geologist Ignatz Venetz (b. 1788) proposed that glaciers formerly occurred throughout Europe.
1825	E.F. August developed the psychrometer.
1827	Baron Joseph Fourier proposed the possibility of CO_2-induced global warming.
1835	Gaspard Gustav de Coriolis demonstrated quantitatively the effects of the Earth's rotation on large-scale motion.
1837	The Swiss-born American naturalist Jean Louis Agassiz (1807-1873) used the term *Eiszeit* (Ice Age) for the first time.
1840	Agassiz publishes his *Etudes des glaciers* (Studies on Glaciers).
1842	Johann Christian Doppler first explained the Doppler effect.
1843	Samuel Heinrich Schwabe discovered the sunspot cycle.
1844	Aneroid barometer invented.
1853	James Coffin suggested that there are three distinct wind zones in the Northern Hemisphere.
1856	William Ferrel proposed a scheme of the general circulation of the atmosphere consisting of three cells.
1857	Lorin Blodget published *Climatology of the United States*.
1857	The Dutch meteorologist Christopher Buys Ballot (1817-90) relates wind direction to the location of a low pressure center (*Buys Ballot's law*).
1862	James Glaisher and Henry Coxwell set manned balloon altitude record of 9000 m (29,500 ft) over Wolverhampton, England.
1869	In Cincinnati, OH Cleveland Abbe (1838-1916) prepared the first regular weather maps for part of the United States.
1870	President Ulysses S. Grant signed legislation establishing a national weather service operated by the U.S. Army Signal Corps.
1871	Cleveland Abbe appointed chief meteorologist of the new national weather service.
1874	American geologist Thomas Chowder Chamberlin (1843-1928) suggested that there were several Ice Ages separated by nonglacial epochs.

1878	The International Meteorological Organization (IMO) was founded.
1884	Ludwig Boltzmann derived the Stefan-Boltzmann law.
1885	First reported sighting of a noctilucent clouds.
1888	Gustavus Hinrichs, an Iowa weather researcher, first used the word *derecho* for a straight-line windstorm.
1891	Luis Carranza, President of the Lima Geographical Society, describes a countercurrent flowing north to south from Paita to Pacasmayo and called El Niño by the local fishermen.
1891	The national weather service was transferred from military to civilian control in a new Weather Bureau within the U.S. Department of Agriculture.
1893	British astronomer E. Walter Maunder (1851-1928) identified a period of low solar activity between 1645 and 1715 (now known as the *Maunder minimum*).
1894	Wilhelm Wien developed his *displacement law of radiation*.
1894	First sounding of the atmosphere using a self-recording thermometer attached to a kite at Blue Hill Observatory, Milton, MA.
1896	Swedish chemist Svante Arrhenius (1859-1927) points out that increasing carbon dioxide in the atmosphere can enhance the greenhouse effect.
1900	Russian-born German climatologist Wladimir Peter Köppen (1846-1940) first published his original climate classification scheme.
1902	French meteorologist Léon Philippe Teisserenc de Bort (1855-1913) identified and named the troposphere and stratosphere as layers of the atmosphere.
1902	Swedish physicist Vagn Walfrid Ekman (1874-1954) produced a mathematical model describing the interaction of wind on ocean water (*Ekman spiral*).
1904	Vilhelm Bjerknes laid the foundations of synoptic meteorology.
1905	Ekman explained flow in the atmospheric boundary layer.
1916	Pyranometer for measuring global radiation developed.
1917	Vilhelm Bjerknes formulated polar front theory.
1920	Serbian astrophysicist Milutin Milankovich (1879-1958) proposed a theory of long-term cyclic climate change based on regular changes in Earth-sun geometry (*Milankovich cycles*).
1920	Theory of atmospheric front was developed by V. Bjerknes, H. Solberg, and J. Bjerknes.
1924	Sir Gilbert Walker discovered the southern oscillation.
1927	G. Stüve introduces a type of thermodynamic diagram (the *Stüve diagram*).
ca. 1928	First radiosonde developed.
1929	Robert H. Goddard conducted the first rocket probe of the atmosphere.

1933	Tor Bergeron published his paper on *Physics of Clouds and Precipitation*.
1935	Radar invented.
1937	Carl-Gustaf Rossby introduced techniques for forecasting upper-level westerly waves.
1939	Work by Walter Findeisen supported Bergeron's theory of precipitation development in cold clouds (later known as the *Bergeron-Findeisen process*).
1941	Radar applied to weather systems for the first time.
1944	Hurd Curtis Willett produced atmospheric cross sections showing the jet stream.
1946	Vincent J. Schaefer and Irving Langmuir performed the first cloud-seeding experiments.
1946	John Von Neumann began mathematical modeling of the atmosphere.
1947	World Meteorological Organization (WMO) founded.
1947	American chemist Willard Frank Libby (1908-1980) invents the radiocarbon dating techniques, a valuable tool in reconstructing late-glacial climates.
1947	First photographs of the Earth's cloud cover obtained by a V2 rocket at altitudes of 110 to 165 km (68 to 102 mi).
1948	First rocket probes of the upper atmosphere.
1950	A team led by the Hungarian-born American mathematician John Von Neumann (1903-1957) produced the first computer-generated weather forecasts using the ENIAC computer.
1951	The World Meteorological Organization (WMO) commences operation.
1951	The American geographer and climatologist Arthur N. Strahler (1918-) introduced a climate classification scheme based on air masses.
1956	The British government attempted to reduce smog by limiting smoke emissions.
1957	CO_2 measurements began at Mauna Loa and Antarctica.
1959	A temperature-humidity index first introduced by the U.S. Weather Bureau.
1960	The United States sent the first weather satellite, TIROS-I, into orbit.
1961	American meteorologist Edward N. Lorenz (1917-) observed that computer predictions of weather are highly sensitive to small differences in initial conditions, an aspect of chaos theory.
1963	Edward N. Lorenz applied chaos theory to meteorology.
1964	U.S. government introduced the first Clean Air Act to set national air pollution standards.
1966	Jacob Bjerknes demonstrated a relationship between El Niño and the southern oscillation.
1967	Verner E. Suomi of the University of Wisconsin-Madison processed the first geostationary satellite image.

1970s	American engineer Herbert Saffir (1917-) and meteorologist Robert Simpson (1912-) developed their hurricane intensity scale (the *Saffir-Simpson scale*).
1971	Japanese-born American meteorologist Tetsuya Fujita (1920-1998) with Allen Pearson introduced their tornado intensity scale (*Fujita* or *Fujita-Pearson scale*).
1972	Earth Resources Technology Satellite 1 (later called Landsat-1) was launched.
1974	The Global Atmospheric Research Program (GARP) conducted the GARP Atlantic Tropical Experiment (GATE) to improve understanding of the tropical atmosphere.
1974	M. J. Molina and F. S. Rowland warned of the CFC threat to the stratospheric ozone layer.
1975	First GOES (Geostationary Operational Environmental Satellite) was launched.
1975	The European Space Agency was formed.
1977	The first Meteosat weather satellite was launched by the European Space Agency.
1984	The British Antarctic Survey obtained evidence of a seasonal hole in the ozone layer over Antarctica.
1985	The Vienna Convention for the Protection of the Ozone Layer was convened.
1985	The Tropical Ocean-Global Atmosphere Program (TOGA) began.
1987	International adoption of the *Montreal Protocol on Substances That Deplete the Ozone Layer*.
1987	*The Joint Global Ocean Flux Study (JGOFS)* initiated to study the fluxes of carbon in the ocean.
1988	The Global Energy and Water Cycle Experiment (GEWEX) launched to examine the global hydrologic cycle.
1990	First Doppler-radar was introduced into meteorological service.
1992	*UN Framework Convention on Climate Change* was ratified.
1997	The Kyoto Protocol addressed reductions in greenhouse gas emissions.
1999	Landsat-7 was launched.

APPENDIX III

CLIMATE CLASSIFICATION

- Tropical Humid Climates
- Dry Climates
- Subtropical Climates
- Snow Forest Climates
- Polar Climates
- Highland Climates

One of the most widely used climate classification systems was designed by German climatologist and plant geographer Wladimir Köppen (1846-1940) and subsequently modified by his students R. Geiger and W. Pohl. The Köppen system is an empirical approach to organizing Earth's myriad of climate types. Recognizing that indigenous vegetation is a natural indicator of regional climate, Köppen and his students looked for patterns in annual and monthly mean temperature and precipitation, which closely correspond to the limits of vegetative communities thereby revealing broad-scale climatic boundaries throughout the world. Records of annual and monthly mean temperature and precipitation are sufficiently long and reliable in many parts of the world that they serve as a good first approximation of climate. Since its introduction early in the 20th century, Köppen's climate classification has undergone numerous and substantial revisions by Köppen himself and by other climatologists and has had a variety of applications.

As shown in the Table, the Köppen climate classification system identifies six main climate groups; four are based on temperature, one is based on precipitation, and one applies to mountainous regions. Köppen's scheme uses letters to symbolize major climatic groups: (A) Tropical Humid, (B) Dry, (C) Subtropical (Mesothermal), (D) Snow Forest

TABLE
Köppen-Based Climate Classification

Tropical humid (A)
  Af   tropical wet
  Am   tropical monsoon
  Aw   tropical wet-and-dry

Dry (B)
  BS   steppe or semiarid (BSh, BSh)
  BW   arid or desert (BWh, BWk)
  BWn   foggy desert

Subtropical (C)
  Cs   subtropical dry summer (Csa, Csb)
  Cs   subtropical dry winter
  Cf   subtropical humid (Cfa, Cfb, Cfc)

Snow forest (D)
  Dw   dry winter (Dwa, Dwb, Dwc, Dwd)
  Df   year-round precip. (Dfa, Dfb, Dfc, Dfd)
  Ds   dry summer

Polar (E)
  Et   tundra
  Ef   ice cap

Highland (H)

(Microthermal), (E) Polar, and (H) Highland. Additional letters further differentiate climate types.

Tropical Humid Climates

Tropical humid climates (A) constitute a discontinuous belt straddling the equator and extending poleward to near the Tropic of Cancer in the Northern Hemisphere and the Tropic of Capricorn in the Southern Hemisphere. Mean monthly temperatures are high and exhibit little variability throughout the year. The mean temperature of the coolest month is no lower than 18 °C (64 °F), and there is no frost. The temperature contrast between the warmest and coolest month is typically less than 10 Celsius degrees (18 Fahrenheit degrees). In fact, the diurnal (day-to-night) temperature range generally exceeds the annual temperature range. This monotonous air temperature regime is the consequence of consistently intense incoming solar radiation associated with a high maximum solar altitude and little variation in the period of daylight throughout the year.

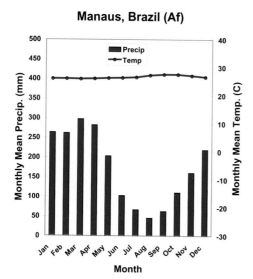

Although tropical humid climate types are not readily distinguishable on the basis of temperature, there are important differences in precipitation regimes. Tropical humid climates are subdivided into tropical wet (Af), tropical monsoon (Am), and tropical wet-and-dry (Aw). Although these climate types generally feature abundant annual rainfall, more than 100 cm (40 in.) on

average, their rainy seasons differ in length and, in the case of Am and Aw, there is a pronounced dry season and wet season. In tropical wet climates, the yearly average rainfall of 175 to 250 cm (70 to 100 in.) supports the world's most luxurious vegetation. Tropical rainforests occupy the Amazon Basin of Brazil, the Congo Basin of Africa, and the islands of Micronesia. For the most part, rainfall is distributed uniformly throughout the year, although some areas experience a brief (one or two month) dry season. Rainfall occurs as heavy downpours in frequent thunderstorms triggered by local convection and the intertropical convergence zone (ITCZ). Convection is largely controlled by solar radiation and rainfall typically peaks in midafternoon, the warmest time of day. Because the water vapor concentration is very high, even the slightest cooling at night leads to saturated air and the formation of dew or radiation fog, giving these regions a sultry, steamy appearance.

Tropical monsoon (Am) climates feature a seasonal rainfall regime with extremely heavy rainfall during several months and a lengthy dry season. The principal control for these climates involves seasonal shifts in wind from land to sea, typified by the so-called Asian monsoon. During the low-sun season, high air pressure over the Asian continent causes dry air to flow southward into parts of southeast Asia and India. During the high-sun season, low air pressure covers the Tibetan Plateau and the winds reverse direction, advecting moisture inland from over the Indian Ocean. Local convection, orographic effects, and a shifting ITCZ conspire to deluge the land with torrential rains. Am climates also occur in western Africa and northeastern Brazil.

For the most part, tropical wet-and-dry climates (Aw) border tropical wet climates (Af) and are transitional to subtropical dry climates in a poleward direction. Aw climates support the savanna, tropical grasslands with scattered deciduous trees. Summers are wet and winters are dry, with the dry season lengthening poleward. This marked seasonality of rainfall is linked to shifts of the intertropical convergence zone (ITCZ) and semipermanent subtropical anticyclones, which follow the seasonal excursions of the sun. In summer (*high-sun* season), surges of the ITCZ trigger convective rainfall; in winter (*low-sun* season), the dry eastern flank of the subtropical anticyclones dominates the weather.

Annual mean temperatures in Aw climates are only slightly lower, and the seasonal temperature range is only slightly greater, than in the tropical wet climates (Af). The diurnal temperature range varies seasonally, however. In summer, frequent cloudy skies and high humidity suppress the diurnal temperature range by reducing both solar heating during the day and radiational cooling at night. In winter, on the other hand, persistent fair skies have the opposite effect on radiational heating and cooling and increase the diurnal temperature contrast. Cloudy, rainy summers plus dry winters also mean that the year's highest temperatures typically occur toward the close of the dry season in late spring.

Dry Climates

Dry climates (B) characterize those regions where average annual potential evaporation exceeds average annual precipitation. *Potential evaporation* is the quantity of water that would be vaporized into the atmosphere from a surface of fresh water during long-term average weather conditions. Air temperature largely governs the rate of evaporation so it is not possible to specify some maximum rainfall amount as the criterion for dry climates. Rainfall is not only limited in B climates but also highly variable and unreliable. As a general rule, the lower the mean annual rainfall, the greater is its variability from one year to the next.

Earth's dry climates encompass a larger land area than any other single climate grouping. Perhaps

30% of the planet's land surface, stretching from the tropics into midlatitudes, experience a moisture deficit of varying degree. These are the climates of the world's deserts and steppes, where vegetation is sparse and equipped with special adaptations that permit survival under conditions of severe moisture stress. Based on the degree of dryness, we distinguish between two dry climate types: steppe or semiarid (BS) and arid or desert (BW). Steppe or semiarid climates are transitional between more humid climates and arid or desert climates. Mean annual temperature is latitude dependent, as is the range in variation of mean monthly temperatures through the year. Hence, a distinction is made between warm dry climates of tropical latitudes (BSh and BWh) and cold, dry climates of higher latitudes (BSk and BWk).

Dryness is the consequence of subtropical anticyclones, cold surface ocean currents, or the rain shadow effect of high mountain ranges. Subsiding stable air on the eastern flanks of subtropical anticyclones gives rise to tropical dry climates (BSh and BWh). These huge semipermanent pressure systems, centered over the ocean basins, dominate the weather year-round near the Tropics of Cancer and Capricorn. Consequently, dry climates characterize North Africa eastward to northwest India, the southwestern United States and northern Mexico, coastal Chile and Peru, southwest Africa, and much of the interior of Australia.

Although persistent and abundant sunshine is generally the rule in dry tropical climates, there are some important exceptions. Where cold ocean waters border a coastal desert, a shallow layer of stable marine air drifts inland. The desert air thus features high relative humidity, persistent low stratus clouds and fog, and considerable dew formation. Examples are the Atacama Desert of Peru and Chile, the Namib Desert of southwest Africa, and portions of the coastal Sonoran Desert of Baja California and stretches of the coastal Sahara Desert of northwest Africa. These anomalous foggy desert climates are designated BWn.

Cold, dry climates of higher latitudes (BWk and BSk) are situated in the rain shadows of great mountain ranges. They occur primarily in the Northern Hemisphere, to the lee of the Sierra Nevada and Cascade ranges in North America and the Himalayan chain in Asia. Because these dry climates are at higher latitudes than their tropical counterparts, mean annual temperatures are lower and the seasonal temperature contrast is greater. Anticyclones dominate winter,

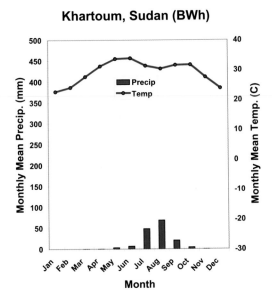

Khartoum, Sudan (BWh)

bringing cold and dry conditions, whereas summers are hot and generally dry. Scattered convective showers, mostly in summer, produce relatively meager precipitation.

Subtropical Climates

Subtropical climates are located just poleward of the Tropics of Cancer and Capricorn and are dominated by seasonal shifts of subtropical anticyclones. There are three basic climate types: subtropical dry summer (or *Mediterranean*) (Cs), subtropical dry winter (Cw), and subtropical humid (Cf), which receive precipitation throughout the year.

Mediterranean climates occur on the western side of continents between about 30 and 45 degrees latitude. In North America, mountain ranges confine this climate to a narrow coastal strip of California. Elsewhere, Cs climates rim the Mediterranean Sea and occur in portions of extreme southern Australia. Summers are dry because at that time of year Cs regions are under the influence of stable subsiding air on the eastern flanks of the semipermanent subtropical highs. Equatorward shift of subtropical highs in autumn allows extratropical cyclones to migrate inland, bringing moderate winter rainfall. Mean annual precipitation varies greatly—ranging from 30 to 300 cm (12 to 80 in.) with the wettest winter month typically receiving at least three times the precipitation of the driest summer month.

Although Mediterranean climates exhibit a pronounced seasonality in precipitation (dry summers and wet winters), the temperature regime is quite variable. In coastal areas, cool onshore breezes prevail, lowering the mean annual temperature and reducing seasonal temperature contrasts. Well inland, however, away from the ocean's moderating influence, summers are considerably warmer; hence, inland mean annual temperatures are higher and seasonal temperature contrasts are greater than in coastal Cs localities. Climatic records of coastal San Francisco and inland Sacramento, CA illustrate the contrast in temperature regime within Cs regions. Although the two cities are separated by only about 145 km (90 mi.), the climate of Sacramento is much more continental (much warmer summers and somewhat cooler winters) than that of San Francisco. The warm climate subtype is designated Csa and the cooler subtype is Csb.

Subtropical dry winter climates (Cw) are transitional between Aw and BS climates and located in South America and Africa between 20 and 30 degrees S. Cw climates also occur between the Aw and H climates of the Himalayas and Tibetan plateau and between the BS and Cfa climates of Southeast and East Asia. Northward shift of the subtropical high pressure systems is responsible for the dry winter in South America and Africa. The narrowness of the two continents between 20 and 30 degrees S means a relatively strong maritime influence and dictates against extreme dryness. In spring, subtropical highs shift southward and rains return. In Asia, winter dryness is caused by winds radiating outward from the massive cold Siberian high. As the continent warms in spring, the Siberian high weakens and is replaced by low pressure. Moist winds then flow inland bringing summer rains. Mean annual precipitation in Cw climates is in the range of 75 to 150 cm (30 to 60 in.).

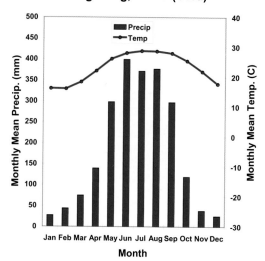

Hong Kong, China (CWa)

Subtropical humid climates (Cf) occur on the eastern side of continents between about 25 and 40 degrees latitude (and even more poleward where the maritime influence is strong). Cfa climates are the most important of the Cf climate subtypes in terms of land area and number of people impacted. Cfa climates are situated primarily in the southeastern United States, a portion of southeastern South America, eastern China, southern Japan, on the extreme southeastern coast of South Africa, and along much of the east coast of Australia. These climates feature abundant precipitation

(75 to 200 cm, or 30 to 80 in., on average annually), which is distributed throughout the year. In summer, Cfa regions are dominated by a flow of sultry maritime tropical air on the western flanks of the subtropical anticyclones. Consequently, summers are hot and humid with frequent thunderstorms, which can produce brief periods of substantial rainfall. Hurricanes and tropical storms contribute significant rainfall (up to 15% to 20% of the annual total) to some North American and Asian Cfa regions, especially from summer through autumn. In winter, after the subtropical highs shift toward the equator, Cfa regions come under the influence of migrating extratropical cyclones and anticyclones.

In Cfa localities, summers are hot and winters are mild. Mean temperatures of the warmest month are typically in the range of 24 to 27 °C (75 to 81 °F). Average temperatures for the coolest months typically range from 4 to 13 °C (39 to 55 °F). Subfreezing temperatures and snowfalls are infrequent.

A strong maritime influence is responsible for the cool summers and mild winters of Cfb climates. These climates occur over much of Northwest Europe, New Zealand, and portions of southeastern South America, southern Africa, and Australia. The coldest subtype, the Cfc, is relegated to coastal areas of southern Alaska, Norway, and the southern half of Iceland. Cfb and Cfc climates are relatively humid with mean annual precipitation ranging between 100 and 200 cm (40 and 80 in.).

Melbourne, Australia (Cfb)

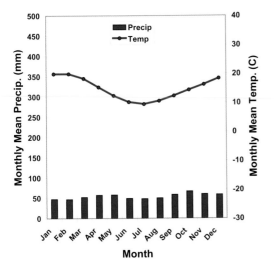

Snow Forest Climates

Snow forest climates (D) occur in the interior and to the leeward sides of large continents. The name emphasizes the link between biogeography and the Köppen climate classification system. These climates feature cold snowy winters (except for the Dw subtype in which the winter is dry) and occur only in the Northern Hemisphere. Snow forest climates are subdivided according to seasonal precipitation regimes with Df climates experiencing year-round precipitation whereas Dw climates have a dry winter. D climates with dry summers (Ds) are rare and small in extent. Additional distinction is made between warmer subtypes (Dwa, Dfa, Dwb, and Dfb) and colder subtypes (Dwc, Dfc, Dwd, Dfd).

The warmer subtypes, sometimes termed *temperate continental*, have warm summers (mean temperature of the warmest month greater than 22 °C or 71 °F) and cold winters. They are located in Eurasia, the northeastern third of the United States, southern Canada, and extreme eastern Asia. Continentality increases inland with maximum temperature contrasts between the coldest and warmest months as great as 25 to 35 Celsius degrees (45 to 63 Fahrenheit degrees). The southerly Dfa climates have cool winters and warm summers and the more northerly Dfb climates have cold winters and mild summers. The freeze-free period varies in length from 7 months in the south to only 3 months in the north. The weather in these regions is very changeable and dynamic because these areas are swept by cyclones and anticyclones and by surges of contrasting air masses. Polar front cyclones dominate winter, bringing episodes of light to moderate frontal precipitation. These storms alternate with incursions of dry polar and arctic air masses. In summer, cyclones are weak and infrequent as the principal storm track shifts poleward. Summer rainfall is mostly convective, and locally amounts can be very heavy in severe thunderstorms and mesoscale convective complexes (MCCs). Although precipitation is distributed rather uniformly throughout the year, most places experience a summer maximum.

In northern portions, winter snowfall becomes an important factor in the climate. Mean annual snowfall and the persistence of a snow cover increase northward. Because of its high albedo for solar radiation and its efficient emission of infrared, a snow cover chills and stabilizes the overlying air. For these reasons, a snow cover tends to be self-sustaining; once established in early winter, an extensive snow cover tends to persist.

Moving poleward, summers get colder and winters are bitterly cold. These so-called *boreal climates* (Dfc, Dfd, Dwc, Dwd) occur only in the Northern Hemisphere as an east-west band between 50 to 55 degrees N and 65 degrees N. It is a region of extreme continentality and very low mean annual temperature. Summers are short and cool, and winters are long and bitterly cold. Because midsummer freezes are possible, the growing season is precariously short. Both continental polar (cP) and arctic (A) air masses originate here, and this area is the site of an extensive coniferous (boreal) forest. In summer, the mean position of the leading edge of arctic air (the arctic front) is located along the northern border of the boreal forest. In winter, the mean position of the arctic front is situated along the southern border of the boreal forest.

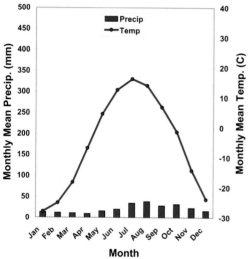

Weak cyclonic activity occurs throughout the year and yields meager annual precipitation (typically less than 50 cm, or 20 in.). Convective activity is rare. A summer precipitation maximum is due to the winter dominance of cold, dry air masses. Snow cover persists throughout the winter and the range in mean temperature between winter and summer is among the greatest in the world.

Polar Climates

Polar climates (E) occur poleward of the Arctic and Antarctic circles. These boundaries correspond roughly to localities where the mean temperature for the warmest month is 10 °C (50 °F). These limits also approximate the tree line, the poleward limit of tree growth. Poleward are tundra and the Greenland and Antarctic ice sheets. A distinction is made between tundra (Et) and ice cap (Ef) climates, with the dividing criterion being 0 °C (32 °F) for the mean temperature of the warmest month. Vegetation is sparse in Et regions and almost nonexistent in Ef areas.

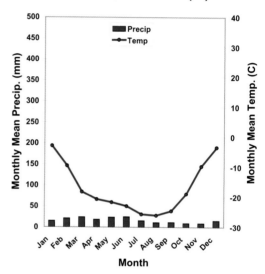

Polar climates are characterized by extreme cold and slight precipitation, which falls mostly in the form of snow (less than 25 cm, or 10 in., melted, per year). Greenland and Antarctica could be considered deserts for lack of significant precipitation, despite the presence of large ice sheets. Although summers are cold, the winters are so extremely cold that polar climates feature a marked seasonal temperature contrast. Mean annual temperatures are the lowest of any place in the world.

Highland Climates

Highland Climates (H) encompass a wide variety of climate types that characterize mountainous terrain. Altitude, latitude, and exposure are among the factors that shape a complexity of climate types. For example, temperature decreases rapidly with increasing altitude and windward slopes tend to be wetter than leeward slopes.   Climate-ecological zones are telescoped in mountainous terrain.   That is, in ascending several thousand meters of altitude, we encounter the same bioclimatic zones that we would experience in traveling several thousand kilometers of latitude. As a general rule, every 300 m (980 ft) of elevation corresponds roughly to a northward advance of 500 km (310 mi.)

GLOSSARY

Absolute humidity   The mass of water vapor per unit volume of air containing the water vapor; usually expressed as grams of water vapor per cubic meter of humid air.

Absolute instability   Property of an ambient air layer that enhances vertical motion for both saturated (cloudy) and unsaturated (clear) air parcels; the *temperature* of the ambient air is dropping more rapidly with altitude than the *dry adiabatic lapse rate*.

Absolute stability   Property of an ambient air layer that suppresses vertical motion for both saturated (cloudy) and unsaturated (clear) air parcels. Occurs when the *temperature* in the ambient air layer drops more slowly with altitude than the *moist adiabatic lapse rate*, the temperature does not change with altitude in the ambient air layer (isothermal), or the temperature increases with altitude in the ambient air layer (temperature inversion).

Absolute zero   The theoretical *temperature* at which a body does not emit *electromagnetic radiation* and all molecular motion ceases (usually some subatomic activity takes place); 0 K (-273.15 °C or -459.57 °F).

Absorption   The energy conversion process whereby a portion of the radiation incident on an object is converted to *heat* energy. Absorption is a function of the surface and composition of the absorbing material and the *wavelength* of the incident radiation.

Adiabatic process   A thermodynamic process in which there no *heat* is exchanged between a mass and its environment. Within the *atmosphere*, *expansional cooling* and *compressional warming* of unsaturated air parcels are essentially adiabatic processes, that is, no heat is exchanged between the air parcels and the surrounding air.

Advanced Weather Interactive Processing System (AWIPS)   A computerized workstation that receives and organizes *Automated Surface Observing System* data and other analysis and guidance products from the *National Centers for Environmental Prediction* and enables *National Weather Service* meteorologists to display, process, and overlay images, graphics, and other data. Inspired by *McIDAS*.

Advection fog   Ground-level *cloud* generated by the cooling of a mild, humid *air mass* (to its *dewpoint*) as it travels over a relatively cool surface.

Aerosols   Tiny liquid or solid particles of various compositions that occur suspended in the *atmosphere*.

Agroclimatic compensation   Poor growing *weather* and decreased crop yields in one area is offset to some extent by better growing weather and increased crop yields in other areas.

Air density   Mass of molecules per unit volume of air; about 1.275 kg per cubic meter at sea-level temperature of 0 °C and pressure of 1000 millibars.

Air mass   A huge volume of air covering thousands of square kilometers that is horizontally relatively uniform in *temperature* and water vapor concentration (*humidity*). The specific characteristics of an air mass depend on the type of surface over which the air mass forms (its source region) and travels.

Air mass advection   Movement of an *air mass* from one place to another; one air mass replaces another air mass having different *temperature* and/or *humidity* characteristics.

Air mass modification   Changes in the *temperature*, *humidity*, and/or stability of an *air mass* as it travels away from its source region. Occurs primarily by exchange of *heat*, moisture, or both with the surface over which the air mass travels, *radiational heating* or *radiational cooling*, and adiabatic heating or cooling associated with large-scale vertical motion.

Air pollutant   A gas or *aerosol* that occurs at a concentration that threatens the wellbeing of living organisms (especially humans) or disrupts the orderly functioning of the environment.

Air pressure   The cumulative force exerted on any surface by the molecules composing air; usually expressed as the weight of a column of air per unit surface area.

Air pressure gradient   Change in *air pressure* with distance.

Air pressure tendency   Change in *air pressure* with time; on a surface *weather* map, the air pressure change over the prior 3 hrs.

Albedo   The fraction or percent of radiation striking a surface (or interface) that is reflected by that surface (or interface); usually applied to the reflectivity of an object to visible radiation.

Altimeter   Any device used to determine altitude such as an *aneroid barometer* calibrated to read in altitude or elevation.

Altocumulus (Ac)   Middle level *clouds* occurring as roll-like patches or puffs forming waves or parallel bands. Altocumulus are distinguished from *cirrocumulus* by the larger size of the cloud patches and by sharper edges indicating the presence of supercooled water droplets rather than ice crystals.

Altocumulus lenticularis   Lens-shaped *altocumulus*; *mountain-wave clouds* generated by the disturbance of horizontal airflow by a prominent mountain range.

Altostratus (As)   Middle level *clouds* occurring as uniformly gray or bluish white layers totally or partially covering the sky. The sun is typically only dimly visible through altostratus.   These clouds are composed of supercooled water droplets and ice crystals.

Aneroid barometer   A portable instrument that utilizes a flexible partially-evacuated metal chamber and spring to measure *air pressure*; may be used as an *altimeter*.

Anomalies   Departures of *temperature*, *precipitation*, or other *weather* elements from long-term averages.

Antarctic Circle   Poleward of this latitude (66 degrees 33 minutes S) there are 24 hs of sunlight on the austral summer *solstice* and 24 hrs of darkness on the austral winter solstice.

Antarctic ozone hole   A large area of significant stratospheric ozone depletion over the Antarctic continent that typically develops annually between late August and early October, and generally ends in mid-November. Ozone thinning is attributed to the action of chlorine (Cl) liberated from a group of chemicals known as chlorofluorocarbons (CFCs).

Anticyclone   A dome of air that exerts relatively high surface *air pressure* compared with surrounding air; same as a *High*. Viewed from above, surface *winds* in an anticyclone blow clockwise and outward in the Northern Hemisphere but counterclockwise and outward in the Southern Hemisphere.

Aphelion   The time of the year when the Earth is farthest (152 million km or 94 million mi) from the sun (currently about 4 July).

Arctic air (A)   Exceptionally cold and dry *air masses* that form primarily in winter over the Arctic Basin, Greenland, and the northern interior of North America north of about 60 degrees N.

Arctic Circle   Poleward of this latitude (66 degrees 33 minutes N) there are 24 hrs of sunlight at the boreal summer *solstice* and 24 hrs of darkness at the boreal winter solstice.

Arctic high   A *cold-core anticyclone* originating in a source region for *Arctic air*; the product of extreme *radiational cooling* over the *snow*-covered continental interior of North America well north of the *polar front*.

Atmosphere   A thin envelope of gasses (also containing suspended solid and liquid particles and *clouds*) that encircles the planet.

Atmospheric boundary layer   The atmospheric zone to which frictional resistance (*eddy viscosity*) is essentially confined; on average the zone extends from the Earth's surface to an altitude of about 1000 m (3300 ft).

Atmospheric stability   Property of ambient air that either enhances (unstable) or suppresses (stable) vertical motion of air parcels; depends on the vertical *temperature* profile or *sounding* of the ambient air and whether air parcels are saturated (cloudy), with cooling or warming at the *moist adiabatic lapse rate*, or unsaturated (clear), with cooling or warming at the *dry adiabatic lapse rate*. See *stable air layer* and *unstable air layer*.

Atmospheric windows   *Wavelength* bands within which atmospheric gases absorb little or no *electromagnetic radiation*.

Aurora   A luminous phenomenon in the night sky consisting of overlapping curtains of greenish-white light, sometimes fringed with pink, appearing in the *ionosphere* of high latitudes. Light is emitted by atoms and molecules that are excited by beams of electrons generated by a complex interaction between the *solar wind* and the Earth's *atmosphere*. Known as the aurora borealis (northern lights) in the Northern Hemisphere and the aurora australis (southern lights) in the Southern Hemisphere.

Automated Surface Observing System (ASOS)   Meteorological sensors, computers, and fully automated communication ports that record and transmit atmospheric conditions automatically; a component of the modernization program of the *National Weather Service*. A network of ASOS units feed observational data to National Weather Service Forecast Offices and local airport control towers 24-hrs per day.

Back-door cold front   A *cold front* that propagates southward or southwestward along the North Atlantic coast east of the Appalachian Mountains. These fronts occur most frequently in the summer and fall and usher in Canadian *continental polar air* or Atlantic *maritime polar air*.

Barograph   A recording instrument that provides a continuous trace of *air pressure* with time; helps determine *air pressure tendency*.

Barometer   An instrument used to measure *air pressure* and monitor its changes. The basic types are the *mercury barometer* and the *aneroid barometer*.

Barrier island   A long, narrow strip of sand parallel to the coast and separated from the mainland by a lagoon or backwater bay. Since a barrier island faces the open ocean, it absorbs the brunt of *tropical cyclone*- and other sea storm-driven waves and *storm surge*, providing some protection for coastal beaches, wetlands, and shoreline structures.

Beaufort scale   A scale of *wind* speed based originally on visual assessment of the effects of wind on seas, and later extended to describe the effects of wind on land-based flexible objects such as trees. The scale ranges from 0 for calm conditions to 12 for *hurricane*-strength winds.

Bergeron-Findeisen process   *Precipitation* formation in a *cold cloud*s whereby ice crystals grow at the expense of supercooled water droplets in response to differences in *vapor pressure* relative to water and ice surfaces. Also known as the ice-crystal process.

Blackbody   A hypothetical "body" that absorbs all *electromagnetic radiation* that is incident on it at every *wavelength* and emits all radiation at every wavelength; no radiation is reflected or transmitted. The "body" must be large compared to the wavelength of incident radiation.

Blizzard warning   Issued when falling or blowing *snow* is accompanied by *winds* of 56 km (35 mi) per hr or higher and visibility reduced to less than 400 m (1300 ft), and when these conditions are expected to persist for at least 3 hrs.

Blocking system   A *cyclone* or *anticyclone* cutoff from the main westerly airflow that blocks the usual west-to-east progression of *weather* systems. A blocking system may be responsible for weather extremes, such as drought or flooding *rains* or excessive *heat* and cold.

Bowen ratio   For a moist surface, the ratio of *heat* energy used for *sensible heating* (*conduction* and *convection*) to the heat energy used for *latent heating* (*evaporation* of water or *sublimation* of *snow*). The Bowen ratio varies from one locality to another depending on the amount of surface moisture, ranging from 0.1 for the ocean surface to more than 5.0 for deserts; negative values are also possible.

British thermal unit (Btu)   The quantity of *heat* needed to raise the *temperature* of one pound of water one Fahrenheit degree (technically, from 62 to 63 °F).

Calorie (cal)   The amount of *heat* needed to raise the *temperature* of one gram of water one Celsius degree (technically, from 14.5 to 15.5 °C).

Centripetal force   An inward-directed force that acts on an object moving in a curved path, confining the object to the curved path; the result of other forces.

Chinook wind   Air that is adiabatically compressed as it is drawn down the leeward slope of a mountain range. As a consequence, the air is warm and dry, causing rapid temperature rises and reduction in snow cover.

Chromosphere   Portion of the sun above the *photosphere*; consists of transparent ionized hydrogen and helium at 4000 to 40,000 °C (7200 to 72,000 °F).

Circumpolar vortex   The planetary-scale circulation regime that surrounds the cold pool of air in the polar regions.

Cirriform cloud   A thin wispy or fibrous cloud composed of ice crystals.

Cirrocumulus (Cc)   High *clouds* that exhibit a wavelike pattern of small, white, rounded patches rarely covering the entire sky. Cirrocumulus are composed of ice crystals.

Cirrostratus (Cs)   High, thin, layered *clouds* composed of ice crystals that form a thin white veil over the sky. Cirrostratus are nearly transparent, so the sun or moon readily shines through them.

Cirrus (Ci)   High, thin, nearly transparent *clouds* composed of ice crystals and occurring as delicate silky strands, which are streaks of falling ice particles blown laterally by strong *winds*.

Climate   *Weather* of some locality averaged over some time period plus extremes in weather behavior observed during the same period or during the entire period of record.

Climatology   Study of *climate*, its controls and spatial and temporal variability. In *weather* forecasting, a forecast derived from climatology is based on the type of weather that occurred on the same day in years past.

Cloud   A visible suspension of minute water droplets and/or ice crystals in the *atmosphere* above the Earth's surface. Clouds differ from *fog* only in that the latter is, by definition, in contact with Earth's surface. Clouds form in the free atmosphere primarily as a result of *condensation* or *deposition* of water vapor in ascending air that nears *saturation*.

Cloud condensation nuclei (CCN)   Tiny solid and liquid particles that promote the *condensation* of water vapor at *temperatures* both above and below the freezing point of water, may include *hygroscopic nuclei*.

Cloud streets   *Clouds* aligned in rows due to strong vertical shear in horizontal *wind* speed or direction; these rows often extend for hundreds of kilometers.

Cold air advection   Flow of air across regional isotherms from relatively cool localities to relatively warm localities.

Cold cloud *Cloud* composed of ice crystals or supercooled water droplets or a mixture of both which have *temperatures* below 0 °C (32 °F).

Cold front *Front* that moves in such a way that relatively cold (more dense) air advances and replaces relatively warm (less dense) air.

Cold-core anticyclone Shallow high-pressure systems that coincide with a dome of *continental polar air* or *Arctic air*; is labeled either a *polar high* or an *arctic high*. The high's shallow anticyclonic circulation weakens rapidly with altitude and often reverses to a cyclonic circulation in the middle and upper *troposphere*.

Cold-core cyclone An occluded low-pressure system in which the lowest *temperatures* occur throughout the column of air above the low-pressure center. Isobaric surfaces dip downward above the low center and the depth of the low increases with altitude, implying that a cyclonic circulation prevails throughout the *troposphere* and is most intense at high altitudes.

Collision-coalescence process Growth of *cloud* droplets into raindrops within a *warm cloud*; droplets merge upon impact. This process takes place in a cloud made up of droplets of different sizes; larger droplets with higher *terminal velocity* overtake, and then collide and coalesce with smaller droplets in their paths.

Comma cloud The pattern of cloudiness associated with a mature *extratropical cyclone*; the head of the comma stretches from the low center to the northwest and its tail follows along the *cold front*. This pattern reflects the strengthening of the system's circulation.

Compressional warming A *temperature* rise that accompanies a pressure increase on and resultant compression of a gas (or mixture of gases). The work of compressing air is converted to *heat*, raising the temperature of the gas. Air parcels that descend within the *atmosphere* undergo compressional warming in response to increasing *air pressure*.

Condensation The process that produces a net gain of liquid water mass at the interface between liquid water and air; more water changes phase from vapor to liquid than liquid to vapor. The phase change from vapor to liquid releases the *latent heat of vaporization*.

Conditional stability Property of an ambient air layer that suppresses vertical motion for unsaturated (clear) air parcels and enhances vertical motion for saturated (cloudy) air parcels. The *sounding* in a conditionally stable layer of air lies between the *dry adiabatic lapse rate* and the *moist adiabatic lapse rate*.

Conduction The transfer of energy (electrical, *heat*) within and through a conductor by means of internal particle or molecular activity; occurs primarily in solids, but also in liquids and gases. Conduction is significant only in a thin layer of air that is in immediate contact with Earth's surface.

Continental drift The slow movement of continents, which are parts of gigantic plates, across the face of the globe.

Continental polar air (cP) Relatively dry *air masses* that develop over the northern interior of North America; these air masses are very cold in winter and relatively mild in summer.

Continental tropical air (cT) Warm, dry *air masses* that form over the subtropical deserts of Mexico and the southwestern United States primarily in summer.

Contrails *Cloud*-like streamers frequently seen forming behind aircraft flying in clear, cold air.

Convection The transport of *heat* within a fluid via motions of the fluid itself; generally occurs only in liquids and gases. Convection is much more important than *conduction* in transporting heat vertically within the *troposphere*.

Convective condensation level (CCL) The altitude at which *condensation* begins to occur through *convection*; coincides with the base of *cumuliform clouds*, typically at altitudes between 1000 and 2000 m (3600 and 6000 ft).

Convergence A wind pattern that has a net inflow of air toward a central point either because of wind direction or changes in wind speed.

Converging winds *Winds* that blow toward a central point of a column of air.

Conveyor-belt model A three-dimensional depiction of a mature *extratropical cyclone* and its corresponding *fronts* in terms of three interacting airstreams, often referred to as conveyor belts. This model developed out of a better understanding of upper-air circulation in the years since the *Norwegian cyclone model*.

Coriolis effect A deflective force arising from the rotation of the Earth on its axis; affects principally synoptic-scale and planetary-scale *winds*. Winds are deflected to the right of their initial direction in the Northern Hemisphere and to the left in the Southern Hemisphere. Magnitude depends on latitude and speed of the moving object.

Corona A series of alternating light and dark colored rings that surround the moon or sun; due to *diffraction* of light by spherical *cloud* droplets.

Crepuscular rays Alternating light and dark bands (solar rays and shadows) that appear to diverge in a fanlike pattern from the solar disk. They can be observed at the beginning of *twilight* or during the day as the sun's rays pass through holes in *clouds* or between clouds in a hazy sky.

Cumuliform clouds *Clouds* that exhibit significant vertical development; often produced by updrafts in *convection* currents.

Cumulonimbus (Cb) *Thunderstorm clouds* that form as a consequence of deep *convection* in the *atmosphere*; have tops that sometimes reach altitudes of 20,000 m (60,000 ft) or higher. The upper portions of these clouds are anvil-shaped and composed of mostly ice crystals, the middle portions are composed of supercooled water droplets or a mixture of supercooled water droplets and ice crystals, and the lower portions are often composed of ordinary water droplets.

Cumulus (Cu) Vertically developed *clouds* that form as a consequence of the updraft in *convection* currents; resemble balls of cotton floating the sky. Cumulus are normally fair-*weather* clouds and consist of mostly water droplets, which may be supercooled.

Cumulus congestus Upward-building convective *clouds* with vertical development between those of *cumulus* and *cumulonimbus*; having a cauliflower appearance. Cumulus congestus can occur when the ambient air in the middle to upper *troposphere* is unstable for saturated air. These clouds consist of mostly water droplets, which are frequently supercooled.

Cumulus stage Initial stage in the life cycle of a *thunderstorm* cell when *cumulus* build both vertical and laterally over a period of about 10 to 15 minutes. The updraft throughout the cell is sufficiently strong to keep water droplets and ice crystals suspended in the *cloud*, therefore *precipitation* does not occur in this stage.

Curvature effect The effect a curved water surface has on the ability of water molecules to vaporize from the surface. The smaller (radius) a droplet, the greater the concentration of surrounding water vapor that is necessary for the droplet to grow. Due to differences in the bond strength of the water molecules on the droplet surface, water molecules more readily vaporize from a small droplet than a large droplet, explaining why the *saturation vapor pressure* over a small droplet is greater than over a large droplet at the same *temperature*. Tiny droplets cannot form or exist without a supersaturated environment.

Cup anemometer An instrument used to monitor *wind* speed. Wind rotates the 3 or 4 open hemispheric cups, which spin horizontally on a vertical shaft, and that motion is calibrated in wind speed.

Cyclogenesis The birth and development of a *cyclone*; for an *extratropical cyclone*, cyclogenesis usually takes place along the *polar front* directly under an area of strong horizontal *divergence* in the upper *troposphere*.

Cyclolysis Process whereby a *cyclone* weakens; usually takes place when the central pressure rises and winds slacken. Upper-air support for the cyclone diminishes to the point where horizontal *divergence* aloft becomes weaker than horizontal *convergence* near the surface.

Cyclone   A *weather* system characterized by relatively low surface *air pressure* compared with the surrounding air; same as a *Low*. Viewed from above, surface *winds* blow counterclockwise and inward in the Northern Hemisphere but clockwise and inward in the Southern Hemisphere.

Dalton's law   A scientific law that states that the total pressure exerted by a mixture of gases is equal to the sum of the partial pressures of each constituent gas. Each gas species in the mixture acts independently of the other molecules.

Deposition   The process which produces a net gain of ice mass at the interface between ice and air; more water changes phase from vapor to ice than ice to vapor; phase change of water from vapor to ice (without first becoming liquid).

Derecho   A family of straight-line *downburst winds* in excess of 94 km (58 mi) per hr that impacts a path up to hundreds of kilometers long; may be produced by a *squall line* or a *mesoscale convective complex*.

Dew   Tiny water droplets formed by *condensation* of water vapor on a relatively cold surface. For water vapor to condense as dew on the surface of an object, the *temperature* of that surface must cool below the *dewpoint*.

Dewpoint   *Temperature* to which air must be cooled (at constant *air pressure* and constant water vapor content) to achieve *saturation* of the air relative to liquid water (if at or above 0 °C or 32 °F).

Dewpoint hygrometer   An instrument that measures the *dewpoint* of the air. In one design, air passes over the surface of a metallic mirror that is cooled electronically. An electronic sensor continually monitors the *temperature* of the mirror at the same time that a beam of *infrared radiation* is pointed at the mirror. When a thin *condensation* film forms on the mirror, the *reflectivity* of the infrared beam changes, and the mirror temperature is automatically recorded as the dewpoint.

Diffraction   The slight bending of a light wave as it moves along a boundary of an object such as a water droplet. As light waves bend, they interfere with each other, either constructively, producing a larger wave, or destructively, canceling each other out. When interference of light waves is constructive, a ring of bright light is seen, and when interference is destructive, darkness is perceived.

Dissipating stage   The final phase in the life cycle of a *thunderstorm* cell; features downdrafts throughout the system, the tapering off and ending of *precipitation*, and vaporization of *clouds*.

Distillation   The purification of water through phase changes (e.g., *evaporation* followed by *condensation*). When water vaporizes, all suspended and dissolved substances are left behind.

Divergence   A *wind* pattern that has a net outflow of air from some point either because of wind direction or changes in wind speed.

Diverging winds   *Winds* that blow away from a central point in a column of air.

Doldrums   An east-west equatorial belt of light and variable surface *winds* where the *trade winds* of the two hemispheres converge.

Doppler effect   A shift in the frequency (or phase) of an electromagnetic or sound wave due to the relative movement of the source or the observer.

Doppler radar   *Weather radar* that determines the velocity of targets (*precipitation*, dust particles) moving directly toward or away from the radar unit based on the difference in frequency (or phase) between the outgoing and returning radar signal.

Downburst   A strong and potentially destructive *thunderstorm* downdraft; depending on size, classified as either a *microburst* or a *macroburst*.

Drizzle   A form of liquid *precipitation* consisting of water droplets having diameters between 0.2 and 0.5 mm (0.01 and 0.02 in); falls very slowly from low *stratus*.

Dropwindsonde   A small instrument package equipped with a radio transmitter and parachute that is dropped from an aircraft to take altitude readings of *temperature, air pressure,* and *dewpoint*. This instrument package is also tracked from the air by a direction-finding antenna or global positioning system (GPS) to measure variations in horizontal *wind* direction and speed with altitude.

Dry adiabatic lapse rate   The adiabatic expansional cooling of ascending, clear (unsaturated) air parcels at the rate of about 9.8 Celsius degrees per 1000 m of uplift (or 5.5 Fahrenheit degrees per 1000 ft). If clear (unsaturated) air parcels descend in the *atmosphere*, the parcels warm adiabatically at the same rate.

Dryline   A boundary between *continental tropical air* and *maritime tropical air* in the southeast sector of a mature *extratropical cyclone*; likely site for *squall line* and *severe thunderstorm* development.

Dust devil   A swirling mass of dust triggered by intense solar heating of dry surface areas. The most common dust devil is less than 1 m (about 3 ft) in diameter, lasts less than a minute, and is too weak to cause serious property damage.

Earth-atmosphere system   The Earth's surface (i.e., continents, ocean, ice sheets) plus the *atmosphere*.

Easterly wave   A ripple in the tropical easterlies (*trade winds*) featuring a weak trough of low pressure; propagates from east to west. Lower-tropospheric convergence on the east side of the wave can help organize convective activity into a *tropical disturbance*.

Eddy viscosity   Fluid *friction* (*viscosity*) arising from eddies (irregular whirls) within a fluid such as air or water.

Ekman spiral   A spiral in water motion in the top 100 m (330 ft) or so of the ocean produced by the directional change and decreasing horizontal motion of successively lower layers of water. Due to a balance between the *Coriolis effect* and frictional drag, the surface ocean layer moves at an angle of up to 45 degrees to the right or left (depending on the hemisphere) of the surface *wind* direction. Due to the same balance of forces, the layer immediately below the surface ocean layer moves slower and at an angle to the motion of the layer above. This spiraling effect is responsible for the *Ekman transport*.

Ekman transport   The net horizontal movement of water in the top 100 m (330 ft) or so of the ocean induced by the coupling of surface *winds* with ocean surface waters. The *Ekman spiral* causes this net transport to be about 90 degrees to the right of the surface wind direction in the Northern Hemisphere and to the left of the wind direction in the Southern Hemisphere.

El Niño   An anomalous warming of surface ocean waters in the eastern tropical Pacific; accompanied by suppression of *upwelling* off the coasts of Peru and Ecuador.

Electromagnetic radiation   Energy in the form of waves that have both electrical and magnetic properties; these waves travel through gases, liquids, and solids, and occur even in a vacuum. All objects emit all forms of electromagnetic radiation, although each object emits its peak radiation at a certain *wavelength* within the *electromagnetic spectrum*.

Electromagnetic spectrum   Range of *electromagnetic radiation* types arranged by *wavelength* or by *wave frequency* or both. Although the electromagnetic spectrum is continuous, different names are assigned to different segments. These segments extend from the longest-wavelength (lowest frequency) *radio waves* through *microwaves, infrared radiation, visible radiation, ultraviolet radiation, X-rays,* to the shortest wavelength (highest frequency) *gamma radiation*.

Electronic hygrometer   Measures the amount of water vapor in the air based on changes in the electrical resistance of certain chemicals as they adsorb water vapor from the air. Variations in electrical resistance are calibrated in terms of *relative humidity* or *dewpoint*. An electronic hygrometer is flown aboard *radiosondes*.

Emissivity   A measure of how close a radiating object approximates a *blackbody*, a perfect radiator. The emissivity of a blackbody is 1.0 and less than 1.0 for all other objects.

ENSO (El Niño/Southern Oscillation) The relationship between *El Niño* and the *southern oscillation*. An El Niño episode begins when the weakening of the surface *air pressure gradient* between the western and central tropical Pacific heralds the slackening of the *trade winds*.

Equinoxes The first days of spring and autumn when day and night are of approximately equal length at all latitudes (except the poles) and the noon sun is directly over the equator (*solar altitude* of 90 degrees).

Evaporation The process which produces a net gain of water vapor mass at the interface between liquid water and air; more water changes phase from liquid to vapor than vapor to liquid. The phase change from liquid to vapor requires the *latent heat of vaporization* and occurs at a *temperature* below the boiling temperature of water.

Evapotranspiration Direct *evaporation* from Earth's surface plus the release of water vapor by vegetation (*transpiration*).

Expansional cooling A *temperature* drop that accompanies a pressure reduction on and resultant expansion of a gas (or mixture of gases). The work of expansion requires energy, which is drawn from the internal energy of the gas, reducing its temperature. Air parcels that ascend within the *atmosphere* undergo expansional cooling in response to falling *air pressure*; expansional cooling is the principal means of *cloud* formation in the *atmosphere*.

Extratropical cyclone A *synoptic-scale* low pressure system that occurs in midlatitudes, often forming along the *polar front*. This low, characterized by *fronts* and a *comma cloud* pattern, becomes a *cold-core cyclone* especially in the later stage of its life cycle.

Eye An area of almost cloudless skies, subsiding air, and light *winds* (less than 25 km or 16 mi per hr) at the center of a *tropical cyclone*. The eye diameter ranges from 10 to 65 km (6 to 40 mi) across and typically shrinks as the system strengthens.

Eye wall A circle of *cumulonimbus* that surrounds the *eye* of a mature *hurricane* and produces heavy *rains* and very strong *winds*. The most dangerous and potentially most destructive part of a hurricane is the portion of the eye wall where the wind blows in the same direction as the hurricane's forward motion.

Faculae Relatively bright and hot areas in the solar *photosphere* that form a veined network near *sunspots*.

Fair-weather bias The observation that fair-*weather* days outnumber stormy days almost everywhere.

Flash flood A short-term, localized, and often unexpected rise in river or stream level causing flooding. Usually occurs in response to torrential *rain* falling over a small geographical area.

Fog A visibility-restricting suspension of tiny water droplets or ice crystals in an air layer next to Earth's surface; *stratus* in contact with the ground.

Forced convection *Convection* aided by topographic uplift or converging surface *winds*.

Free convection *Convection* triggered by intense solar heating of Earth's surface.

Freezing rain Supercooled raindrops that at least partially freeze on contact with cold surfaces.

Friction The resistance an object encounters as it comes into contact with other objects; the friction of fluid flow is known as *viscosity*.

Front A narrow zone of transition between *air masses* of contrasting *air density*, that is, air masses of different *temperature*, *humidity*, or both. Fronts are classified as stationary, warm, cold, and occluded.

Frontal fog *Fog* formed when *precipitation* falling from relatively warm air aloft into a wedge of relatively cool air near the Earth's surface evaporates and raises the *vapor pressure* in the cool air to saturation; occurs either just ahead of a *warm front* or just behind a *cold front*.

Frontogenesis   The development or strengthening of a *front*.

Frontolysis   The dissipation or weakening of a *front*.

Frost   Ice crystals that are formed by *deposition* of water vapor on a relatively cold surface. For water vapor to deposit as frost on the surface of an object, the *temperature* of that surface must cool below the *frost point*.

Frost point   The *temperature* to which air must be cooled at constant *air pressure* to achieve saturation at or below 0°C (32 °F).

F-scale (Fujita scale)   A six-category *tornado* intensity scale developed by T. Theodore Fujita that rates tornadoes from F0 to F5 on the basis of rotational *wind* speed estimated from property damage. Categories are weak (F0, F1), strong (F2, F3), and violent (F4, F5).

Funnel cloud   A tornadic circulation extending below *cloud* base but not reaching the ground; made visible by a cone-shaped cloud.

Gamma radiation   *Electromagnetic radiation* having very short *wavelength* (0.001 to 0.1 micrometer) and great penetrating power; gamma radiation is the shortest wavelength portion of the *electromagnetic spectrum*. The Earth's *atmosphere* blocks out virtually all incoming gamma radiation; overexposure to this type of radiation is dangerous to living organisms.

Gas law   The relationship between the *variables of state* of a gas (or mixture of gases) stating that the pressure exerted by a gas is directly proportional to the product of its density and *temperature*.

Geologic time   A span of millions or billions of years in the past. The geological time scale is a standard division of time on Earth into eons, eras, periods, and epochs based on large-scale geological events.

Geostationary Operational Environmental Satellite (GOES)   A *geostationary satellite*, operated by NOAA, designed to remotely monitor *weather* systems.

Geostationary satellite   A satellite that orbits the Earth at the same rate and in the same direction as the Earth rotates so that the satellite is always directly over the same point (the *sub-satellite point*) on Earth's equator and has the same field of view. The altitude of the satellite's orbit is about 36,000 km (22,300 mi).

Geostrophic wind   A hypothetical, unaccelerated horizontal *wind* that flows along a straight path parallel to *isobars* or height contours above the *atmospheric boundary layer*; results from a balance between the horizontal *pressure gradient force* and the *Coriolis* force.

Glacial climate   Long-term average conditions favorable to the initiation and growth of glacial ice.

Global climate model   A *numerical model* that describes the physical interactions among the various components of the *climate* system: the *atmosphere*, ocean, land, ice-cover, and biosphere. Predicts broad regions of expected positive and negative *temperature* and *precipitation anomalies*, and the mean location of circulation features such as jet streams and principal storm tracks.

Global radiative equilibrium   The balance between net incoming solar radiation and *infrared radiation* emitted to space by the *Earth-atmosphere system*.

Global water budget   Balance sheet for the inputs and outputs of water to and from the various global water reservoirs. The global water budget indicates an annual net gain of water mass on land and an annual net loss of water mass from the ocean; the excess water mass on land flows to the sea and balances the budget.

Global water cycle   The ceaseless movement of water among the atmospheric, oceanic, terrestrial, and biospheric reservoirs on a planetary scale; assumes an essentially fixed amount of water in the *Earth-atmosphere system*.

Glory Concentric rings of color about the shadow of an observer's head that appear on top of a *warm cloud* situated below the observer. A glory is caused by much the same optics as a primary *rainbow*, two important differences being the size of the reflecting and refracting particles, and the direction of reflected and refracted light. A glory is the consequence of sunlight interacting with a mass of much smaller cloud droplets that compose a *warm cloud*.

Gradient wind A hypothetical horizontal *wind* that blows parallel to curved *isobars* or height contours above the *atmospheric boundary layer*. It differs from the *geostrophic wind* in that the path of the gradient wind is curved.

Granules A network of huge, irregularly shaped convective cells in the sun's *photosphere*.

Gravity The force that accelerates air downward to the Earth's surface. It is the net result of gravitation, the force of attraction between Earth and all other objects, and the *centripetal force* arising from Earth's rotation on its axis.

Graupel *Ice pellets*, generally 2 to 5 mm (0.1 to 0.2 in.) in diameter, formed in a *cloud* when supercooled water droplets collide and freeze on impact.

Green flash A brilliant, thin, green rim that occasionally appears on the upper limb of the sun as it rises or sets (at *twilight*). A green flash is primarily the consequence of atmospheric *refraction* and *scattering* of light from a low sun; it is best seen on a distant horizon when the *atmosphere* is clear.

Greenhouse effect Heating of Earth's surface and lower atmosphere as a consequence of differences in atmospheric transparency to *electromagnetic radiation*. The *atmosphere* is nearly transparent to incoming solar radiation, but much less transparent to outgoing *infrared radiation*. Terrestrial infrared radiation is absorbed and radiated primarily by water vapor and, to a lesser extent, by carbon dioxide and other trace gases, thereby slowing the loss of *heat* to space from the *Earth-atmosphere system*.

Ground clutter A pattern of *radar echoes* produced by *reflection* of radar signals by fixed objects such as buildings on Earth's surface.

Gust front Leading edge of a mass of relatively cool gusty air that flows out of the base of a *thunderstorm cloud* (downdraft) and spreads along the ground well in advance of the parent thunderstorm cell; a mesoscale *cold front*. Uplift along the gust front may produce additional *cumulus* that may evolve into secondary thunderstorms cells tens of kilometers ahead of the parent cell.

Haboob A dust- or sandstorm caused by the strong, gusty downdraft of a desert *thunderstorm*. The mass of dust or sand rolls along the ground as a huge ominous black *cloud* that may be more than 100 km (60 mi) wide and may reach altitudes of several thousand meters.

Hadley cell Thermally-driven air circulation in tropical and subtropical latitudes of both hemispheres resembling a huge convective cell with rising air near the equator in the *intertropical convergence zone* and sinking air in the *subtropical anticyclones*. Equatorward blowing surface winds and poleward directed upper-level *winds* complete the circulation.

Hail *Precipitation* in the form of jagged to nearly spherical chunks of ice often characterized by internal concentric layering. Hail is associated with *thunderstorm* cells that have strong updrafts, an abundant supply of supercooled water droplets, and great vertical *cloud* development.

Hailstreak Accumulation of *hail* in a long narrow path along the ground typically around 2 km (1.2 mi) wide and 10 km (6.2 mi) long.

Hair hygrometer An instrument designed to monitor *relative humidity* by measuring the changes in the length of human hair that accompany *humidity* variations.

Halo A whitish ring of light surrounding the sun or moon formed when the tiny ice crystals that compose high, thin *clouds* (such as *cirrus* or *cirrostratus*) refract the sun's rays.

Heat A form of energy transferred between systems or components of a system in response to differences in *temperature*. Heat energy is always transferred from a warmer system to a colder system.

Heat equator The latitude (about 10 degrees N) of highest mean annual surface air *temperature*. The mean position of the *intertropical convergence zone* approximately corresponds to the heat equator.

Heterosphere The *atmosphere* above 80 km (50 mi) and the *homosphere* where gases are stratified, with concentrations of the heavier gases decreasing more rapidly with altitude than concentrations of the lighter gases.

Holocene The interval since the end of the Pleistocene glaciation (approximately 10,500 years ago) that represents the present interglacial. This epoch has been characterized by spatially and temporally variable *temperature* and *precipitation*.

Homosphere The *atmosphere* up to 80 km (50 mi) where the relative proportions of principal gases, such as nitrogen and oxygen, are constant.

Hook echo A distinctive hook-shaped reflectivity pattern in a *radar echo* that often indicates the presence of a *severe thunderstorm* cell and perhaps tornadic circulation. This pattern is produced by rainfall being drawn around the *mesocyclone*.

Horse latitudes Areas of persistent light *winds* or calm air between about 30 and 35 degrees N and S under *subtropical anticyclones*.

Hot-wire anemometer An instrument that measures *wind* speed based on the rate of *heat* loss to air flowing passed a heated wire. The rate of heat loss from the wire to the air increases as wind speed increases.

Humidity A general term referring to any one of many ways of describing the amount or concentration of water vapor in the air.

Hurricane An intense *tropical cyclone* that originates over tropical ocean basins; called a typhoon in the western Pacific Ocean. Maximum near-surface sustained *winds* are 119 km (74 mi) per hr or higher.

Hydrostatic equilibrium A balance of the *atmosphere's* vertical *pressure gradient force* and the equal, but oppositely directed force of *gravity*.

Hygrograph An instrument that records a continuous trace of *relative humidity* variations with time. The most common hygrograph is a *hair hygrometer* designed to move a pen on a clock-driven drum.

Hygrometer An instrument that measures the amount of water vapor in the air. See *dewpoint hygrometer, hair hygrometer*, and *electronic hygrometer*.

Hygroscopic nuclei A special category of *cloud condensation nuclei* that have a special chemical affinity for water molecules, so that *condensation* may take place on these nuclei at *relative humidities* under 100%.

Hypothesis A proposed explanation for some observation or phenomenon which is tested through the *scientific method*.

Ice pellets *Precipitation* consisting of frozen raindrops 5 mm (0.2 in.) or less in diameter that bounce on impact with the ground; also called sleet.

Ice-forming nuclei (IN) Tiny particles that promote the formation of ice crystals at *temperatures* well below freezing; include *freezing nuclei* and *deposition nuclei*.

Infrared radiation (IR) *Electromagnetic radiation* at *wavelengths* ranging from 0.8 micrometer (near-infrared) to about 0.1 mm (far infrared). Infrared radiation has wavelengths shorter than *microwaves* and longer than visible red light; most objects in the *Earth-atmosphere system* have their peak emission in the infrared.

Infrared satellite image   Picture or image processed from radiometers onboard a satellite that sense thermal (or infrared) radiation (typically, from *wavelengths* of approximately 8 to 12 micrometers) emitted by earth and *cloud* surfaces of the *Earth-atmosphere system*. Infrared radiation signals are routinely calibrated to give the surface *temperature* of objects in the sensor's field of view.

Intertropical convergence zone (ITCZ)   Discontinuous low-pressure belt of *thunderstorms* paralleling the equator and marking the convergence of the Northern and Southern Hemisphere surface *trade winds*.

Inverse square law   Intensity of radiation emitted by a point source (e.g., the sun) decreases as the inverse square of distance traveled.

Ion   An atomic-scale particle that caries an electric charge.

Ionosphere   Region of the upper *atmosphere* above 80 (50 mi) that contains a relatively high concentration of *ions* (electrically charged particles). The ionosphere is located primarily in the *thermosphere*.

Iridescent cloud   A *cloud* (usually *altocumulus*, *cirrostratus*, or *cirrocumulus*) having bright spots, bands, or borders of color, usually red or green. Iridescent clouds are produced by *diffraction* and typically appear up to about 30 degrees from the sun.

Isobar   A line plotted on a map joining locations reporting the same *air pressure*. Drawing an isobar on a map of sea-level air pressures requires interpolation between reporting *weather* stations.

Jet streak   An area of accelerated air flow within a jet stream; the *wind* may strengthen by as much as an additional 100 km (62 mi) per hr. Jet streaks occur where surface horizontal *temperature gradients* are particularly steep and play an important role in the generation and maintenance of synoptic-scale *cyclones*. The strongest jet streaks develop during winter in the *polar front jet stream* along the East Coasts of North America and Asia.

Kinetic energy   The energy within a body that is a result of its motion.

La Niña   A period of particularly strong *trade winds* and unusually low sea-surface *temperatures* (SSTs) in the central and eastern tropical Pacific; opposite of *El Niño*. The strong trade winds induce exceptionally vigorous *upwelling* in the eastern tropical Pacific, which causes unusually low SSTs.

Lake breeze   A relatively cool surface *wind* directed from a large lake toward land in response to differential heating between land and lake; develops during daylight hours.

Land breeze   A relatively cool surface *wind* directed from land to sea or land to lake in response to differential cooling between land and a body of water; develops at night.

Latent heat   The quantity of *heat* involved in the phase changes of water.

Latent heat of fusion   *Heat* released to the environment when water changes phase from liquid to solid; 80 *calories* per gm.

Latent heat of melting   *Heat* required to change the phase of water from solid to liquid; 80 *calories* per gram.

Latent heat of vaporization   *Heat* required to change the phase of water from liquid to vapor; 540 to 600 *calories* per gram, depending on the *temperature* of the water.

Latent heating   Transport of *heat* from one place to another within the *atmosphere* as a consequence of phase changes of water. Heat is supplied for melting, *evaporation*, and *sublimation* of water at the Earth's surface, and heat is released during fusion, *condensation*, and *deposition* within the atmosphere.

Law of energy conservation Energy is neither created nor destroyed but can change from one form to another; same as the first law of thermodynamics.

Law of reflection The angle of incident radiation is to equal to the angle of reflected radiation.

Lee-wave clouds Lens-shaped (lenticular) *clouds* that form in the crests of a standing wave downwind from a prominent mountain range.

Lifting condensation level (LCL) The altitude to which air must be lifted so that *expansional cooling* leads to *condensation* (or *deposition*) and *cloud* development; corresponds to the base of clouds.

Lightning A brilliant flash of light produced by an electrical discharge in response to the buildup of an electrical potential between a *cloud* and Earth's surface, between different clouds, or between different portions of the same cloud.

Lightning detection network (LDN) System that provides real-time information on the location and severity of *lightning* strokes.

Little Ice Age The interval of the *Holocene* from about 1400 to 1850 when average global *temperatures* were lower, and alpine glaciers increased in size and advanced down mountain valleys. The Little Ice Age followed the *Medieval Warm Period*.

Macroburst A *downburst* that affects a path longer than 4.0 km (2.5 mi), has maximum surface *winds* that may top 210 km (130 mi) per hr, and has a life expectancy of up to 30 minutes. The leading edge of a macroburst may be marked by a *gust front*.

Magnetosphere Region of the upper *atmosphere* above an altitude of about 150 km (95 mi) encompassed by the Earth's magnetic field; deflected by *solar wind* into a teardrop-shaped cavity.

Mammatus clouds *Clouds* that form on the underside of a *thunderstorm* anvil and exhibit pouchlike, downward protuberances; may indicate turbulent air.

Maritime polar air (mP) Cool, humid *air masses* that form over the cold ocean waters of the North Pacific and North Atlantic, especially north of 40 degrees N. Along the West Coast, maritime polar air contributes to heavy winter *rains* and *snows* and persistent summer coastal *fogs*.

Maritime tropical air (mT) Warm, humid *air masses* that form over tropical and subtropical ocean waters (e.g., the Gulf of Mexico). Maritime tropical air is responsible for oppressive summer *heat* and *humidity* east of the Rocky Mountains.

Mature stage The middle and most intense phase in the life cycle of a *thunderstorm* cell; begins when *precipitation* reaches Earth's surface and is characterized by both updrafts and downdrafts. *Rain* is heaviest, *lightning* is most frequent, and *hail*, strong surface *winds*, and *tornadoes* may develop during this stage, which typically lasts about 10 to 20 minutes.

Maunder minimum A 70-year period from 1645 to 1715 when *sunspots* were rare.

McIDAS (Man-computer Interactive Data Access System) A computerized data management system for *weather* information; ingests weather data and satellite images and integrates and organizes those data into guidance products for potential users. Developed by scientists at the Space Science and Engineering Center (SSEC) at the University of Wisconsin-Madison.

Medieval Warm Period A relatively mild episode of the *Holocene* between about 950 and 1250 A.D.

Mercury barometer A mercury-filled tube used to measure *air pressure*; the standard barometric instrument, which features great precision.

Meridional flow pattern   Flow of the *planetary-scale* westerlies in a series of deep troughs and sharp ridges; westerlies exhibit considerable amplitude. In this pattern, cold *air masses* surge southward and warm air masses stream northward, leading to strong *temperature gradients*.

Mesocyclone   A vertical column of cyclonically rotating air 3 to 10 km (2 to 6 mi) across that develops in the updraft of a *severe thunderstorm* cell; an early stage in the development of a *tornado*. A mesocyclone forms a tornado about 10% of the time.

Mesopause   Narrow zone of transition between the *mesosphere* below and the *thermosphere* above; the top of the mesosphere. Has an average altitude of 80 km (50 mi) and features the lowest average *temperature* in the *atmosphere*.

Mesoscale convective complex (MCC)   A nearly circular organized cluster of many interacting *thunderstorm* cells covering an area of many thousands of square kilometers. MCCs occur mostly in the warm-season over the eastern two-thirds of the United States and develop at night during weak synoptic-scale flow. Lasting from 6 to 12 hrs, they are driven by a flow of warm, humid air at low levels and *radiational cooling* at upper levels, which work together to destabilize the *troposphere*.

Mesoscale systems   *Weather* phenomena that may influence the weather in only a portion of a large city or county; includes *thunderstorms*, *sea breezes*, and *lake breezes*. These systems have dimensions of 1 to 100 km (1 to 60 mi) and last from hours to a day or so.

Mesosphere   The atmospheric layer between the *stratosphere* and the *thermosphere*; *temperature* falls with increasing altitude. Located between average altitudes of 50 to 80 km (31 to 50 mi).

Meteorology   The scientific study of the *atmosphere* and atmospheric processes.

Microburst   A *downburst* that affects a path on the ground that is 4.0 km (2.5 mi) or shorter, has maximum surface *winds* as high as 270 km (170 mi) per hr, and has a life expectancy of less than 10 minutes.

Microscale systems   *Weather* phenomena that represent the smallest spatial subdivision of atmospheric circulation, such as a weak *tornado*. These systems have dimensions of 1 m to 1 km (3 ft to 1 mi) and last from seconds to an hour or so.

Microwave radiation   A form of *earth-atmosphere system* having *wavelengths* in the 0.1 to 300 mm ranges, that is, between *infrared radiation* and *radio waves*. Some microwave frequencies are used in *weather radar*.

Midlatitude westerlies   Prevailing planetary-scale *winds* in the middle and upper *troposphere* between about 30 and 60 degrees of latitude; blow on average from the southwest in the Northern Hemisphere and from the northwest in the Southern Hemisphere out of the poleward flanks of the *subtropical anticyclones*.

Mie scattering   The optical effect whereby light is scattered equally at all *wavelengths*. It is produced by spherical particles having about the same diameter as the wavelength of *visible radiation;* responsible for a milky white sky when large numbers of *aerosols* are present.

Milankovich cycles   Systematic changes in three elements of Earth-sun geometry: precession of the *solstices* and *equinoxes*, tilt of Earth's rotational axis, and orbital eccentricity; affect the seasonal and latitudinal distribution of incoming solar radiation and influence climatic fluctuations over tens to hundreds of thousands of years.

Mirage   An optical phenomenon that makes an object appear to be displaced from its true position. A mirage is caused by *refraction* of light rays due to the change in *air density* with altitude within the lower *atmosphere*. When an object appears higher that it actually is, it is called a superior mirage. When an object appears lower that it actually is, it is called an inferior mirage.

Mist   Very thin *fog* in which visibility is greater than 1.0 km (0.62 mi). Mist is also known as light *drizzle*.

Mixing ratio   Mass of water vapor per mass of the remaining dry air; usually expressed as so many grams of water vapor per kilogram of dry air.

Moist adiabatic lapse rate   A variable rate of cooling applicable to saturated (cloudy) air parcels that are ascending within the *atmosphere*. This rate is less than the *dry adiabatic lapse rate* because some of the *expansional cooling* is compensated by *latent heat* released during *condensation* or *deposition* of water vapor.

Molecular viscosity   Fluid *friction* (*viscosity*) arising from the random motions and interactions of molecules composing a fluid such as air or water.

Monsoon circulation   Seasonal reversals in prevailing *winds* that cause wet summers and relatively dry winters. The most vigorous monsoon circulation occurs over Africa and southern Asia.

Mountain breeze   A shallow, gusty downslope flow of cool air that develops at night in some mountain valleys in response to differential heating between the air adjacent to the valley wall and air at the same altitude out over the valley floor.

Mountain-wave clouds   Stationary lenticular (lens-shaped) *clouds* situated over and downwind of a prominent mountain range; formed by the disturbance of the large-scale horizontal *winds* by the mountain range.

Nacreous clouds   Rarely seen *clouds* with a soft, pearly luster that form in the upper *stratosphere*; may be composed of ice crystals or supercooled water droplets. Nacreous clouds are *cirrus* and *altocumulus lenticularis* that form on sulfuric acid nuclei. Also called mother-of-pearl clouds or polar stratospheric clouds.

National Centers for Environmental Prediction (NCEP)   Centers responsible for the interpretation of weather maps and charts, and the generation of forecasts and other guidance products for the nation and for exchange with other nations.

National Climatic Data Center (NCDC)   An agency of the *National Oceanographic and Atmospheric Administration* that archives climatic data of the United States; located in Asheville, NC.

National Hurricane Center (NHC)   A branch of the Tropical Prediction Center (part of the *National Centers for Environmental Prediction*) located in south Florida that is responsible for forecasting *tropical cyclones* in the Atlantic Basin, Gulf of Mexico, and the eastern tropical Pacific Ocean.

National Oceanic and Atmospheric Administration (NOAA)   The administrative unit within the U.S. Department of Commerce that oversees the *National Weather Service*.

NOAA Weather Radio   Low power, VHF-high-band FM radio transmitters operated by the *National Oceanic and Atmospheric Administration* that broadcast continuous *weather* information (e.g., regional conditions, local forecasts, marine warnings) directly from *National Weather Service* Forecast Offices 24 hrs per day.

National Weather Service (NWS)   The agency of the *National Oceanic and Atmospheric Administration* responsible for *weather* data acquisition, data analysis, weather forecasting and dissemination, and storm watches and warnings.

NWS Cooperative Observer Network   Consists of more than 8000 *weather* stations across the United States that record daily *precipitation* and maximum/minimum *temperatures* for hydrologic, agricultural, and climatic purposes. These manned stations complement observations from *Automated Surface Observing System* units.

Neutral air   An ambient air layer in which an ascending or descending air parcel always has the same *temperature* (and *air density*) as its surroundings.

Newton's first law of motion   An object at rest or in straight-line, unaccelerated motion remains that way unless acted upon by a net external force.

Newton's second law of motion   A net force is required to cause a unit mass of a substance to accelerate; force = (mass) × (acceleration).

Nimbostratus (Ns)   Low, gray, layered *clouds* that resemble *stratus* but are thicker, appear darker gray, have a less uniform base, and yield more substantial *precipitation*. Nimbostratus are composed of mostly water droplets.

Noctilucent clouds   Wavy, thin *clouds* resembling *cirrus*, but usually bluish-white or silvery; best seen at high latitudes just before sunrise or just after sunset. These rare clouds occur in the upper *mesosphere* at altitudes above about 80 km (50 mi) and may be composed of ice deposited on meteoric dust.

Non-glacial climate   Conditions favorable to no glaciers or the shrinkage of existing glacial ice.

Norwegian cyclone model   The original description of the structure and life cycle of an extratropical low-pressure system based mostly on surface observations, first proposed during World War I by researchers at the Norwegian School of Meteorology at Bergen.

Nuclei   Tiny solid or liquid particles suspended in the *atmosphere* that provide surfaces on which water vapor condenses into droplets or deposits into ice crystals; essential for *cloud* formation.

Number density   Number of molecules of a gas (or mixture of gases) per unit volume.

Occluded front   A *front* formed late in the life cycle of an extratropical *cyclone*; its behavior depends upon the characteristics of air behind the *cold front* and ahead of the *warm front*. Also known as an occlusion.

Orographic lifting   The forced rising of air up the slopes of a hill or mountain. Air that is forced to ascend the slopes facing the oncoming *wind* (windward slopes) expands and cools, which increases its *relative humidity*. With sufficient cooling, *clouds* and *precipitation* develop.

Outgassing   Release of gasses to the *atmosphere* from hot, molten rock during volcanic activity and from impact of meteorites on the rocky surface of the planet; the origins of most atmospheric gases.

Overrunning   The process whereby less dense air flows up and over denser air; occurs along a *warm front* and may occur along a *stationary front*. Ascending air cools by expansion, leading to the formation of *clouds* and perhaps *precipitation* over a widespread area.

Pacific air   Term used to describe cool, humid *maritime polar air* air swept inland from the Pacific Ocean that experiences *air mass modification* over the Rocky Mountains, emerging milder and drier to the east of the mountains. During a *zonal flow pattern*, Pacific air floods the eastern two-thirds of the United States and southern Canada, causing mild and generally dry *weather*.

Parhelia   Two bright spots of light appearing on either side of the sun; each is separated from the sun by an angle of 22 degrees. Parhelia are caused by *refraction* of sunlight by ice crystals; also called mock suns and *sundogs*.

Perihelion   The time of the year when the Earth's orbital path brings it closest (147 million km or 91 million mi) to the sun (at present, about 3 January).

Persistence   Tendency for *weather* episodes to continue for some period of time.

Photodissociation   The process by which radiation breaks down molecules into their smallest components.

Photosphere   The intensely bright portion of the sun visible to the unaided eye; this several-hundred-kilometer-thick layer is what we perceive as the surface of the sun. Features such as *sunspots* and *faculae* are observed on the photosphere.

Photosynthesis   The process whereby plants use sunlight, water, and carbon dioxide to manufacture their food and generate oxygen as a byproduct.

Planetary albedo The fraction (or percent) of incident solar radiation that is scattered and reflected back into space by the *Earth-atmosphere system*; measurements by satellite sensors indicate a planetary albedo of about 31%.

Planetary-scale systems *Weather* phenomena operating at the largest spatial scale of atmospheric circulation; includes the global *wind* belts and *semipermanent pressure systems*. These systems have dimensions of 10,000 to 40,000 km (6000 to 24,000 mi) and exhibit patterns that last from weeks to months.

Plate tectonics Concept that the outer 100 km (60 mi) of solid Earth is divided into a dozen gigantic rigid plates that move relative to one another across the surface of the planet. The drift of these plates move continents and open and close ocean basins over the vast expanse of *geologic time*; mountain building and most volcanic activity occur at plate boundaries.

Polar amplification The tendency of a major *temperature* change to increase in magnitude with latitude.

Polar front Narrow transition zone where the relatively mild *midlatitude westerlies* meet and override the relatively cold polar easterlies. When the *temperature gradient* across the *front* is steep, the front is well defined and is a potential site for development of *extratropical cyclones*.

Polar front jet stream A corridor of strong westerlies in the upper *troposphere* between the midlatitude *tropopause* and the polar tropopause and directly over the *polar front*.

Polar high A *cold-core anticyclone* that originates in a source region for *continental polar air*; this shallow system is the product of intense *radiational cooling* over the *snow*-covered continental interior of North America well north of the *polar front*.

Polar-orbiting satellite A satellite in relatively low orbit that travels near the geographical poles on meridional trajectories. Earth rotates through the plane of the satellite. The altitude of the satellite's orbit is about 800 to 1000 km (500 to 620 mi).

Poleward heat transport Flow of *heat* from tropical to middle and high latitudes in response to latitudinal imbalances in net *radiational heating* and *radiational cooling*. Poleward heat transport is accomplished primarily by *air mass* exchange, but also by storms, and ocean circulation.

Precipitable water The depth of water that would be produced if all the water vapor in a vertical column of air were condensed; usually the column of air extends from Earth's surface to the top of the *troposphere*.

Precipitation Water in solid, liquid, or freezing form that falls to Earth's surface from *clouds*; can be in the form of *rain*, *drizzle*, *snow*, *ice pellets*, *hail* or *freezing rain*.

Pressure gradient force A force operating in the *atmosphere* that accelerates air parcels away from regions of high *air pressure* directly across *isobars* toward regions of low air pressure in response to an *air pressure gradient*.

Pressure systems Individual synoptic-scale features of atmospheric circulation; commonly denoted as highs (or *anticyclones*) or lows (or *cyclones*), less frequently ridges or troughs.

Psychrometer An instrument used to measure the amount of water vapor in the air. It consists of two identical liquid-in-glass *thermometers* mounted side by side with the bulb of one thermometer wrapped in a muslin wick. To take a reading, the wick-covered bulb is first soaked in distilled water and then the instrument is ventilated, either by being whirled about (*sling psychrometer*) or with a small fan (*aspirated psychrometer*). The dry bulb thermometer measures the actual air *temperature*. Water vaporizes from the muslin wick into the air streaming past the wet-bulb thermometer and evaporative cooling lowers the *wet-bulb temperature* to a steady value. By referring to special psychrometric tables, the *relative humidity* or the *dewpoint* can be determined from the difference between the dry-bulb temperature and wet-bulb temperature (*wet-bulb depression*).

Pyranometer The standard instrument for measuring solar radiation incident on a horizontal surface; calibrates the *temperature* response of a special sensor in units of radiation flux, such as watts per square meter.

Radar echo *Microwave* signals emitted by *weather radar* that are scattered or reflected by *rain* or *snow* back to a receiver where they are electronically processed and displayed on a cathode ray tube.

Radiation fog Ground-level *cloud* formed by nocturnal *radiational cooling* of a humid air layer near the ground so that its *relative humidity* approaches 100 %; sometimes called *ground fog*,

Radiational cooling The drop in *temperature* of an object or a surface accomplished whenever the object or surface undergoes a net loss of *heat* due to a greater rate of emission of *electromagnetic radiation* than *absorption*.

Radiational heating The rise in *temperature* of an object or a surface accomplished whenever the object or surface undergoes a net gain of *heat* due to a greater rate of *absorption* of *electromagnetic radiation* than emission.

Radio waves Long-*wavelength* (1000 mm to hundreds of kilometers), low-frequency *electromagnetic radiation*; the longest wavelength portion of the *electromagnetic spectrum*.

Radiosonde A small balloon-borne instrument package equipped with a radio transmitter that takes altitude readings (*soundings*) of *temperature*, *air pressure*, and *humidity* in the *atmosphere*.

Rain Form of *precipitation* consisting of liquid water drops having diameters generally between 0.5 and 6.0 mm (0.02 and 0.2 in); falls mostly from *nimbostratus* and *cumulonimbus*.

Rain gauge A device for collecting and measuring rainfall (or melted snowfall); a standard rain gauge consists of a cylindrical container equipped with a cone-shaped funnel at the top.

Rain shadow A region situated downwind (often hundreds of kilometers) of a prominent mountain range and characterized by descending air and, as a consequence, a relatively dry *climate*.

Rainbow An arc of concentric colored bands formed by *refraction* and internal *reflection* of sunlight by raindrops. To see a rainbow, an observer must be looking at a distant *rain* shower with the sun at his/her back.

Rawinsonde A *radiosonde* tracked from the ground by a direction-finding antenna or global positioning system (GPS) to measure variations in horizontal *wind* direction and wind speed with altitude.

Rayleigh scattering The optical effect where light at the short-*wavelength* end of the visible *electromagnetic spectrum* is scattered much more efficiently than light at the long-wavelength end of the visible spectrum. This wavelength-dependent *scattering* is caused by spherical scattering particles having diameters much smaller than the wavelength of scattered radiation. Rayleigh scattering is responsible for the blue of the daytime sky.

Reflection The process whereby a portion of the radiation that strikes an interface between two different media is redirected such that the angle of reflection equals the angle of incidence. Reflection is a special case of *scattering*.

Refraction The bending of a light ray as it passes from one transparent medium into another (from air to water, for example). Bending is due to the differing speeds of light in the two media.

Relative humidity Compares the amount of water vapor in the air with the amount of water vapor in the same air at *saturation*. Relative humidity is a measure of how close air is to *saturation* at a specific *temperature*, always expressed as a percentage. It can be computed from either the ratio of the *vapor pressure* to the *saturation vapor pressure* or the ratio of the *mixing ratio* to the *saturation mixing ratio*.

Roll cloud A low, cylindrically-shaped and elongated *cloud* occurring behind a *gust front*; associated with but detached from a *cumulonimbus* cloud. The cloud appears to rotate slowly about its horizontal axis.

Rossby waves Series of long-*wavelength* troughs and ridges that characterize the planetary-scale westerlies (above the 500-mb level) as they encircle the globe; also called *long waves*. Typically, between 2 and 5 waves encircle the hemisphere at one time.

Saffir-Simpson Hurricane Intensity Scale   A five-category *hurricane* intensity scale developed by H. S. Saffir and R. H. Simpson that rates hurricanes from 1 to 5 on the basis of maximum sustained *wind* speed. Categories are 1 (weak), 2 (moderate), 3 (strong), 4 (very strong), and 5 (devastating).

Santa Ana wind   A hot, dry *chinook wind* that blows from the desert plateaus of Utah and Nevada across the Sierra Nevada and downslope toward coastal southern California. This *wind* desiccates vegetation and contributes to outbreaks of forest and brush fires.

Saturation absolute humidity   The value of *absolute humidity* when air is saturated with respect to water vapor.

Saturation mixing ratio   The value of the *mixing ratio* when air is saturated with respect to water vapor; varies directly with *temperature*, and to a lesser extent upon *air pressure*.

Saturation specific humidity   The value of the *specific humidity* when air is saturated with respect to water vapor; varies directly with *temperature*.

Saturation vapor pressure   The value of the *vapor pressure* when air is saturated with respect to water vapor; varies directly with *temperature*.

Scattering   The process by which *aerosols* and molecules disperse radiation in all directions. Scattering is a function of the *wavelength* of the incident radiation and the size, surface, and composition of the scattering aerosol or molecule, where size is defined relative to the wavelength. No energy transformation results from scattering.

Scientific method   A systematic form of inquiry that involves observation, speculation, and formulation and testing of hypotheses.

Scientific model   An approximate representation or simulation of a real system that omits all but the most essential variables of the system.

Scientific theory   A *hypothesis* that has stood the test of time and is generally accepted by the scientific community.

Sea breeze   A relatively cool mesoscale surface *wind* directed from the ocean toward land in response to differential heating of land and sea; develops during daylight hours.

Second law of thermodynamics   All systems tend toward a state of disorder. A gradient in a system, such as a *temperature gradient*, signals order in the system. As a system tends towards disorder, gradients are eliminated.

Semipermanent pressure systems   Persistent areas of high and low *air pressure* that are *planetary-scale systems*. They undergo some important seasonal changes in location and surface air pressures and include *subtropical anticyclones*, the *intertropical convergence zone*, *subpolar lows*, and *polar highs*.

Sensible heating   Transport of *heat* from one location or object to another via *conduction*, *convection* or both.

Severe thunderstorm   *Thunderstorms* accompanied by locally damaging surface *winds*, frequent *lightning*, or large *hail*. The official *National Weather Service* criterion for the severe thunderstorm designation includes any one or a combination of the following: hailstones with diameters 0.75 in. (1.9 cm) or larger; *tornadoes* or *funnel clouds*; surface winds stronger than 58 mi (93 km) per hr.

Shelf cloud   A low, wedge-shaped, and elongated *cloud* that occurs along a *gust front*; associated with and attached to a *cumulonimbus* cloud. Damaging surface *winds* may occur under a shelf cloud, and this cloud may be associated with a *severe thunderstorm*. Also known as an arcus cloud.

Short waves   Relatively small short-*wavelength* ripples (troughs and ridges) superimposed on *Rossby waves* in the planetary-scale westerlies; they propagate with the airflow in the middle and upper *troposphere*. Typically, a dozen or more short waves encircle the hemisphere at one time.

Single-station forecasts   *Weather* forecasts based on observations at one location.

Snow   A frozen form of *precipitation* consisting of an assemblage of ice crystals in the form of flakes.   Although snowflakes vary in shape and size, the constituent ice crystals are hexagonal (six-sided) and may consist of needles, dendrites, plates, or columns.

Snow grains   Frozen form of *precipitation* consisting of flat particles of opaque, white ice having diameters less than 1 mm (0.04 in.); originates in the same way as *drizzle* except that particles freeze prior to reaching the ground.   Also known as granular *snow*.

Snow pellets   Frozen form of *precipitation* consisting of soft spherical (or sometimes conical) particles of opaque, white ice having diameters of 2 to 5 mm (0.08 to 0.2 in.).   They often break up when striking a hard surface and differ from *snow grains* in being softer and larger.   Formally called soft *hail* or *graupel*.

Solar altitude   The angle of the sun 90 degrees or less above the horizon; influences the intensity of solar radiation that strikes Earth's surface.   At a maximum possible solar altitude of 90 degrees, the solar rays are most intense; the intensity declines with decreasing solar altitude (towards 0 degrees).

Solar constant   The flux of solar radiational energy falling on a surface that is positioned at the outer edge of the *atmosphere* and oriented perpendicular to the solar beam when Earth is at its average distance from the sun.

Solar corona   Outermost portion of the solar atmosphere (above the *chromosphere*) that is a region of extremely hot (1 to 4 million °C or 1.8 to 7.2 million °F), highly rarefied gases extending millions of kilometers in space, to the outer limits of the solar system.   The *solar wind* originates in the solar corona.

Solar wind   A stream of super-hot electrically charged subatomic particles (mainly protons and electrons) flowing into space from the sun.   The solar wind originates in the *solar corona*.

Solstice   A time during the year when the sun is at its maximum poleward location relative to the Earth (23 degrees, 30 minutes, North or South); the first days of astronomical summer and winter.

Sounding   Continuous altitude measurements that provide vertical profiles of such variables as *temperature*, *humidity*, and *wind* speed.

Southern oscillation   Opposing swings of surface *air pressure* between the western and central tropical Pacific Ocean. When air pressure is low at Darwin (Australia) it is high at Tahiti (a south Pacific island) and when air pressure is high at Darwin it is low at Tahiti.

Specific heat   The amount of *heat* required to raise the *temperature* of 1 gram of a substance by 1 Celsius degree.

Specific humidity   The ratio of the mass of water vapor (in grams) to the mass (in kilograms) of air containing the water vapor, that is, the combined mass of dry air plus water vapor.

Split flow pattern   Wave pattern in the *planetary-scale circulation* regime where westerlies to the north have a wave configuration that differs from that of westerlies to the south.

Spörer minimum   A period of reduced *sunspot* activity between about 1450 and 1550 A.D.

Squall line   An elongated cluster of intense *thunderstorm* cells that is accompanied by a continuous *gust front* at the cluster's leading edge.   A squall line is located parallel to and up to 300 km (180 mi) ahead of a fast-moving well-defined *cold front*.

Stable air layer An ambient air layer characterized by a vertical *temperature* profile such that air parcels return to their original altitudes following any upward or downward displacement. An ascending air parcel becomes cooler (denser) than the ambient air and a descending air parcel becomes warmer (less dense) than the ambient air, returning in both cases to the original altitude.

Standard atmosphere A model that represents the state of the *atmosphere* averaged for all latitudes and seasons. It features a fixed sea-level air *temperature, air pressure,* and *air density* plus fixed vertical profiles of temperature and air pressure.

Station model A conventional representation of *weather* station observations on a weather map using standard symbols for weather conditions.

Stationary front A *front* that exhibits essentially no lateral movement; *winds* blow parallel but in opposite directions on either side of the *front*.

Steam fog The general name for *fog* produced when extremely cold, dry air comes in contact with a relatively warm (unfrozen) water surface; has the appearance of rising streamers. Also known as Arctic sea smoke.

Stefan-Boltzmann law A radiation law that states that the total energy radiated by a *blackbody* at all *wavelengths* is directly proportional to the fourth power of the absolute *temperature* (in kelvins) of the body. For example, the sun's energy output per square meter is about 190,000 times that of the *Earth-atmosphere system*.

Storm surge A dome of water perhaps 80 to 160 km (50 to 100 mi) wide that sweeps over the coastline near the landfall of a tropical or extratropical *cyclone*, often causing considerable coastal erosion and flooding. The dome of water is caused primarily by strong onshore *winds* and, to a lesser extent, low *air pressure* associated with the storm system.

Stratiform clouds Layered *clouds*, such as *altostratus*, often produced by *overrunning*.

Stratocumulus (Sc) Low *clouds* occurring as large, irregular puffs or rolls arranged in a layer. Stratocumulus are composed of mostly water droplets.

Stratopause Transition zone between the *stratosphere* below and the *mesosphere* above; the top of the stratosphere at an average altitude of 50 km (31 mi).

Stratosphere The *atmosphere's* thermal subdivision situated between the *troposphere* and *mesosphere* and the primary site of ozone formation. Air *temperature* in the lower portion of the stratosphere is constant with altitude, and then the temperature increases with altitude to the *stratopause*. Located between the *tropopause* and an average altitude of 50 km (31 mi).

Stratospheric ozone shield Ozone in the *stratosphere* that shields organisms at the Earth's surface from exposure to potentially lethal intensities of solar *ultraviolet radiation*.

Stratus (St) Low *clouds* that occur as a uniform gray layer stretching from horizon to horizon. They may produce *drizzle*, and where they intersect the ground, they are classified as *fog*. Stratus are composed of water droplets.

Streamline A line that graphically portrays the flow of *wind* and can be used to identify regions of *divergence* and convergence.

Stüve thermodynamic diagram A graphical representation of *soundings* and *adiabatic processes* in the *atmosphere*.

Sublimation The process which produces a net loss of water mass at the interface between ice and air; more water changes phase from ice to vapor than vapor to ice. The phase change of water from ice to vapor (without first becoming liquid) requires the addition of the *latent heat of melting* plus the *latent heat of vaporization*.

Subpolar lows High-latitude, semipermanent *cyclones* marking the convergence of planetary-scale surface southwesterlies of midlatitudes with surface northeasterlies of polar latitudes in the Northern Hemisphere or midlatitude northwesterlies and polar southeasterlies in the Southern Hemisphere. The Icelandic low and Aleutian low are Northern Hemisphere examples.

Subtropical anticyclones Semipermanent high-pressure systems centered over subtropical latitudes (on average, near 30 degrees N and S) of the Atlantic, Pacific, and Indian Oceans. These warm-core systems extend from the ocean surface up to the *tropopause*.

Subtropical jet stream A zone of relatively strong *winds* aloft situated between the tropical *tropopause* and the midlatitude tropopause, on the poleward side of the *Hadley cell*.

Sundog A colored luminous spot at the same altitude as the sun produced by *refraction* of light by ice crystals. In some cases, sundogs are part of a *halo* surrounding the sun. Also called *parhelia*.

Sunspot Relatively dark, cool area that develops on the surface of the sun's *photosphere* where an intense magnetic field suppresses the flow of gases transporting *heat* from the sun's interior.

Supercell thunderstorm A relatively long-lived, large and intense *thunderstorm* cell characterized by an exceptionally strong updraft sometimes in excess of 240 km (150 mi) per hr; may produce a *tornado*.

Synoptic-scale systems *Weather* phenomena operating at the continental or oceanic spatial scale; includes migrating *cyclones* and *anticyclones*, *hurricanes*, *air masses* and *fronts*. These systems have dimensions of 100 to 10,000 km (60 to 6000 mi) and last from days to a weeks or so.

Teleconnection A linkage between changes in atmospheric circulation occurring in widely separated regions of the globe, often many thousands of kilometers apart.

Temperature A measure of the average *kinetic energy* of the individual atoms or molecules composing a substance.

Temperature gradient *Temperature* change with distance.

Terminal velocity Constant downward-directed speed of a particle within a fluid due to a balance between *gravity* (acting downward) and fluid resistance (directed upward).

Thermal inertia Resistance to a change in *temperature*.

Thermograph A recording instrument that provides a continuous trace of *temperature* variations with time.

Thermometer An instrument used for measuring *temperature* by incorporating a thermal sensor that utilizes the variation of the physical properties of substances according to their thermal states. Thermal sensors include liquid-in-glass, deformation, and electronic thermometers, along with radiometers that sense *electromagnetic radiation*.

Thermosphere The outermost thermal subdivision of the *atmosphere* above the *mesopause* in which the air *temperature* is isothermal in the lower reaches and then increases with altitude. The thermosphere starts at an average altitude of 80 km (50 mi) and includes most or all of the *ionosphere*.

Thunder Sound accompanying *lightning*; produced by violent expansion of air due to intense heating by a lightning discharge.

Thunderstorm A mesoscale *weather* system produced by strong *convection* currents that reach to great altitudes within the *troposphere*, sometimes reaching the *tropopause* or higher. Consists of *cumulonimbus* accompanied by *lightning* and *thunder* and often, locally heavy rainfall (or snowfall) and gusty surface *winds*. A thunderstorm consists of one or more convective cells, each of which progresses through the life cycle of *cumulus stage*, *mature stage*, and *dissipating stage*.

Tipping-bucket rain gauge   A recording *rain gauge* that collects rainfall in increments of 0.01 in. by containers that alternately fill, tip and empty.   A heated tipping-bucket rain gauge is a standard component of the *Automated Surface Observing System*.

Tornado   A small mass of air in contact with the ground that whirls rapidly about an almost vertical axis; made visible by water droplets formed by *condensation* and by dust and debris sucked into the system.

Tornado alley   Region of maximum *tornado* frequency in North America; a corridor stretching from eastern Texas and the Texas panhandle northward into Oklahoma, Kansas and Nebraska, and into southeastern South Dakota.

Trade wind inversion   An elevated *stable air layer* that occurs on the eastern flank of *subtropical anticyclones* in the vicinity of the *trade winds*; a persistent and climatically significant feature.   Formed when the subsiding and compressionally warmed air in the subtropical anticyclones encounters the marine air layer, a layer of cool, humid, and stable air formed where sea-surface *temperatures* are relatively low.   A temperature inversion develops at the altitude where air subsiding from above meets the top of the marine air layer.

Trade winds   Prevailing planetary-scale surface *winds* in tropical latitudes; blow from the northeast in the Northern Hemisphere and from the southeast in the Southern Hemisphere out of the equatorward flanks of the *subtropical anticyclones*.

Transpiration   Process by which water that is taken up by plant roots escapes as vapor through tiny leaf pores. Measurements of direct *evaporation* from Earth's surface plus transpiration are usually combined as *evapotranspiration*.

Triple point   The point of occlusion in an *extratropical cyclone* where *cold front*, *warm front*, and *occluded front* all come together.   Conditions at this location are favorable for the formation of a new extratropical *cyclone*.

Tropic of Cancer   A *solstice* position of the sun with a latitude 23 degrees 27 minutes N.   On 21 June, the sun's noon rays are vertical (*solar altitude* of 90 degrees) at this latitude.

Tropic of Capricorn   A *solstice* position of the sun with a latitude of 23 degrees, 27 minutes S.   On 21 December, the sun's noon rays are vertical (*solar altitude* of 90 degrees) at this latitude.

Tropical cyclone   A *synoptic-scale system* of low pressure that originates over the tropical ocean; includes *tropical depression*, *tropical storm*, *hurricane*, and *typhoon*.   Example of a *warm-core cyclone*.

Tropical depression   A *tropical cyclone* with sustained *wind* speeds of at least 37 km (23 mi) per hr but less than 63 km (39 mi) per hr; an early stage in the development of a *hurricane*.

Tropical disturbance   A region of convective activity over tropical seas with a detectable center of low *air pressure* at the surface; the initial stage in the development of a *hurricane*.   A tropical disturbance is typically triggered by the *intertropical convergence zone*, by a trough in the westerlies intruding into the tropics, or by a wave in the easterly *trade winds* (an *easterly wave*).

Tropical storm   A *tropical cyclone* having sustained *wind* speeds of 63 to 118 km (39 to 73 mi) per hr; a storm at pre-*hurricane* stage.   When a tropical cyclone reaches tropical storm strength, it is assigned a name.

Tropopause   Zone of transition between the *troposphere* below and the *stratosphere* above; the top of the troposphere. Has an average altitude ranging from 6 km (3.7 mi) at the poles to about 20 km (12 mi) at the equator.

Troposphere   Lowest thermal subdivision of the *atmosphere* in which air *temperature* normally drops with altitude; the site of most *weather*.   Located between Earth's surface and the *tropopause* at an average altitude ranging from 6 km (3.7 mi) at the poles to about 20 km (12 mi) at the equator.

Turbulence   A state of fluid flow characterized by irregular (eddy) motion.

Twilight A period after sunset or before sunrise when the sky is illuminated by sunlight scattered by constituents of the upper *atmosphere*. Twilight is divided into the stages of civil twilight, nautical twilight, and astronomical twilight.

Ultraviolet radiation Short-wave, energetic *electromagnetic radiation* with *wavelengths* longer than *X-rays*, but shorter than *visible radiation*. The sun emits most of the ultraviolet radiation incident on the Earth but much of the solar ultraviolet radiation is absorbed in the *stratosphere*, where it is involved in the formation and destruction of ozone.

Universal Coordinated Time (UTC) A worldwide time reference used for synchronizing *weather* observations; the time at 0 degrees longitude, the prime meridian.

Unstable air layer An ambient air layer characterized by a vertical *temperature* profile such that air parcels accelerate upward or downward and away from their original altitudes. An ascending air parcel remains warmer (less dense) than the ambient air and continues to ascend and a descending air parcel remains cooler (denser) than the ambient air and continues to descend.

Upslope fog *Fog* formed as a consequence of the *expansional cooling* of humid air that is forced to ascend a mountain slope.

Upwelling The upward movement of cold, nutrient-rich water from depths of 200 to 1000 m (about 650 to 3300 ft) toward the ocean surface associated with the offshore flow of near-surface waters.

Urban heat island An area of higher air *temperatures* in a city setting compared to the air temperatures of the suburban and rural surroundings; shows up as an island in the pattern of isotherms on a surface map.

Valley breeze A shallow, upslope flow of relatively warm air that develops during daylight hours within *snow*-free mountain valleys in response to differential heating between the air adjacent to the valley wall and air at the same altitude out over the valley floor.

Vapor pressure The portion of the total *air pressure* exerted by the water vapor component of air; increases as the content of water vapor in air increases.

Variables of state The minimum number of descriptors of the physical state of a thermodynamic system; for a gas (and mixtures of gas) these include *temperature*, pressure and density. The variables of state are interrelated through the *gas law*.

Vertical wind shear The change in *wind* speed or direction with increasing altitude.

Virga Streaks of water and ice particles falling from a *cloud* that vaporize before reaching Earth's surface.

Viscosity *Friction* within fluids such as air and water.

Visible radiation (light) *Electromagnetic radiation* that is perceptible to the human eye; *wavelengths* of visible light range from about 0.40 (violet) to 0.70 (red) micrometers. Visible radiation has wavelengths longer than *infrared radiation* but shorter than *ultraviolet radiation*.

Visible satellite image Image processed from radiometers onboard a satellite that sense visible solar radiation reflected or back-scattered from surfaces in the *Earth-atmosphere system*.

Wall cloud A roughly circular lowered portion of the *rain*-free base of a *cumulonimbus cloud* about 3 km (2 mi) in diameter associated with a humid updraft; may develop into a *mesocyclone* and *tornado*. A wall cloud forms in the region of strongest *thunderstorm* updraft and often accompanies a mesocyclone.

Warm air advection The flow of air across regional isotherms from a relatively warm locality to a relatively cool locality.

Warm cloud   *Cloud* composed of liquid water droplets having a *temperatures* above 0°C (32°F).

Warm front   A *front* that moves in such a way that the cold (more dense) air retreats, allowing relatively warm (less dense) air to advance. May be associated with a broad band of cloudiness and light to moderate *precipitation*.

Warm-core anticyclone   High-pressure system occupying a thick column of subsiding warm, dry air. Isobaric surface bulge upward above the high center and the anticyclonic circulation is most intense at high altitudes. *Subtropical anticyclones* are examples.

Warm-core cyclone   A surface, synoptic-scale stationary *cyclone* that develops as a consequence of intense solar heating of a large, relatively dry geographical area; same as a thermal low. The thermal low's shallow cyclonic circulation weakens rapidly with altitude and often reverses to an anticyclonic circulation in the middle and upper *troposphere*.

Water vapor plume   Extensive stream of water vapor that often originates over tropical seas and is transported horizontally; easily identified using *water vapor satellite imagery*. A typical water vapor plume can be several hundred kilometers wide and thousands of kilometers long. Plumes supply moisture to *hurricanes*, *thunderstorm* clusters, and winter storms.

Water vapor satellite imagery   Picture or image processed from a satellite radiometer that senses *infrared radiation* at those *wavelengths* (typically near 6.7 micrometers) emitted by *clouds* and water vapor in the *atmosphere*; this imagery displays flow patterns in the middle *troposphere* near the top of the bulk of atmospheric water.

Wave frequency   Number of crests or troughs of a wave that pass a given point in a specified period of time, usually 1 second.

Wavelength   The distance between successive wave crests (or equivalently, wave troughs).

Weather   The state of the *atmosphere* at some place and time described in terms of such variables as *temperature*, *humidity*, cloudiness, *precipitation*, and *wind* speed and direction.

Weather radar   An adaptation of radar for meteorological purposes. The *scattering* of *microwave* radiation (*wavelengths* of a few millimeters to several centimeters) by raindrops, snowflakes, or hailstones is used for locating and tracking the movement of areas of *precipitation* and monitoring the circulation within small-scale *weather* systems such as *thunderstorms*.

Weather warning   Issued by the *National Weather Service* when hazardous *weather* is observed or imminent.

Weather watch   Issued by the *National Weather Service* when hazardous *weather* is considered possible based on current or anticipated atmospheric conditions.

Weighing-bucket rain gauge   A recording *rain gauge* that is calibrated so that the weight of cumulative rainfall is recorded directly in terms of water depth (in millimeters or inches).

Wet-bulb depression   On a *psychrometer*, the difference between the dry-bulb *temperature* and the *wet-bulb temperature*; used to determine *relative humidity*.

Wet-bulb temperature   The *temperature* an air parcel would have if cooled adiabatically to saturation at constant *air pressure* by *evaporation* of water into it; measured using a *psychrometer* in which the wick-covered bulb is first soaked in distilled water and the instrument is ventilated to promote evaporation.

Wien's displacement law   A radiation law whereby the *wavelength* of maximum emission by a *blackbody* is inversely to its absolute *temperature* (in kelvins). For example, hot objects (such as the sun) emit peak radiation at relatively short wavelengths, whereas cold objects (such as the *Earth-atmosphere system*) emit peak radiation at longer wavelengths.

Wind   Air in motion measured relative to the Earth's rotating surface.

Wind shear Change in *wind* speed or direction with distance.

Wind vane An instrument used to monitor *wind* direction; consists of a free-swinging shaft with a vertical plate at one end and a counterweight (arrowhead) at the other end; the counterweight always points into the wind.

Windsock An instrument used to monitor *wind* direction; consists of a cone-shaped cloth bag that is open at both ends. The larger end is held open by a metal ring attached to a pole; air enters the larger opening, rotating the sock and stretching it downwind.

World Meteorological Organization (WMO) The agency of the United Nations that coordinates *weather* data collection and analysis by 185 member nations and territories; based in Geneva, Switzerland.

World Weather Watch (WWW) Standardized international *weather*-monitoring network coordinated by the *World Meteorological Organization*.

X-rays Highly energetic short-*wavelength electromagnetic radiation* with wavelengths longer than *gamma radiation*, but shorter than *ultraviolet radiation*. Earth's upper *atmosphere* blocks out virtually all incoming X-rays; overexposure to this type of radiation can be dangerous to living organisms. Artificially produced X-rays are used as a powerful medical diagnostic tool and both X-rays and *gamma radiation* are used to treat cancer.

Younger Dryas A relatively cold period from about 12,900 to 11,600 year ago.

Zonal flow pattern Flow of the *planetary-scale* westerlies almost directly from west to east; westerlies exhibit little amplitude. In this pattern, the north-south exchange of *air masses* is minimal.

Index

N

O

P

X

Y-Z